ILLUSTRATED FLORA *of* BRITISH COLUMBIA

Volume 2

Dicotyledons

(Balsaminaceae through Cuscutaceae)

edited by

George W. Douglas[1], Gerald B. Straley[2], Del Meidinger[3] and Jim Pojar[4]

[1] Conservation Data Centre
B.C. Ministry of Environment, Lands and Parks
PO Box 9344 Stn Prov Govt
Victoria BC V8W 9M1

[2] Botanical Garden
University of British Columbia
6501 N.W. Marine Drive
Vancouver BC V6T 1W5

[3] Research Branch
B.C. Ministry of Forests
PO Box 9519 Stn Prov Govt
Victoria BC V8W 9C2

[4] B.C. Forest Service
PO Box 5000
Smithers, BC V0J 2N0

September 1998

Canadian Cataloguing in Publication Data
Main entry under title:
Illustrated flora of British Columbia

Revision in part of: The vascular plants of British Columbia, 1989–1994
Co-published by British Columbia Ministry of Forests.
Includes bibliographical references.
Contents: v. 1. Gymnosperms and dicotyledons (Aceraceae through Asteraceae)—
v. 2. Dicotyledons (Balsaminaceae through Cuscutaceae)
ISBN 0-7726-3685-0 (set)

1. Botany - British Columbia. I. Douglas, George W., 1938- . II. British Columbia. Ministry of Environment, Lands and Parks. III. British Columbia. Ministry of Forests. IV. Title: Vascular plants of British Columbia.

QK203.B7I44 1998 581.9711 C98-960275-3

Contributors

Ms. E.M. Campbell, 780 Rue Andre Jobin, Ile Bizard PQ H9C 1W8 – Glossary.

Dr. G.W. Douglas, Conservation Data Centre, B.C. Ministry of Environment, Lands and Parks, PO Box 9344, Stn Prov Govt, Victoria BC V8W 9M1 – Balsaminaceae through Betulaceae, Brassicaceae (except *Arabis*) through Cuscutaceae.

Mr. F. Lomer, 304-315 Hospital Street, New Westminster BC V3L 3L5 – Clusiaceae.

Ms. L. Milnes, 2124 McLaren Street, Victoria BC V8S 2P8 – Technical Descriptions.

Mr. D. Meidinger, Research Branch, B.C. Ministry of Forests, 712 Yates Street, 3rd Floor, Victoria, BC V8W 9C2 – Brassicaceae (except *Arabis*).

Mr. A. MacKinnon, Research Branch, B.C. Ministry of Forests, 712 Yates Street, 3rd Floor, Victoria, BC V8W 9C2 – Caryophyllaceae.

Dr. J. Pojar, Ministry of Forests and Lands, PO Box 5000, Smithers BC V0J 2N0 – Boraginaceae, Caprifoliaceae.

Dr. G. B. Straley, Botanical Garden, University of British Columbia, Vancouver BC V6T 1Z4 – *Arabis*.

Illustrated Flora of British Columbia - Volumes 1 and 2

ERRATA

Illustration scales for floral parts (both volumes):

These should be used with caution since a number of errors were introduced to the final copy. Almost all 0.5 mm scales should read 5 mm. Description measurements should be checked on the facing page for each part.

Volume 1

Page 2 - U.S. state abbreviations - at Missouri and Montana read MO and MT, respectively
Page 5 - end of 2nd paragraph - remove "constricted between the seeds".
Page 30 - line 2 down - after "bigleaf" add "maple"
Page 38 - line 13 down- after "sometimes" add "sharp-pointed at the tips"
Page 59 - illustration of *Glehnia littoralis* should read "*Glehnia littoralis* ssp. *leiocarpa*"
Page 74 - line 8 down- "30-1.8 m" should read "30-180 cm"
Page 74 - line 10 down- add "times" after "2-4"
Page 78 - line 10 down- add "cm" after "0.5-4"
Page 85 - at *Torilis arvensis* -" Habitat/Range" is missing -it should read as follows:

> Habitat/Range:
>
> Mesic to dry roadsides and waste places in the lowland zone; rare in SW BC, known only from the lower Fraser Valley; introduced from Eurasia.

Page 98 - at 2nd lead #8 of key - after "orange" add", red or pink (in *Agoseris lackschewitzii)*
Page 100 - line 9 up - "*Megalodontia*" should read *"Megalodonta"*
Page 114 - at *Ambrosia artemisiifolia* - at "General"- "0.1-1.0 cm" should read "10-100 cm tall
Page 140 - line 1 -"*chammissoni*" should read "*chamissonis*"
Page 141 - top two illustrations - replace "var." with "ssp."
Page 144 - at *Arnica gracilis* common name should read "tall mountain arnica"
Page 152 - line 10 up - "4.0-1.2 m" should read "40-120 cm"
Page 166 - 2nd lead #1 in key - replace "annuals" with "perennials"
Page 174 - line 1 down-"*Seriocarpos*" should read "*Sericocarpus*"
Page 176 - at *Aster ericoides* - at "General" - "3-20 cm" should read "30-200 cm"
Page 180 - line 14 down -"5-7 cm" should read "5-7 mm"
Page 184 - line 20 down - replace Lac La Biche" with "La Biche River"
Page 188- replace lines 13 to 19 with the following:

> **Flowers:** Heads with ray and disk flowers, few to more commonly many in a round-topped inflorescence; involucres 5-6 mm tall; involucral bracts often strongly graduated, linear to lanceolate, abruptly sharp-pointed, usually with a conspicuous, yellowish or brownish basal portion and with an evidently thin, dry, papery or sometimes minutely fringed margin; ray flowers 20-30, violet, 10-15 mm long; disk flowers yellow.
>
> **Fruits:** Achenes faintly nerved, more or less hairy; pappus usually reddish at maturity.

Page 194- at *Bidens tripartita* - at "General" -"0.2-1.0 cm" should read "20-100 cm"
Page 198- line 16 down -"1.5-2.5 mm" should read "1.5-2.5 cm"
Page 198- line 6 up - " 2-8 mm" should read "2-8 cm"
Page 202- line 6 down - "*Centaurea debauxii*" should read "*Centaurea debeauxii*"
Page 204- line 17 down - "*C. debauxii*" should read "*C. debeauxii*
Page 205- at illustration on upper right - "*Centaurea debauxii*" should read "*Centaurea debeauxii*

Page 208- at *Centaurea nigrescens* - common name - replace " knotweed" with "knapweed"
Page 216- line 14 up - "YK" should read "YT"
Page 232- line 9 down - "3-1.1 m" should read "30-110 cm"
Page 236- line 10 up - place "Fraser" after "lower"
Page 238- at *Erigeron* key, 1st lead #2 -"0.6-1.5 cm" should read 60-150 cm"
Page 258- line 17 down - replace "female" with "ray"
Page 260- line 13 up -*Eupatorium* key needed as follows:

1. Leaves glabrous, saw-toothed, the teeth several; disk flowers 5-6, reddish-mauve or whitish. .... *E. cannabinum*
1. Leaves with spreading, curly hairs below, saw-toothed, the teeth numerous; disk flowers 9-22, purple.......... *E. maculatum*

Page 262- line 7 down - "0.6-2.0 cm" should read "60-200 cm"
Page 262- line 3 up - remove "7-22 mm" and insert "1.5-2.5 cm"
Page 264- line 14 down - place "up to" in front of "4 cm"
Page 265- at *Filago minima* - the illustration for this species is incorrect -
Page 266- line 5 down - replace "River" with "Fraser Valley"
Page 268- line 1- replace "*C. ciliata*" with "*G. ciliatata*"
Page 270- line 2 up - replace "1 mm" with "1-3 mm"
Page 282- at 2nd key lead #13 "*lachenali*" should read "*lachenalii*"
Page 284- line 5 down - replace "River" with "Fraser Valley"
Page 288- line 11 up - replace "River" with "Fraser Valley"
Page 289- at *Hieracium pilosella* - the illustration for this species is incorrect -
Page 304- line 17 up - replace "River" with "Fraser Valley"
Page 307- illustration at bottom of page - "*Leontodon taraxacoides* should read "*Leucanthemum vulgare*"
Page 318- at 2nd key lead # 2 "galbrous" should read "glabrous"
Page 321- at lower illustration - replace "*nutan*" with "*nutans*"
Page 332- line 17 up - replace "0.6-1.2 long" with "0.6-1.2 mm long"
Page 336- line 4 down - place "Fraser" after "lower"
Page 344- line 17 up "*S. ogorukensis*" should read "*S. ogotorukensis*"
Page 346- line 13 up - add "Fraser" between "the" and "and"
Page 358- line 18 up - replace the "L." after *Senecio pseudarnica* with "Less."
Page 373- illustration of *Solidago spathulata* should read "*Solidago spathulata* var. *neomexicana*"
Page 375- at *Sonchus arvensis* var. *glabrescens* -the illustration for this species is incorrect
Page 378- line 2 up - replace "**Silicles**" with "**Fruits**"
Page 384- line 2 down "*Haplopappus lyalli*" should read "*Haplopappus lyallii*"
Page 384- line 7 up -replace "**Silicles**" with "**Fruits"**
Page 388- line 8 up - replace "narrower" with "broader"
Page 394- line 4 down - replace "CT" with "CA"
Page 395- lines 11 down and 13 up - replace "CT" with "CA"
Page 396- line 6 up - replace "MA" with "England"
Page 397- lines 18 down and line 8 up "*asiocarpa*" should read "*lasiocarpa*"
Page 399- line 14 down "ASCLEPADACEAE" should read "ASCLEPIADACEAE"
Page 401- line 3 down "*riophyllum*" should read "*Eriophyllum*"
Page 401- lines 7 and 10 down "Inulae" should read "Inuleae"
Page 401- line 20 up "Anthernidae" and "Eupatoreae" should read "Anthemidae" and "Eupatorieae"
Page 401- line 5 up -add "No." after "University" and replace "CLVII.I" with "CLVIII"
Page 402- line 11 down "flavinoids" should read "flavonoids"

Page 402- line 19 down - replace "MA" with "England"
Page 403- line 7 down - replace "MA" with "England"
Page 403- line 18 down - replace "CT" with "CA"
Page 405- line 7 down "*Anthriscus cerefolium*" should not appear in the EXCLUDED
Page 407- line 14 down "*Galinsoga parviflora*" should not appear in the EXCLUDED
Page 407- line 11 up "*Helianthus maximillianii*" should read "*Helianthus maximilianii* and "has not been collected since 1954"
Page 407- line 9 up - replace "Bolvin" with "Boivin"
Page 408- first line "*Hieracium caespitosum*" should not appear in the EXCLUDED
Page 409- line 13 down "*bierbersteinii*' should read "*biebersteinii*"
Page 409- line 16 down - replace "*Ericameria nauseosa* var. *speciosa*" with "*Ericameria nauseosus* var. *speciosa*"
Page 409- line 17 down - replace "*Conoselinum*" with "*Conioselinum*"
Page 409- line 8 up - replace "*lyalli*" with "*lyallii*" in both genera
Page 425- line 29 down right column - add "**tall mountain** 144" after "**streambank** 136"
Page 430- line 13 up (right side) - replace "*lyalli*" with "*lyallii*"
Page 433- line 24 down left column - remove "**mountain, tall** 136"
Page 435- line 21 up left column "*Seriocarpus*" should read "*Sericocarpus*"

Volume 2

Page 2 - U.S. state abbreviations - at Missouri and Montana read MO and MT, respectively
Page 6 - line 22 down - replace "0.1-1.5 cm" with "1.0-1.5 cm"
Page 6 - line 10 up - replace "2.5 mm" with "10-20 mm"
Page 8 - line 13 down - replace "2. 5 cm" with "1.5-2.5 cm"
Page 8 - line 11 up - replace "2.5 cm" with "10-20 mm"
Page 8 - last line - "2.5 cm" should read "10-20 mm"
Page 13 - *Mahonia repens* flower and fruit drawings are incorrect
Page 44 - line 1 - "Opiz Bercht" should be "Opiz ex Bercht"
Page 66 - Footnote, "Hitchcock *et al.* (1994)" should read "Hitchcock *et al.* (1964)"
Page 69 - 1st lead #31 (*Cardaria*) - replace "leaves 1-0 cm long" with "leaves 4-10 cm long"
Page 71 - 2nd lead #68 - *Aphragmus*, which was omitted from the Brassicaceae key, will key here. It differs from *Arabis* in having spoon-shaped leaves.
Page 95 - Illustration of *Barbarea vulgaris* is incorrect
Page 98 - line 22 down - replace "stems 3.0-1.0 m tall' with "stems 30-100 cm tall"
Page 114 - line 22 down - replace "3-15 mm long" with "3-15 cm long"
Page 124 - 1st lead #4 - "*D. nemerosa*" should read "*D. nemorosa*"
Page 150 - line 12 down - replace "3-10 mm" with "3-10 cm"
Page 153 - Illustration of *E. cheiri* is incorrect
Page 157 - Illustration of *Hutchinsia procumbens* is incorrect
Page 174 - 2nd lead #6 - remove "silicles" appearing before "pimpled"
Page 174 - at leads #7 - the species should be reversed
Page 182 - line 4 up - replace "8-15 long" with "8-15 mm long"
Page 188 - line 3 down "var. *lacinatum*" should read "var. *laciniatum*"
Page 188 - both leads in key (*Thelypodium*) - replace all "silicles" with "siliques"
Page 194 - line 21 up - replace "less than 1 mm" with "1-2 mm"
Page 222 - 2nd lead #2 in *Sambucus racemosa* key - replace "In Cronq. & Hitchcock" with "*In* Hitchcock & Cronquist"

Page 231 - add *Dianthus* after lead #18

Page 272 - last line - replace "var. *subcaulescens*" with "var. *subacaulescens*"

Page 273 - upper left illustrations - replace " var. *subcaulescens*" with "var. *subacaulescens*"

Page 300 - line 24 down - replace "Chinnapappa" with "Chinnappa"

Page 304 - 1st lead #1 in key - replace "*Pachistima*" with "*Paxistima*"

Page 316 - line 21 down - replace "cm" with "m"

Page 328 - lines 18 and 19 down - replace "cm" with "m" in each line

Page 330 - lines 14 and 15 up "*Polycnemon*" should read "*Polycnemum*"

Page 331 - Illustration at bottom "*Polycnemon*" should read "*Polycnemum*"

Page 344 - line 12 down - replace "4-7 mm" with "4-7 cm"

Page 344 - line 17 up - replace "3-6 mm" with "3-6 cm"

Page 362 - line 4 down - replace "CT" with "CA"

Page 363 - line 11 down - replace "CT" with "CA"

Page 363 - line 13 up - replace "CT" with "CA"

Page 366 - line 16 up - remove Mulligan1971c reference

Page 370 - line 3 down - replace "*C. septum*" with "*C. sepium*"

Page 372 - *Lobularia maritima* should be removed here (It should be included)

Page 373 - line 13 down - replace "*chalapensis*" with "*chalepensis*" in both instances

Page 373 - line 17 down - replace "*Descurainea*" with "*Descurainia*"

Page 373 - lines 11, 13 and 14 up - replace "*Rorripa*" with "*Rorippa*"

Page 373 - line 3 up - replace "*Thelluginiella*" with "*Thellungiella*"

Page 373 - line 2 up - replace "*Thellypodium*" with "*Thelypodium*"

Page 373 - line 2 up - replace "*lacinatum*" with "*laciniatum*"

Page 397 - line 22 down, right column - replace "*Polycnemon*" with "*Polycnemum*"

Gerald Bane Straley
(1945–1997)

Prior to the publication of this manual, Gerald B. Straley passed away after a lengthy illness. With Gerald's passing, British Columbia lost an excellent botanist and an extremely knowledgable plantsman.

Gerald's poor health in recent years made it very difficult for him to work, especially when the research required intense concentration. As we began to revise our earlier work, *The Vascular Plants of British Columbia*, and produce the *Illustrated Flora*, we realized that Gerald would not be able to carry the research load of the former manual. Despite this, he continued to work, and true to form, Gerald chose one of our most difficult genera, *Arabis*. This genus has always been perplexing and a continual headache for botanists. To make matters even more difficult two very competent systematic botanists, R.C. Rollins and G.A. Mulligan, produced respectively, a North American treatment of *Arabis* in 1993 and a Canadian treatment of *Arabis* in 1995. These treatments varied markedly in their taxonomic concepts. Gerald was forced to re-examine most of our British Columbia material in light of each author's description and interpetation of each species. The result is an impressive, practical treatment for our 20 *Arabis* taxa that will be with us for years to come.

Gerald's persistence, patience and mellow approach during the frustrating months of working on *Arabis*, reflected his remarkable outlook towards life. We have indeed lost an irreplaceable friend and scientist.

George W. Douglas

ACKNOWLEDGEMENTS

Gerry A. Allen provided valuable suggestions during the initial stages of preparation for *The Vascular Plants of British Columbia* project. Gerry A. Allen, Adolf Ceska and Fred Ganders reviewed taxonomically difficult groups for the latter project.

The present flora was meticulously reviewed by Frank Lomer, Ernie McNaughton, Malcolm Martin and Hans Roemer. The latter four botanists and Anna Roberts also provided the senior editor with manuscript corrections and range information during the past several years. W. (Bill) J. Cody and Gerry Mulligan kindly searched the DAO herbarium and provided information on several families.

Special thanks are due to our illustrators: Donald Gunn, Gail F. Harcombe, Jane Lee Ling, Sherry Mitchell, Lora M. Richards and Elizabeth J. Stephen.

Many of the initial write-ups for the technical descriptions were contributed by Lynne Milnes. Technical descriptions for Brassicaceae were researched by Tina Lee. Andy MacKinnon undertook the rather large task of completing technical descriptions for the Caryophylliaceae. Marie Fontaine and Jenifer Penny made numerous corrections and provided aid on late drafts. Sharon Hartwell proofread the final draft of the manuscript.

Elizabeth Campbell prepared the glossary and Marie Fontaine revised it with respect to terms used the present text. Art prepress and final typesetting were done by Frances Hunter, and Gail F. Harcombe coordinated the prepress and print production of the flora. Cover design was created by Annie Weeks.

The curators of all of the herbaria visited during this study deserve a special thanks for providing help whenever needed. The herbaria curators and staff include: George Argus and Michael Shchepanek (The National Herbarium, National Museums, Ottawa); T. Christopher Brayshaw, John Pinder-Moss, Jim Cosgrove, Joan Kerik, Lesley Kennes and Leon Pavlick (Herbarium, Royal British Columbia Museum, Victoria); William Cody, Gisele Mitrow and Paul Catling (Vascular Plant Herbarium, Department of Agriculture, Ottawa); Olivia Lee and Jack Maze (Herbarium, Department of Botany, University of British Columbia, Vancouver); Brenda Constanzo and Gerry Allen (Herbarium, Department of Biology, University of Victoria); Ray Coupé and Anna Roberts (Herbarium, B.C. Ministry of Forests, Williams Lake) and Jim Pojar and Karen McKeown (Herbarium, B.C. Ministry of Forests, Smithers).

The production of this extension project was made possible through the funding support of Forest Renewal British Columbia (FRBC) and the B.C. Ministry of Environment, Lands and Parks, Wildlife Branch and Resources Inventory Branch.

TABLE OF CONTENTS

INTRODUCTION

This flora, in eight volumes, keys, describes and illustrates all of the flowering plants and vascular cryptogams, both native and naturalized, known to occur at the time of writing within the borders of British Columbia. In part, it is a revision of *The Vascular Plants of British Columbia* (Douglas et al. 1989, 1990, 1991, 1994). Keys are included for all genera, species, subspecies and varieties. Species previously recorded for the province but excluded from this treatment appear in an annotated list in Appendix I.

With respect to this volume, 3 new species taxa, discovered in the province in recent years, have been added. Recent taxonomical or nomenclatural research has also resulted in new name changes for 40 taxa since Douglas (1989). These new name changes are listed in Appendix II.

FORMAT

The floristic treatment is in alphabetical order by families, genera and species since most users may not be familiar with the more traditional "Englerian system". These users (e.g., foresters, wildlife specialists, park managers, amateur naturalists, students, etc.) often require specific information about a species. Thus we have attempted to include the most frequently required information. Each entry includes the following:

Synonyms
Synonyms are given only when there are references in major or pertinent floristic treatments to a given species by another name or names.

English Names
An English name for each species is included. These are based mainly on names appearing most frequently in the western North American literature.

Habitat and Moisture Regime
A brief description of each species habitat preference, including moisture regime, is given. The latter uses a simple (wet - moist - mesic - dry) scale.

Vegetation Zones
The elevational limits of a species are presented using vegetation zones. The zonal system used here, unlike some of the regional zonation systems (Krajina 1965; Rowe 1959, 1972; Franklin and Dyrness 1973), has wide application in the northern hemisphere and differs only slightly from that proposed by Danserau (1975) and Meusel et al. (1965) and recommended by Löve (1970) in a review of the subject. The *montane zone* includes all continuous forests in British Columbia except for the coastal lowlands and some islands, which are included in the *lowland zone*. The upper forests in southern British Columbia, included by some ecologists (e.g., Krajina 1965) in the subalpine zone, are considered the *upper montane zone* in this work. The *subalpine zone* is defined as that area above the montane zone and below the upper limit of conifers as an upright tree form (Douglas l972). In the southern part of the province, subalpine vegetation consists of a meadow/tree-clump complex (Douglas 1971, 1972), while in the northern part of the province the subalpine zone is dominated by tall shrubs (mainly two-to four-metre *Salix* and *Betula*) and scattered trees (Douglas 1974, Krajina 1975). Above the subalpine zone is the *alpine zone* where trees occur only in krummholz (or dwarfed) form and the vegetation is extremely short (less than one metre) and commonly referred to as alpine tundra (Krajina 1969, Douglas 1972, Douglas and Bliss 1977). The *steppe zone* (Daubenmire 1970) occurs in the interior of the province and includes what is often referred to as sagebrush or grassland vegetation (Tisdale 1947). This zone occurs at lower elevations in the Thompson, Okanagan, Kootenay and Columbia River valleys in southern British Columbia. Parts of the Fraser and Thompson plateaus also have grassland or steppe vegetation.

Abundance

The abundance of a species is given by a general scale: rare – infrequent – frequent – common. The rare plants are well documented in Douglas et al. (1998). Most of the common plants are also well known from numerous other sources. However, limited collecting or research enabled us to make only general assumptions about the abundance of some of the infrequent or frequent species.

Range

The general distribution of a species, both within and outside the province, is presented (Figure 1). Maps compiled at the Royal British Columbia Museum were extremely helpful for the more common species. Rare plant distributions were provided by the maps in Douglas et al. (1998). Abbreviations used within the province are as follows:

E - Eastern	S - Southern
EC - Eastcentral	SC - Southcentral
N - Northern	SE - Southeastern
NE - Northeastern	SW - Southwestern
NW - Northwestern	WC - Westcentral

For ranges outside of B.C. the outer peripheries (i.e., provinces/states) of the ranges are given using the following abbreviations:

Alberta	AB	Maine	ME
British Columbia	BC	Maryland	MD
Manitoba	MB	Massachusetts	MA
New Brunswick	NB	Michigan	MI
Newfoundland	NF	Minnesota	MN
Northwest Territories	NT	Mississippi	MS
Nova Scotia	NS	Missouri	M
Ontario	ON	Montana	M
Prince Edward Island	PE	Nebraska	NE
Quebec	PQ	Nevada	NV
Saskatchewan	SK	New Hampshire	NH
Yukon	YT	New Jersey	NJ
		New Mexico	NM
Alabama	AL	New York	NY
Alaska	AK	North Carolina	NC
Arizona	AZ	North Dakota	ND
Arkansas	AR	Ohio	OH
California	CA	Oklahoma	OK
Colorado	CO	Oregon	OR
Connecticut	CT	Pennsylvania	PA
Delaware	DE	Rhode Island	RI
District of Columbia	DC	South Carolina	SC
Florida	FL	South Dakota	SD
Georgia	GA	Tennessee	TN
Hawaii	HI	Texas	TX
Idaho	ID	Utah	UT
Illinois	IL	Vermont	VT
Indiana	IN	Virginia	VA
Iowa	IA	Washington	WA
Kansas	KS	West Virginia	WV
Kentucky	KY	Wisconsin	WI
Louisiana	LA	Wyoming	WY
Mexico	MX		

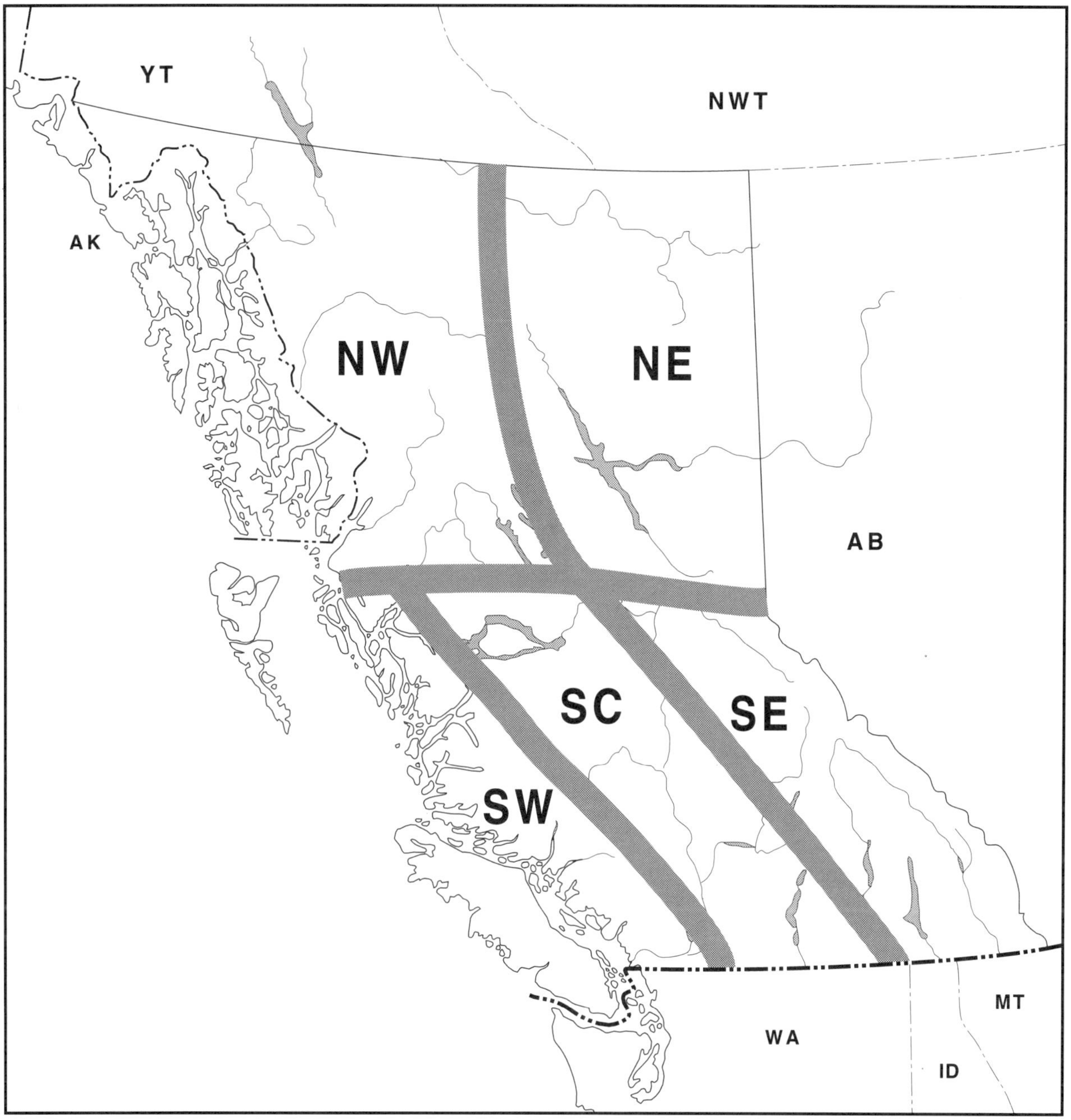

Figure 1. Geographical Regions of British Columbia.

Descriptions

Major morphological characters are given for each taxon. In most cases technical terminology has been kept to a minimum since most of our users have a limited botanical background. All terms, both technical and simplified appear in the glossary.

Notes

Specific items of importance, or problems, especially those related to taxonomy and nomenclature, are noted. Infraspecific taxa worthy of recognition are keyed here.

A floristic bibliography, arranged by families, is also appended for users wishing to study a species further.

DATABASE

The present manual is based on a wealth of regional publications dating back over a hundred years. In addition, most of the species (especially the rare and infrequent ones) have been examined in various Canadian herbaria. Historically, the floristic documentation of British Columbia's flowering plants and vascular cryptogams began with the enumeration of the extensive collections of J. Macoun (1883-1890). This was soon followed by the botanical contributions of his son, J.M. Macoun (1889, 1894-1906). In 1915, the first provincial flora appeared (Henry 1915), treating the southern part of the province. A supplement to this work was later provided by Eastham (1947). The southern part of the province was also included in a treatment of Pacific Northwest vascular plants by Hitchcock et al. (1955-1969) – later edited and revised by Hitchcock and Cronquist (1973). During this same period several other major works also appeared. Boivin (1966-1967), after numerous visits to Canadian and American herbaria, produced a checklist of Canadian vascular plants. In his *Flora of Alaska and Neighboring Territories*, Hultén (1968) provided brief descriptions with range maps that, in many cases, included British Columbian species. Meanwhile, in a treatment of the Queen Charlotte Island's vascular plants, Calder and Taylor (1968) contributed one of the few floras dealing solely with a part of British Columbia. British Columbia was again included in a regional flora of northwestern North America when Welsh (1974) treated the plants of Alaska and adjacent parts of Canada, including northern British Columbia. A literature review by Taylor and MacBryde (1977), mainly of North American works, resulted in a resource inventory of vascular plants for British Columbia. (Unfortunately, this extensive review treated several hundred erroneous citations as valid records for the province. It also ignored many herbaria records that had never been published.) Recently, a modern Canadian flora was produced by Scoggan (1978-1979). All of the province's rare or infrequent plants were examined and verified in various Canadian herbaria by Straley et al. (1985) and, more recently, by Douglas et al. (1998). Finally, the B.C. flora was documented, annotated and all plants keyed in *The Vascular Plants of B.C.* (Douglas et al. 1989, 1990, 1991, 1994).

In addition to the major works mentioned above, there have been numerous taxonomic, floristic and related research papers dealing with British Columbia's vascular plants. Many of these are cited in the floristic bibliography by Douglas et al. (1983) and many are specifically referenced in this manual.

This study was supplemented by the examination of specimens in all major Canadian herbaria and some US herbaria (e.g., UC, CAS, WTU), as well as a complete review of the pertinent literature.

Although decriptions and illustrations are provided for all of British Columbia's plant species in this manual, additional descriptions and/or illustrations may also be found in a number of floras (Abrams 1923-1951, Argus 1973, Cody 1996, Fernald 1950, Ferris 1960, Hickman 1993, Hitchcock *et al.* 1955-1969, Packer 1983, Porsild and Cody 1980, Tutin et al. 1964-1980, Welsh 1974) or in treatments of some British Columbian plant families (Brayshaw 1976, 1985, 1996a, 1996b; Douglas 1982, 1995; Szczawinski 1959, 1962; Taylor 1963, 1966, 1970, 1973a, 1974a, 1974b, 1983).

TAXONOMIC CONCEPTS

Our view of species limits is a practical one. Therefore, species had to be readily characterized morphologically in order to produce user-friendly keys. Many of the keys were newly constructed while others were adapted for our British Columbia plants from well-tested manuals.

Wherever possible, geographically separate, or largely separate, infraspecific taxa are treated at the subspecies level, while those with sympatric ranges are treated at the varietal level. Since we have not made any nomenclatural changes associated with the production of the manual, this distinction between subspecies and variety was not always possible. With the exception of *Sambucus racemosa,* we have not used more than one infraspecific rank in this work for a single species. Taxa are not recognized at the *forma* level.

BALSAMINACEAE

IMPATIENS

1. Pouched sepals not spurred .... *I. ecalcarata*
1. Pouched sepals spurred.
   2. Leaves opposite or whorled (at least some of them, finely saw-toothed; flowers blue to purple or purplish-pink or sometimes white .... *I. glandulifera*
   2. Leaves alternate, coarsely saw-toothed; flowers orange or yellowish.
      3. Spurs of the sepals straight, directed backward .... *I. parviflora*
      3. Spurs of the sepals curved or hooked.
         4. Flowers less than 2 cm long, without spots .... *I. aurella*
         4. Flowers more than 2 cm long, spotted, mottled or sparingly flecked.
            5. Flowers orange, brown-spotted or -mottled .... *I. capensis*
            5. Flowers yellow, sparingly flecked with brownish-purple .... *I. noli-tangere*

Impatiens aurella **Rydb.**
orange touch-me-not
HABITAT/RANGE:
Moist streambanks and meadows in the steppe or lower montane zones; rare in S BC east of the Coast-Cascade Mountains S of 55° N; S to ID and MT.
DESCRIPTION:
General: Annual succulent herb; stems erect to ascending, often branched, glabrous, 0.5-1.0 m tall.
Leaves: Alternate, egg-shaped to elliptic, coarsely saw-toothed, 2-8 cm long.
Flowers: One to several in the leaf axils, orange, unspotted, mostly 0.1-1.5 cm long; sepals pouched with spur curved or deflexed forward.
Fruits: Capsules, elastically dehiscent, 1-2 cm long; many-seeded.

Impatiens capensis **Meerb.** (*I. biflora* Walt, *I. noli-tangere* ssp. *biflora* [Walt.] Hult.)
spotted touch-me-not
HABITAT/RANGE:
Moist forests and lake edges in the lowland, steppe and lower montane zones; frequent in SW and SC BC; E to MB and MN and S to NE WA, MT, AL and FL.
DESCRIPTION:
General: Annual succulent herb; stems erect to ascending, often branched, glabrous, 20-80 cm tall.
Leaves: Alternate, stalked, egg-shaped to elliptic, shallowly and remotely saw-toothed, 3-12 cm long.
Flowers: One to several in leaf axils, orange with mottled reddish-brown spots; sepals pouched, with spur curved or hooked.
Fruits: Capsules, elastically dehiscent, 2.5 mm long, many-seeded.

Impatiens ecalcarata **Blank.** (*I. biflora* Walt. var. *ecalcarata* [Blank.] M. E. Jones)
spurless touch-me-not; western touch-me-not
HABITAT/RANGE:
Moist forests in the montane zone; rare in SC and SE BC; S to MT and N OR.
DESCRIPTION:
General: Annual succulent herb; stems erect to ascending, freely branched, glabrous, 0.4-1.0 m tall.
Leaves: Alternate, stalked, elliptic to elliptic-lanceolate, coarsely saw-toothed, 3-10 cm long.
Flowers: Mostly paired in leaf axils, pale yellow to orange, unspotted; sepals pouched but not spurred.
Fruits: Capsules, elastically dehiscent, 2.5 cm long, many-seeded.

Impatiens aurella

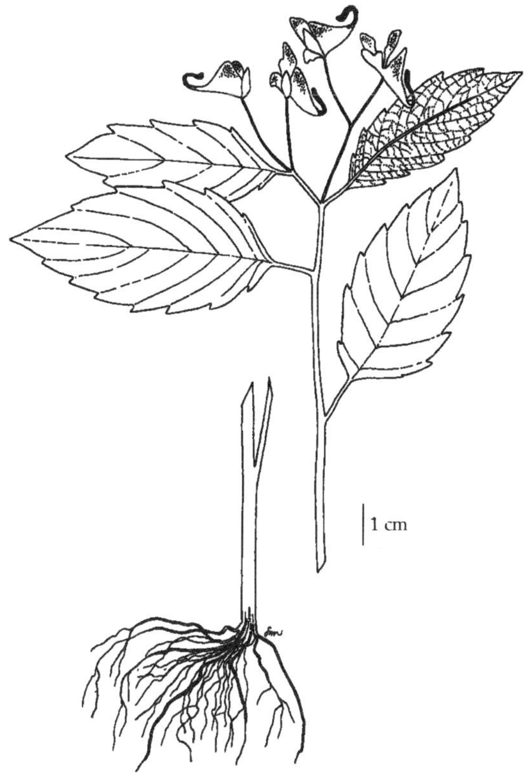

Impatiens capensis

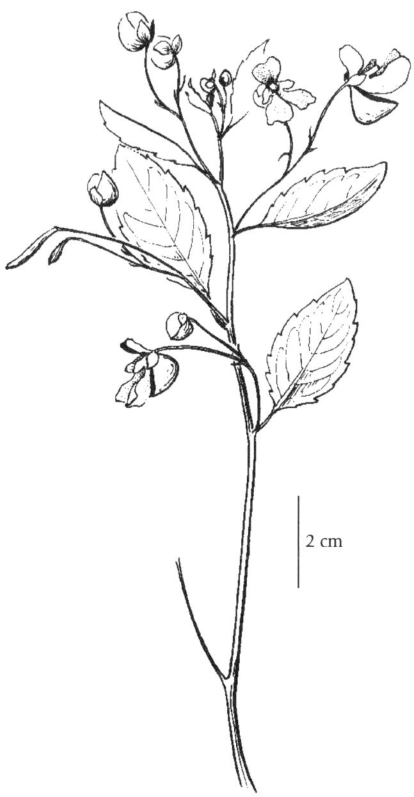

Impatiens ecalcarata

Impatiens glandulifera **Royle** (*I. roylei* Walp.)

policeman's helmet

HABITAT/RANGE:

Moist to mesic streambanks, meadows, roadsides and waste places in the lowland and steppe zones; infrequent in the lower Fraser Valley and extreme SC BC; introduced from Asia.

DESCRIPTION:

General: Annual succulent herb; stems erect to ascending, often branched, hairy, often purplished-tinged, 0.6-2.0 m tall.

Leaves: Opposite to whorled (at least some of them), stalked, egg-shaped to elliptic, sharply and closely saw-toothed, 6-15 cm long.

Flowers: One to several in leaf axils, whitish to red, usually purplish-spotted; sepals pouched, with a short recurved spur.

Fruits: Capsules, elastically dehiscent, 2.5 cm long, many-seeded.

Impatiens noli-tangere **L.** (*I. occidentalis* Rydb.)

common touch-me-not

HABITAT/RANGE:

Moist streambanks and forests in the lowland, steppe and montane zones; infrequent throughout BC south of 57° N; circumboreal, N to SE AK, E to MB, and S to ID and OR; Eurasia.

DESCRIPTION:

General: Annual succulent herb; stems erect to ascending, often branched, hairy, 20-60 cm tall.

Leaves: Alternate, stalked, elliptic to egg-shaped, coarsely saw-toothed, 3-12 cm long.

Flowers: One to several in leaf axils, yellow, sparingly flecked with brownish-purple; sepals pouched, with spurs curved or deflexed forward.

Fruits: Capsules, elastically dehiscent, 2.5 cm long, many-seeded.

Impatiens parviflora **DC.**

small touch-me-not; balsam

HABITAT/RANGE:

Moist disturbed areas in the lowland zone; infrequent in SW BC, known only from the Vancouver area; introduced from Asia.

DESCRIPTION:

General: Annual succulent herb; stems erect to ascending, often branched, glabrous, 20-80 cm tall.

Leaves: Alternate, stalked, egg-shaped to elliptic, finely and sharply saw-toothed, 3-12 cm long.

Flowers: One to several in leaf axils, pale yellow; sepals pouched, with spurs straight and directed backward.

Fruits: Capsules, elastically dehiscent, 2.5 cm long, many-seeded.

Impatiens glandulifera

Impatiens noli-tangere

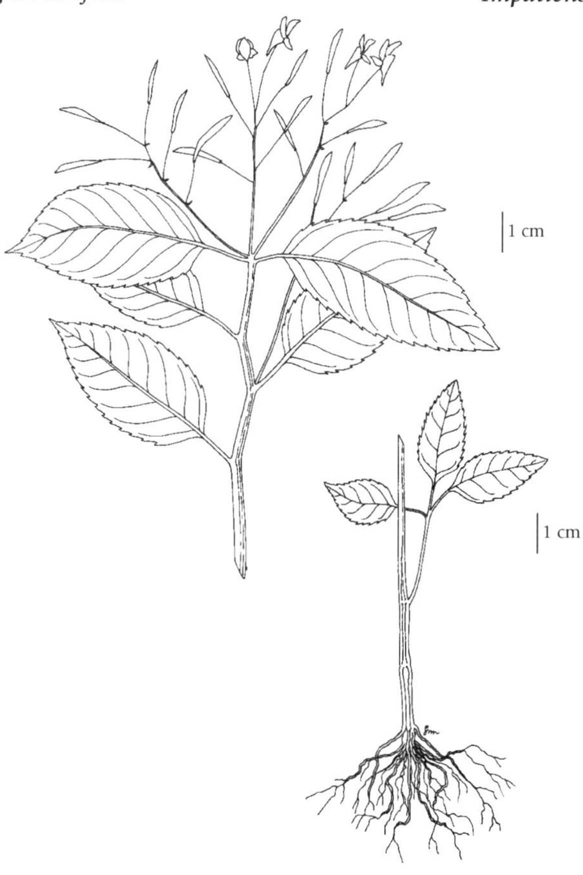

Impatiens parviflora

BERBERIDACEAE

1. Leaves simple .... *Berberis*
1. Leaves pinnate or compound with 3 leaflets.
    2. Leaves pinnate, margins with small spines, evergreen; shrubs .... *Mahonia*
    2. Leaves compound with 3 leaflets, these 3-9 lobed, deciduous; herbs .... *Achlys*

ACHLYS

Achlys triphylla **(J.E. Smith) DC.** (*A. californica* Fukada & Baker)
vanilla-leaf; deer foot

Habitat/Range:

Moist to mesic open or closed forests in the lowland and montane zones; frequent in SW BC; known from Vancouver Island, the Gulf Islands and the adjacent lower mainland; S to NW CA.

Description:

General: Deciduous perennial herb from a slender rhizome, spreading; aerial stems absent but single leaves are sent up along the rhizome.

Leaves: Basal, palmately compound with 3 leaflets, blades fan-shaped, 3-9 lobed, deciduous, with a vanilla-like fragrance when dry.

Flowers: Inflorescence in compact spikes, 2-5 cm long; sepals and petals lacking; stamens white, 8-20 mm long.

Fruits: Achenes dark brown to reddish-purple, 3-4 mm long, finely hairy, with a somewhat fleshy to leathery thickened ridge.

Notes: Recently some taxonomists (e.g., Williams 1993, Kartesz 1994) have recognized a second species (*A. californica*) in western North America. The latter is characterized by having a central leaflet with 3 lobes, whereas *A. triphylla* has a central leaflet with 6-8 lobes. Examination of herbaria material reveals that many sheets have mixed collections of plants with central leaflets with both 3 lobes (actually the 3-lobed ones usually have 2 additional, much smaller lobes, on the sides) and 7-9 lobes. This has also been noted in the field. At this time, we are not convinced that *A. californica* is worthy of recognition.

BERBERIS

Berberis vulgaris **L.**
common barberry

Habitat/Range:

Mesic to dry disturbed areas in the lowland and steppe zones; infrequent in SW and SC BC; introduced from Eurasia.

Description:

General: Clump-forming shrub; stems erect, freely branched, 0.5-2.5 m tall; branches angled, grey or yellowish-grey.

Leaves: Deciduous, appearing simple, in fascicles in axils of branched spines, egg-shaped, 1-6 cm long, coarsely spine-toothed, prominently veined beneath.

Flowers: Inflorescence of hanging or arching racemes; racemes 2-6 cm long, arising in the leaf axils; petals yellow, entire.

Fruits: Oblong berries, 1 cm long, red or purple.

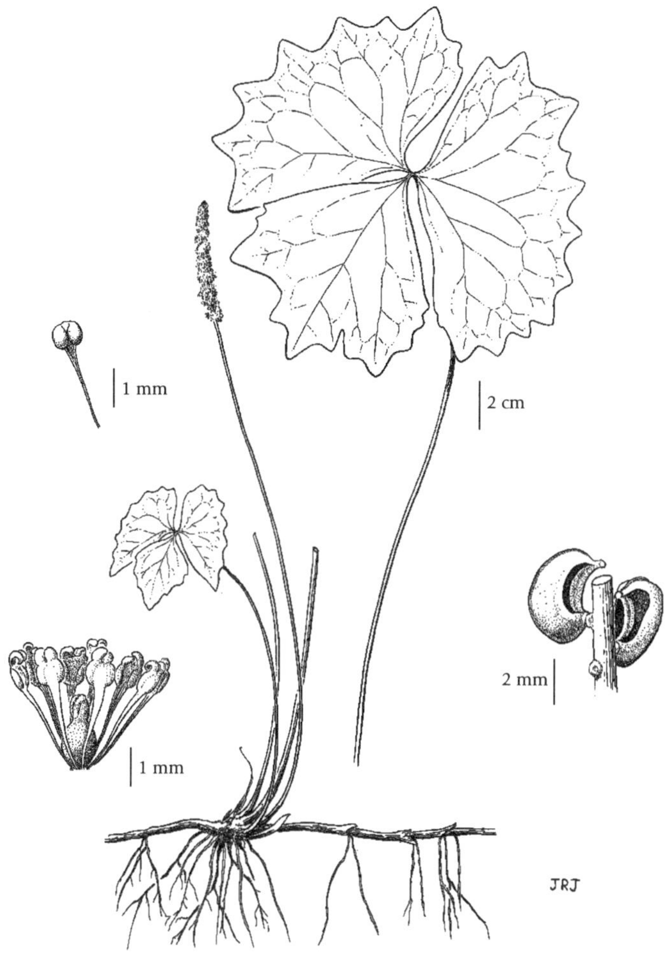

Achlys triphylla

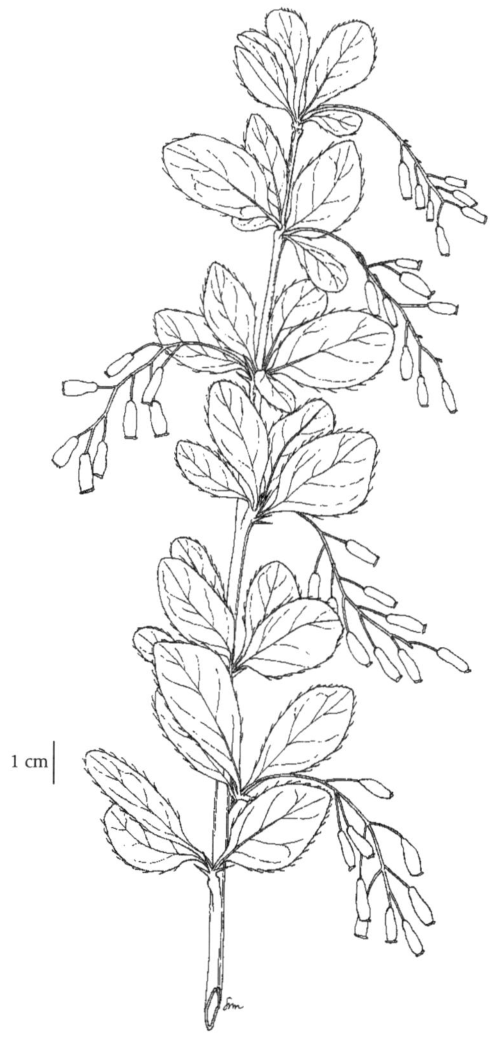

Berberis vulgaris

MAHONIA

1. Leaflets usually 9-19, palmately nerved .... *M. nervosa*
1. Leaflets usually 3-11, pinnately nerved.
   2. Leaves more than twice as long as broad, mostly with 12-29 prominent spiny teeth; leaflets 5-11; plants 0.5-4.5 m tall .... *M. aquifolium*
   2. Leaves less than twice as long as broad, with 15-43 inconspicuous, spiny teeth; leaflets 3-7; plants 0.1-0.6 m tall .... *M. repens*

***Mahonia aquifolium* (Pursh) Nutt.** (*Berberis aquifolium* Pursh)
tall Oregon-grape

HABITAT/RANGE:

Mesic to dry open or closed forests in the lowland, steppe and montane zones; common in S BC; E to AB and S to ID and OR.

DESCRIPTION:

General: Evergreen shrub from a rhizome; stems branched, stoloniferous to stiffly erect, 0.5-2.5 (4.5) m tall; bark and wood yellowish.

Leaves: Evergreen, hollylike, alternate, pinnate; leaflets usually 5-11, pinnately nerved, more than twice as long as broad, mostly with 12-29 prominent spiny teeth, shiny above but less so beneath.

Flowers: Inflorescence of clustered racemes 3-8 cm long; flowers yellow, the segments in 6's.

Fruits: Clustered berries 7-14 mm in diameter, blue, glaucous, with a few large seeds, edible.

***Mahonia nervosa* (Pursh) Nutt.** (*Berberis nervosa* Pursh)
dull Oregon-grape

HABITAT/RANGE:

Mesic to dry open slopes and open forests in the lowland, steppe and montane zones; common in SW BC, infrequent eastward in S BC; S to CA.

DESCRIPTION:

General: Evergreen shrub from a rhizome; stems branched, erect, 10-60 cm tall, leafy; bark and wood yellowish.

Leaves: Evergreen, hollylike, turning reddish in the fall, alternate, pinnate; leaflets 9-19, palmately nerved, with spiny teeth.

Flowers: Inflorescence of clustered racemes up to 20 cm long; flowers yellow, the segments in 6's.

Fruits: Clustered berries 8-11 mm in diameter, blue, glaucous, with a few large seeds, edible.

***Mahonia repens* (Lindl.) G. Don** (*Berberis aquifolium* Pursh var. *repens* [Lindl.] Scoggan, *B. repens* Lindl.)
creeping Oregon-grape

HABITAT/RANGE:

Dry rocky slopes and open forests in the steppe and montane zones; common in SE BC, rare in C and SC BC; E to AB and S to SD, TX, NM, NV and NE CA.

DESCRIPTION:

General: Evergreen shrub from a rhizome; stems branched, more or less procumbent, widely stoloniferous, 15-100 cm tall; bark and wood yellowish.

Leaves: Evergreen, hollylike, alternate, pinnate; leaflets 3-7, pinnately nerved, less than twice as long as broad, with 15-43 inconspicuous spiny teeth.

Flowers: Inflorescence of clustered racemes 3-8 cm long; flowers yellow, the segments in 6's.

Fruits: Clustered berries 8-14 mm in diameter, blue, glaucous, with a few large seeds, edible.

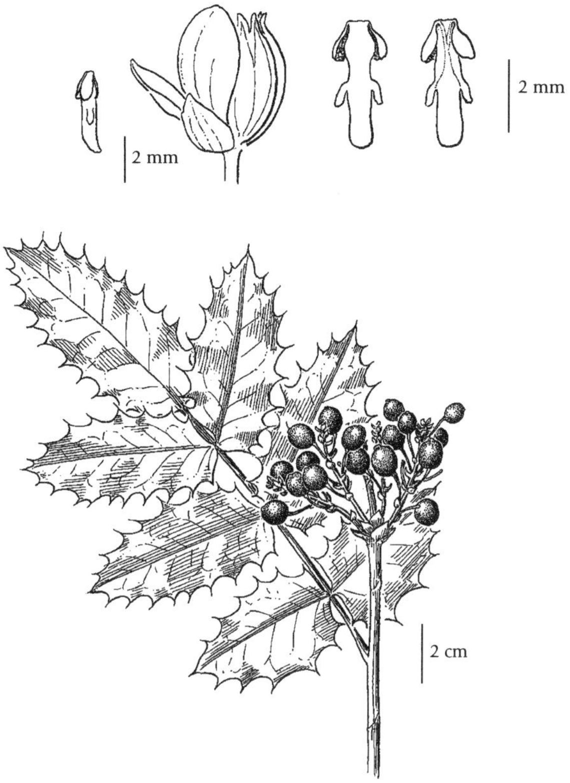

Mahonia aquifolium

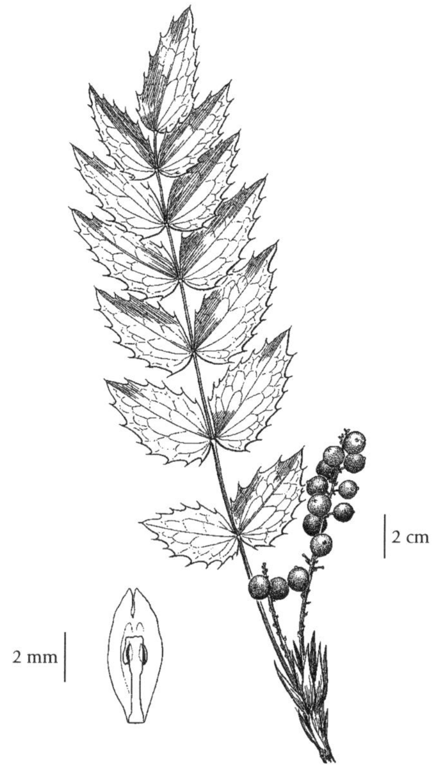

Mahonia nervosa

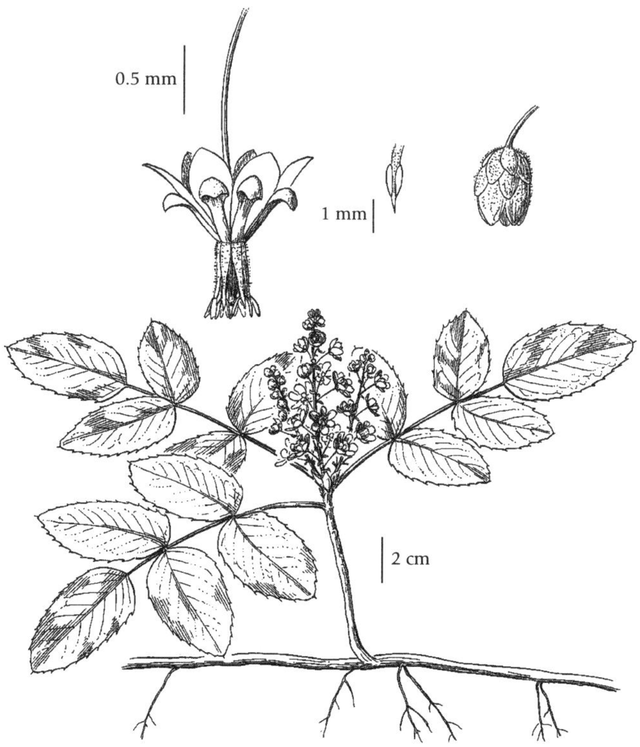

Mahonia repens

BETULACEAE

1. Fruit a rounded nut, borne in a flasklike involucre of fused bractlets; leaves usually heart-shaped ... *Corylus*
1. Fruit a flattened nutlet, without an involucre; leaves rarely heart-shaped.
    2. Fruiting female catkin conelike, hardened and persistent after the release of the nutlets; pith usually 3-angled in cross section .... *Alnus*
    2. Fruiting female catkin cylindric, deciduous with the release of the nutlets; pith usually flattened in cross section .... *Betula*

ALNUS

1. Leaves finely once or twice saw-toothed; axillary buds unstalked, pointed; male catkins unstalked; stalks as long as or longer than the conelike silicles .... *A. viridis*
1. Leaves coarsely to irregularly round-toothed; axillary buds pedunculate, blunt or short-pointed; male catkins stalked; stalks shorter than the conelike silicles.
    2. Leaf margins rolled under; nutlets with narrow-winged margins .... *A. rubra*
    2. Leaf margins not rolled under; nutlets wingless .... *A. incana*

***Alnus incana* [L.] Moench ssp. *tenuifolia* [Nutt.] Breit.** (*A. incana* [L.] Moench ssp. *rugosa* [DuRoi] Clausen var. *occidentalis* [Dippel] C.L. Hitchc., *A. tenuifolia* Nutt. var. *occidentalis* [Dippel] Collier, *A. tenuifolia* Nutt.)
mountain alder

Habitat/Range:

Moist forests, streamsides, bogs and fens in the montane zone; common in BC east of the Coast-Cascade Mountains; N to AK, YT and NT, E to SK, and S to NM, AZ and CA.

Description:

General: Deciduous shrub or tree, up to 12 m tall, usually 2-5 m tall, new growth short-hairy; axillary buds with short stalks; bark scaly, often lichen-covered, yellowish-brown or grey.

Leaves: Alternate, deciduous, smooth, coarsely to irregularly toothed, the teeth pointing outwards, leaf margins not rolled under, brownish in the fall.

Flowers: Inflorescence of male and female catkins, which open before the leaves on previous year's growth; male catkins with stalks.

Fruits: Small nutlets, without wings; female cones 1-1.5 cm long, egg-shaped.

***Alnus rubra* Bong.** (*A. oregona* Nutt.)
red alder

Habitat/Range:

Moist woodlands, forests, floodplains and clearcuts in the lowland and montane zones; common in coastal BC; N to SE AK and S to CA.

Description:

General: Deciduous shrub or tree, up to 25 m tall; axillary buds with stalks; bark scaly, often lichen-covered, yellowish-brown or grey-splotched with white.

Leaves: Alternate, deciduous, smooth, coarsely to irregularly toothed, the teeth pointing outwards, leaf margins rolled under, brownish in the fall.

Flowers: Inflorescence of male and female catkins which open before the leaves enlarge; male catkins with stalks.

Fruits: Small nutlets, with narrow-winged margins; female cones 1.5-2.5 cm long, egg-shaped.

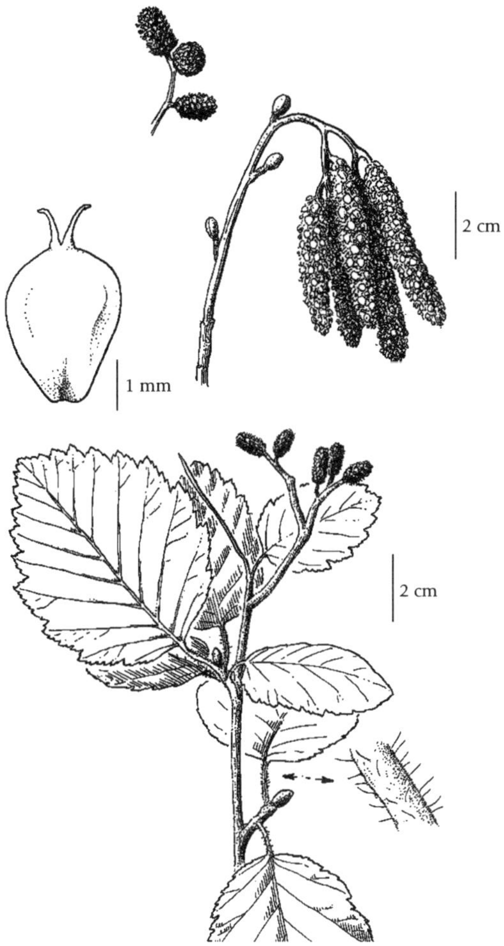

Alnus incana ssp. *tenuifolia*

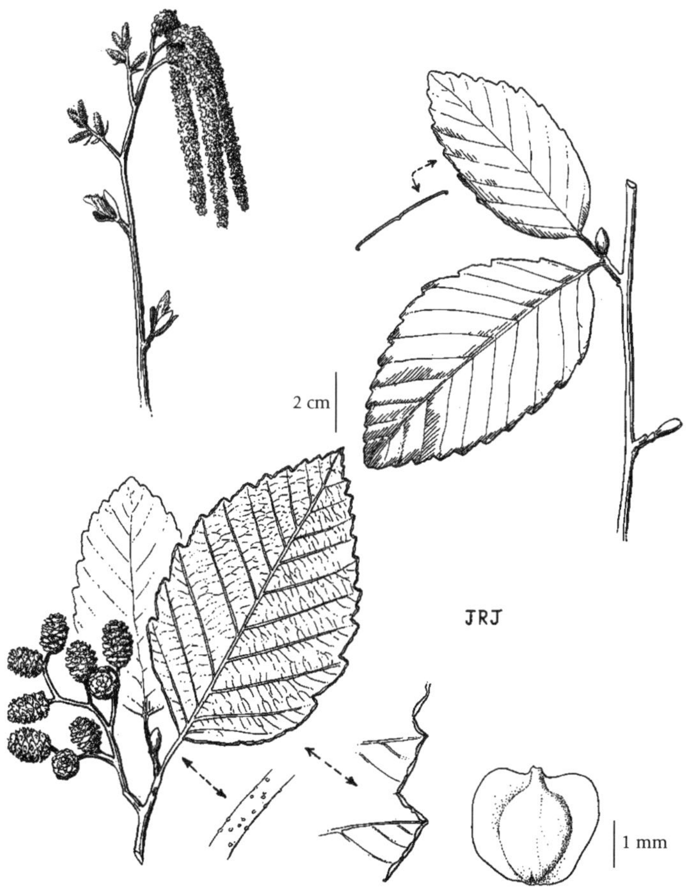

Alnus rubra

***Alnus viridis* [Chaix] DC.** (*A. crispa* [Ait.] Pursh ssp. *laciniata* Hult., *A. crispa* var. *sinuata* Regel, *A. sinuata* [Regel] Rydb. = ssp. *sinuata; A. crispa* = ssp. *crispa*)
green alder (ssp. *crispa*); **Sitka alder** (ssp. *sinuata)*

HABITAT/RANGE:

Moist slopes, streambanks, avalanche tracks, bogs and fens in all zones; ssp. *crispa* frequent in N BC, ssp. *sinuata* common throughout S BC, becoming less frequent and transitional to ssp. *crispa* in N BC; circumboreal, E to PQ and S to NC, MN, CO and CA.

DESCRIPTION:

General: Deciduous shrub or tree, 1-5 m tall; pointed axillary buds without stalks; bark scaly, sometimes lichen-covered, yellowish-brown or grey.

Leaves: Alternate, deciduous, smooth, finely toothed 1-2 times, oval with pointed tips, 4-10 cm long, brownish in the fall.

Flowers: Inflorescence of male and female catkins which open at the same time as the leaves on current year's growth; male catkins unstalked.

Fruits: Small nutlets, with broad wings; female cones 1.5-2 cm long, egg-shaped.

Notes: Two subspecies occur in BC.

1. Leaf margins shallowly lobed as well as finely saw-toothed .... ssp. *sinuata* (Regel) A.& D. Love
1. Leaf margins not at all or only slightly lobed, mostly merely finely saw-toothed .... ssp. *crispa* [Ait.] Turrill

BETULA

1. Nutlet wings narrower than the body of the nutlet; leaves relatively thick and leathery in texture, usually less than 3 cm long, stalks less than 1 cm long; shrubs up to about 2 m tall.
   2. Twigs densely glandular and short-hairy; leaves with 3 or fewer lateral veins and 10 or fewer teeth on each side; nutlet wings relatively narrow, less than half the width of the nutlet body .... *B. nana*
   2. Twigs densely hairy and slightly glandular; leaves with 4 or more lateral veins and 10 or more teeth on each side; nutlet wings equal to about half the width of the nutlet body .... *B. pumila*
1. Nutlet wings at least as wide as the body of the nutlet; leaves relatively thin and membranous, usually more than 3 cm long, stalks more than 1 cm long; trees or tall shrubs 3 m tall or taller.
   3. Small trees or shrubs 3-10 m tall; bark dark reddish-brown to black; fruiting bracts with pointed lateral lobes .... *B. occidentalis*
   3. Medium to large trees 10-30 m tall; bark white or brown, often peeling; fruiting bracts with rounded lateral lobes.
      4. Leaves essentially rounded or broadly egg-shaped in outline and hairy below, rarely glandular; twigs short-hairy or long-hairy but usually not glandular.
         5. Leaves 4-10 cm long; twigs hairy with both short and long hairs; native species found throughout most of BC .... *B. papyrifera*
         5. Leaves 2-5 cm long; twigs short-hairy; introduced horticultural species of extreme SW BC .... *B. pubescens*
      4. Leaves essentially wedge-shaped in outline and glabrous below; twigs glandular.
         6. Bark thin, smooth, pale and papery; twigs with wartlike resin glands; native species of NE BC .... *B. neoalaskana*
         6. Bark thick, rough, dark and furrowed; twigs with numerous small glands, sometimes sticky; introduced horticultural species of extreme SW BC .... *B. pendula*

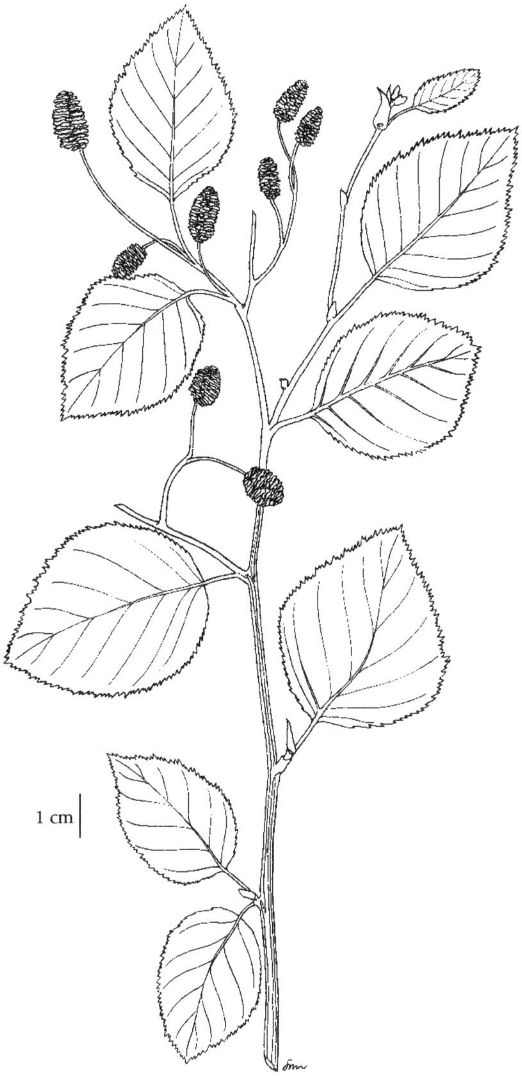

Alnus viridis ssp. *crispa*

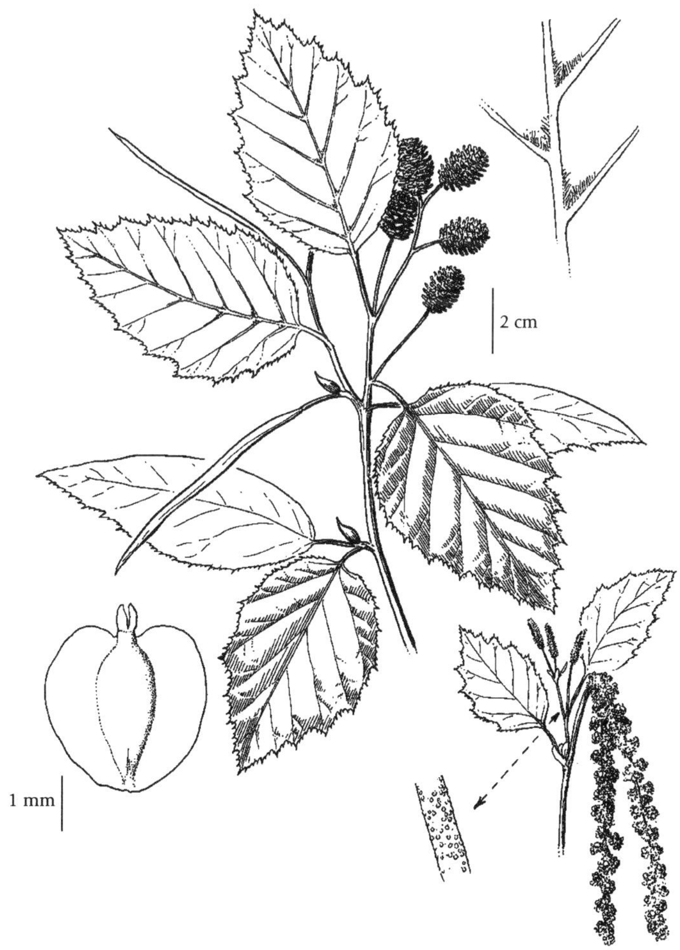

Alnus viridis ssp. *sinuata*

Betula nana **L.** (*Betula glandulosa* Michx., *B. glandulosa* var. *sibirica* [Ledeb.] Blake, *B. nana* L. ssp. *exilis* [Sukats.] Hult.)

scrub birch; bog birch; dwarf birch; glandular birch

HABITAT/RANGE:

Wet to moist bogs, fens, streamsides, marshes, meadows and rocky slopes from the montane to the alpine zones; common throughout BC east of the Coast-Cascade Mountains; N to AK, YT and NT, E to NB and NS, and S to MN, CO and N CA.

DESCRIPTION:

General: Deciduous shrub, 0.3-2 m tall; young twigs densely glandular and distinctly short-hairy.

Leaves: Alternate, deciduous, oval to round, usually coarsely toothed, thick and leathery in texture, 0.5-3 cm long; leaf stalks less than 1 cm long.

Flowers: Male and female flowers in separate catkins 1-2.5 cm long; flowers emerging before or with the leaves; catkins breaking up at maturity.

Fruits: Nutlets with wings relatively narrow, less than half the width of the body of the nutlets.

Betula neoalaskana **Sarg.** (*B. alaskana* Sarg., *B. papyrifera* Marsh. var. *neoalaskana* [Sarg.] Raup)

Alaska paper birch; northwestern white birch

HABITAT/RANGE:

Wet bogs, fens, swamps or moist forest sites in the montane zone; frequent in NE BC; N to AK, YT and NT and E to ON.

DESCRIPTION:

General: Deciduous tree, 10-30 m tall; young twigs with yellowish, wartlike resin glands; bark thin, smooth, pale and papery.

Leaves: Alternate, deciduous, wedge-shaped in outline, doubly saw-toothed, thin and membranous, glabrous below, usually 2.5-8 cm long; leaf stalks more than 1 cm long.

Flowers: Male and female flowers in separate catkins 2-4 cm long; flowers emerging before or with the leaves; catkins breaking up at maturity.

Fruits: Nutlets with broad wings, at least twice as wide as the body of the nutlet; bracts with rounded, lateral lobes.

Betula occidentalis **Hook.** (*B. papyrifera* Marsh. ssp. *occidentalis* [Hook.] Hult., *B. papyrifera* Marsh. var. *occidentalis* [Hook.] Sarg.)

water birch; black birch; red birch; mountain birch

HABITAT/RANGE:

Wet to moist streambanks, forests and marshes in the steppe and montane zones; frequent in S BC east of the Coast-Cascade Mountains, rare northward; N to AK, YT and NT, E to NF, and S to CO, UT and CA.

DESCRIPTION:

General: Deciduous small tree or large shrub, 3-10 m tall; bark dark reddish-brown to black.

Leaves: Alternate, deciduous, rounded to sharp-pointed at the tip, usually doubly saw-toothed, thin and membranous, 1-7 cm long, leaf stalks more than 1 cm long.

Flowers: Male and female flowers in separate catkins 2-4 cm long; flowers emerging before or with the leaves; catkins breaking up at maturity.

Fruits: Nutlets with broad wings, at least twice as wide as the nutlets; bracts with pointed, lateral lobes.

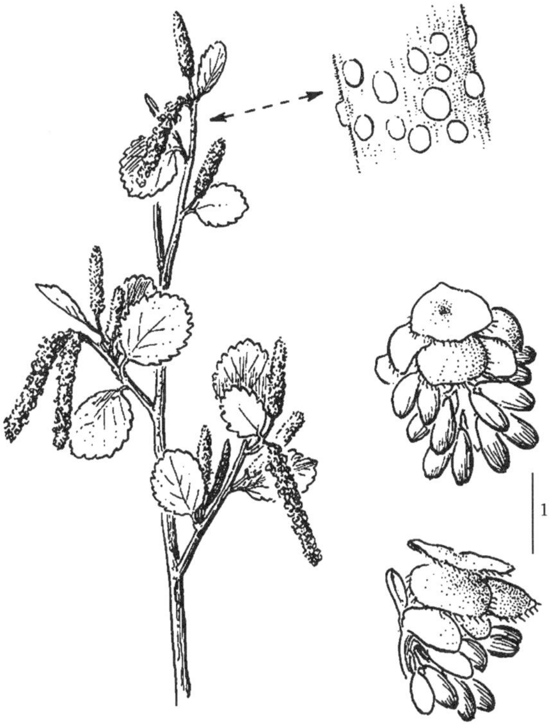

Betula nana

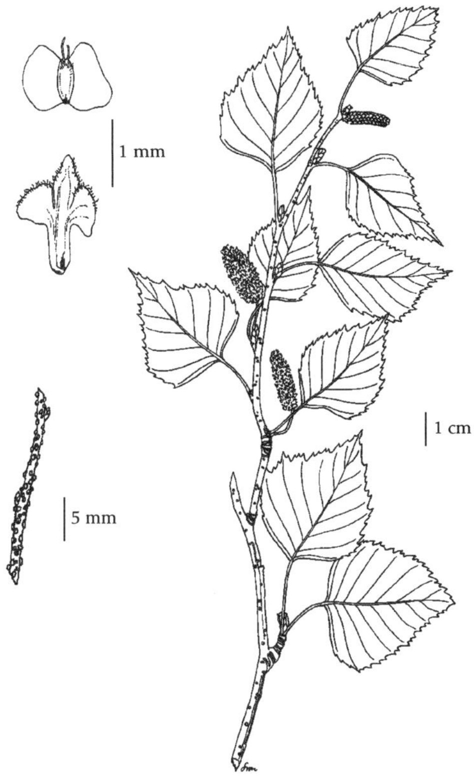

Betula neoalaskana

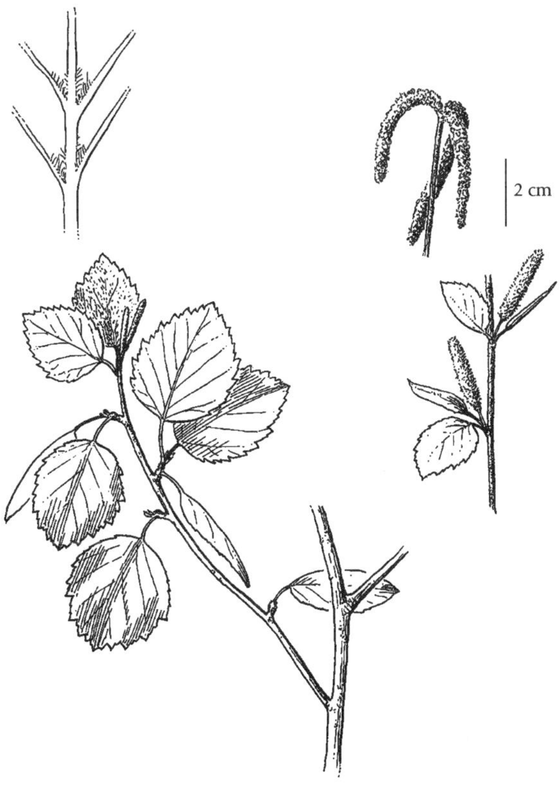

Betula occidentalis

***Betula papyrifera* Marsh.**

paper birch; white birch; canoe birch

Habitat/Range:

Moist to mesic woodlands, forests, clearcuts, burns and open areas in the lowland, steppe and montane zones; var. *commutata* – frequent in extreme SW BC, var. *papyrifera* – common throughout BC east of the Coast-Cascade Mountains; N to AK, YT and NT, E to NF, and S to MN, PA, CO and NE OR.

Description:

General: Deciduous tree, 10-30 m tall; bark white, yellowish- to reddish-brown or dark grey, often peeling; twigs hairy with both short and long hairs.

Leaves: Alternate, deciduous, broadly egg-shaped in outline, the tips sharp-pointed, hairy below, rarely glandular, 4-10 cm long, leaf stalks more than 1 cm long.

Flowers: Male and female flowers in separate catkins 2-4 cm long; flowers emerging before or with the leaves; catkins breaking up at maturity.

Fruits: Nutlets with broad wings, at least twice as wide as the body of the nutlets; bracts with rounded, ascending or diverging lateral lobes.

Notes: Numerous apparent hybrids between *B. papyrifera* and *B. neoalaskana* or *B. occidentalis* (e.g., *B.* x *utahensis* Britt. and *B. subcordata* Rydb.) have been formally named (see Brayshaw 1996b, Scoggan 1979). Two varieties occur in BC.

1. Mature bark white; fruiting bracts with ascending lateral lobes .... var. *papyrifera*

1. Mature bark white, yellowish- to reddish-brown or dark grey; fruiting bracts with lateral lobes diverging at right angles .... var. *commutata* (Regel) Fern.

***Betula pendula* Roth**

European birch

Habitat/Range:

Bogs or marshes in the lowland zone; frequent horticultural escape in the lower Fraser Valley, less common on extreme SE Vancouver Island; introduced from Eurasia.

Description:

General: Deciduous tree, 10-30 m tall; bark thick, rough, dark brown, furrowed, often peeling; twigs glandular, numerous small glands, sometimes sticky; branch ends often drooping.

Leaves: Alternate, deciduous, wedge-shaped in outline, saw-toothed, glabrous below, 1-6 cm long.

Flowers: Male and female flowers in separate catkins, the male 3-6 cm long, the female 1.5-3.5 cm long; flowers emerging before or with the leaves; catkins breaking up at maturity.

Fruits: Nutlets with broad wings, at least twice as wide as the body of the nutlets; bracts with rounded lateral lobes.

***Betula pubescens* Ehrh.** (*B. alba* L.)

silver birch

Habitat/Range:

Bogs in the lowland zone; rare in SW BC, known from the lower Fraser Valley and extreme SE Vancouver Island; introduced from Europe.

Description:

General: Deciduous tree, 10-30 m tall; bark white or brown, often peeling; twigs short-hairy, not glandular.

Leaves: Alternate, deciduous, rounded in outline, hairy below, rarely glandular, 1-5 cm long.

Flowers: Male and female flowers in separate catkins, the male 3-6 cm long, the female 1-4 cm long; flowers emerging before or with the leaves; catkins breaking up at maturity.

Fruits: Nutlets with broad wings, at least 1-1 1/2 times as wide as the body of the nutlets; bracts with rounded lateral lobes.

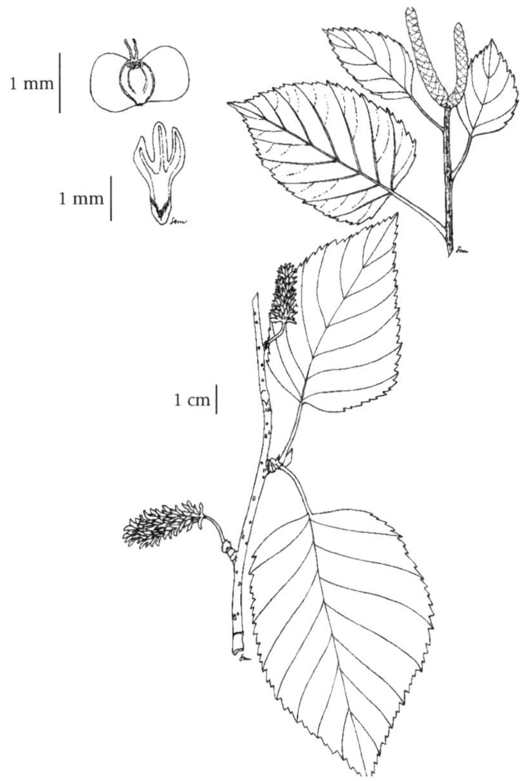

Betula papyrifera **var.** ***papyrifera***

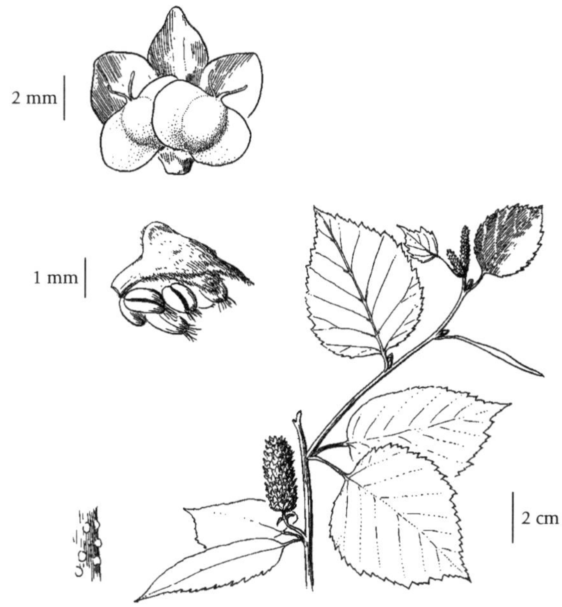

Betula papyrifera **var.** ***commutata***

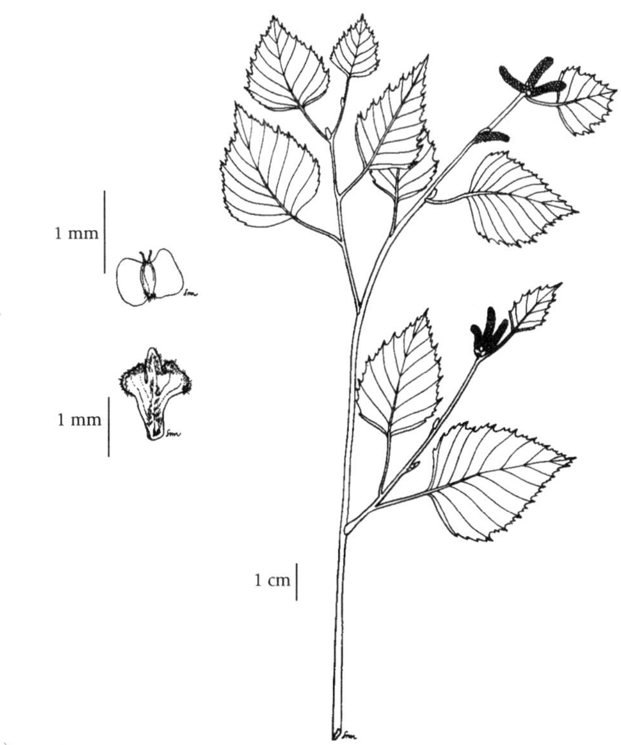

Betula pendula

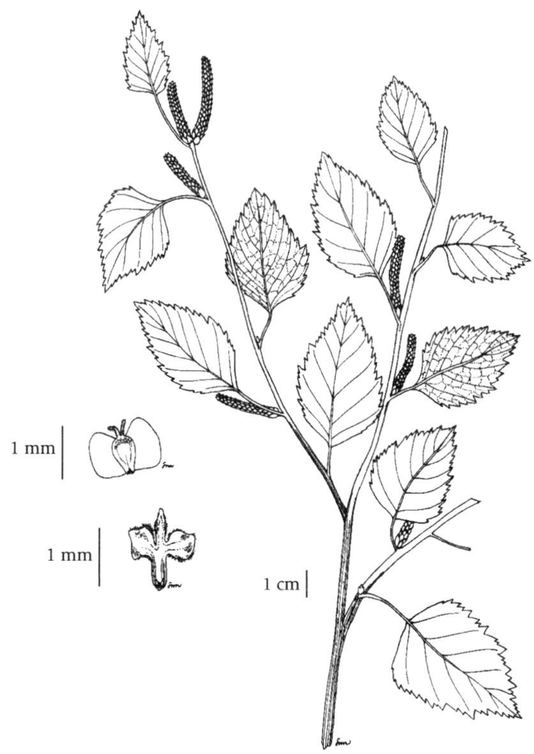

Betula pubescens

***Betula pumila* L. var. *glandulifera* Regel** (*B. glandulosa* var. *glandulifera* [Regel] Gleason, *B. pumila* var. *glandulifera* f. *hallii* [Howell] Brayshaw, *B. glandulosa* var. *hallii* [Howell] Hitchc.)
low birch; swamp birch; scrub birch; glandular birch
Habitat/Range:
Wet to moist bogs, fens, marshes, meadows and seepage slopes in all vegetation zones; frequent throughout BC east of the Coast-Cascade Mountains, rare in extreme SW BC; N to AK, YT and NT, E to PQ, and S to WY and OR.
Description:
General: Deciduous shrub, 0.5-3 m tall; twigs densely hairy, slightly glandular.

Leaves: Alternate, deciduous, thick and leathery in texture, 1-4 cm long, stalks less than 1 cm long, 4 or more lateral veins, with 10 or more coarse teeth on each side.

Flowers: Male and female flowers in separate catkins 1.5-2 cm long; flowers emerging before or with the leaves; catkins breaking up at maturity.

Fruits: Nutlets with narrow wings, 1/2 the width of the body of the nutlets.

Notes: This species apparently hybridizes with *B. pendula* and *B. papyrifera* in our area (Brayshaw 1996b).

CORYLUS

1. Twigs sparsely to moderately hairy, sometimes slightly glandular; silicles completely enclosed by bristly bractlets .... *C. cornuta*
1. Twigs both hairy and glandular; silicles not completely enclosed by the thinly downy, lacerated bractlets .... *C. avellana*

***Corylus avellana* L.**
filbert; hazelnut
Habitat/Range:
Moist meadows and thickets in the lowland zone; rare on Saltspring Island and in the Lower Fraser Valley; introduced from Europe.
Description:
General: Deciduous shrub, 1-5 m tall with many stems, densely clumped or spreading by suckers; twigs hairy, glandular.

Leaves: Alternate, deciduous, elliptic to oval, leaf blades heart-shaped with sharp-pointed tips, doubly saw-toothed, paler below than above, 5-10 cm long, turning yellow in the fall.

Flowers: Male flowers in catkins appearing before the leaves in spring; female flowers in a very small catkin with protruding red stigmas.

Fruits: Edible, hard-shelled nuts not completely enclosed by thinly downy, lacerated bractlets, in 2's or 3's at the ends of the branches.

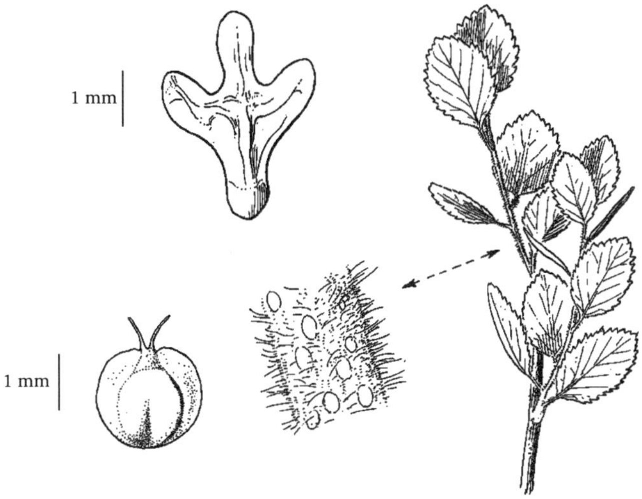

Betula pumila var. *glandulifera*

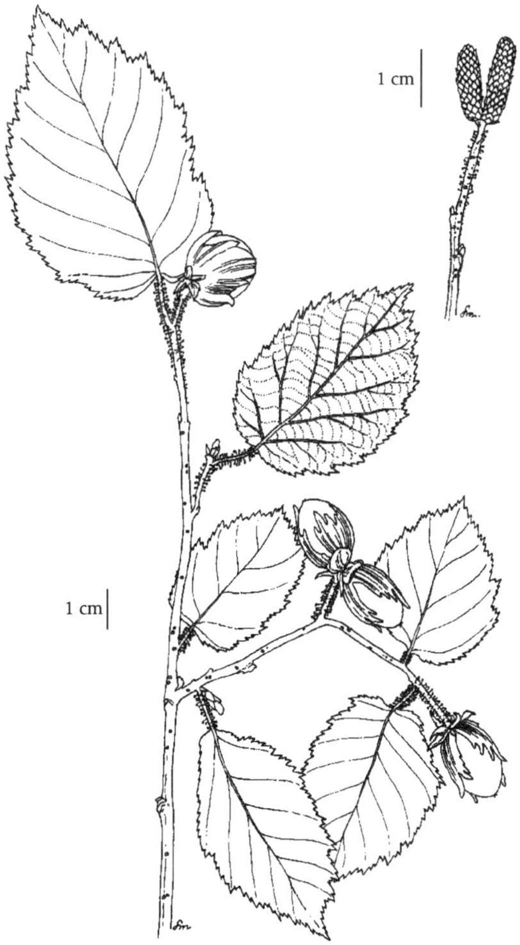

Corylus avellana

***Corylus cornuta* Marsh.** (*C. californica* [A. DC.] Rose, *C. cornuta* var. *glandulosa* Boivin = *C. cornuta* var. *californica)*
beaked hazelnut; California hazelnut

HABITAT/RANGE:

Mesic sites in the lowland and montane zones; var. *cornuta* - common south of 57° N east of the Coast-Cascade Mountains, var. *californica* - frequent on S Vancouver Island and the lower Fraser Valley, becoming rare to the east; E to NF and S to GA (var. *cornuta)* and E to ID and S to CA (var. *californica)*.

DESCRIPTION:

General: Deciduous shrub, 1-3 m tall with many stems, densely clumped or spreading by suckers; twigs sparsely to moderately hairy, sometimes glandular.

Leaves: Alternate, deciduous, elliptic to oval, leaf blades heart-shaped with a sharp-pointed tip, doubly saw-toothed, paler below than above, 4-10 cm long, turning yellow in the fall.

Flowers: Male flowers in catkins appearing before the leaves in spring; female flowers in a very small catkin with protruding red stigmas.

Fruits: Edible hard-shelled nuts completely enclosed by bristly bractlets, in 2's or 3's at the end of branches, barely 1.5 cm long, thinly hairy or glabrous.

Notes: Two varieties occur in BC.

1. Involucral beaks about twice as long as the fruit; silicles thinly short-hairy; twigs sparsely hairy .... var. *cornuta*
1. Involucral beaks about equal in length to the fruit; silicles glabrous; twigs hairy, sometimes glandular .... var. *californica* (A. DC.) Sharp

BORAGINACEAE

1. Calyces greatly enlarged in fruit; flowers in axils of leaves; plants scrambling, with backward-pointing bristles on prickly stems .... *Asperugo*
1. Calyces not enlarged in fruit; flowers in terminal clusters called clusters (typical of the family); habit and stems various.
   2. Corollas flat, star-shaped, greater than 1 cm wide; stamens held together in a cone projecting from the petals; rare garden escape .... *Borago*
   2. Corollas shaped like funnels, trumpets or deep bells, less than 1 cm wide; stamens not held together in a cone.
      3. Corollas irregular (bilaterally symmetric); stamens sticking out of the corolla tube .... *Echium*
      3. Corollas regular (radially symmetric); stamens included within the corolla tube.
         4. Nutlets armed with prickles.
            5. Nutlets flattened, with sunken centre and hooked marginal . bristles; low, white-flowered annuals .... *Pectocarya*
            5. Nutlets rounded or angled with harpoonlike prickles.
               6. Nutlets spreading, broadest near apex, uniformly prickly all over .... *Cynoglossum*
               6. Nutlets erect, broadest near base, predominant or largest prickles marginal.
                  7. Fruiting stalks reflexed; inflorescences mostly lacking bracts beneath each flowe .... *Hackelia*
                  7. Fruiting stalks erect; inflorescences with bracts beneath each flower .... *Lappula*
         4. Nutlets not armed with prickles.
            8. Corollas tubular, flaring at top to cuplike limbs (the expanded part of a fused orolla, above the throat).
               9. Stems glabrous or with few, relatively soft hairs; hairy bulges at the corolla throat nearly absent or much shorter than anthers; native .... *Mertensia*
               9. Stems with stiff, spreading hairs; hairy bulges at throat equalling anthers; corollas purplish to white; introduced .... *Symphytum*

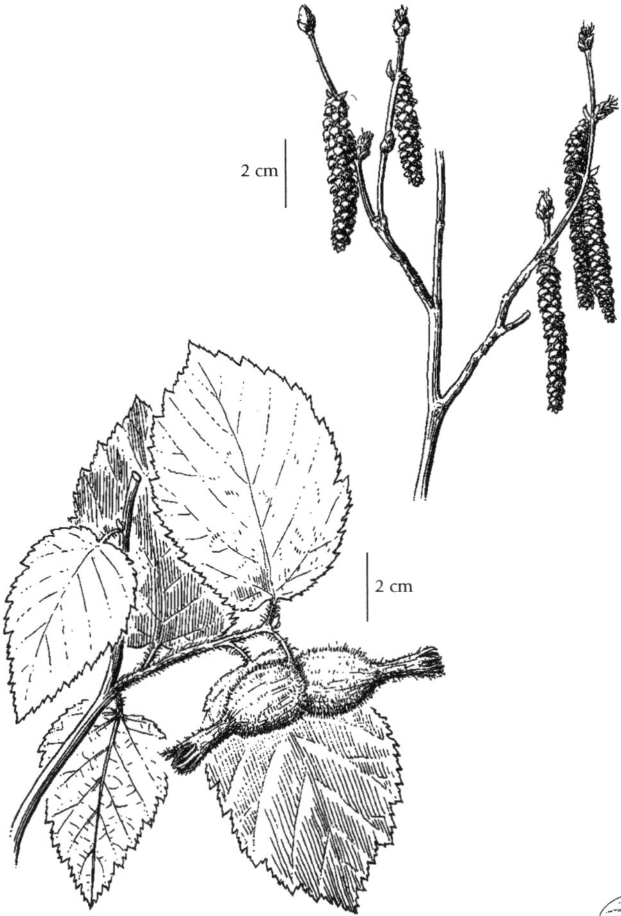

Corylus cornuta var. *californica*

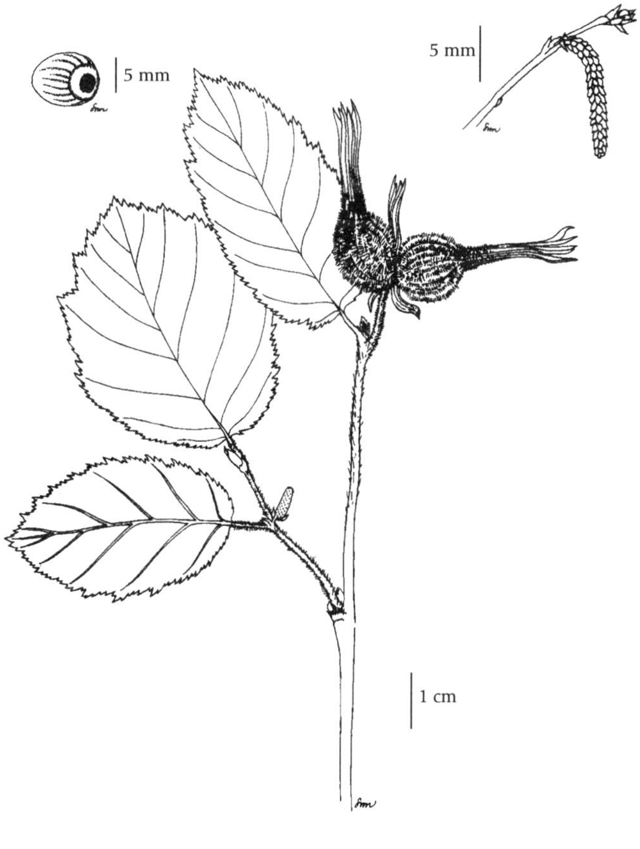

Corylus cornuta var. *cornuta*

8. Corollas tubular, flaring to spreading limbs.
   10. Nutlets nearly round, smooth and shining, with small basal scar; flowers usually blue .... *Myosotis*
   10. Nutlets usually pointed, dull and wrinkled or if smooth, with broad flat base or with lateral scar; flower colour various.
       11. Nutlets attached basally, with broad, flat base.
           12. Flowers bright blue, with prominent fringed bulges at the corolla throat; rare garden escape .... *Anchusa*
           12. Flowers white, yellow or greenish, without throat bulges and without fringes .... *Lithospermum*
       11. Nutlets attached along inside edge.
           13. Flowers bright orange-yellow .... *Amsinckia*
           13. Flowers white.
               14. Nutlets with a long groove, mostly smooth; leaves mostly alternate; annuals or perennials .... *Cryptantha*
               14. Nutlets with a raised scar, mostly rough; basal leaves usually opposite; annuals .... *Plagiobothrys*

AMSINCKIA

1. Leaves toothed; one pair of calyx lobes partially united (like 2 webbed fingers); plants maritime .... *A. spectabilis*
1. Leaves entire; calyx lobes all free; plants not usually maritime.
   2. Corolla tube with hairy bulges at the throat .... *A. lycopsoides*
   2. Corolla throats open, without hairy bulges.
      3. Stems coarsely stiff-hairy, but also with shorter, softer, downswept hairs; leaf hairs more ascending than wide-spreading .... *A. retrorsa*
      3. Stems stiff-hairy, mostly without shorter, softer hairs below the inflorescence.
         4. Corollas 4-7 mm wide .... *A. menziesii*
         4. Corollas 7-11 mm wide .... *A. intermedia*

Amsinckia intermedia **Fisch. & C.A. Mey.** (*A. menziesii* [Lehm.] A. Nels. & J.F. Macbr. var. *intermedia* [Fischer & C. Mey.] Ganders)
common fiddleneck
HABITAT/RANGE:
Dry roadsides and waste places in the lowland and lower montane zones; frequent in BC east of the Coast-Cascade Mountains, probably introduced in Vancouver and on S Vancouver Island; S to ID, NM and Baja CA.
DESCRIPTION:
General: Annual herb from a taproot; stems simple or branched, spreading stiff-hairy, 20-80 cm tall.

Leaves: Basal or lower leaves soon deciduous; stem leaves linear to oblong or lanceolate, up to 15 cm long and 2 cm wide, alternate, entire.

Flowers: Inflorescence of terminal, flat-topped flower clusters (clusters); petals orange to orange-yellow, fused basally into a tube 7-11 mmm long that spreads to 5 free lobes; calyx lobes 5, silky-hairy, also with some longer, stiff hairs.

Fruits: Nutlets 4, clustered together, 2.5-4 mm long, egg-shaped, wrinkled, pointed, dull, greenish or grey to brown or black.

Notes: The differences (corolla size and color) between this species and *A. menziesii* are slight and some taxonomists (i.e., Ganders [1993]) treat them at the varietal level.

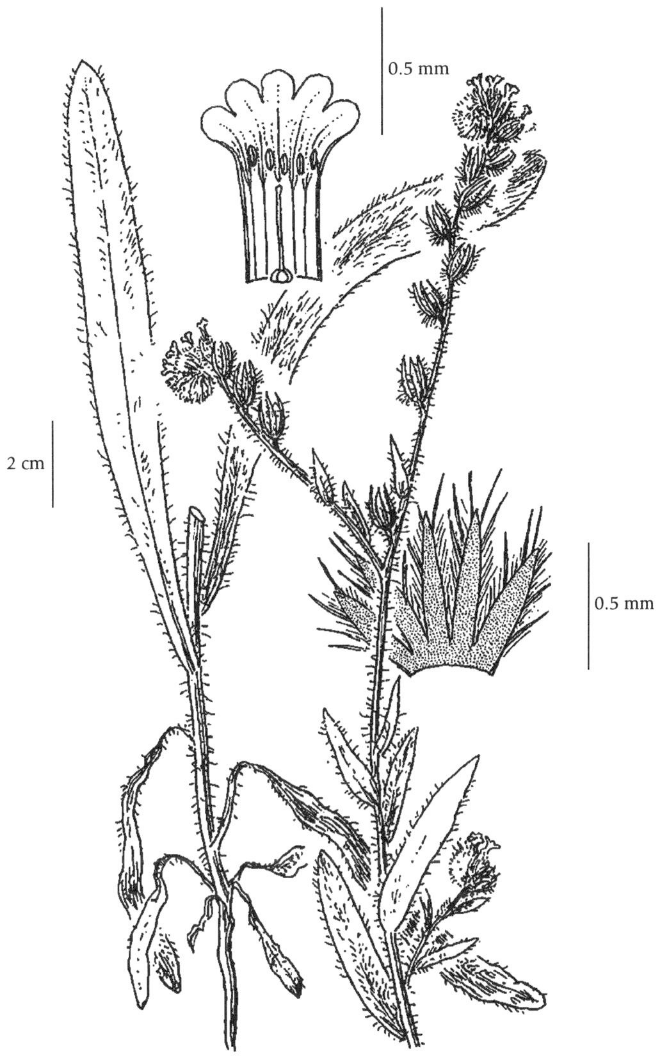

Amsinckia intermedia

Amsinckia lycopsoides **Lehm.**

bugloss fiddleneck

HABITAT/RANGE:

Dry disturbed areas in the lowland and lower montane zones; infrequent in extreme S BC; N to S AK, S to MT and CA.

DESCRIPTION:

General: Annual herb from a taproot; stems simple or few-branched, spreading stiff-hairy and also with softer, often downswept hairs, 10-60 cm tall.

Leaves: Basal leaves persistent, often crowded; stem leaves linear to oblong or oblanceolate, to 10 cm long and 1.5 cm wide, alternate, entire.

Flowers: Inflorescence a coiled cluster; petals yellow to orange, fused basally into a tube 6-9 mm long that spreads (3-6 mm wide) at the top to 5 stubby lobes; calyx lobes 5, with both silky and longer, stiff hairs.

Fruits: Nutlets 4, clustered together, 2.5-3 mm long, egg-shaped, warty, greenish to dark brown.

Amsinckia menziesii **(Lehm.) A. Nels. & J.F. Macbr.**

small-flowered fiddleneck

HABITAT/RANGE:

Moist to dry disturbed areas in the steppe and montane zones; common in S BC east of the Coast-Cascade Mountains, rare northward, introduced in Vancouver; N to AK and YK, S to CA.

DESCRIPTION:

General: Annual herb from a taproot; stem simple to much-branched, somewhat flimsy, sparingly spreading stiff-hairy, without shorter softer hairs except often in the inflorescence, 15-80 cm tall.

Leaves: Basal leaves not forming a persistent tuft; stem leaves lance-oblong, elliptic, or linear, 2-12 cm long and 0.5-2.5 cm wide, alternate, entire.

Flowers: Inflorescence a coiled cluster; petals yellow, fused into a tube 4-7 mm long that spreads (2-4 mm wide) at top to 5 shallow lobes, the throat open; calyx lobes 5, with both silky and longer, stiff hairs.

Fruits: Nutlets 4, clustered together, 2-3.5 mm long, egg-shaped, warty or pimply and often wrinkled, greenish or grey to brownish-black.

Amsinckia retrorsa **Suksd.**

rigid fiddleneck

HABITAT/RANGE:

Dry roadsides, fields and disturbed areas in the steppe zone; rare in SC BC, known only from the Kelowna area; S to UT and CA.

DESCRIPTION:

General: Annual herb from a taproot; stems simple to few-branched, spreading stiff-hairy and also with shorter, softer, downswept hairs, 10-60 cm tall.

Leaves: Basal leaves often crowded, larger than stem leaves, which are linear to linear-oblong, to 12 cm long and to 1 cm wide, alternate, entire.

Flowers: Inflorescence a coiled cluster; petals orange or orange-yellow with vermilion markings in the open throat, fused into a tube 5-8 mm long that spreads (1.5-3 mm wide) at top to 5 short lobes; calyx lobes 5, hairy.

Fruits: Nutlets 4, clustered together, 2-3 mm long, egg-shaped, pimply and often wrinkled.

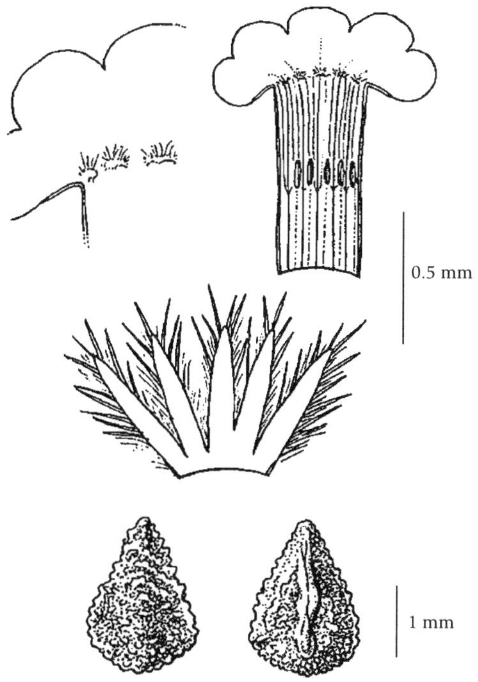

Amsinckia lycopsoides

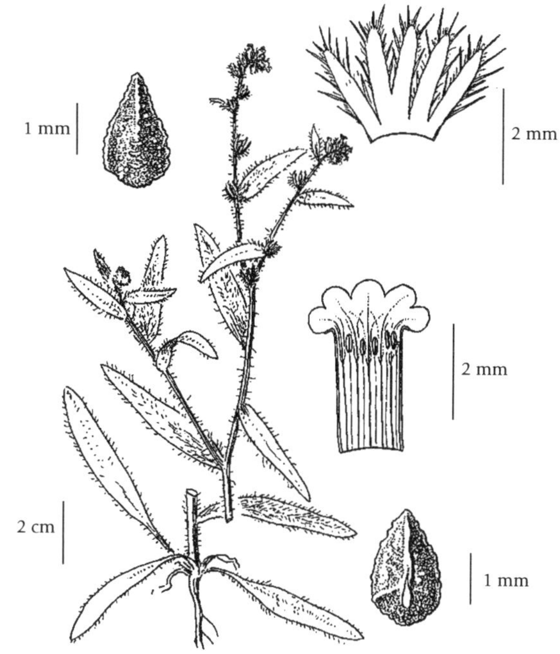

Amsinckia menziesii

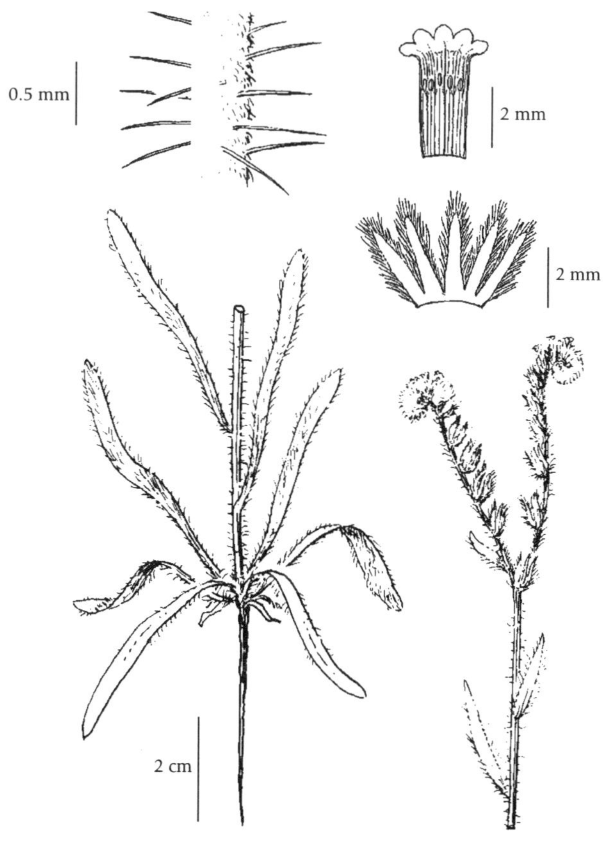

Amsinckia retrorsa

Amsinckia spectabilis **Fisch. & C.A. Mey.**

seaside fiddleneck

Habitat/Range:

Moist sandy coastal beaches in the lowland zone; infrequent on the Queen Charlotte Islands, Vancouver Island the Gulf Islands and the lower Fraser Valley; S to N Baja CA.

Description:

General: Annual herb from a taproot; stems erect or often reclining to prostrate, spreading stiff-hairy, simple or branched, to 40 cm tall.

Leaves: Basal leaves not persistent; stem leaves irregularly fine-toothed, lanceolate to lance-linear or lance-elliptic, to 5 cm long and about 1 cm wide, alternate.

Flowers: Inflorescence a coiled cluster; petals orange, fused into a tube 4-8 mm long that spreads (3-5 mm wide) at the top to 5 short lobes; calyx lobes 5, but 2 of them half-fused, with long stiff hairs and a few shorter, softer hairs.

Fruits: Nutlets 4, clustered together, 1.5-2.5 mm long, egg-shaped, weakly wrinkled or pimply, blackish.

ANCHUSA

1. Corolla tube bent, lobes unequal; leaves toothed .... *A. arvensis*
1. Corolla tube straight, lobes equal; leaves entire.
    2. Limb (flared part) of corollas 12-25 mm wide; calyx lobes linear; nutlets 5-10 mm long .... *A. azurea*
    2. Limb of corollas 6-11 mm wide; calyx lobes lanceolate; nutlets 2-4 mm long .... *A. officinalis*

Anchusa arvensis **(L.) Bieb.**

European bugloss

Habitat/Range:

Dry roadsides and fields in the montane zone; rare in SE BC, known only from Grand Forks; introduced from Europe.

Description:

General: Annual herb from a taproot, spreading stiff-hairy, the hairs with bulbous bases; stems simple to few-branched, 30-80 cm tall.

Leaves: Basal and lower stem leaves stalked; middle and upper stem leaves becoming unstalked upwards, irregularly and distantly toothed or wavy-edged, lanceolate to oblong or egg-shaped, to 15 cm long and 1-2 cm wide, alternate.

Flowers: Inflorescence axillary and terminal, of elongating, often forked clusters; petals blue, fused into a curved tube 4-7 mm long that spreads (4-6 mm wide) at the top to 5 unequal lobes, throat closed by 5 white, hairy bulges; calyx lobes 5, with long, abruptly sharp-pointed.

Fruits: Nutlets 4, clustered together, squat, asymmetric, 2-3 mm tall, 3-4 mm wide at thickened, concave base, irregularly angled and wrinkled.

Anchusa azurea **P. Mill.**

Italian bugloss

Habitat/Range:

Mesic to dry waste places, forest edges and shrubby banks in the lowland zone; rare in SW BC, known only from the Capilano River and Burnaby; introduced from Europe.

Description:

General: Perennial herb from a taproot, spreading stiff-hairy; stems mostly single, 0.4-1.5 m tall.

Leaves: Basal leaves stalked, oblanceolate; stem leaves unstalked, lanceolate or narrowly egg-shaped to oblong, to 30 cm long and 8 cm wide, alternate, entire.

Flowers: Inflorescence a terminal cluster; petals blue, fused into a tube 10-12 mm long that spreads (12-25 mm wide) at the top to 5 peg-shaped lobes, throat closed by 5 white, frilly bulges; calyx lobes 5, linear, white-bristly.

Fruits: Nutlets 4, clustered together, erect, asymmetric, 5-9 mm tall, 3-5 mm thick, irregularly ridged, concave and with thickened rim at the base.

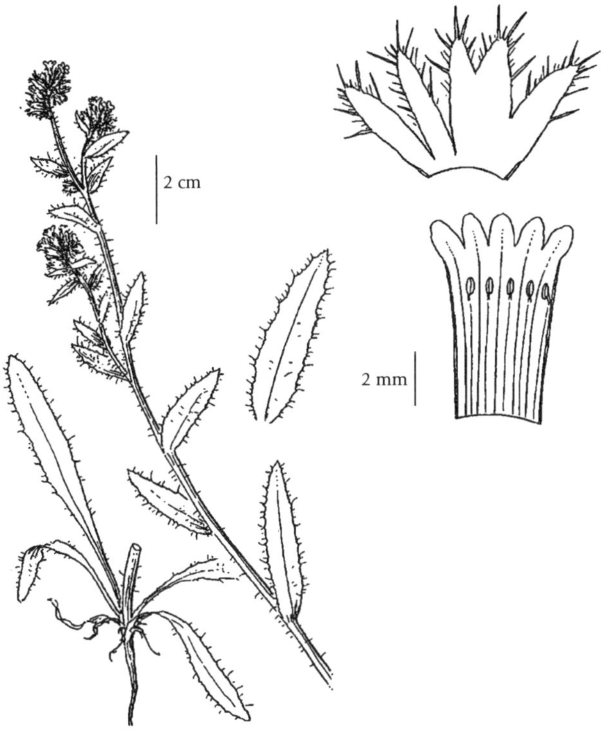

Amsinckia spectabilis

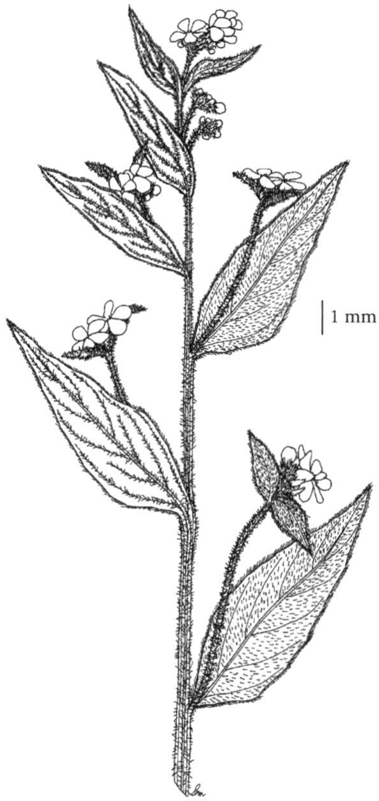

Anchusa arvensis

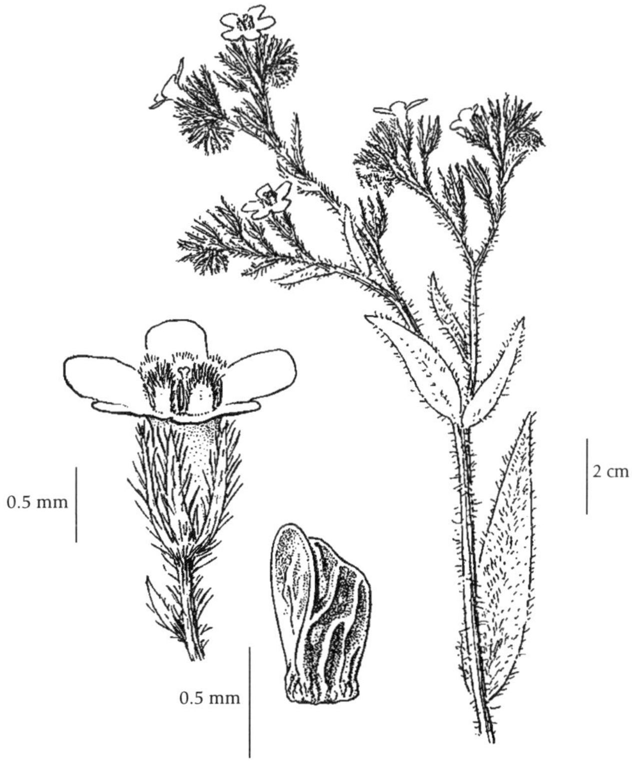

Anchusa azurea

Anchusa officinalis **L.**

alkanet; common bugloss

Habitat/Range:

Dry roadsides and waste places in the lowland, steppe and lower montane zones; rare in SC BC and S Vancouver Island; introduced from Europe.

Description:

General: Biennial or perennial herb from a taproot, softly spreading-hairy; stems often several, 30-80 cm tall.

Leaves: Basal and lower leaves stalked, oblanceolate, 6-20 cm long (stalk included) and 1-2.5 cm wide, the others gradually reduced upwards, becoming unstalked and more lanceolate, alternate, entire.

Flowers: Inflorescence a coiled cluster; petals blue-purple, fused into a tube 6-11 mm long that spreads (6-11 mm wide) at the top to 5 tooth-shaped lobes, throat closed by 5 white, hairy bulges; calyx lobes 5, lanceolate, stiff-hairy.

Fruits: Nutlets 4, clustered together, horizontal, squat, asymmetric or conical, concave and with a thickened rim at the base, 2 mm high, 3-4 mm long.

ASPERUGO

Asperugo procumbens **L.**

madwort; catchweed

Habitat/Range:

Dry to moist disturbed areas and waste places in the steppe and lower montane zones; infrequent in SC and SE BC; introduced from Eurasia.

Description:

General: Annual herb from a taproot; stems weak, branched, climbing-sprawling, angled, with short backward-pointing prickles along the angles, 0.3-1.0 m long.

Leaves: Lower leaves often deciduous, oblanceolate, stalked, 2-10 cm long, to 2.5 cm wide, roughly stiff-hairy and irregularly stiff-fringed, opposite, more or less entire, the others gradually reduced upwards, becoming more elliptic or somewhat diamond-shaped and unstalked, partly whorled or alternate, often widely spaced.

Flowers: Single or in clusters of 2-3 on short, thick, downward-curving stalks from the axils of leaves or bracts or forks of branches; corollas narrowly bell-shaped; petals at first purplish then blue, fused at base into a short tube that flares (2-3 mm wide) at top to 5 tiny lobes, with 5 white bulges at the throat; calyces with 5 deltoid lobes, each lobe with a tooth on each side near the base; fruiting calyces enlarging (to 1-2 cm wide), becoming compressed, firm-papery, net-veined, prickly with stiff hooked or curved hairs, and forming a 2-lipped covering around the silicles.

Fruits: Nutlets 4, clustered together, enveloped by the calyx, narrowly egg-shaped, 2.5 mm long, with shiny, cobblestone-patterned surface and small, off-centre groove.

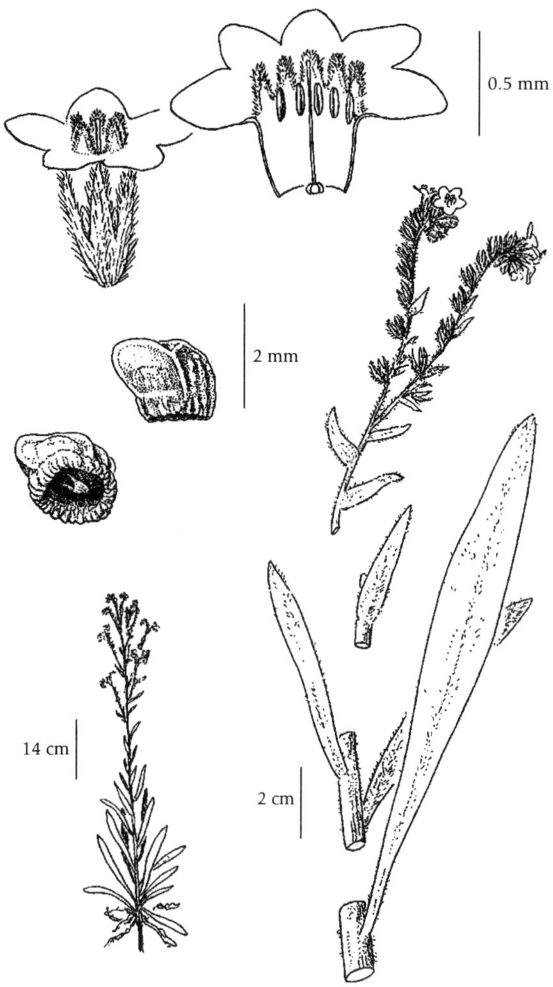

Anchusa officinalis

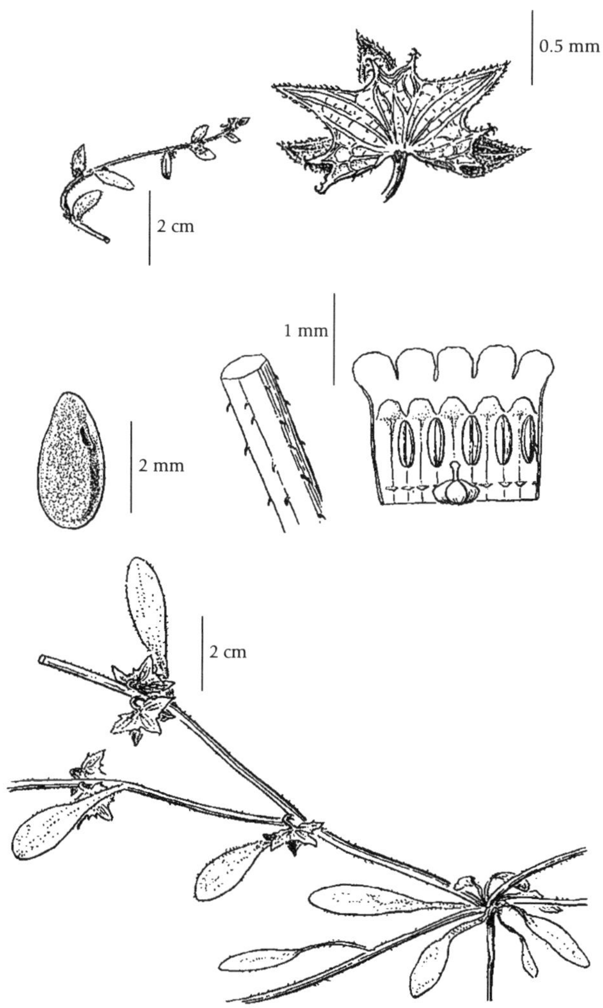

Asperugo procumbens

BORAGO

***Borago officinalis* L.**
common borage
HABITAT/RANGE:
Dry to moist waste places in the steppe and lower montane zones; rare in SC and SE BC; introduced from Europe.
DESCRIPTION:
General: Annual herb from a taproot, stiff-hairy or bristly, with cucumber-scented juice; stems ascending to erect, usually branching above, 20-70 cm tall.

Leaves: Lower leaves stalked, blades broadly elliptic or egg-shaped to oblanceolate, 5-15 cm long and 2-6 cm wide, the others reduced and shorter-stalked upwards, the uppermost often unstalked and clasping, alternate, entire or wavy-edged.

Flowers: Inflorescence of several loose, terminal clusters with leafy bracts below; flower stalks spreading to drooping; corollas flat or somewhat backswept, star-shaped; petals bright blue, fused at base, spreading wide (to 2 cm across) to 5 sharp-pointed lobes, with 5 erect, notched, white, glabrous bulges at the throat; stamens purplish-brown, sticking out from the flower and forming a cone; fruiting calyces 1-1.5 cm long, densely bristly.

Fruits: Nutlets 4, clustered together, oblong or egg-shaped, 4-6 mm long, irregularly ridged and pimply, with a swollen budlike basal attachment surrounded by a thickened rim.

CRYPTANTHA

1. Plants perennials with a distinct tuft of basal leaves .... *C. celosioides*
1. Plants annuals, lacking a tuft of basal leaves.
    2. Flowers large, 4-8 mm wide .... *C. intermedia*
    2. Flowers small, not over 2 mm wide.
        3. Nutlets warty .... *C. ambigua*
        3. Nutlets smooth.
            4. Nutlets lanceolate, with groove open at the base.
                5. Margins of nutlets prominent, sharply angled .... *C. watsonii*
                5. Margins of nutlets rounded .... *C. fendleri*
            4. Nutlets egg-shaped, with groove closed at the base.
                6. Nutlets compressed, with groove toward one margin .... *C. affinis*
                6. Nutlets not compressed, groove medial .... *C. torreyana*

***Cryptantha affinis* (A. Gray) Greene**
common cryptantha
HABITAT/RANGE:
Dry to mesic sites in the lowland and steppe zones; infrequent in S BC; E to SW AB and S to WY and CA.
DESCRIPTION:
General: Annual herb from a taproot; stems simple or branched, with stiff, straight, appressed to somewhat coarse, long hairs, 5-30 cm tall.

Leaves: Scattered along the stem, linear to oblong, 1-4 cm long, short-bristly, alternate, entire.

Flowers: Inflorescence of several coiled, often paired,1-sided, spikelike arms; petals white, fused into a short tube that spreads (1-2 mm wide) at the top to 5 lobes, with yellow bulges at the throat; calyces 2.5-4 mm long in fruit, with long, coarse hairs or bristles.

Fruits: Nutlets 4, clustered together, to 2 mm long, egg-shaped, smooth or finely sandy on surface (granular), compressed, groove on inner face off to one edge, more less closed at the base.

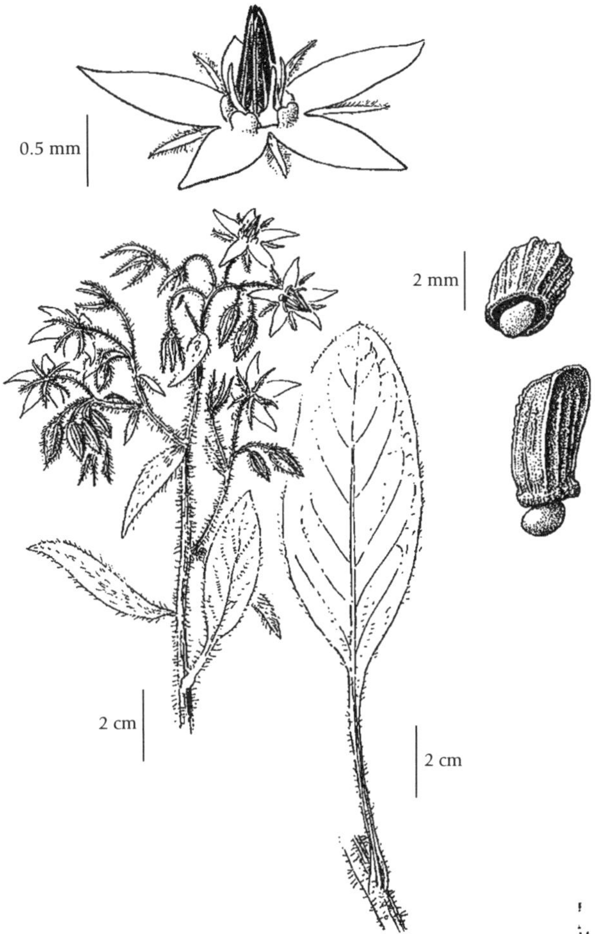

Borago officinalis

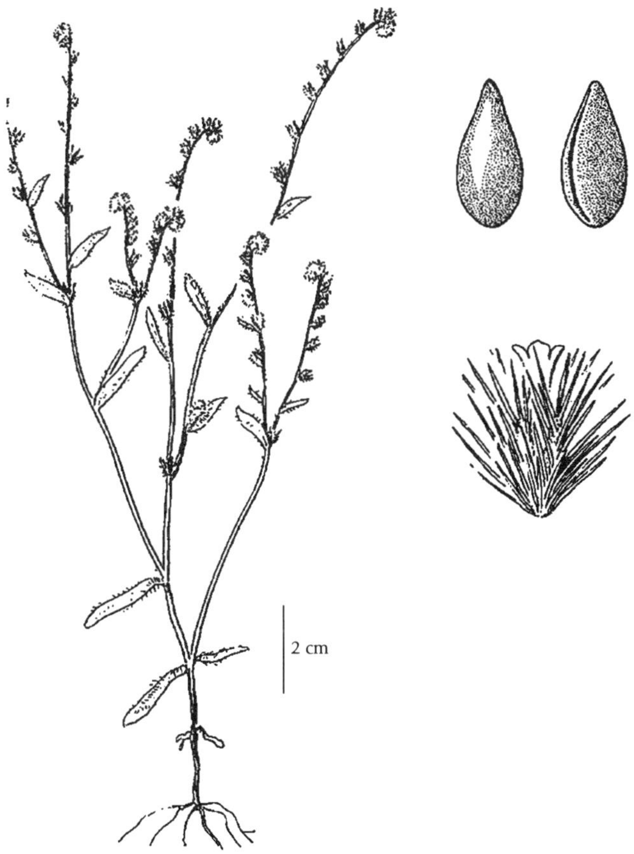

Cryptantha affinis

Cryptantha ambigua **(A. Gray) Greene**
obscure cryptantha
HABITAT/RANGE:
Dry grasslands and shrublands in the steppe zone; rare in SC and SE BC; S to CO, NV and NE CA.
DESCRIPTION:
General: Annual herb from a taproot, with long, spreading hairs and also shorter, straight, appressed hairs; stems usually with several branches from near the base, 5-30 cm tall.

Leaves: Basal leaves few and small or absent; stem leaves linear to linear-oblong, 1-4 cm long, alternate, entire.

Flowers: Inflorescence of several coiled,1-sided, spikelike arms; petals white, fused into a short tube that spreads (2-4 mm wide) at the top to 5 lobes, with 5 yellow bulges at the throat; calyces 4-6 mm long in fruit, densely straight, appressed-hairy and spreading-bristly.

Fruits: Nutlets 4, clustered together, 1-2 mm long, egg-shaped, surface sandy (granular) and with scattered bumps or pimples, with a medial groove essentially closed or flared open at the base.

Cryptantha celosioides **(Eastw.) Payson** *(C. nubigena* [Greene] Payson var. *celosioides* [Eastw.] Boivin)
cockscomb cryptantha
HABITAT/RANGE:
Dry grasslands and shrublands in the steppe zone; rare in extreme SC BC; S to NE and NV.
DESCRIPTION:
General: Biennial or short-lived perennial herb from a sometimes branched stem-base, silky throughout with fine, straight, appressed hairs and also with bulbous-based bristles; stems 1-several, 10-50 cm tall.

Leaves: Basal leaves several, tufted, conspicuous, spoon-shaped or broadly oblanceolate, rounded at tip, 2-8 cm long, 4-15 mm wide; stem leaves somewhat reduced, entire, alternate.

Flowers: Inflorescence of short "spikes" aggregated into a long cluster (thyrse); petals white, fused into a tube that spreads (8-12 mm wide) at top to 5 lobes, with 5 yellow bulges at throat; calyces 4-8 mm long in fruit, coarsely stiff-hairy and bristly.

Fruits: Nutlets 4, clustered together, 3-5 mm long, egg-shaped or lance-egg-shaped, wrinkled and bumpy, with medial groove closed at base.

Cryptantha fendleri **(A. Gray) Greene**
Fendler's cryptantha
HABITAT/RANGE:
Dry grasslands and shrublands in the steppe zone; rare in extreme SC BC; S to NE and AZ.
DESCRIPTION:
General: Annual herb from a taproot; stems often branched, with straight, appressed and spreading, stiff hairs, 10-40 cm tall.

Leaves: Basal leaves smaller, soon withering; stem leaves fairly numerous, linear, sharp-pointed, to 4 cm long, rough/stiff long-hairy with bulbous-based hairs of varying length, entire, alternate.

Flowers: Inflorescence of several naked, elongating, coiled spikes; petals white, fused into a tube that spreads at top merely 1 mm wide into 5 tiny lobes, with yellow bulges at the throat; fruiting calyces 4-6 mm long, coarsely long-hairy and with pustule-based hairs.

Fruits: Nutlets 4, clustered together, 1.5-2 mm long, lanceolate, smooth and shining, with medial groove open at base.

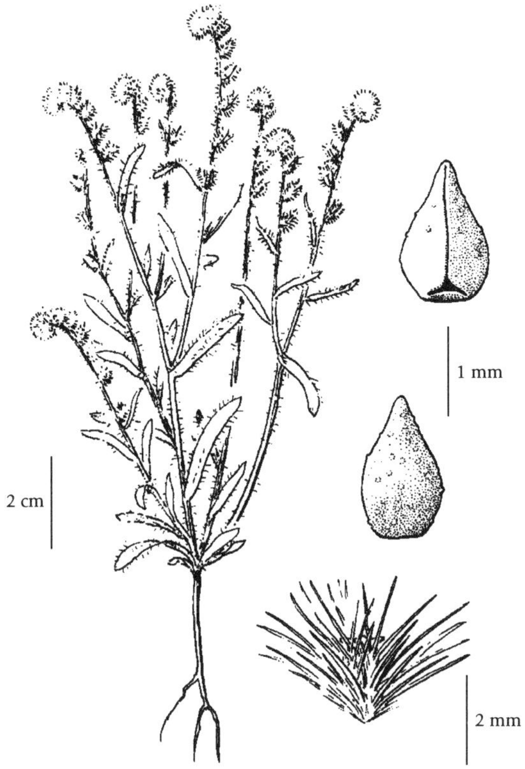

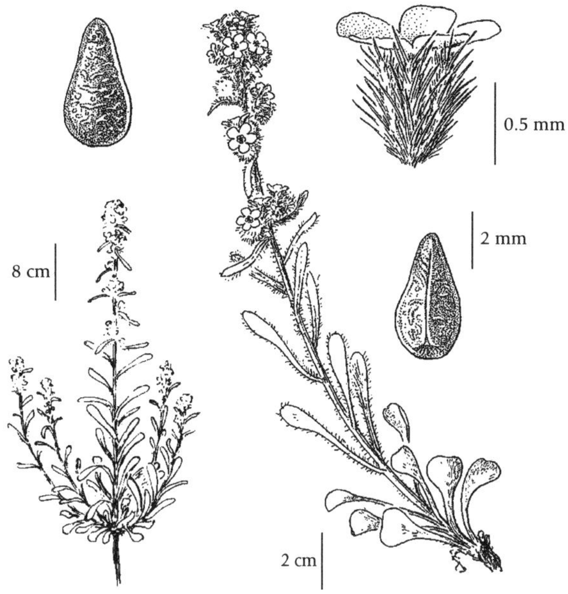

Cryptantha celosioides

Cryptantha ambigua

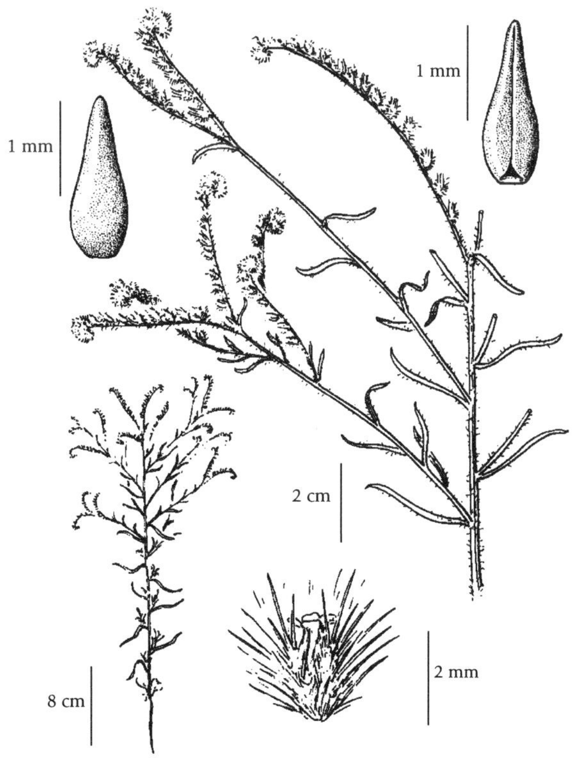

Cryptantha fendleri

Cryptantha intermedia **(A. Gray) Greene var. *grandiflora* (Rydb.) Cronq.**
large-flowered cryptantha

HABITAT/RANGE:

Dry to mesic disturbed areas in the lowland and montane zones; rare on S Vancouver Island (introduced?) and NE BC, known recently only from Nevis Creek; S to ID and CA.

DESCRIPTION:

General: Annual herb from a taproot, with straight, appressed and stiff, spreading hairs; stems simple or branched, 5-50 cm tall.

Leaves: Basal leaves more or less lacking; stem leaves scattered, linear to narrowly oblong or linear-oblanceolate, 1.5-5 cm long.

Flowers: Inflorescence of naked, elongating, coiled "spikes", tending to be paired or in 3s; petals white, fused at base into a tube that spreads (4-8 mm wide) at top to 5 lobes, with 5 yellow bulges at the throat; fruiting calyces 3-6 mm long, spreading long-hairy to bristly.

Fruits: Nutlets 1-4, narrowly egg-shaped, 2-3 mm long, typically sandy and sparingly bumpy, less often smooth and shining, with medial groove open at base.

Cryptantha torreyana **(A. Gray) Greene**
Torrey's cryptantha

HABITAT/RANGE:

Dry to mesic rocky slopes, grasslands and shrublands in the lowland and steppe zones; infrequent in SC BC and on S Vancouver Island; S to UT, WY and CA.

DESCRIPTION:

General: Annual herb from a taproot; stems simple or branched, with straight, appressed and spreading, stiff hairs, 10-40 cm tall.

Leaves: Basal leaves smaller, withering; stem leaves linear to narrowly oblong, 0.3-3 cm long, uniformly hairy with somewhat appressed to spreading hairs, entire, alternate.

Flowers: Inflorescence of naked, coiled "spikes", tending to be paired, typically raised somewhat above the leafier part of the plant; petals white, fused at base into a tube that spreads (about 1 mm wide) at the top to 5 tiny lobes, with 5 bulges at the throat; fruiting calyces 4-8 mm long, stiffly appressed-hairy and spreading-bristly.

Fruits: Nutlets 4, clustered together, 1.5-2.3 mm long, egg-shaped, usually smooth and shining, with medial groove forked and closed at base.

Cryptantha watsonii **(A. Gray) Greene**
Watson's cryptantha

HABITAT/RANGE:

Dry grasslands, shrublands and open forests in the steppe and lower montane zones; rare in SC BC; S to MT, CO and NV.

DESCRIPTION:

General: Annual herb from a taproot; stems usually branched, spreading rough-hairy and sometimes also appressed straight-hairy, 10-30 cm tall.

Leaves: Basal leaves mostly lacking; stem leaves linear, blunt-tipped, 0.5-3 cm long, densely straight-appressed to spreading long-hairy.

Flowers: Inflorescence of naked, coiled "spikes", single or often paired; petals white, fused at base into a tube spreading (1-2 mm wide) at top to 5 small lobes, with 5 bulges at the throat; fruiting calyces 2-4 mm long, long, straight, appressed-hairy and usually also with spreading bristles.

Fruits: Nutlets 4, clustered together, 1.5-2 mm long, lanceolate, with prominent sharp-angled margins and medial groove opening at base.

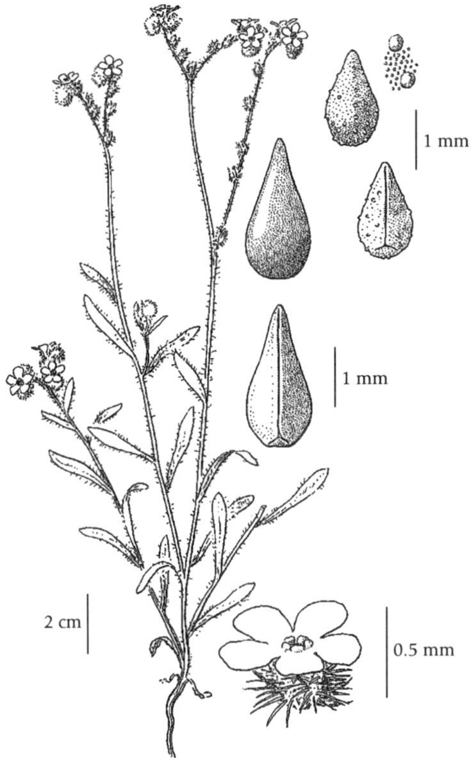

Cryptantha intermedia

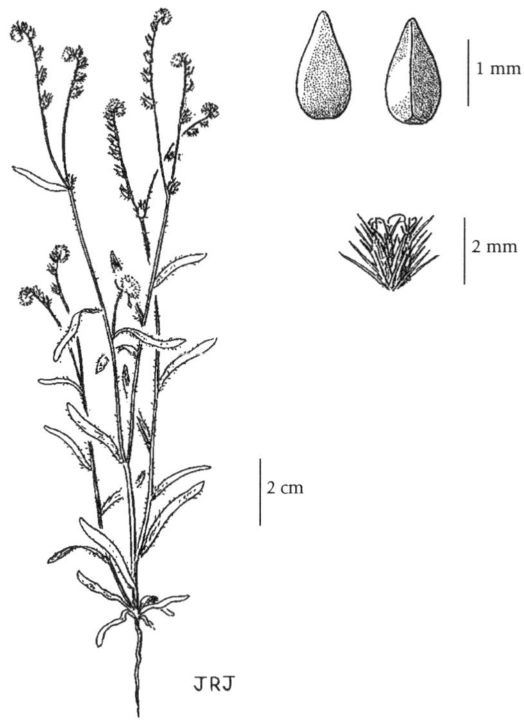

Cryptantha torreyana

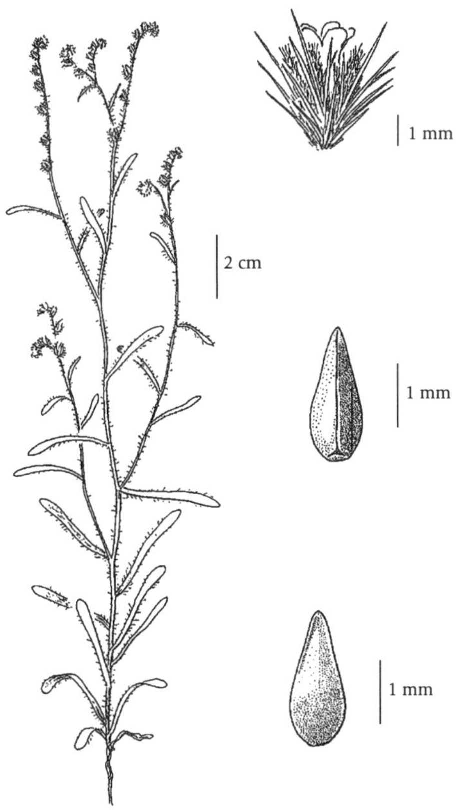

Cryptantha watsonii

CYNOGLOSSUM

1. Flowers blue; stems naked below inflorescences; native perennials .... *C. boreale*
1. Flowers dull red-purple (rarely white); stems leafy at least to base of inflorescences; weedy introduced biennials .... *C. officinale*

Cynoglossum boreale **Fern.**

northern hound's-tongue; northern wild comfrey

Habitat/Range:

Mesic forest openings in the lower montane zone; infrequent in C and N BC; E to NF and S to N MI, IN and CT.

Description:

General: Perennial herb from a taproot; stems single, spreading long-hairy, 40-80 cm tall.

Leaves: Basal and lowermost stem leaves long-stalked, 7-20 cm long, 2.5-7 cm wide, the blades elliptic or egg-shaped, straight, appressed-hairy; middle and upper leaves reduced upwards, unstalked, clasping, becoming more oblong or egg-shaped, alternate, entire.

Flowers: Inflorescence of 3-several wands (false racemes) atop a long, naked stalk; corollas funnel-shaped, blue; petals fused at base into a squat (to 3 mm long) tube that spreads (5-8 mm wide) at top to 5 rounded lobes, with 5 oblong, yellow bulges at the throat; mature calyces 1-2.5 mm long.

Fruits: Nutlets 4, clustered together but spreading apart at maturity, 3.5-5 mm long, egg-shaped, covered with barb-tipped prickles.

Cynoglossum officinale **L.**

common hound's-tongue

Habitat/Range:

Dry disturbed areas in the lowland, steppe and lower montane zones; common E of the Coast-Cascade Mountains; introduced from Europe.

Description:

General: Biennial herb from a taproot, long soft-hairy or greyish-downy throughout; stems single, leafy to the top, 0.3-1.2 m tall.

Leaves: Basal and lowermost stem leaves oblanceolate or narrowly elliptic, tapering to stalks, 10-30 cm long, 2-5 cm wide; middle and upper leaves numerous, unstalked and more oblong or lanceolate, gradually reduced upward, alternate, entire.

Flowers: Inflorescence of numerous curved wands (false racemes) from upper leaf axils; corollas dull reddish-purple (maroon); petals fused at base into short (5-8 mm long) broad tube that spreads (about 1 cm wide) at top to 5 rounded lobes, with 5 rounded bulges sticking out at the throat; mature calyces 5-8 mm long at maturity.

Fruits: Nutlets 4, clustered together but spreading wide when ripe, egg-shaped, 5-7 mm long, covered with barb-tipped prickles.

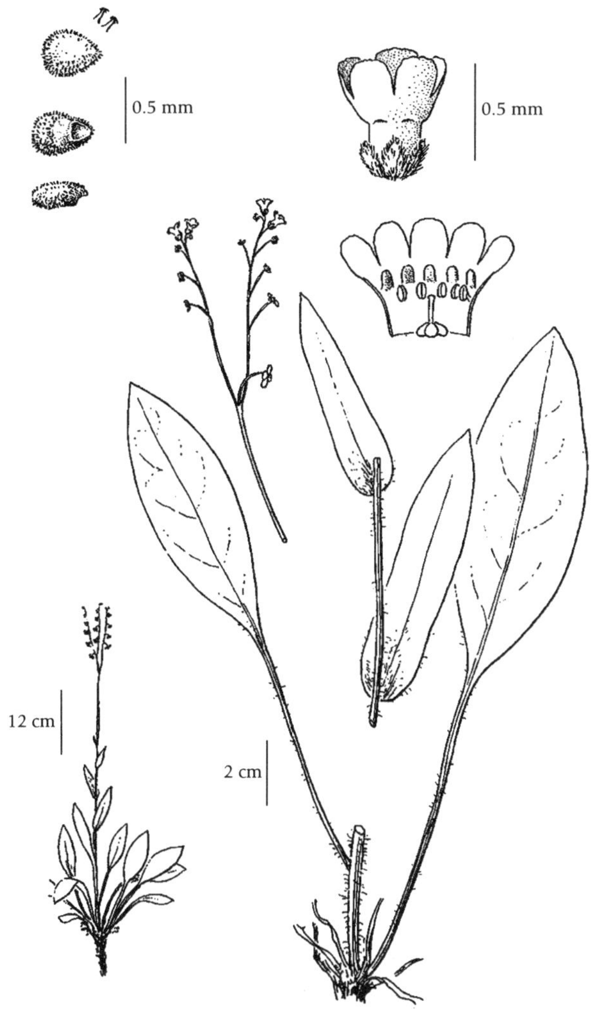

Cynoglossum boreale

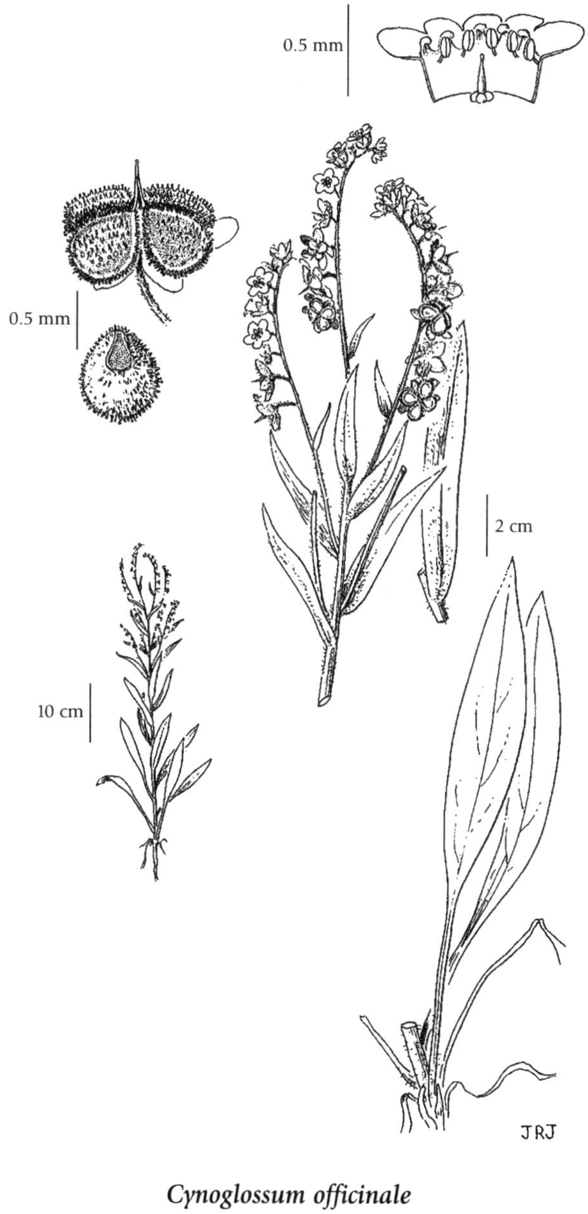

Cynoglossum officinale

ECHIUM

***Echium vulgare* L.**

viper's bugloss; blueweed

HABITAT/RANGE:

Dry roadsides and waste places in the lowland, steppe and montane zones; frequent in C and S BC east of the Coast-Cascade Mountains, infrequent on the coast; introduced from Europe.

DESCRIPTION:

General: Biennial herb from a taproot, roughly long- or short-hairy and spreading-bristly, the bristles often bulbous-based; stems single, 30-80 cm tall.

Leaves: Basal leaves oblanceolate, narrowed to stalk, 6-25 cm long (stalk included), 0.5-3 cm wide; stem leaves becoming smaller and unstalked upward, alternate, entire.

Flowers: Inflorescence of numerous short, coiled clusters in a long, narrow (often wandlike) cluster; corollas funnel-shaped, asymmetric (upper part longer); petals blue (sometimes pink or white), fused at base into a short tube that flares at top into 5 unequal lobes, the throat open without bulges; 4 of 5 stamens exserted from the corolla.

Fruits: Nutlets 4, clustered together, angular, roughened and wrinkled.

HACKELIA

1. Flowers 1-3 mm wide; annuals or biennials .... *H. deflexa*
1. Flowers 4-20 mm wide; biennials or perennials.
    2. Flowers white .... *H. diffusa*
    2. Flowers blue.
        3. Nutlets with marginal prickles only .... *H. floribunda*
        3. Nutlets with central (facial) prickles in addition to marginal ones.
            4. Marginal prickles united for at least 1/3 their length, forming a cup .... *H. ciliata*
            4. Marginal prickles mostly distinct to base, not forming a cup .... *H. micrantha*

***Hackelia ciliata* (Dougl. *ex* Lehm.) I.M. Johnst.**

Okanogan stickseed; Okanogan hackelia

HABITAT/RANGE:

Dry grassy slopes in the steppe zone; rare in SC BC, known only from Rock Creek and Kettle Valley; S to WA.

DESCRIPTION:

General: Perennial herb from a taproot; stems several, straight, appressed-hairy, downswept below, upswept above, and often with coarse, spreading hairs as well, 30-90 cm tall.

Leaves: Basal leaves well-developed, oblanceolate, stalked, 5-15 cm long, 3-12 mm wide; stem leaves almost linear, mostly unstalked, to 10 cm long and 6 mm wide; closely appressed straight-hairy to often densely and softly short-hairy, and frequently with some coarse bristles, alternate, entire.

Flowers: Inflorescence of several short clusters with leafy bracts at the base; corollas light blue (fading pink); petals fused at base into a short, narrow tube that spreads flat (5-12 mm wide) abruptly at top to 5 lobes, with 5 pimply bulges at the throat.

Fruits: Nutlets 4, clustered together, the margins with prominent, barb-tipped prickles, the prickles united at base for 1/3-1/2 their length to form a shallow cup; intramarginal prickles several, smaller than marginal ones.

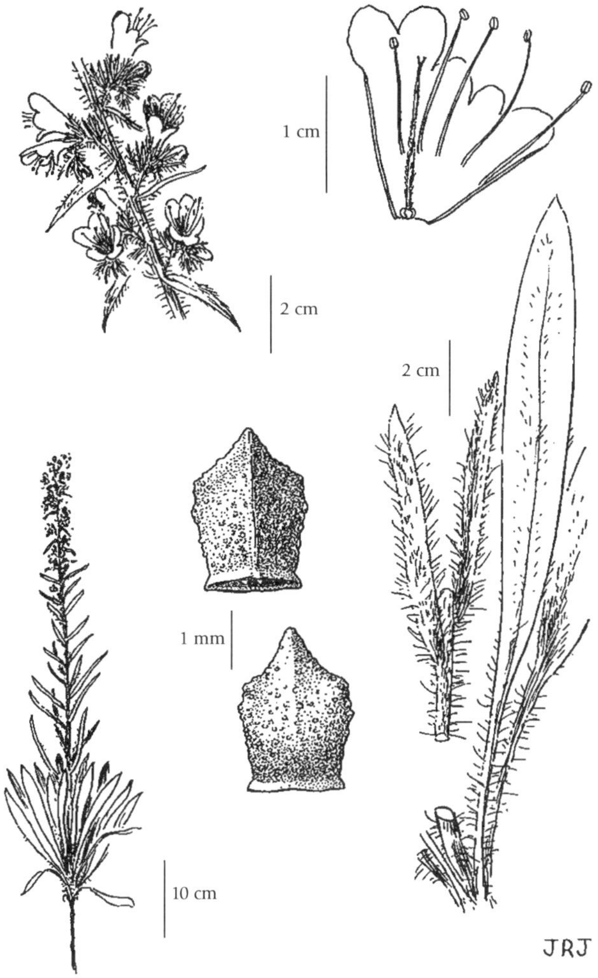

Echium vulgare

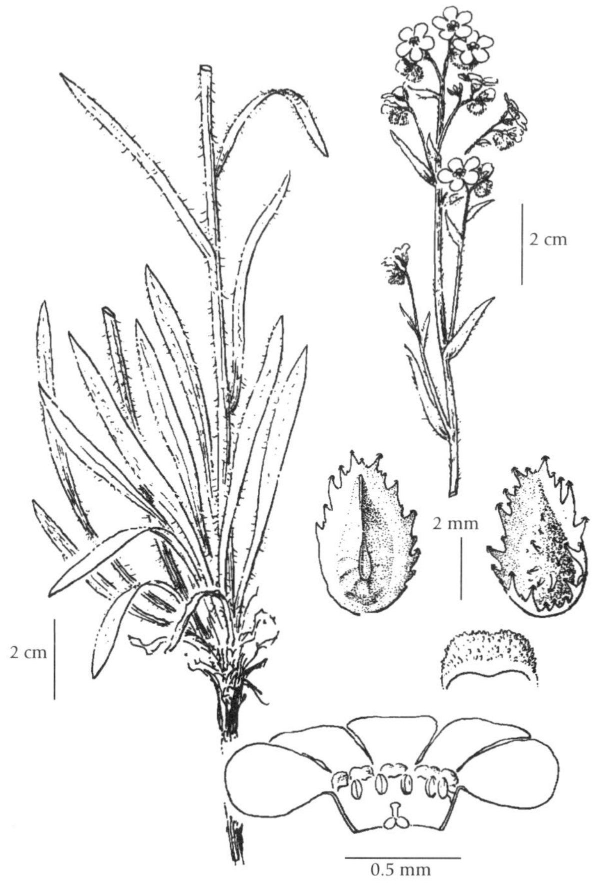

Hackelia ciliata

Hackelia deflexa **(Wahl.) Opiz Bercht. ssp. *americana* (A. Gray) Hult.**
nodding stickseed; nodding hackelia
HABITAT/RANGE:
Dry grasslands, shrublands, forest openings and disturbed areas in the steppe and lower montane zones; frequent throughout BC east of the Coast-Cascade Mountains, circumboreal, E to PQ and S to WA, CO and IA.
DESCRIPTION:
General: Annual or biennial herb from a taproot; stems usually single, sometimes few, spreading-hairy below, appressed straight-hairy above, 15-100 cm tall.

Leaves: Basal leaves few and withering or absent; stem leaves 2-15 cm long, 0.5-3 cm wide, the lower oblanceolate and somewhat stalked, the upper lanceolate or narrowly elliptic and unstalked, coarsely stiff-hairy or appressed straight-hairy, entire, alternate.

Flowers: Inflorescence of several elongating, narrow clusters, naked or with leafy bracts at the base, the flower stalks curved downward in fruit; corollas blue or sometimes white, tiny; petals fused at base into short narrow tube that spreads (1.5-3 mm wide) to 5 lobes, with 5 bulges at the throat.

Fruits: Nutlets 4, clustered together, 2-3 mm long, with marginal barb-tipped prickles distinct to base, facial prickles 0-3.

Hackelia diffusa **(Dougl. ex Lehm.) I.M. Johnst.**
spreading stickseed; spreading hackelia
HABITAT/RANGE:
Dry grasslands, shrublands and forest openings in the steppe and lower montane zones; rare in SC and SE BC; S to WA and OR.
DESCRIPTION:
General: Perennial herb from a taproot; stems several or sometimes single, spreading stiff-hairy on lower half, more appressed, upswept, straight-hairy above, 20-70 cm tall.

Leaves: Basal leaves well-developed, persistent, stalked, oblanceolate or narrowly elliptic, 6-18 cm long and 8-25 mm wide; stem leaves ample, mostly unstalked and sometimes also clasping upward, lanceolate or lance-oblong to lance-elliptic, 5-12 cm long, 8-18 mm wide; stiff-hairy or appressed, coarsely straight-hairy, entire, alternate.

Flowers: Inflorescence of several elongating, narrow clusters with leafy bracts at the base; petals white (rarely blue), fused at base into a short, narrow tube that spreads (7-12 mm wide) at top to 5 lobes, with 5 pimply, fuzzy, yellow bulges at the throat.

Fruits: Nutlets 4, clustered together, with marginal, barb-tipped prickles distinct nearly to the base, facial prickles 8-30, much shorter than marginal ones.

Hackelia floribunda **(Lehm.) I.M. Johnst.**
many-flowered stickseed; many-flowered hackelia
HABITAT/RANGE:
Moist to mesic rocky slopes, grasslands and shrublands in the lowland and steppe zones; frequent in S BC east of the Coast-Cascade Mountains; E to W ON and S to CA and NM.
DESCRIPTION:
General: Biennial or short-lived perennial herb from a taproot; stems single to few, spreading long-hairy on lower part, appressed, upswept, straight-hairy above, 0.3-1 m tall.

Leaves: Basal leaves sometimes persistent but usually smaller than stem leaves and soon withering, oblanceolate, stalked; stem leaves ample, 4-20 cm long and 5-30 mm wide, the lower oblanceolate and stalked, the upper more lanceolate or lance-elliptic and unstalked, reduced upward, appressed straight-hairy or coarsely stiff-hairy, entire, alternate.

Flowers: Inflorescence of several narrow, elongating, many-flowered clusters with leafy bracts at the base; petals blue, fused at base into a short,narrow tube that spreads (4-8 mm wide) at top to 5 lobes, with 5 minutely pimply, yellow bulges at the throat.

Fruits: Nutlets 4, clustered together, 2-4 mm long, with marginal, barb-tipped prickles mostly free (but sometimes fused) to near base, facial prickles 0-3.

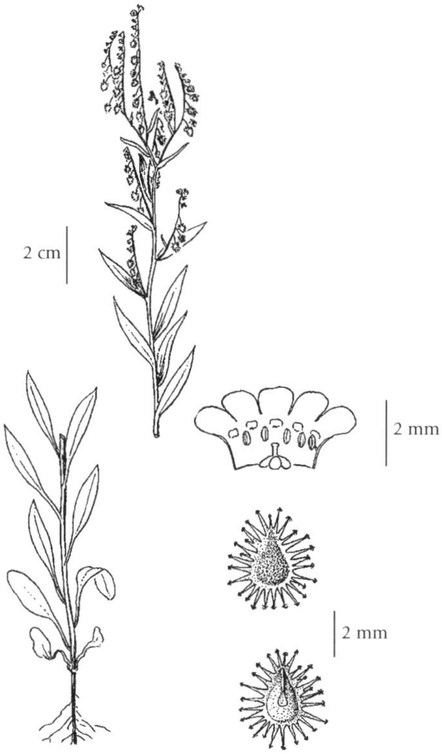

Hackelia deflexa

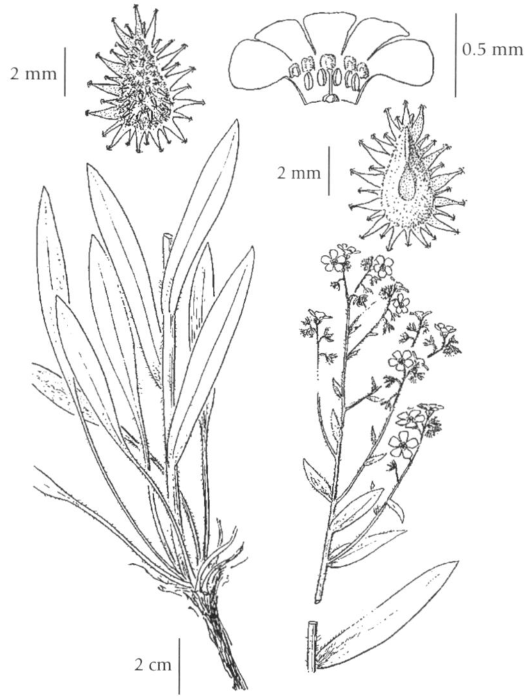

Hackelia diffusa

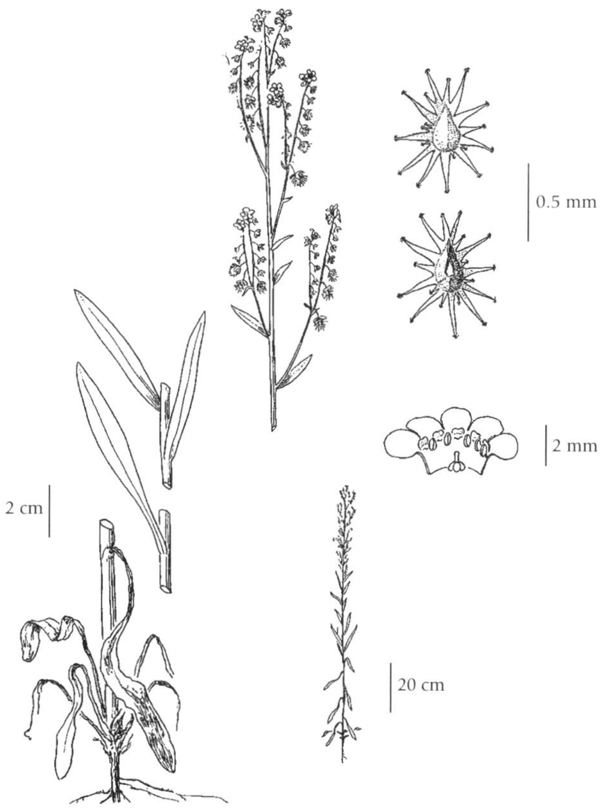

Hackelia floribunda

***Hackelia micrantha* (Eastw.) J.L. Gentry** *(H. jessicae* [McGregor] Brand)

blue stickseed; blue hackelia

HABITAT/RANGE:

Dry to mesic rocky slopes, grasslands, shrublands and forest openings in the lowland, steppe and montane zones; frequent in C and S BC east of the Coast-Cascade Mountains; E to AB and S to UT and CA.

DESCRIPTION:

General: Perennial herb from a taproot and branching base; stems several to numerous, relatively sparsely hairy with coarse, stiff-spreading and (especially upward) short, appressed, straight hairs, 0.3-1 m tall.

Leaves: Basal leaves usually well-developed, persistent, oblanceolate or narrowly elliptic, stalked, 6-35 cm long and to 4 cm wide; stem leaves also well-developed, 5-20 cm long, 7-20 mm wide, the lower oblanceolate and stalked, the upper narrowly elliptic to oblong and unstalked, reduced upward, short appressed-hairy or coarsely stiff-hairy, entire, alternate.

Flowers: Inflorescence of several clusters with leafy bracts at the base, tending to be shorter, wider, and looser than that of *H. floribunda*; petals blue, fused at base into a short, narrow tube that spreads (5-11 mm wide) abruptly at top to 5 lobes, with 5 minutely pimply, yellow or whitish bulges.

Fruits: Nutlets 4, with marginal, barb-tipped prickles mostly distinct to the base, facial prickles several (usually 4-10), much shorter than marginal.

LAPPULA

1. Nutlets with single row of marginal prickles .... *L. occidentalis*
1. Nutlets with 2 rows of marginal prickles .... *L. squarrosa*

***Lappula occidentalis* (S. Wats.) Greene** (*L. redowskii auct. non* (Hornem.) Greene = var. *occidentalis*, *L. redowskii* var. *cupulata* [A. Gray] M.E. Jones, *L. texana* [Scheele] Britt. = var. *cupulata)*

western stickseed

HABITAT/RANGE:

Dry to mesic roadsides, disturbed areas, grasslands and shrublands in the lowland, steppe and lower montane zones; common in S BC east of the Coast-Cascade Mountains, less common northward; circumpolar, S to NM, AZ and CA; S America, Eurasia.

DESCRIPTION:

General: Annual herb from a taproot, short curly- or stiff-hairy (or appressed-hairy) throughout; stems simple or branched above, 5-50 cm tall.

Leaves: Basal leaves oblanceolate, often deciduous, 2-6 cm long, to 1 cm wide; stem leaves linear or linear-oblong to oblanceolate, reduced upward to the leafy, lanceolate bracts of the inflorescence, unstalked or nearly so, entire, alternate.

Flowers: Inflorescence of narrow, elongating clusters that have bracts throughout; flower stalks ascending to erect when in fruit; petals blue or white, fused at base into small (2-4 mm long) tube that flares a bit (1.5-2.5 mm wide) at the top to 5 lobes, with 5 yellow bulges at the throat; fruiting calyces 3-3.5 mm long, the lobes erect.

Fruits: Nutlets 4, clustered together, egg-shaped, 2-3 mm long, with a crown of marginal, barb-tipped prickles in a single row, prickles distinct or sometimes fused at the base to form a cuplike rim.

Note: Two varieties occur in BC.

1. Prickles of nutlets distinct to base; the common variety .... var. *occidentalis*
1. Prickles of nutlets fused at base, forming a cup; rare .... var. *cupulata* (A. Gray) Higgins

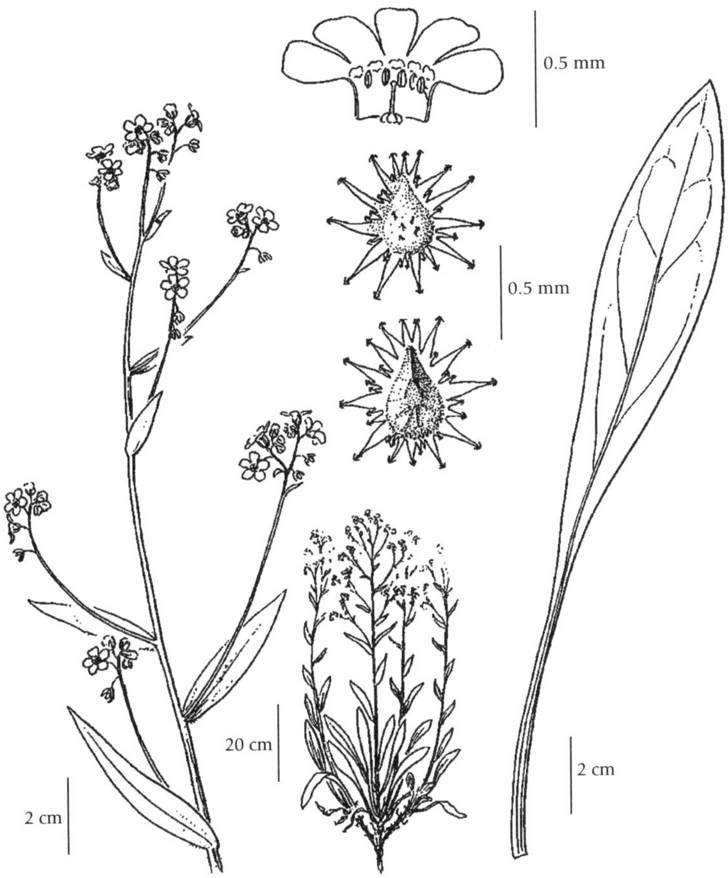

Hackelia micrantha

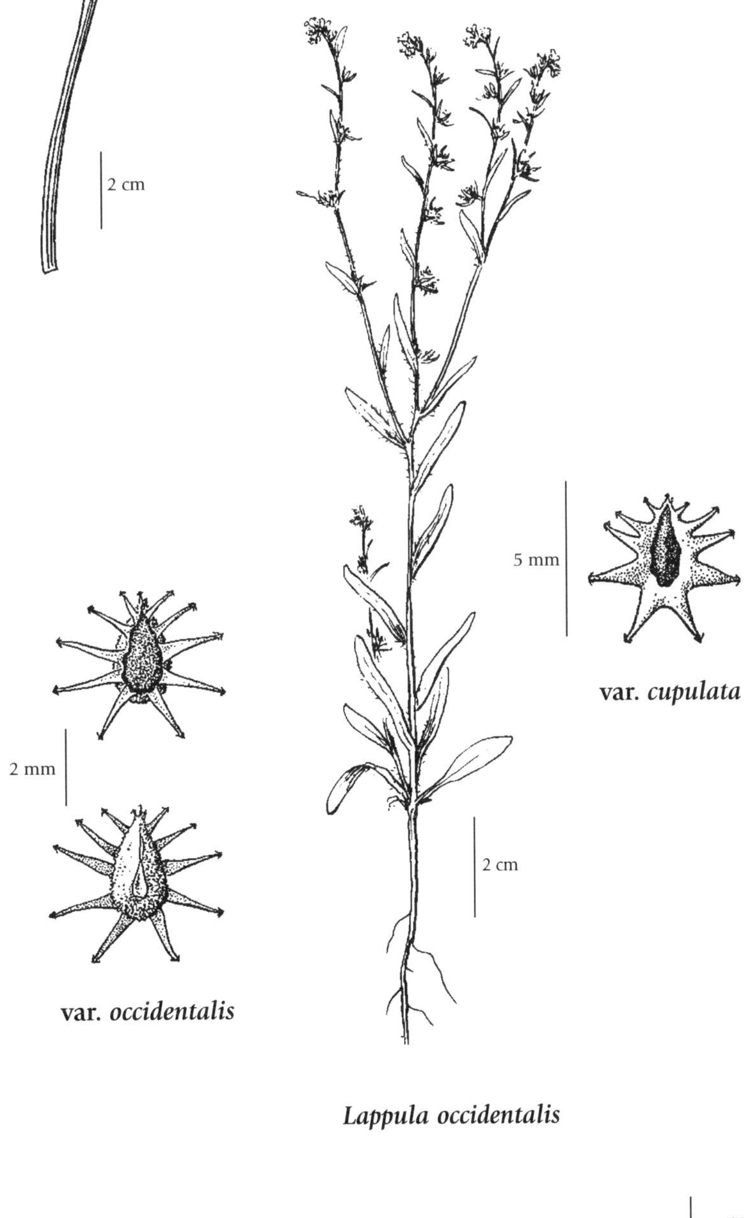

Lappula occidentalis

***Lappula squarrosa* [Retz.] Dumort.** (*Lappula echinata* Gilib., *L. myosotis* Moench)

bristly stickseed; common stickseed

HABITAT/RANGE:

Dry to mesic grasslands, shrublands, forest openings and disturbed areas in the steppe and montane zones; common in S BC east of the Coast-Cascade Mountains, less frequent northward; introduced from Eurasia.

DESCRIPTION:

General: Annual herb from a taproot, short, straight, appressed-hairy (and sometimes curly- or spreading-hairy) throughout; stems usually branched, 10-80 cm tall.

Leaves: Basal leaves oblanceolate, to 10 cm long, often deciduous; stem leaves linear to narrowly oblong or oblanceolate, reduced upwards to the leafy, lanceolate bracts of the inflorescence, unstalked, entire, alternate.

Flowers: Inflorescence of narrow, elongating clusters with bracts throughout; flower stalks ascending to erect when in fruit; petals blue (sometimes white), fused at base into a small tube that flares (2-4 mm wide) at top to 5 lobes, with 5 yellow bulges at throat; fruiting calyces 2.5-3 mm long, the lobes spreading

Fruits: Nutlets 4, clustered together, egg-shaped, 3-4 mm long, with 2(-3) marginal rows of slender, barb-tipped prickles distinct to the base, those of the outer row usually shorter than those of the inner.

LITHOSPERMUM

1. Plants taprooted annuals, flowers white .... *L. arvense*
1. Plants perennials; flowers greenish to yellow.
   2. Corollas yellow-green, lobes entire .... *L. ruderale*
   2. Corollas bright yellow, lobes ragged- or wavy-edged .... *L. incisum*

***Lithospermum arvense* L.** (*Buglossoides arvense* [L.] I.M. Johnston)

corn gromwell

HABITAT/RANGE:

Dry roadsides and disturbed areas in the lowland, steppe and lower montane zones; frequent in S BC; introduced from Europe.

DESCRIPTION:

General: Annual herb from a taproot, short, stiff, appressed-hairy throughout; stems 1-several, sometimes branched, 10-70 cm tall.

Leaves: Lowermost leaves oblanceolate and soon deciduous, others linear to lanceolate or oblong, unstalked, 1.5-6 cm long and 2-15 mm wide, entire, alternate.

Flowers: Inflorescence of several narrow clusters (compact at first, elongate and open in fruit); flowers solitary in axils of upper leaves, on short (1 mm) stalks that are erect to ascending in fruit; corollas funnel-shaped, 5-8 mm long, white or bluish-white; petals fused at base into a narrow tube that flares (2-4 mm wide) at top to 5 lobes, lacking bulges but with 5 hairy lines in the throat.

Fruits: Nutlets usually 4, clustered together, brownish, wrinkled, pitted, sometimes pimply.

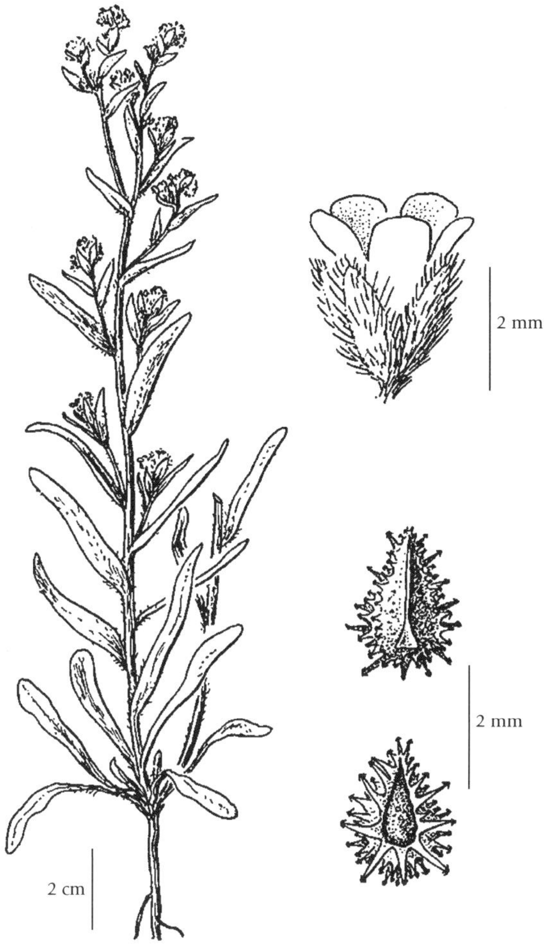

Lappula squarrosa

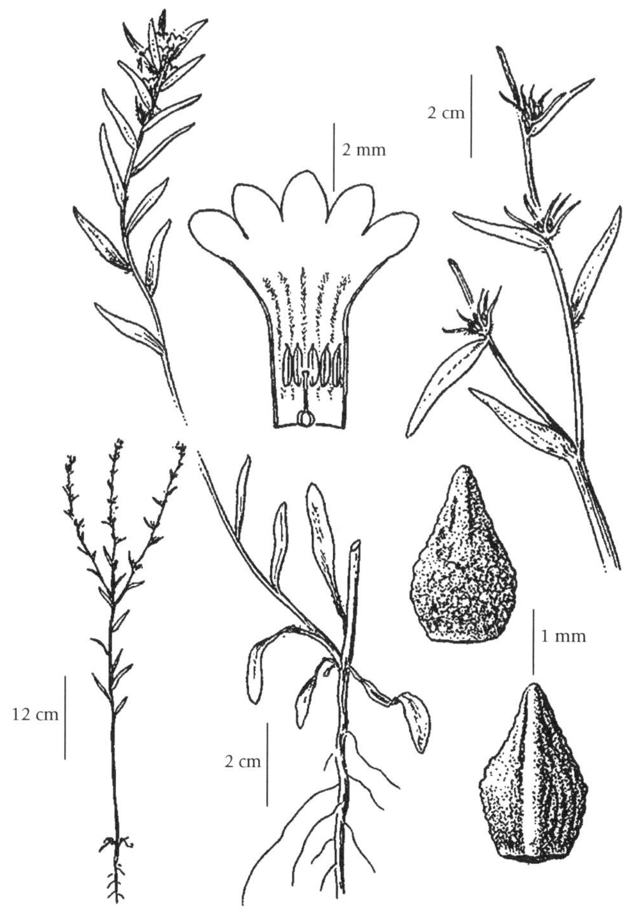

Lithospermum arvense

Lithospermum incisum **Lehm.**
yellow gromwell
Habitat/Range:
Dry grasslands and shrublands in the steppe zone; infrequent in S BC east of the Coast-Cascade Mountains; E to S ON and S to IN, MO, TX, NM, AZ and MX.
Description:
General: Perennial herb from heavy, woody taproot and branched stem-base; stems few to several, 5-30 cm tall, short, stiff appressed-hairy.
Leaves: Numerous, all along the stem, unstalked, linear to oblong or narrowly lanceolate, 2-6 cm long and 2-6 mm wide, entire, alternate, short, stiff appressed-hairy.
Flowers: Well-developed flowers in small clusters in upper leaf axils, the flower stalks bent down somewhat in fruit; corollas bright yellow, 15-35 mm long, fused at base into long (12-30 mm) tube that spreads (10-20 mm wide) at top to 5 ragged-margined lobes, with 5 bulges at the throat; smaller unopened flowers often appear later in the season, in branching clusters lower on the stem.
Fruits: Nutlets usually 4, clustered together, grey, shining, sparingly pitted, 3-3.5 mm long.

Lithospermum ruderale **Dougl. *ex* Lehm.**
lemonweed; Columbia gromwell
Habitat/Range:
Dry grasslands and shrublands in the steppe zone; common in S BC east of the Coast-Cascade Mountains; E to SW SK and S to CO and N CA.
Description:
General: Perennial herb from a stout, woody taproot and branching stem-base; spreading stiff-hairy throughout but especially on the stem; stems 1 to several, clustered, more or less unbranched, 20-60 cm tall.
Leaves: Numerous, all along the stem, lowermost reduced, others unstalked, lanceolate to linear, 3-10 cm long, 2-10 mm wide, entire, alternate.
Flowers: Inflorescence of small clusters with leafy bracts, in the axils of upper leaves; flower stalks ascending in fruit; corollas pale yellow, often tinged with green, 8-13 mm long; petals fused at base into a 4-6-mm long tube that spreads (7-13 mm wide) to 5 entire lobes, the throat glandular within but lacking bulges.
Fruits: Nutlets often 1-2, sometimes 4, clustered together, grey, smooth, shining, 3.5-6(8) mm long.

MERTENSIA

1. Plants decumbent, glaucous, succulent, maritime .... *M. maritima*
1. Plants erect, not glaucous or succulent, not maritime.
    2. Leaves with distinct lateral veins; plants usually much taller than 15 cm .... *M. paniculata*
    2. Leaves lacking distinct lateral veins; plants usually less than 15 cm tall .... *M. longiflora*

Mertensia longiflora **Greene** (*M. oblongifolia auct. non* [Nutt.] G. Donn)
long-flowered bluebells; long-flowered mertensia
Habitat/Range:
Dry to mesic grasslands, shrublands and open forests in the steppe and montane zones; frequent in S BC east of the Coast-Cascade Mountains; S to W MT, ID and N CA.
Description:
General: Perennial herb from tuberlike, easily detached root; glabrous or straight appressed-hairy; stems usually 1-2, 5-25 cm tall.
Leaves: Basal leaves rare on flowering plants, if present then stalked, with broadly elliptic, rounded blades; stem leaves few, lowermost strongly reduced, the others well-developed, mostly unstalked, 2-6 cm long and 0.5-3 cm wide, rounded at tip, entire, alternate; lateral veins obscure.
Flowers: Inflorescence compact, somewhat headlike; corollas tubular funnel-shaped, blue, 15-25 mm long; petals fused at base into a tube that flares slightly at top to the funnel-shaped limb and 5 lobes, the tube 2-3 times as long as the limb, and glabrous within but with 5 bulges at the throat.
Fruits: Nutlets 4, clustered together, wrinkled.

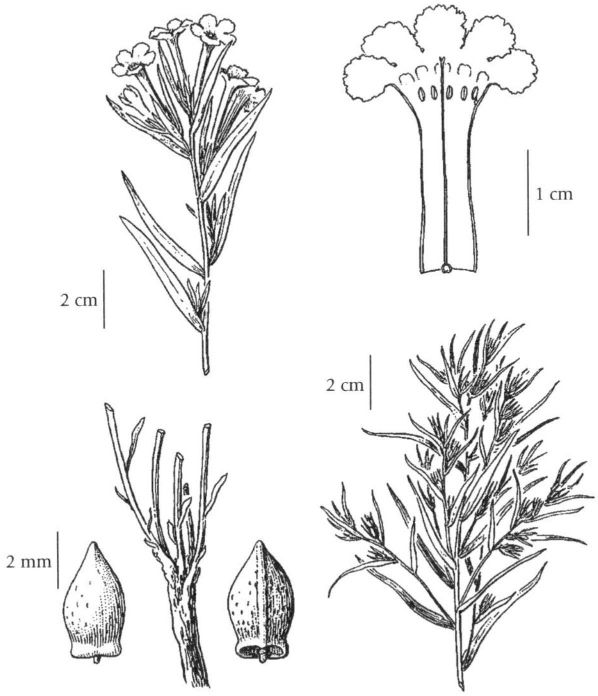

Lithospermum incisum

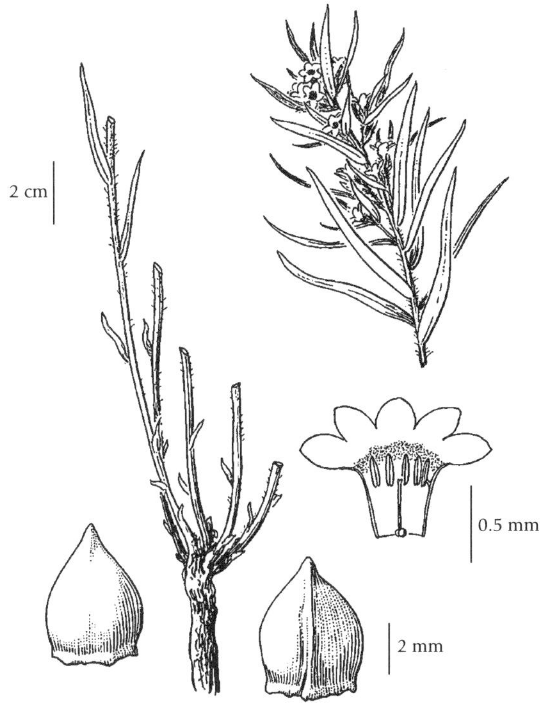

Lithospermum ruderale

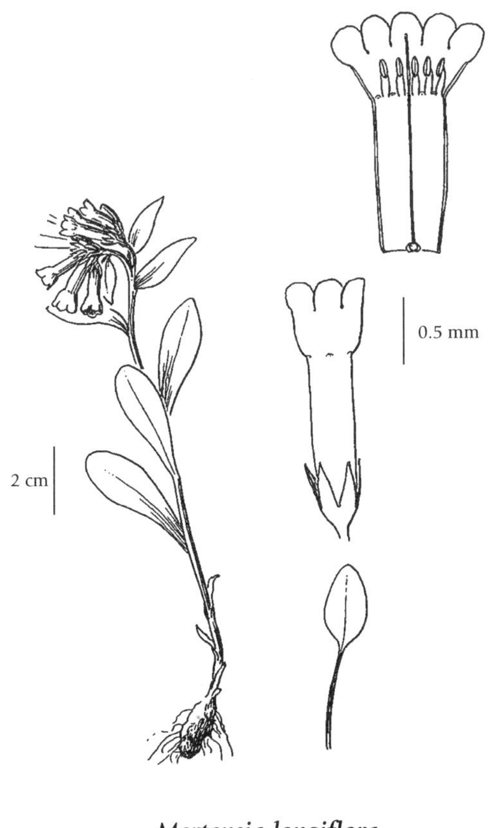

Mertensia longiflora

Mertensia maritima **(L) S.F. Gray**

sea bluebells; sea-mertensia; sea-lungwort; oysterleaf

HABITAT/RANGE:

Moist sandy or gravelly beaches in the lowland zone; rare in W BC, known only from the Queen Charlotte Islands; circumpolar, N to AK, YT and NT and E to NF, N Europe, E Asia.

DESCRIPTION:

General: Perennial herb from a taproot and branched, often subterranean stem-base; glabrous throughout; stems several to many, sprawling to decumbent, 10-70 cm long.

Leaves: Fleshy, with whitish bloom, spoon-shaped to elliptic or egg-shaped, abruptly sharp-pointed, the lower leaves stalked, the others becoming smaller and nearly unstalked upward, entire, mostly alternate in two rows.

Flowers: Several to many in compact to open clusters, or solitary in leaf axils; corollas tubular bell-shaped, blue or blue-purple, rarely white, sometimes pink in bud, 3-8 mm long; petals fused at base into a short (1-3 mm) tube, with 5 bulges at the throat, flaring at top to a limb and 5 lobes, the limb longer than the tube.

Fruits: Nutlets 4, clusterd together, 2.5-4 (6) mm long, smooth or slightly wrinkled, flattened, somewhat fleshy, the outer coat becoming inflated and spongy.

Mertensia paniculata **(W. Ait.) G. Don**

tall bluebells; panicled mertensia

HABITAT/RANGE:

Wet to mesic meadows, streambanks and open forests in the montane, subalpine and alpine zones; frequent in N BC, infrequent southward; N to AK, YT, and NT, E to PQ and S to IA, MT, ID and OR.

DESCRIPTION:

General: Perennial herb from a taproot and stout rhizome or branched base; stems several, erect or ascending, glabrous or hairy, 0.2-1.5 m tall.

Leaves: Basal leaves long-stalked and more or less heart-shaped; stem leaves numerous, well-developed, mostly short-stalked, lanceolate to broadly egg-shaped with abruptly sharp-pointed tips, 3-15 cm long, 1-7 cm wide, becoming smaller and unstalked upward, with short, straight, appressed- or stiff-spreading hairs; entire, alternate.

Flowers: Few to many in generally branched, compact to open terminal clusters, or with some solitary in leaf axils; corollas tubular bell-shaped, blue to pink (pink in bud), 9-19 mm long; petals fused below into a tube 5-8 mm long, flaring at top to bell-shaped limb (a bit longer than the tube) and 5 short lobes, with 5 bulges at the throat.

Fruits: Nutlets 4, clustered together, 2.5-5 mm long, wrinkled.

Notes: Two varieties occur in BC.

1. Leaves straight appressed-hairy to spreading coarse-hairy on both surfaces; calyces hairy; plants infrequent to frequent throughout BC .... var. *paniculata*

1. Leaves glabrous above; calyces glabrous; plants rare, known only from extreme SE BC .... var. *borealis* (J.F. Macbr.) L.O. Williams

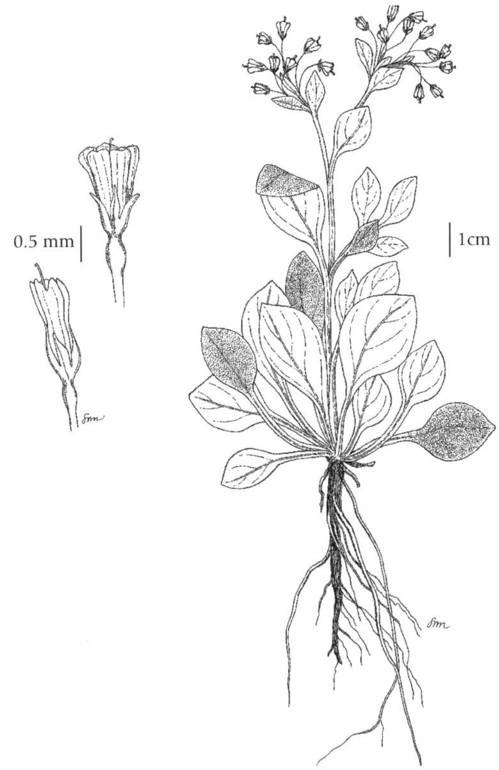

Mertensia maritima

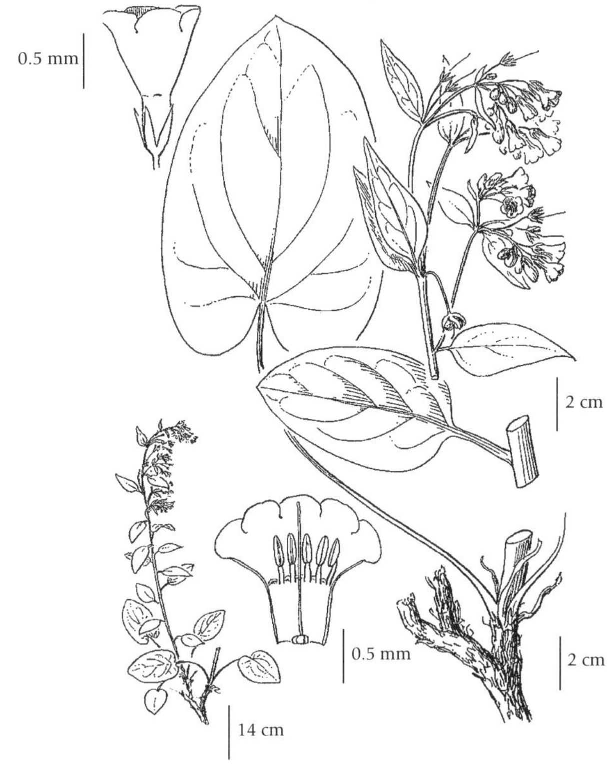

Mertensia paniculata var. *paniculata*

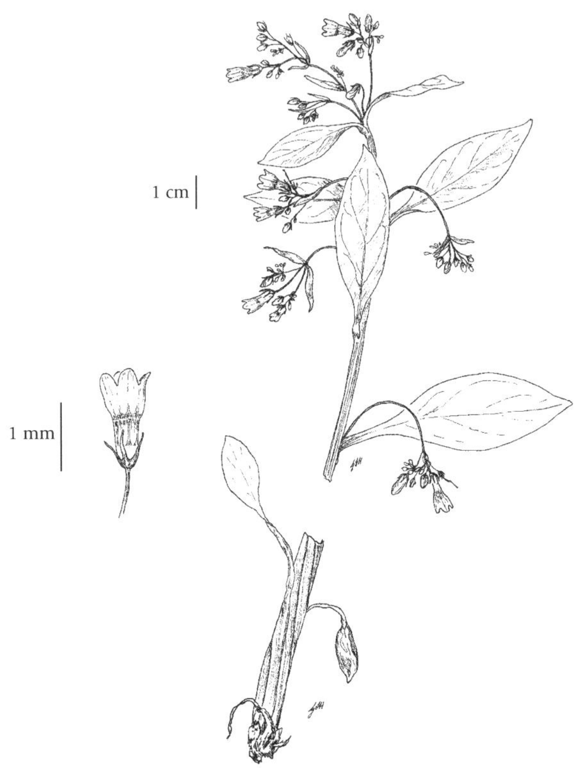

Mertensia paniculata var. *borealis*

MYOSOTIS

1. Calyces with appressed hairs, lacking hooks.
   2. Corollas 2-5 mm wide, the tube equal to the calyces .... *M. laxa*
   2. Corollas 5-10 mm wide, the tube longer than calyces .... *M. scorpioides*
1. Calyces with spreading hairs, some of which are hooked.
   3. Corollas 4-10 mm wide.
      4. Stems lax, usually branching above ground; more or less uniformly leafy throughout; introduced at low elevations .... *M. sylvatica*
      4. Stems erect, from branching stem-base or short rhizome,forming tight, erect clumps; leaves largest basally, gradually reduced upwards; native, common at high elevations .... *M. asiatica*
   3. Corollas 1-3 mm wide.
      5. Calyces asymmetric, 2 lobes longer than other 3 .... *M. verna*
      5. Calyces symmetric, all 5 lobes equal.
         6. Fruiting stalks equal to or longer than calyces .... *M. arvensis*
         6. Fruiting stalks shorter than calyces.
            7. Corollas blue, tube equal to calyces .... *M. stricta*
            7. Corollas yellow, aging blue, tube surpassing calyces .... *M. discolor*

***Myosotis arvensis* (L.) Hill**
field forget-me-not

Habitat/Range:

Moist to mesic roadsides and waste places in the lowland zone; frequent in SW BC; introduced from Europe.

Description:

General: Biennial or annual herb from fibrous roots, short soft spreading-hairy to straight appressed-hairy throughout; stems simple or branched, 10-40 cm tall.

Leaves: Lower leaves in a loose rosette, stalked, oblanceolate; upper leaves unstalked, more oblong or lanceolate; 1-6 cm long, 3-16 cm wide, entire, alternate.

Flowers: Inflorescence of lax, coiled, elongating, bractless, terminal clusters, not much longer than leafy part of plant; fruiting stalks ascending or spreading, as long as or longer than the calyces; petals grey-blue, sometimes white, fused at the base into a tube that spreads (2-4 mm wide) at top to 5 lobes, with 5 bulges at the throat; fruiting calyces 3-5 mm long, short stiff-hairy and also with numerous, short, coarse, hooked hairs.

Fruits: Nutlets 4, clustered together, egg-shaped, smooth, dark brown, shining.

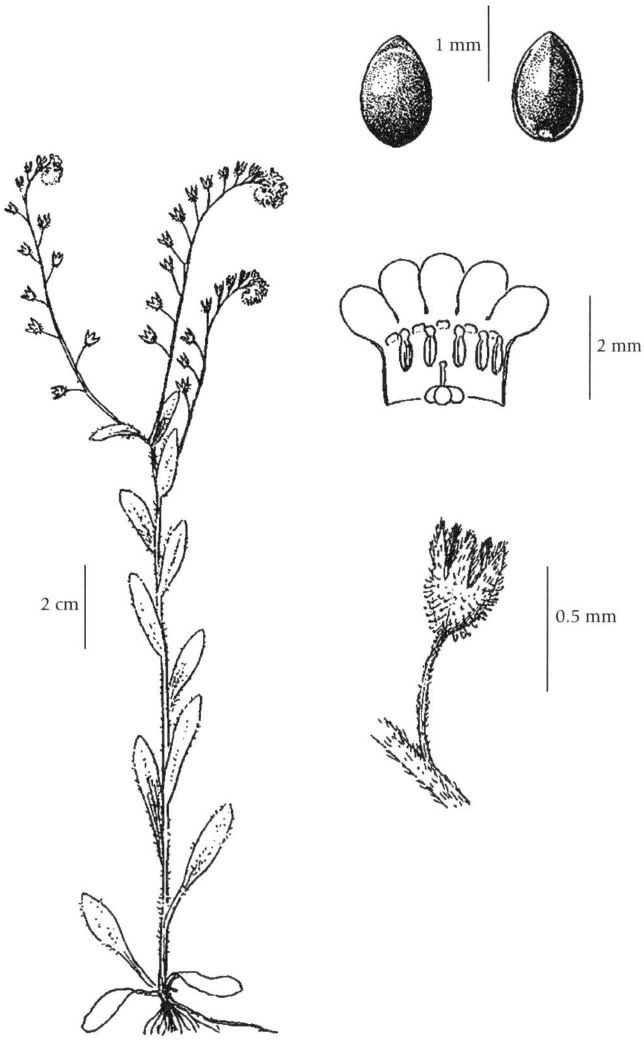

Myosotis arvensis

Myosotis asiatica **[Vesterg.] Schischkin & Sergievskaja** (*Myosotis alpestris auct. non* F.W. Schmidt, *M. alpestris* F.W. Schmidt. ssp. *asiatica* Vesterg., *M. sylvatica* Hoffm. var. *alpestris* [F.W. Schmidt] Koch)

mountain forget-me-not

HABITAT/RANGE:

Mesic meadows, streambanks, rocky slopes and forest openings in the montane to alpine zones; common in N BC, less common southward; circumpolar, N to AK, YT and NT, E to AB and S to ID, SD and WY; Eurasia.

DESCRIPTION:

General: Perennial herb from fibrous roots and a branching stem-base or short rhizome; spreading-hairy throughout; stems several to many, tufted, erect, 3-50 cm tall.

Leaves: Basal leaves oblanceolate or elliptic, stalked, 5-13 cm long, 2-12 mm wide; stem leaves several, oblong to lance-elliptic, smaller, to 6 cm long, gradually reduced upward, unstalked, entire, alternate.

Flowers: Several to many in coiled, bractless, not much elongating clusters (clusters); flowers showy, the flower stalks erect then spreading to ascending, about as long as calyces; petals intensely blue, rarely white, fused at base into a tube that spreads flat (4-10 mm wide) at top to 5 lobes, with 5 yellow bulges at the throat; fruiting calyces 2-5 mm long, with spreading to ascending, often hooked hairs.

Fruits: Nutlets 4, clustered together, egg-shaped, 1-2 mm long, smooth black, shining.

Myosotis discolor **Pers.**

common forget-me-not

HABITAT/RANGE:

Moist to mesic roadsides, ditches and disturbed areas in the lowland and lower montane zones; frequent on S Vancouver Island, the Gulf Islands and adjacent coast, rare E of the Coast-Cascade Mountains; introduced from Europe.

DESCRIPTION:

General: Annual herb from fibrous roots; stems simple or branched, erect, slender, with spreading hairs below and appressed hairs above, 8-50 cm tall.

Leaves: Lowermost leaves oblanceolate, somewhat stalked, 1-4 cm long and 2-8 mm wide; middle and upper leaves oblong to narrowly elliptic or linear, more or less unstalked; straight-hairy on both surfaces, entire, alternate.

Flowers: Several to many in lax, coiled, elongating, terminal clusters, the bracts 0 or 1-2 near the base; inflorescence not much longer than leafy part of stem; fruiting stalks ascending or spreading, 1-3 mm long, shorter than calyces; corollas at first yellow or white, becoming blue; petals fused at base into a tube that flares (1-2 mm wide) at top to 5 small lobes, the throat with 5 bulges; fruiting calyces 3-5 mm long, sepal-tube covered with short, spreading, hooked hairs, the lobes short appressed-hairy.

Fruits: Nutlets 4, clustered together, egg-shaped, dark brown to black, smooth, shining.

Myosotis laxa **Lehm.**

small-flowered forget-me-not

HABITAT/RANGE:

Moist to wet meadows, ditches, pond margins and swamps in the lowland, steppe and montane zones; common along the coast, less common in S BC; circumboreal, E to AB and S to N CA; S. America, Eurasia.

DESCRIPTION:

General: Short-lived perennial, sometimes annual herb from fibrous roots; inconspicuously straight appressed-hairy; stems slender, weak, often decumbent but base not creeping or stoloniferous, 10-40 cm tall.

Leaves: Lower leaves oblanceolate; middle and upper leaves more oblong or narrowly elliptic to lanceolate; 1.5-8 cm long, 3-15 mm wide, entire, alternate.

Flowers: Inflorescence of lax, coiled, elongating, terminal clusters, not much longer than leafy part of stem, bracts 0 to few at base; fruiting stalks spreading, mostly longer than the calyces; petals blue, fused at base into a tube that spreads flat (2-5 mm wide) at the top to 5 lobes, with 5 hairy bulges at the throat; fruiting calyces 3-7 mm long, closely appressed-hairy.

Fruits: Nutlets 4, clustered together, egg-shaped, brown to black, smooth, shining.

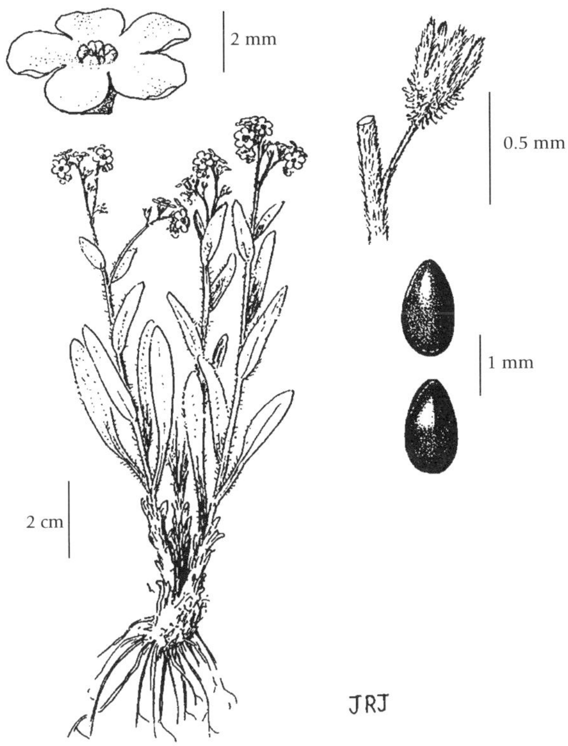

Myosotis asiatica

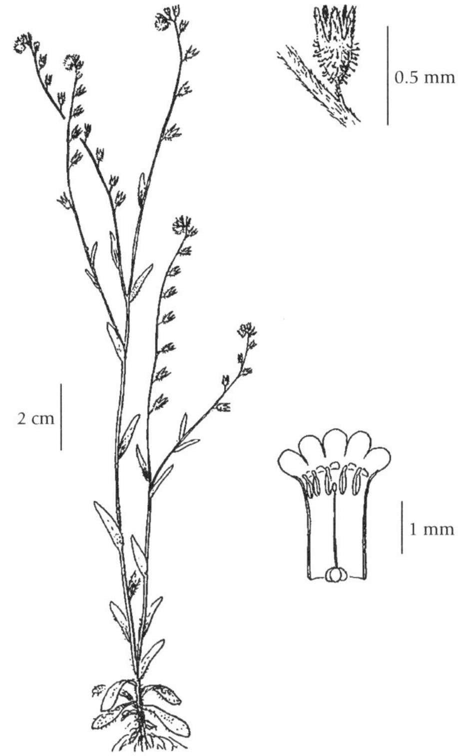

Myosotis discolor

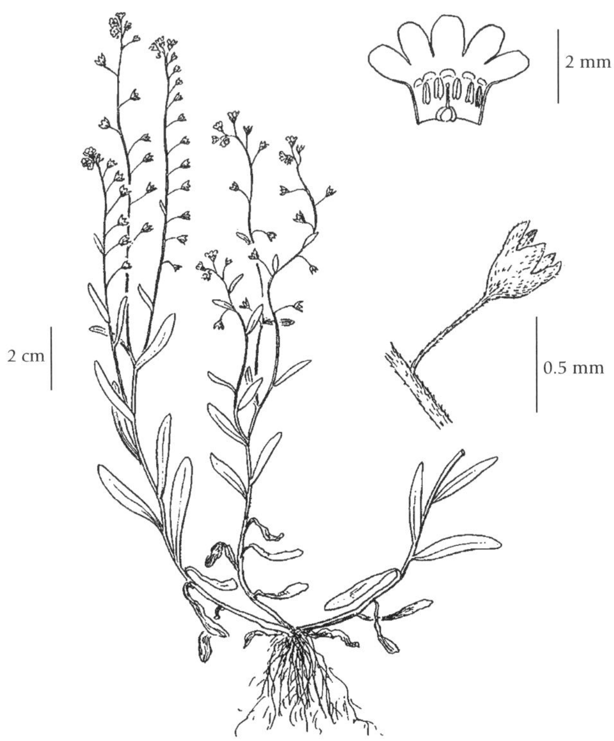

Myosotis laxa

Myosotis scorpioides **L.**
European forget-me-not
HABITAT/RANGE:
Moist to wet meadows, ditches, pond margins and swamps in the lowland, steppe and montane zones; frequent in S BC; introduced from Europe.
DESCRIPTION:
General: Perennial herb from fibrous roots, often creeping at the base and frequently stoloniferous as well; stems mostly unbranched, angled, inconspicuously appressed-hairy, 15-60 cm long.

Leaves: Lower leaves oblanceolate, narrowed at base but scarcely stalked; middle and upper leaves more oblong or elliptic to lance-elliptic; 2.5-8 cm long, 7-20 mm wide, nearly glabrous or with short appressed hairs, entire, alternate.

Flowers: Inflorescence of lax, coiled, narrow, elongating, bractless, terminal clusters, not much longer than leafy part of stem; fruiting stalks spreading, 1-2 times as long as calyces; petals sky-blue, rarely white, fused at base into a tube that spreads flat (4-10 mm wide) at top to 5 lobes, with 5 bulges at the throat; fruiting calyces 3-5 mm long, closely appressed-hairy.

Fruits: Nutlets 4, clustered together, narrowly egg-shaped, 2-2.5 mm long, blackish, smooth, shining.

Myosotis stricta **Link *ex* Roemer & Schultes** (*M. micrantha auct. non* Pallas)
blue forget-me-not
HABITAT/RANGE:
Mesic to dry meadows, ditches and disturbed areas in the lowland, steppe and lower montane zones; common in S BC; introduced from Europe.
DESCRIPTION:
General: Annual herb from fibrous roots, with short, soft, spreading hairs throughout; stems simple or commonly branched from near the base, to 20 cm tall.

Leaves: Lowermost leaves oblanceolate or wider; middle and upper leaves more oblong or elliptic; to 2 cm long and 7 mm wide, entire, alternate.

Flowers: Inflorescence of lax, coiled, narrow, elongating clusters, with some flowers nearly to the base of the stems, with leafy bracts; fruiting stalks ascending to spreading, shorter than calyces; petals blue, fused at base into a tube that flares (1-2 mm wide) at top to 5 small lobes; fruiting calyces 3-5 mm long, with hooked hairs on the sepal-tubes and appressed hairs on the lobes.

Fruits: Nutlets 4, clustered together, egg-shaped, brown, smooth, shining.

Myosotis sylvatica **Ehrh. *ex* G.F. Hoffm.**
wood forget-me-not
HABITAT/RANGE:
Mesic grasslands, shrublands, disturbed areas and open forests in the lowland, steppe and lower montane zones; frequent in S BC, rare elsewhere; introduced from Europe.
DESCRIPTION:
General: Perennial herb from fibrous roots and short rhizome, spreading-hairy throughout; stems simple or much-branched, 15-45 cm tall.

Leaves: Lower leaves in basal rosette, somewhat stalked, egg- or spoon-shaped; stem leaves lanceolate to oblong, unstalked, 5-12 cm long, 2-12 mm wide, entire, alternate.

Flowers: Several to many in lax, coiled, elongating, bractless, terminal clusters; fruiting stalks spreading, 1 1/2 to 2 times as long as the calyces; petals bright blue, rarely white, fused at base into a tube that spreads flat (5-10 mm wide) at the top to 5 lobes, with 5 yellow bulges at the throat; fruiting calyces 3-5 mm long, with short, curly and hooked hairs on the sepal tube.

Fruits: Nutlets 4, clustered, egg-shaped, 1.5-2 mm long, dark brown, smooth, shining.

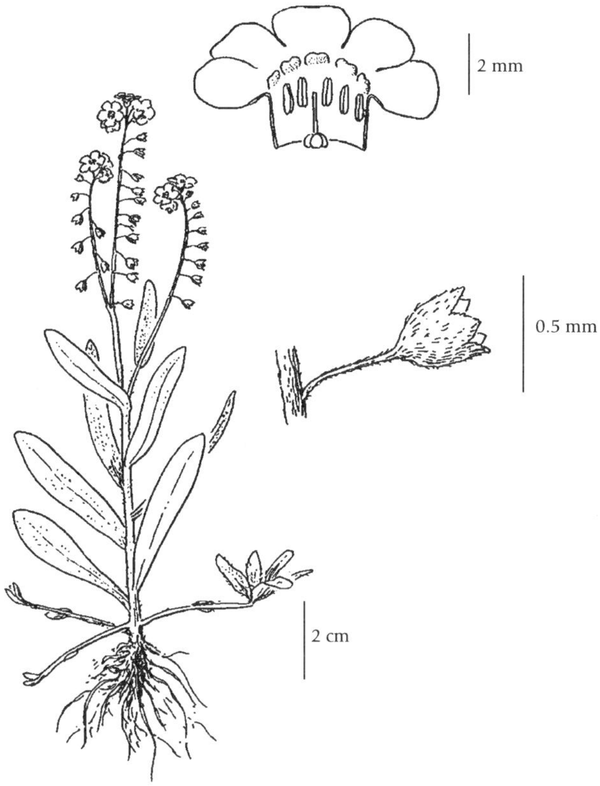

Myosotis scorpioides

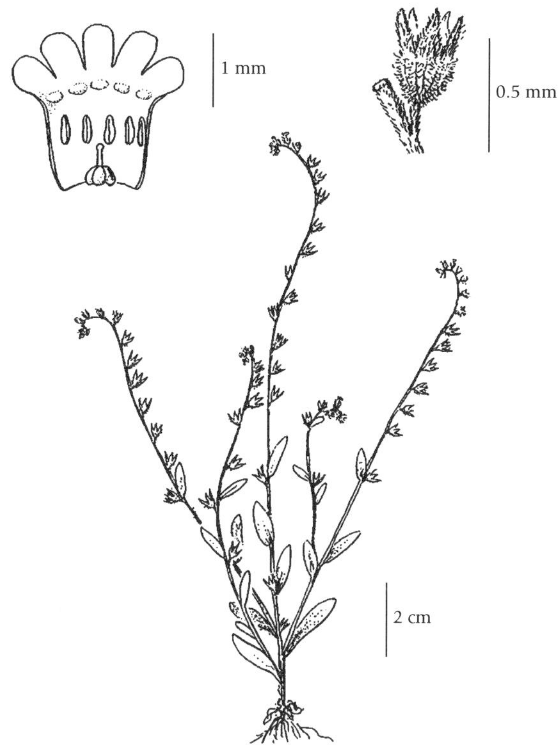

Myosotis stricta

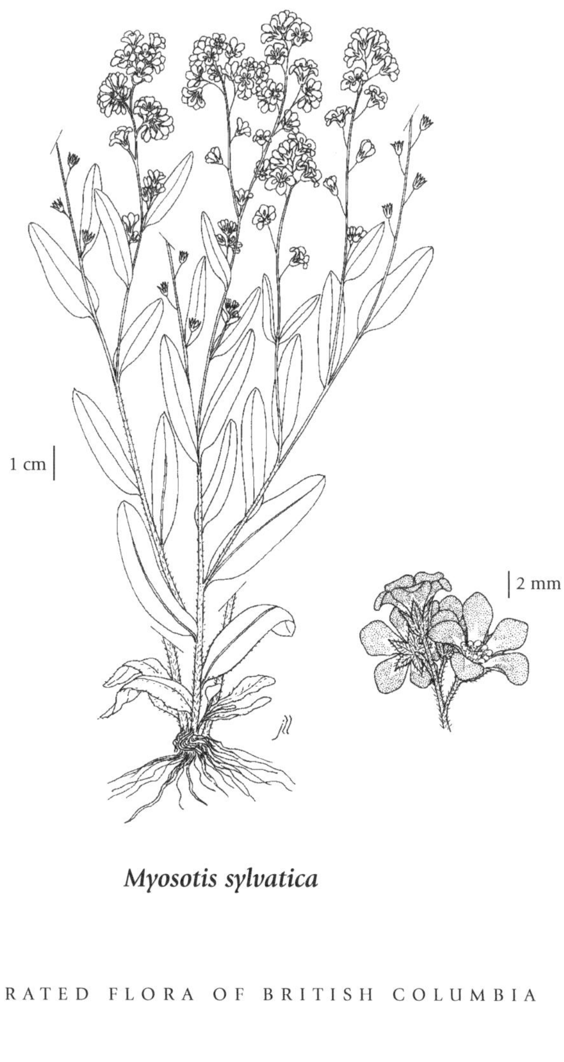

Myosotis sylvatica

Myosotis verna **Nutt.**
spring forget-me-not
HABITAT/RANGE:
Moist to dry meadows, grasslands, shrublands and forest openings in the lowland, steppe and lower montane zones; infrequent throughout BC; E to ON and S to FL, TX and OR.
DESCRIPTION:
General: Annual herb from fibrous roots, with short, coarse, spreading or appressed hairs throughout; stems simple or branched, 5-40 cm tall.

Leaves: Lowermost leaves oblanceolate and somewhat stalked; middle and upper leaves more oblong or narrowly elliptic to linear and nearly unstalked, 1-5 cm long, 2-10 mm wide, entire, alternate.

Flowers: Inflorescence of coiled, narrow, elongating clusters, with some leafy bracts on lower flowers, not much longer than the leafy portion of the plant; fruiting stalks ascending, usually shorter than calyces; corollas inconspicuous; petals white, fused at base into a tube that flares (1-2 mm wide) at the top to 5 small lobes; fruiting calyces 4-7 mm long, covered with short hooked hairs on the sepal-tube, stiff ascending-hairy on the lobes, asymmetric, 3 of the lobes shorter than the other 2.

Fruits: Nutlets 4, clustered, egg-shaped, brown, commonly with pale stippling.

PECTOCARYA

Pectocarya penicillata **(H.& A.) DC.** (*P. linearis* [R.& P.] DC. var. *penicillata* [H.& A.] M. E. Jones)
winged combseed
HABITAT/RANGE:
Dry grasslands and shrublands in the steppe zone; rare in SC BC; S to ID, UT and MX; Argentina, Chile.
DESCRIPTION:
General: Annual herb from a taproot; stems prostrate to decumbent, often branched at base, straight appressed-hairy, 2-25 cm long.

Leaves: Basal and stem leaves linear, to 2 cm long and 2 mm wide, appressed-hairy to spreading-hairy, entire, mostly alternate, reduced towards the branch tips.

Flowers: In small clusters along and at tips of stems; corollas tiny (to 3 mm long), white, tubular at base, flaring merely 1 mm wide or less at top.

Fruits: Nutlets usually 4, spreading in pairs, oblong, 1.5-2 mm long, flattened and sunken in the centre, with prominent, upturned or incurved margins beset (mostly above the middle) with hooked prickles.

PLAGIOBOTHRYS

1. Flowering plants with basal rosette of leaves; stem leaves alternate; nutlets cross-shaped with warty surface .... *P. tenellus*
1. Flowering plants lacking distinct basal rosette; at least some stem leaves opposite; nutlets egg-shaped.
    2. Corollas large, 5-10 mm wide .... *P. figuratus*
    2. Corollas small, 1-4 mm wide .... *P. scouleri*

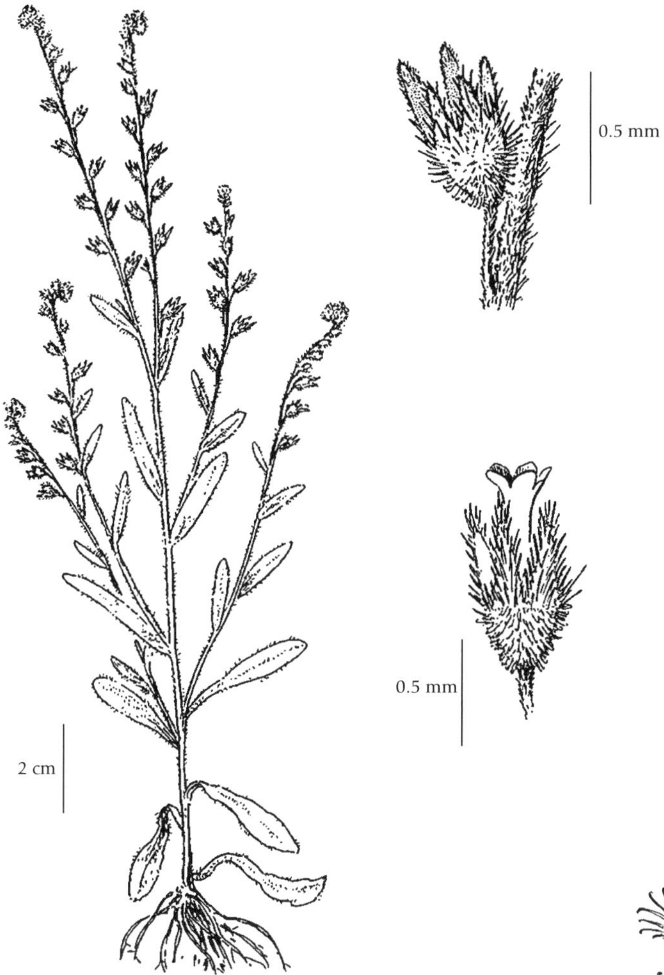

Myosotis verna

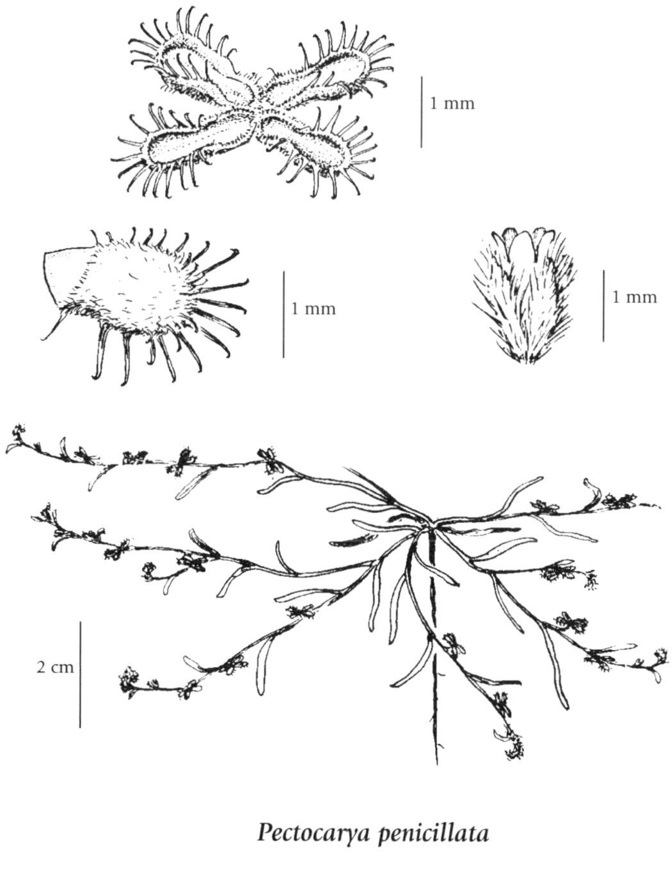

Pectocarya penicillata

Plagiobothrys figuratus **(Piper) Johnst. *ex* M.E. Peck**
fragrant popcornflower

Habitat/Range:
Moist to mesic coastal bluffs in the lowland zone; rare on SE Vancouver Island and the Gulf Islands, known recently only from Whaling Station Bay (Hornby Island); S to OR.

Description:
General: Annual herb from fibrous roots; sparsely to moderately straight appressed-hairy, also sometimes with slender spreading hairs; stems simple or branched, 10-40 cm tall.

Leaves: Basal leaves nearly lacking; stem leaves oblong to linear, entire, 0.5-8 cm long, 3-8 mm wide, the lower 2-4 pairs opposite, the others alternate.

Flowers: Inflorescence of several coiled, narrow, elongating, often paired spikes that lack bracts; flowers relatively large and showy; petals white, fused at base into a tube that spreads (5-10 mm wide) abruptly at top to 5 lobes, the throat with 5 yellow bulges; fruiting calyces 3-4 mm long, densely ascending-hairy.

Fruits: Nutlets 4 (or fewer by abortion), egg-shaped, 1.2-1.7 mm long, wrinkled-pimply on both sides.

Plagiobothrys scouleri **(H.& A.) I.M. Johnst. var. *scouleri***
Scouler's popcornflower

Habitat/Range:
Moist meadows, grasslands and shrublands in the lowland, steppe and lower montane zones; frequent on S Vancouver Island and SC BC; E to MB and S to NM and OR.

Description:
General: Annual herb from a taproot or fibrous roots; straight, slender, appressed-hairy throughout; stems numerous and prostrate to ascending, or single and erect, to 20 cm long/tall.

Leaves: Basal leaves virtually absent; stem leaves linear to oblong, entire, 2-6.5 cm long, to 5 mm wide, lower 1-4 pairs opposite, the others alternate.

Flowers: Inflorescence of terminal, coiled, elongating, loosely-flowered spikes, often with leafy bracts on the lower part; flowers small; petals white, fused at base into a tube that spreads (1-4 mm wide) at top to 5 lobes, with 5 yellow or white bulges at the throat; fruiting calyces 2-4 mm long, ascending-hairy.

Fruits: Nutlets usually 4, clustered together, egg-shaped or narrowly egg-shaped, 1.5-2 mm long, more or less wrinkled and pimply and sometimes also bristly.

Plagiobothrys tenellus **(Nutt. *ex* Hook.) A. Gray**
slender popcornflower

Habitat/Range:
Mesic to dry meadows and coastal bluffs in the lowland zone; rare on S Vancouver Island and the Gulf Islands; S to ID, UT and Baja CA.

Description:
General: Annual herb from a taproot; stems several, simple or few-branched, erect to ascending, 5-25 cm tall, spreading-hairy and often also with backward-pointing, appressed hairs.

Leaves: Basal leaves in a peristent rosette, 0.7-4 cm long, 2-8 mm wide; stem leaves few, lanceolate, usually smaller than basal leaves, spreading-hairy, entire, alternate.

Flowers: Inflorescence terminal, coiled, narrow, elongating, sometimes paired clusters (racemes); flowers small; petals fused at the base into a tube that flares (2-4 mm wide) at top to 5 lobes, with 5 bulges at the throat; fruiting calyces 3-5 mm long, ascending-hairy.

Fruits: Nutlets usually 4, cross-shaped, 1.5-2.5 mm long, warty in rows.

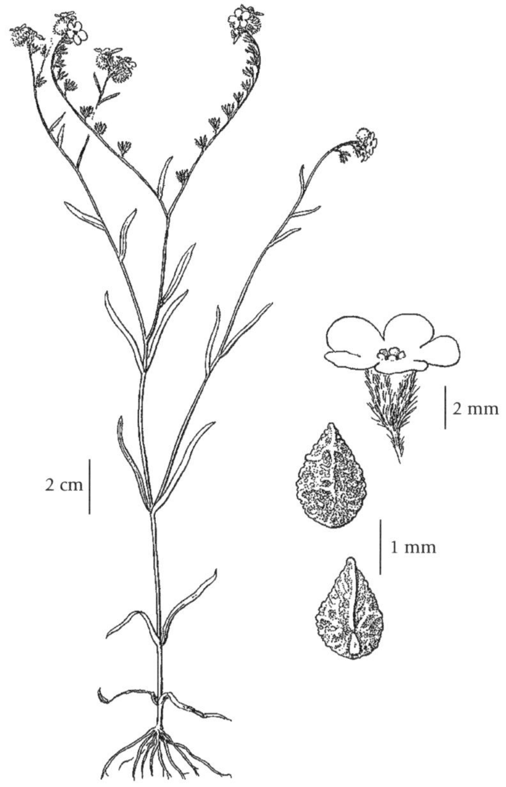

Plagiobothrys figuratus

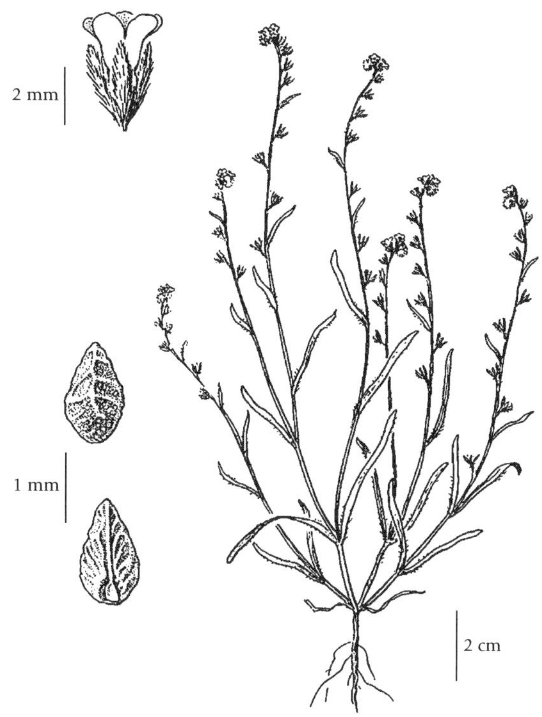

Plagiobothrys scouleri* var. *scouleri

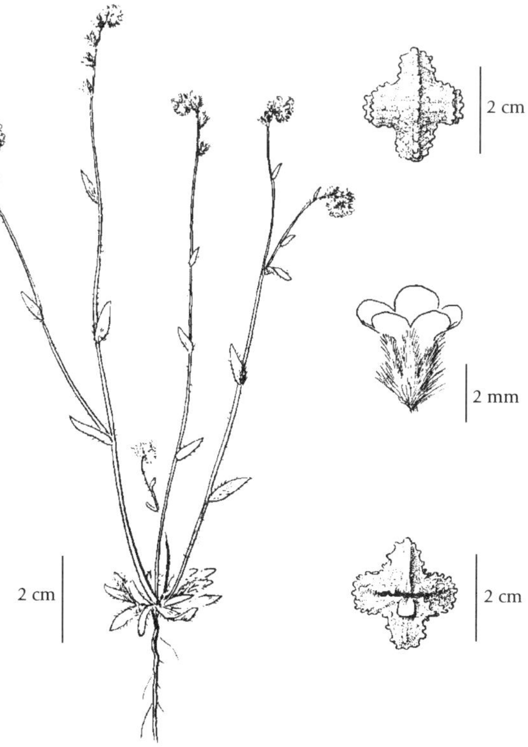

Plagiobothrys tenellus

SYMPHYTUM

1. Stems not winged; corolla lobes erect .... *S. asperum*
1. Stems winged from ridges that run down the stem from the leaf base; corolla lobes recurved at tip.
   2. Corollas cream, sometimes purplish; nutlets smooth, shining .... *S. officinale*
   2. Corollas blue or purplish-blue; nutlets wrinkled, more or less dull .... S. x *uplandicum*

Symphytum asperum **Lepech.**

rough comfrey; prickly comfrey

Habitat/Range:

Wet to mesic roadsides and waste places in the lowland, steppe and lower montane zones; rare in S BC, infrequent northward; introduced from Europe.

Description:

General: Perennial herb from a thick taproot; stems branched, not winged, 0.3-1.2 m tall; stem and inflorescence with coarse, spreading, bristly hairs.

Leaves: Basal leaves stalked, large, the blades elliptic to egg-shaped or narrowly egg-shaped, heart-shaped or rounded at the base, 15-20 cm long and 5-10 cm wide; stem leaves ample, gradually becoming smaller and shorter-stalked upward, with long-tapering, pointed tips, with coarse bristly hairs on both surfaces, entire, alternate.

Flowers: Inflorescence of several to many small, coiled, bractless clusters; corollas blue (pink in bud), tubular bell-shaped, 1-1.5 cm long; petals fused at base into a tube that flares at top to bell-shaped limb (longer than tube) and 5 lobes, the lobe tips erect, the throat with 5 narrow, hairy bulges; calyx teeth blunt-tipped.

Fruits: Nutlets 4, clustered together, curved inward, granular, finely wrinkled.

Symphytum officinale **L.**

common comfrey

Habitat/Range:

Wet to mesic roadsides and waste places in the lowland, steppe and lower montane zones; rare in S BC; introduced from Europe.

Description:

General: Perennial herb from a thick, branched taproot; stems branched, winged from ridges that run down from the leaf bases, 0.5-1.5 m tall; stem and inflorescence with spreading and downward-curving, conical, bristly hairs.

Leaves: Basal leaves stalked, large, the blades egg-shaped or narrowly egg-shaped to oblong, 15-30 cm long and 7-12 cm wide; stem leaves ample, oblong-lanceolate, with long-tapering, pointed tips, gradually becoming smaller and shorter-stalked upwards, spreading-hairy, entire, alternate.

Flowers: Inflorescence of several to many small, coiled, drooping, bractless clusters; corollas cream or dull blue to purplish, 1.5-1.8 cm long; petals fused at base into a tube that flares at top to a bell-shaped limb (longer than tube) and 5 lobes with recurved tips, the throat with 5 narrow hairy bulges; calyx teeth sharp-tipped.

Fruits: Nutlets 4, clustered together, curved inward, smooth, shining.

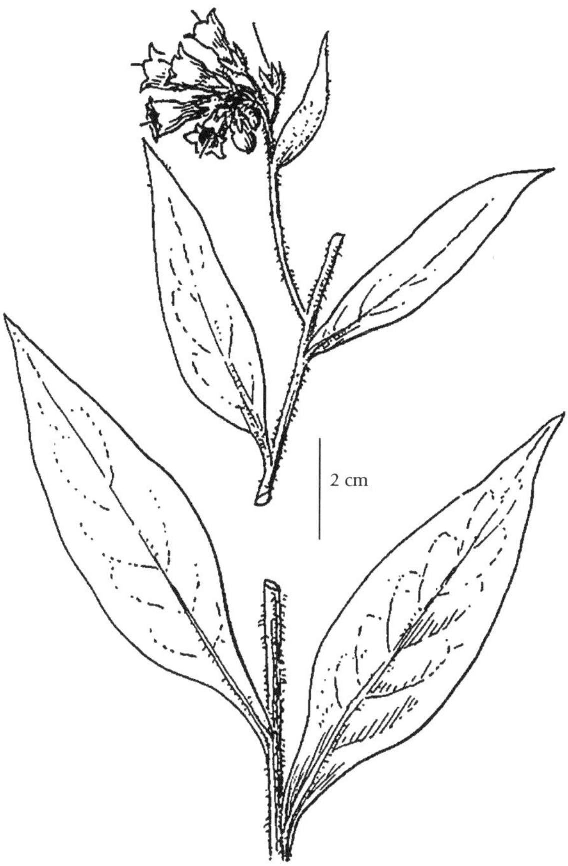

Symphytum asperum

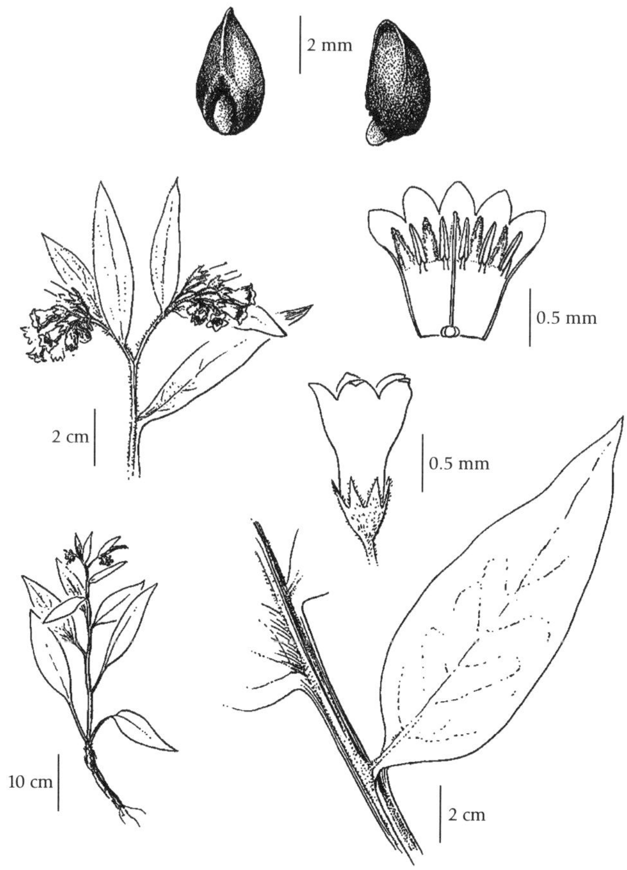

Symphytum officinale

***Symphytum* x *uplandicum* Nyman**
hybrid comfrey
HABITAT/RANGE:
Wet to mesic roadsides, waste places, ditches and forest edges in the lowland zone; frequent in SW BC; introduced from Europe.
DESCRIPTION:
General: Perennial herb from a thick, branched taproot; stems branched, winged from ridges that run down from the leaf bases, 0.5-1.7 m tall; stem and inflorescence with spreading and downward-curving, conical, bristly hairs.

Leaves: Basal leaves stalked, large, the blades egg-shaped or narrowly egg-shaped to oblong, 15-35 cm long and 7-14 cm wide; stem leaves ample, oblong-lanceolate, with long-tapering, pointed tips, gradually becoming smaller and shorter-stalked upwards, spreading-hairy, entire, alternate.

Flowers: Inflorescence of several to many small, coiled, drooping, bractless clusters; corollas blue to purplish-blue, 1.5-1.8 cm long; petals fused at base into a tube that flares at top to a bell-shaped limb (longer than tube) and 5 lobes with recurved tips, the throat with 5 narrow, hairy bulges; calyx teeth sharp-tipped.

Fruits: Nutlets 4, clustered together, curved inward, wrinkled, dull.

BRASSICACEAE [18, 19]

1. Silicles stalked, the stalks at least 1 mm long.
    2. Petals purple (rarely white); Fruits elliptic to oval silicles; stalks 7-12 mm long .... *Lunaria*
    2. Petals white; fruits slender siliques; stalks 1-4 mm long .... *Thelypodium*
1. Silicles unstalked or the stalks less than 1 mm long.
    3. Silicles composed of 2 distinct segments, the basal portion usually of 2 segments, although often indehiscent, usually seed-bearing, sometimes sterile, the upper segment often not of 2 segments, usually sterile, sometimes consisting entirely of the scarcely narrowed, often flattened or otherwise beaklike style.
        4. Siliques of 2 rather similar segments, each 1-locular, usually 1-seeded, sometimes 0 or 2-seeded, indehiscent, finally breaking apart; style absent .... *Cakile*
        4. Siliques of 2 dissimilar segments; seeds sometimes several, usually in the lower part or if in the upper then the style usually conspicuous and beaklike or the siliques constricted between the seeds.
            5. Plants stalked-glandular and soft-hairy; petals pink to magenta; siliques breaking crosswise, not dehiscent; styles beaklike .... *Chorispora*
            5. Plants not stalked-glandular; petals mostly yellow or white, rarely purplish-tinged; siliques often dehiscent; styles beaklike or beakless.
                6. Styles beakless, not differentiated from the upper portion of the beaked silique; siliques with a very short, sterile, basal segment distinct from the 2-parted fertile segment .... *Raphanus*
                6. Styles beaklike and usually sharply differentiated from the body of the silique; siliques without a lower sterile segment.
                    7. Beaks stout, often much flattened, not styliform, usually well over 5 mm long.
                        8. Seeds in two rows; petals with distinct reddish-purple veins .... *Eruca*
                        8. Seeds in one row; petals without distinct veins .... *Brassica*

[18] Brassicaceae, except for *Arabis*, contributed by G.W. Douglas and D. Meidinger.
[19] Key adapted from Hitchcock *et al.* (1994) and Scoggan (1978).

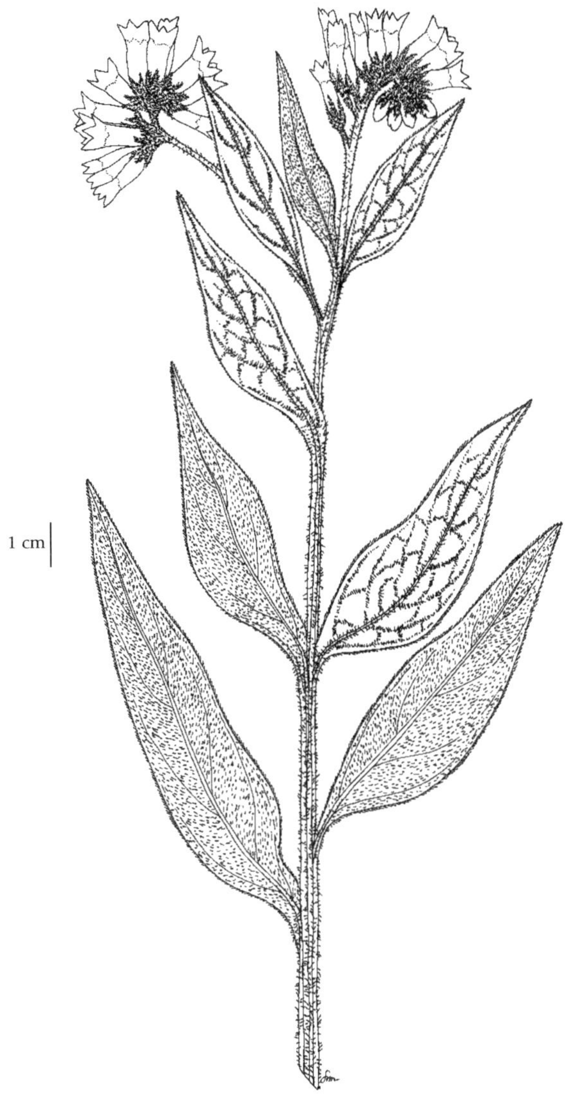

Symphytum x *uplandicum*

7. Beaks usually not flattened, more obviously styliform, less than 5 mm long .... *Erucastrum*

3. Silicles composed of only 1 segment, this seed-bearing; styles often conspicuous but smaller in diameter than the silicle and usually not beaklike.

9. Fruits silicles, varying from heart-shaped, with the notch at the base oval or somewhat oblong, elliptic, or heart-shaped and the notch at the tip, sometimes paired or only 1-seeded.

10. Silicles from greatly compressed to somewhat inflated, but never globose or rounded in cross section, not paired.

11. Seeds 1 per each chamber of the silicle.

12. Silicles broadly winged, oblong, more or less dry and indehiscent; glabrous biennial or perennial herbs; stem leaves clasping and ear-like at the bases .... *Isatis*

12. Silicles more nearly oval or rotund; usually hairy annuals without clasping leaves that are ear-like at the bases.

13. Silicles inflated, broadly elliptic in section, very strongly net-veined and pitted; petals yellow; upper stem leaves ear-like at the bases .... *Neslia*

13. Silicles strongly compressed, or inflated but not conspicuously net-veined and pitted or the stem leaves not ear-like at the bases; petals white or purplish-tinged.

14. Silicles 2-chambered.

15. Stems leafless; stem hairs not starlike or pick-shaped .... *Draba*

15. Stems with leaves; stem hairs starlike or pick-shaped.

16. Silicles nerveless; hairs starlike .... *Alyssum*

16. Silicles l-nerved; hairs pick-shaped .... *Lobularia*

14. Silicles 1-chambered.

17. Silicles 2.0-2.5 mm long and hairy with hooked hairs .... *Athysanus*

17. Silicles usually more than 5.0-8.0 mm long and glabrous, or if hairy then without hooked hairs .... *Thysanocarpus*

11. Seeds 2 or more per chamber.

18. Stems leafless, bearing a solitary terminal flower; seeds conspicuously wing-margined .... *Idahoa*

18. Stems with leaves or supporting 2- or more-flowered racemes; seeds wingless.

19. Silicles oblong to elliptic, usually over twice as long as wide .... *Draba*

19. Silicles oval to egg-shaped, not over twice as long as wide.

20. Petals yellow; plants perennials, densely covered with appressed, starlike hairs .... *Lesquerella*

20. Petals usually white or if pale yellow then either annuals or not with appressed, starlike hairs.

21. Styles not over 1 mm long; silicles oval in outline, less than 5 mm long; petals not bilobed .... *Alyssum*

21. Styles usually 2-3 mm long; silicles oval-elliptic to egg-shaped, usually at least 5 mm long; petals often bilobed.

22. Petals white, deeply bilobed; seeds slightly winged; silicles compressed, not thickly biconvex in section .... *Berteroa*

22. Petals pale yellow, not deeply bilobed; seeds unwinged; silicles somewhat swollen, thickly biconvex in section .... *Camelina*

10. Silicles from conspicuously to only slightly obcompressed, or so greatly inflated as to be rounded in cross section and often paired.

23. Silicles 1-chambered and 1-seeded; plants glabrous, glaucous, ear-like at the leaf bases, biennial or perennial .... *Isatis*

23. Silicles 2-chambered and usually several-seeded; plants sometimes annual, or hairy, or not earlike at the leaf bases.

24. Plants aquatic, the leaves all basal, linear, more or less rounded in cross section and needle-shaped .... *Subularia*

24. Plants not aquatic or if appearing so then the leaves not all basal or needle-shaped.

25. Plants fleshy maritime herbs; basal leaves with slender stalks that are considerably longer than the kidney-shaped to oval, simple blades; seeds numerous, in 2 rows .... *Cochlearia*

25. Plants various but not usually maritime; basal leaves various; seeds commonly few and usually not in 2 rows.

26. Silicles inflated, sometimes developing in pairs.

27. Petals yellow.

28. Plants glandular-warty in the inflorescence .... *Bunias*

28. Plants glabrous in the inflorescence, or if hairy, then not glandular.

29. Hairs starlike; plants usually of dry sites .... *Physaria*

29. Hairs, if any, unbranched; plants usually of moist sites ..... *Rorippa*

27. Petals white or purple.

30. Plants annuals; leaves pinnately cleft .... *Coronopus*

30. Plants perennials; leaves simple.

31. Petals 2-4 mm long; leaves 1-0 cm long .... *Cardaria*

31. Petals 5-7 mm long; leaves 30-50 cm long...................... *Armoracia*

26. Silicles more or less obcompressed, not inflated.

32. Silicles heart-shaped to triangular, with more than 1 seed per chamber; stem leaves clasping and ear-like at the bases .... *Capsella*

32. Silicles not heart-shaped or triangular, or seeds only 1 per chamber, or stem leaves not clasping and ear-like at the bases.

33. Plants with starlike hairs.

34. Silicles usually more than 7 mm long, usually developing in pairs or at least with a prominent apical notch .... *Physaria*

34. Silicles less than 7 mm long, never developing in pairs, usually without an apical notch or the notch not over 1 mm deep .... *Lesquerella*

33. Plants glabrous to hairy with simple hairs.

35. Flowers with the lower 2 petals much larger than the others *Iberis*

35. Flowers with petals about same size or sometimes petals lacking.

36. Seeds 1 per chamber .... *Lepidium*

36. Seeds 2-several per chamber.

37. Stem leaves more or less ear-like at the bases; silicles usually more than 4 mm long .... *Thlaspi*

37. Stem leaves, if any, not ear-like at the bases; silicles not more than 4 mm long.

38. Plants leafless or few-eaved; filaments of the stamens with small, scalelike basal appendages; seeds 2 per chamber .... *Teesdalia*

38. Plants leafy-stemmed; filaments not appendaged at the base; seeds more than 2 per chamber .... *Hutchinsia*

9. Fruits elliptic to linear siliques, in normal specimens the more than 1-seeded.

39. Leaves (at least the basal and lower stem leaves) more or less pinnately cleft.

40. Petals pale to deep yellow.

41. Plants more or less hairy with forked or starlike hairs .... *Descurainia*

41. Plants glabrous to hairy with simple hairs.

42. Racemes bracted nearly or quite throughout; stems with downswept, straight, soft hairs .... *Erucastrum*

42. Racemes bractless except sometimes at the base; stems glabrous or with hairs spreading and relatively stiff (except sometimes downswept in *Sisymbrium loeselii*).

43. Styles (or what appear to be the styles) beaklike, nearly as wide as the silique segments.

44. Siliques 1-celled, indehiscent, usually constricted between the seeds .... *Raphanus*

44. Siliques 2-celled, tardily dehiscent, not constricted between the seeds .... *Barbarea*

43. Styles not beaklike, if prominent then much narrower than the silique segments.

45. Leaves linear, 4 mm wide or less; silique segments with an obscure midnerve .... *Schoenocrambe*

45. Leaves broad (at least below), over 4 mm wide; silique segments with a prominent midnerve.

46. Stem leaves strongly ear-like at the bases, clasping .... *Barbarea*

46. Stem leaves not ear-like at the bases or clasping.

47. Seeds in 1 row; silique segments nerved .... *Sisymbrium*

47. Seeds in 2 rows; silique segments nerveless .... *Rorippa*

40. Petals white or pinkish to purple.

48. Seeds in 2 rows (more or less in 1 row in *Rorippa microphylla*) .... *Rorippa*

48. Seeds in 1 row.

49. Plants copiously greyish-hairy with both long, soft, simple, forking hairs and shorter freely branched hairs .... *Smelowskia*

49. Plants glabrous or hairy with simple hairs only.

50. Siliques indehiscent, usually constricted between the seeds .... *Raphanus*

50. Siliques dehiscent, not constricted between the seeds.

51. Plants annuals or biennials; upper stem leaves not pinnately cleft .... *Thelypodium*

51. Plants perennials, or if annuals or biennials then the upper stem leaves pinnately cleft .... *Cardamine*

39. Leaves entire to coarsely toothed or sometimes lobed or lyrate, rarely distinctly pinnately cleft.

52. Petals creamy, yellow, dark orange or reddish.

53. Stem leaves strongly heart-shaped, clasping with the notch at the base .... *Conringia*

53. Stem leaves not clasping.

54. Plants glabrous and somewhat glaucous .... *Schoenocrambe*

54. Plants hairy with forked or branched hairs and often simple hairs.

55. Seeds in 1 row; plants with appressed, 2-pronged hairs .... *Erysimum*

55. Seeds in 2 rows; plants with mostly starlike or branched, stalked hairs and also with or without simple hairs .... *Draba*

52. Petals white to pink or purple.

56. Plants with forked, starlike, branched or treelike hairs and with or without simple hairs.

57. Petals 15-25 mm long .... *Hesperis*

57. Petals usually less than 15 mm long.

58. Siliques moderately to strongly compressed.

59. Siliques less than 8 times as long as wide .... *Draba*

59. Siliques usually at least 8 times as long as wide.

60. Seeds wingless; petals pink to purple .... *Erysimum*

60. Seeds winged, or if wingless, then the petals white or cream, rarely pink .... *Arabis*

58. Siliques rounded, or nearly so, in cross section, or slightly 4-angled, slightly if at all compressed.

61. Stem leaves sagittate-clasping, seeds in 2 rows; siliques glabrous .... *Halimolobos*

61. Stem leaves not clasping at the base; seeds in 1 row or in 2 rows; siliques glabrous or hairy.

62. Siliques constricted between the seeds .... *Braya*

62. Siliques not constricted between the seeds or only slightly so.

63. Siliques glabrous; plants annuals .... *Arabidopsis*

63. Siliques hairy; plants perennials .... *Halimolobos*

56. Plants glabrous or hairy with simple hairs only.

64. Stems leafless or few-leaved.

65. Plants glandular-hairy, rarely glabrous; siliques 4-7 mm wide .... *Parrya*

65. Plants not glandular-hairy; siliques usually less than 4 mm wide.

66. Siliques 1.0-3.5 cm long; plants glabrous .... *Cardamine*

66. Siliques less than 1.0 cm long; plants subglabrous to hairy .... *Braya*

64. Stems leafy.

67. Stem leaves ear-like at the bases, clasping.

68. Plants glabrous and glaucous .... *Arabidopsis*

68. Plants hairy .... *Arabis*

67. Stem leaves not ear-like at the bases or clasping.

69. Siliques indehiscent .... *Chorispora*

69. Siliques dehiscent.

70. Leaves lanceolate to oblanceolate or narrowly egg-shaped.

71. Plants glabrous; leaves entire; seeds in 1 row .... *Eutrema*

71. Plants usually hairy; leaves often toothed, at least below; seeds in 1 row or 2 rows .... *Arabis*

70. Leaves deltoid to heart-shaped (the notch at the base), rounded or kidney-shaped.

72. Plants native perennials with slender, elongate rhizomes; petals 7-9 mm long .... *Cardamine*

72. Plants introduced biennials from a taproot, garlic-scented; petals less than 7 mm long .... *Alliaria*

ALLIARIA

***Alliaria petiolata* (Bieb.) Cav. & Grande** (*A. officinalis* Andrz. *ex* Bieb.)
garlic mustard; hedge garlic

HABITAT/RANGE:

Mesic to dry disturbed sites and waste places in the lowland and lower montane zones; rare in SW and SC BC; introduced from Eurasia.

DESCRIPTION:

General: Biennial herb from a taproot; stem erect and unbranched below inflorescence, sparsely hairy below, glabrous or with whitish bloom above, 20-120 cm tall; garlic smell, especially from root but also from crushed leaves.

Leaves: Basal leaves kidney-shaped, slender-stalked; stem leaves heart-shaped (the notch at the base) to deltoid, margins wavy or coarsely toothed, (1.5) 3-6 cm wide.

Flowers: In simple or compound racemes, often with leafy bracts at base, 6 mm in diameter; petals white, 6 mm long, twice the length of sepals.

Fruits: Siliques, spreading and curving upward to be almost erect, 4-6 cm long, 2 mm wide, somewhat 4-angled, on thick stalks 4-6 mm long; seeds blackish, 3 mm long.

ALYSSUM

1. Ovules and seeds 1 per chamber .... *A. murale*

1. Ovules and seeds 2 per chamber.

2. Silicles glabrous; styles about 1 mm long .... *A. desertorum*

2. Silicles hairy, the hairs starlike; styles 0.5 mm long, or less .... *A. alyssoides*

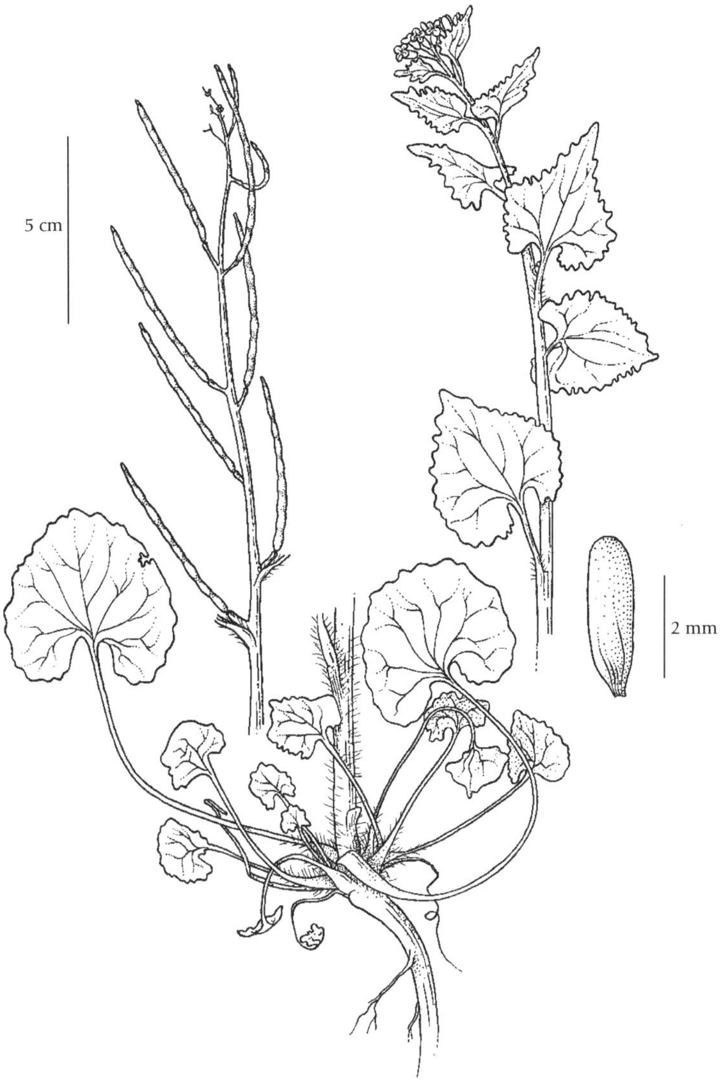

Alliaria petiolata

Alyssum alyssoides **L.**

pale alyssum

Habitat/Range:

Dry roadsides and waste places in the steppe and montane zones; frequent in SC and SE BC, rare on S. Vancouver Island; introduced from Eurasia.

Description:

General: Annual or biennial herb from a slender taproot; stems erect or ascending with several ascending branches near the base, 10-25 (40) cm tall, the shoots grey-hairy with starlike hairs.

Leaves: Basal leaves lacking; stem leaves narrowly oblanceolate, entire, 5-25 mm long, grey with starlike down.

Flowers: Inflorescene elongate; flower stalks spreading, 2-5 mm long; flowers about 3 mm in diameter; petals cream or white, linear, 3-4 mm long; sepals about 2 mm long, soft-hairy and with starlike hairs, persistent in fruit.

Fruits: Silicles, 3-4 mm long, almost circular (oval), starlike hairy, strongly flattened near the edges but bulging in the middle, 2 seeds in each chamber; styles up to 0.5 mm long.

Alyssum desertorum **Stapf**

desert alyssum

Habitat/Range:

Dry disturbed areas and waste places in the steppe zone; infrequent in SC BC, introduced from Eurasia.

Description:

General: Annual herb from a slender taproot; stems erect or ascending with several ascending branches near the base, 10-25 cm tall, the shoots grey-hairy with starlike hairs.

Leaves: Basal leaves lacking; stem leaves narrowly oblanceolate, entire, 5-25 mm long, grey with starlike down.

Flowers: Inflorescene elongate; flower stalks spreading, 2-5 mm long; flowers less than 3 mm in diameter; petals light yellow, very narrow, 2-3 mm long; sepals about 2 mm long, soft-hairy and with starlike hairs, deciduous shortly after fruiting.

Fruits: Silicles, 3-4 mm long, round, shallowly notched, glabrous, strongly flattened near the edges and bulging over seeds, 2 seeds in each chamber; style about 1 mm long.

Alyssum murale **Waldst. & Kit.**

wall alyssum

Habitat/Range:

Mesic to dry disturbed areas and waste places; rare garden escape in SW BC, introduced from Eurasia.

Description:

General: Annual herb from a slender taproot; stems erect or ascending with several ascending branches near the base, 25-70 cm tall, appressed-hairy.

Leaves: Basal leaves egg- or spoon-shaped; stem leaves 10-20 mm long, 3-6 mm wide, lanceolate or oblanceolate, grey with starlike hairs.

Flowers: Inflorescence elongate, flowers less than 2-3.5 mm in diameter; flower stalks spreading, 2-5 mm long; petals yellow, very narrow, 2-3 mm long; sepals about 2 mm long, soft-hairy and with starlike hairs, deciduous shortly after fruiting.

Fruits: Silicles, 2-5.5 mm long, obtuse, nearly glabrous or hairy, the hairs starlike, strongly flattened; seeds 1 in each chamber; style about 0.5-2 mm long.

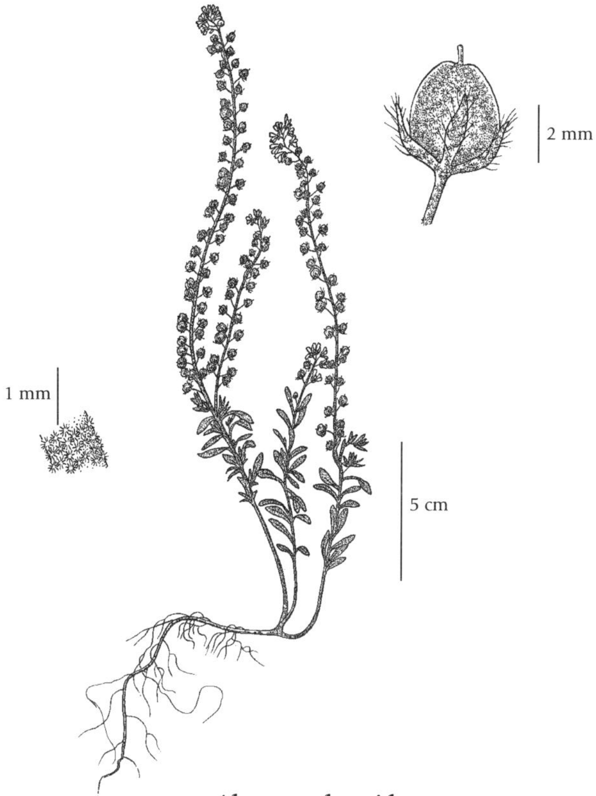

Alyssum alyssoides

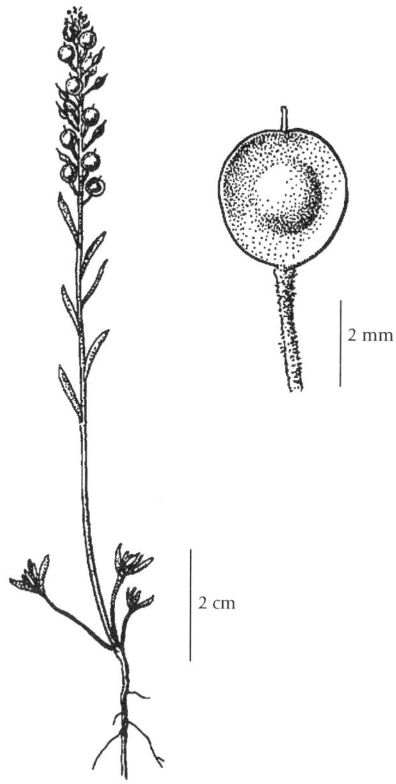

Alyssum desertorum

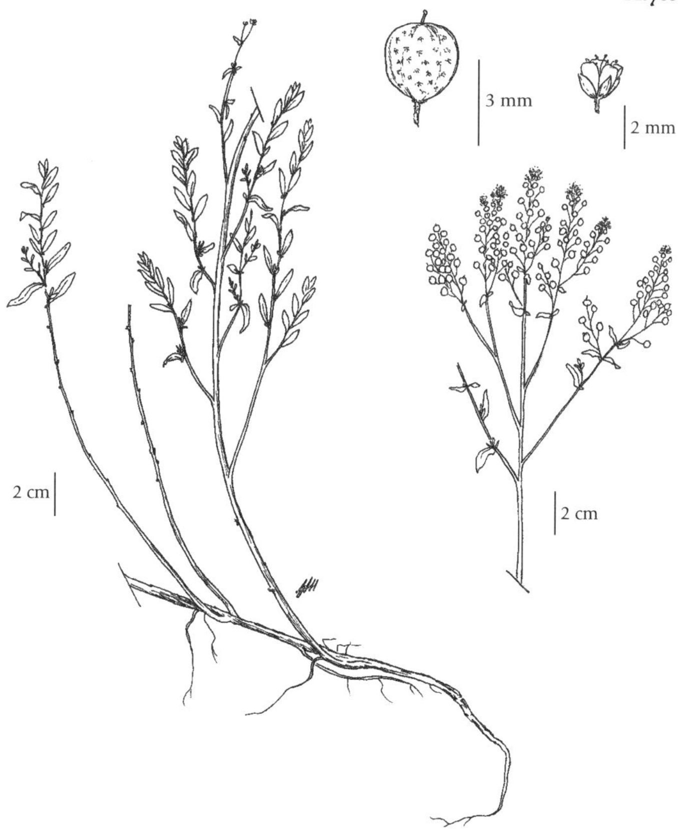

Alyssum murale

APHRAGMUS

Aphragmus eschscholtzianus **Andrz. *ex* DC.**
Eschscholtz's little nightmare
Habitat/Range:
Moist to wet solifluction slopes, scree slopes and rocky ridges in the alpine zone; rare in NW BC, known only from the Tatshenshini River; amphiberingian, N to AK and YT; NE Asia.
Description:
General: Perennial from a short rhizome and stolons; stems ascending, branched above, glabrous or minutely hairy, 2-6 cm tall.
Leaves: Basal leaves simple, spoon-shaped, 1.5-4.0 cm long, 4-8 mm wide, tapering to a slender stalk; stem leaves crowded below the raceme, oblanceolate, 0.5-1.2 cm long.
Flowers: Short clustered racemes; petals 2-3 mm long, longer than the sepals; sepals 1.5-1.8 mm long.
Fruits: Silicles, oblong-ellipsoid, with a short style, 6-15 mm long, 2-3 mm wide; seeds 4-10.

ARABIDOPSIS [20]

1. Stem leaves with ear-like lobes at the base; plants glabrous and glaucous .... *A. salsuginea*
1. Stem leaves without ear-like lobes at the base; plants usually somewhat hairy .... *A. thaliana*

Arabidopsis salsuginea **(Pall.) N. Busch** (*Thellungiella salsuginea* [Pall.] O.E. Schulz)
saltwater cress
Habitat/Range:
Dry saline lakes and meadows in the steppe zone; rare in SE BC; amphiberingian, N to YT and NT, E to MB and S to CO; Eurasia.
Description:
General: Annual or biennial herb from a slender taproot; stems single, 5-20 cm tall, simple or usually branched, glabrous throughout.
Leaves: Basal leaves few, egg-shaped, 5-25 mm long, entire or with 1-2 wide teeth, stalked; stem leaves many, oblong to egg-shaped, 5-20 mm long, 3-10 mm wide, obtuse, entire, clasping with ear-like lobes at the base.
Flowers: Inflorescences dense, few-flowered, elongating in fruit; flower stalks ascending, 3-7 mm long, slender, straight; petals white, 2-3.5 mm long, egg-shaped to spoon-shaped; sepals yellowish to purplish, papery-margined, 1.5-2 mm long, erect, without pouches.
Fruits: Siliques, ascending, 1-2 cm long, about 0.8 mm wide, straight or slightly curved, almost round in cross section; beaks 0.1-0.2 mm long; seeds in 1 row, plump, oblong, about 0.6 mm long.

---

[20] Key from Hitchcock *et al.* (1964)

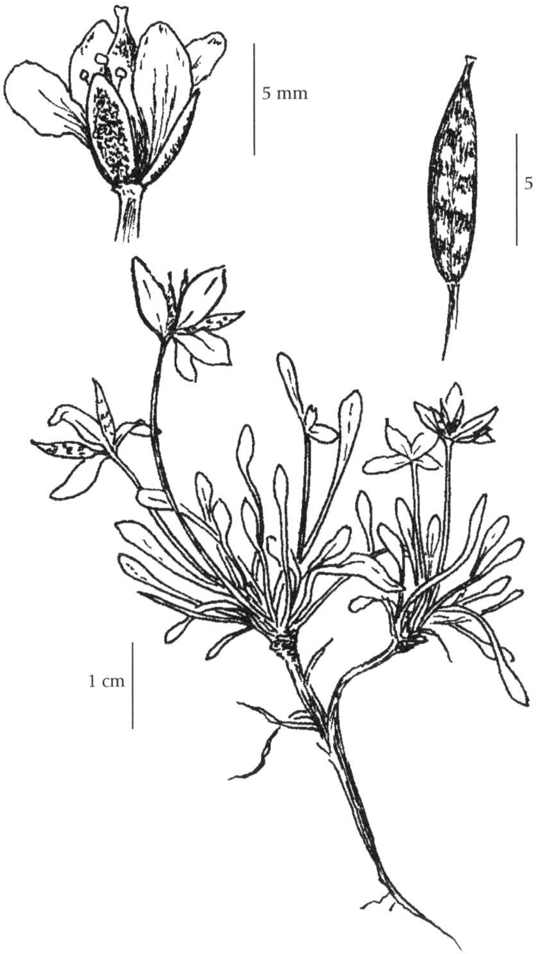

Aphragmus eschscholtzianus

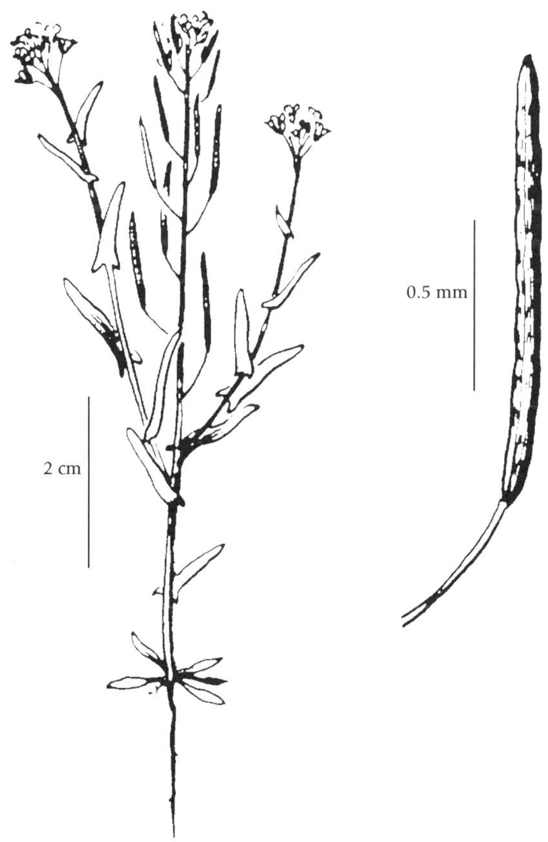

Arabidopsis salsuginea

Arabidopsis thaliana **(L.) Heynh.**

mouse-ear; Thale cress

Habitat/Range:

Mesic to dry roadsides, gardens, fields and waste places; frequent on Vancouver Island, the Gulf Islands and Lower Mainland, rare in SE and SC BC; introduced from Eurasia.

Description:

General: Annual herb from a slender taproot; stems 1-several, erect, 10-40 cm tall, hairy near base with simple, spreading hairs, glabrous above.

Leaves: Mostly in basal rosette, oblanceolate, entire or minutely dentate, 1-4 cm long, with simple and branched hairs; stem leaves lanceolate, nearly unstalked, 5-20 mm long, upper ones usually glabrous.

Flowers: In racemes; flower stalks slender, erect to ascending, 5-10 mm long; petals white, about 3 mm long, narrowly egg-shaped; sepals approximately 1.5 mm long.

Fruits: Siliques, linear, 10-15 mm long, less than 1 mm wide, glabrous; seeds numerous, in one row, less than 1 mm long, oblong.

ARABIS [21]

This is a very difficult genus. Mature silicles and seeds are often necessary for identification. The following treatment is a conservative and practical one, which will allow for identification of many specimens, but there will always be individuals and populations that do not fit the keys. For a recent alternative treatment emphasizing hair types and sizes on leaves and stems, see Mulligan (1995), which includes several new species. However, few people interested in identifying members of this genus will have access to equipment accurate enough to measure hairs less than 1/10 of a mm long, which is needed to use parts of his key.

1. Mature siliques predominantly erect or angled upward.
   2. Stem leaves clasping with ear-like lobes at the bases.
      3. Mature siliques tightly appressed against the axis of the inflorescene.
         4. Stems glabrous or sparsely stiff-hairy at base; siliques 2-3 mm thick; seeds with wings up to 1 mm wide .... *A. drummondii*
         4. Stems densely stiff-hairy at base; siliques 1-2 mm thick; seeds with wings up to 0.3 mm wide.
            5. Seeds in 2 rows; siliques only slightly compressed; sepals not bulging on one side at the base .... *A. glabra*
            5. Seeds in 1 row; siliques noticeably compressed; sepals bulging on one side at the base .... *A. hirsuta*
      3. Mature siliques erect but somewhat spreading away from axis of the inflorescence.
         6. Stems at mid-length less than 1 mm in diameter .... *A. microphylla*
         6. Stems at mid-length more than 1 mm in diameter.
            7. Stems generally glabrous toward base .... *A. lyallii*
            7. Stems generally stiff-hairy toward base.
               8. Basal leaves green and prominent at flowering time; outer sepals bulging on one side at the base; plants from in and W of the Coast-Cascade Mountains .... *A. eschscholtziana*
               8. Basal leaves usually dry and withered at flowering time; outer sepals not bulging on one side at the base; plants from E of the Coast-Cascade Mountains .... *A.* x *divaricarpa*
   2. Stem leaves not clasping and ear-like at the bases.
      9. Basal leaves entire to toothed, with fringed stalks .... *A. nuttallii*

---

[21] Contributed by G.B. Straley.

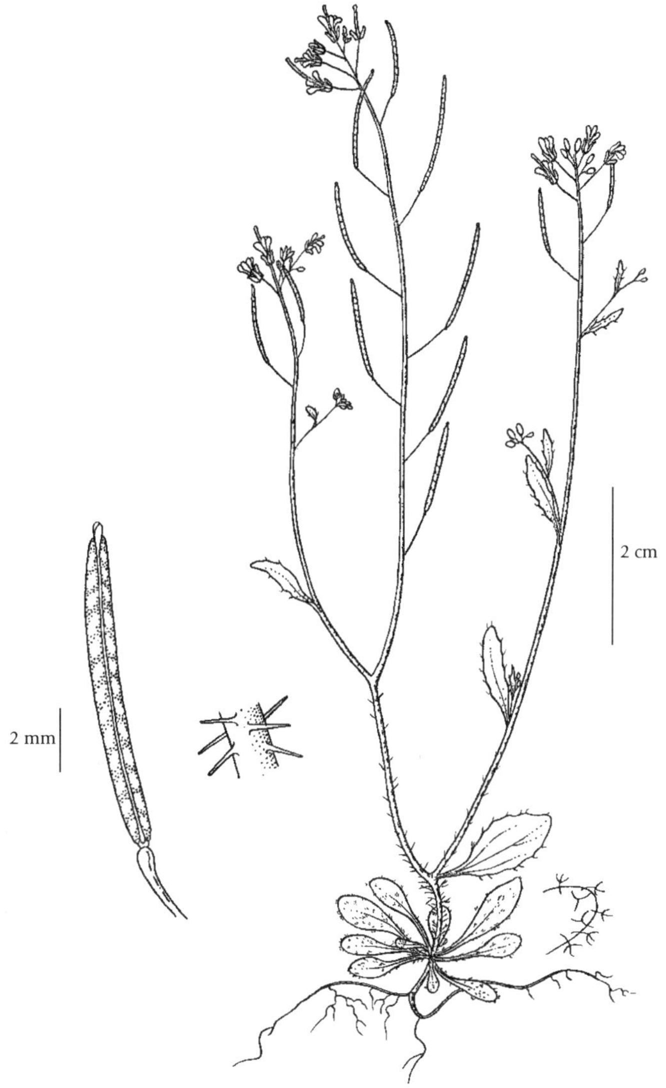

Arabidopsis thaliana

9. Basal leaves, or at least some, pinnately cleft with the terminal lobe the largest, without fringed stalks.
    10. Petals 5-5.5 mm long; mature siliques 1.25-1.5 mm wide .... *A. kamchatica*
    10. Petals 6-8 mm long; mature siliques 0.75-1.0 mm wide .... *A. lyrata*

1. Mature siliques arching, drooping or bent downward.
    11. Stem leaves not ear-like at the bases or clasping .... *A. exilis*
    11. Stem leaves ear-like at the bases, clasping.
        12. Fruiting stalks at right angles to axis of the inflorescence or bent or curved downward.
            13. Basal leaves forming a distinct rosette; fruiting stalks, or at least some of them, bent downward, hairy; siliques usually straight .... *A. holboellii*
            13. Basal leaves usually not in a distinct rosette; fruiting stalks at right angles to axis of the inflorescence or gradually bent downward, glabrous; siliques curved downward .... *A. lignifera*
        12. Fruiting stalks more or less erect, siliques curved downward.
            14. Inflorescences usually symmetric; plants generally more than 30 cm tall .... *A. sparsiflora*
            14. Inflorescences usually 1-sided; plants generally less than 20 cm tall .... *A. lemmonii*

***Arabis* x *divaricarpa* A. Nels.** (*A. divaricarpa* var. *dacotica* [Greene] Boivin)
spreading-pod rockcress

HABITAT/RANGE:

Mesic to dry meadows, gravelly and rocky slopes, rock outcrops, sandy shorelines and roadsides in the montane to alpine zones; frequent in BC east of the Coast-Cascade Mountains; N to AK, YT and NT, E to PQ and NB, S to VT, NE, CO and CA.

DESCRIPTION:

General: Biennial or short-lived perennial herb from a simple or branched stem-base; stems 1-several, simple or branched, 25-80 cm tall, lower part of stem with 3-rayed and often some simple hairs, glabrous above.

Leaves: Basal leaves oblanceolate, entire or slightly toothed (often withered or absent when plant in flower or fruit), 3-6 cm long, 3-8 mm wide; narrowed to a slender stalk; stem leaves unstalked, ear-like at the bases, narrowly oblong to lanceolate, 1.5-3 cm long, 2-5 mm wide.

Flowers: Racemes many-flowered; petals 6-9 mm long, pink to purple; sepals not bulging on one side at the base

Fruits: siliques, erect to spreading, straight to slightly curved, 3-8 cm long, 1.5-2.5 mm wide; seeds in 1 row or sometimes partially in 2 rows, about 2 mm long, with wings 0.5 mm wide.

Notes: Mulligan (1995) recognizes two varieties in BC (var. *divaricarpa* and var. *dacotica* [Greene] Boivin), based mainly on small, but apparently consistent differences in hairs on the leaves. We have chosen not to recognize these in our flora.

***Arabis drummondii* A. Gray** (*A. calderi* G. Mulligan)
Drummond's rockcress

HABITAT/RANGE:

Mesic to dry rock outcrops, talus slopes, gravelly soils, open forests and roadsides in the montane to alpine zones; frequent throughout BC; N to AK, YT and NT, E to NF and S to CA and NJ, AZ and CA.

DESCRIPTION:

General: Short-lived perennial herb from a simple or branched stem-base; stems 1-several, branched, 30-80 cm tall, glabrous throughout or with simple or 2-branched hairs basally.

Leaves: Basal leaves oblanceolate to elliptic, entire, 2-7 mm long, 3-15 mm wide, narrowed to a slender stalk; stem leaves overlapping, unstalked, ear-like at the bases, oblong-lanceolate to linear-lanceolate, 1.5-3 cm long, 3-15 mm wide.

Flowers: Few- to many-flowered racemes; petals 7-12 mm long, white; outer sepals slightly bulging on one side at the base.

Fruits: siliques, erect, straight or slightly curved, glabrous, 3-9 cm long, 2-3 mm wide; seeds in 2 rows, about 2 mm long, with wings to 1 mm wide on one side of seed.

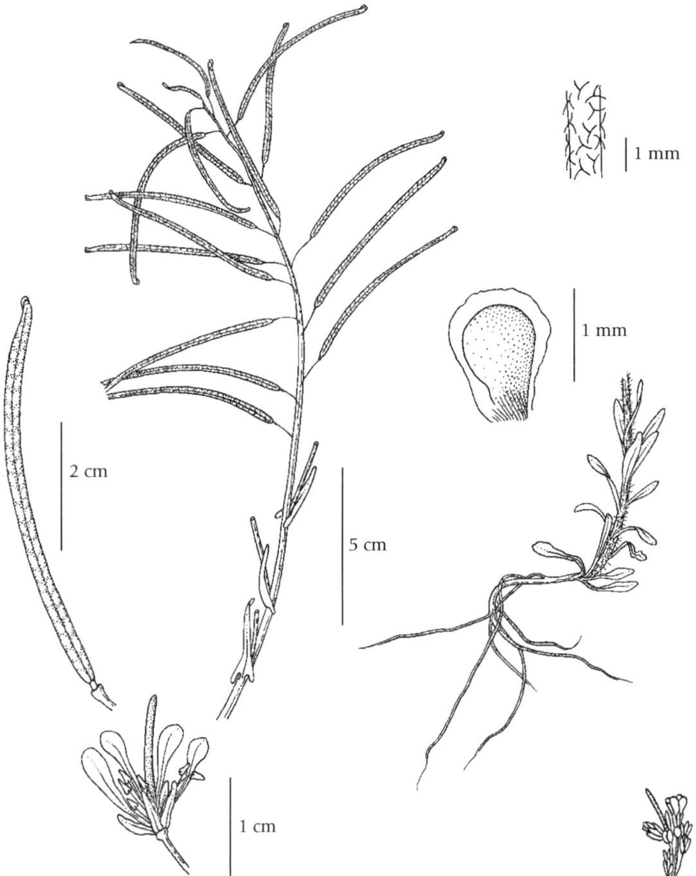

Arabis x *divaricarpa*

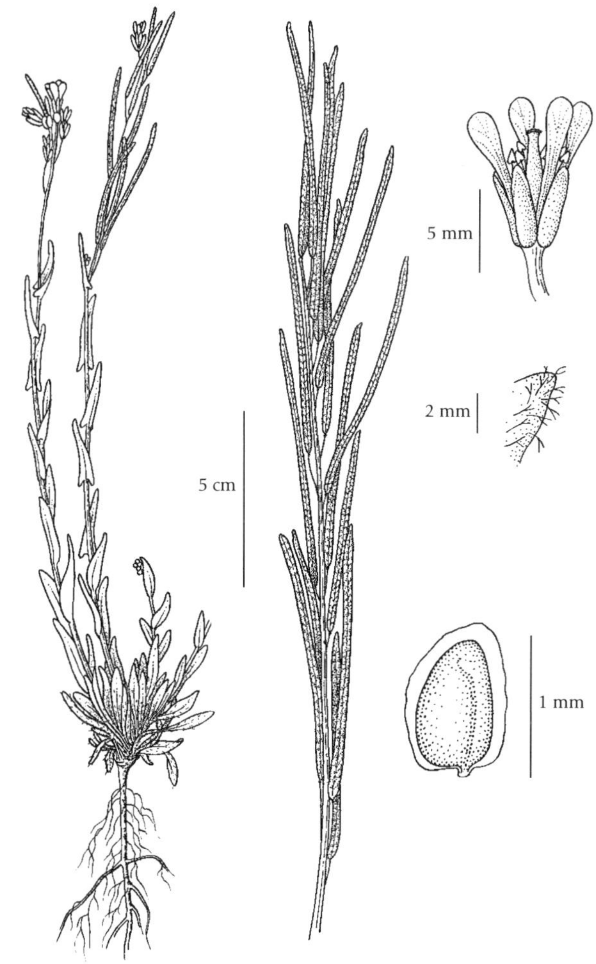

Arabis drummondii

***Arabis eschscholtziana* Andrz.** (*A. hirsuta* [L.] Scop. var. *eschscholtziana* [Andrz.] Rollins)
Eschscholtz's rockcress
HABITAT/RANGE:
Moist to mesic meadows, streambanks, rocky slopes and disturbed areas in the lowland to alpine zones; frequent in BC in and W of the Coast-Cascade Mountains; N to AK and YT and S to W ID and OR.
DESCRIPTION:
General: Biennial or short-lived perennial herb usually from an unbranched stem-base; stems 1-few, 20-80 cm tall, stiff-hairy basally, glabrous above.

Leaves: Basal leaves oblanceolate, 2-8 cm long, 0.5-2 cm wide, narrowed to a slender, fringed stalk, glabrous to sparsely hairy above and below; stem leaves broadly egg-shaped to oblanceolate, unstalked, with clasping bases and often toothed margins, more or less glabrous.

Flowers: Few to many-flowered racemes; petals 6-9 mm long, white to pale pink; sepals bulging on one side at the base.

Fruits: siliques, erect to slightly spreading away from the axis of the inflorescence, straight or slightly curved, 5-8 cm long, 1-2 mm wide, glabrous; seeds in 1 row, 1-2 mm long, wingless or with very narrow wings.

Notes: Often included in *A. hirsuta,* as the coastal form of that species.

***Arabis exilis* A. Nels.** (*A. holboellii* Hornem. var. *pendulocarpa* [A. Nels.] Rollins)
slender rockcress
HABITAT/RANGE:
Mesic to dry grasslands, shrublands, meadows, rocky slopes and rock outcrops in the steppe and montane zones; common throughout BC, N to YT, E to SK and S to CO, NV and CA.
DESCRIPTION:
General: Biennial or short-lived perennial herb from a simple or branched stem-base; stems 1-several, usually simple, 15-30 cm tall, glaucous, with simple to several-branched hairs toward base, glabrous above.

Leaves: Basal leaves in prominent, persistent rosette, lanceolate, 10-15 mm long, 3-5 mm wide, tapering to a slender stalk, densely grey-hairy with branched hairs; stem leaves unstalked, often broadly attached to stem, but not clasping or ear-like at the bases, sparsely hairy, fringed margins, glaucous.

Flowers: Many-flowered racemes; petals 3-5 mm long, white to pale lavender; sepals not bulging on one side at the base.

Fruits: siliques, strongly bent downward, straight, 3-5 cm long, 2 mm wide; seeds 1 mm long with very narrow wings.

Notes: Often included in *A. holboellii,* where it is out of place, without clasping, ear-like lobes at the bases of the stem leaves.

Arabis glabra* (L.) Bernh. var. *glabra
tower mustard
HABITAT/RANGE:
Moist to dry streambanks, meadows, open forests, rock outcrops, roadsides and waste places in the lowland and montane zones; common throughout BC; E to PQ and S to GA and N CA; native status uncertain, also occurs in Eurasia.
DESCRIPTION:
General: Biennial or short-lived perennial herb from a simple stem-base; stems simple or branched above, 0.3-1.2 m tall, densely stiff-hairy at base, usually glabrous above.

Leaves: Basal leaves oblanceolate, 3-14 cm long; stem leaves strongly overlapping, narrowly egg-shaped to lanceolate, 5-15 cm long, to 4 cm wide, unstalked and ear-like at the bases.

Flowers: Many-flowered racemes; petals 5-6 mm long, cream;

sepals not bulging on one side at the base.

Fruits: siliques, erect, appressed to axis of the inflorescence, straight, glabrous, 6-10 cm long, 1-1.5 mm wide; seeds 1-1.5 mm wide, with very narrow wings.

Notes: This taxon occurs in many native habitats but also readily colonizes roadsides and waste places. Its native status is therefore uncertain (Rollins 1993).

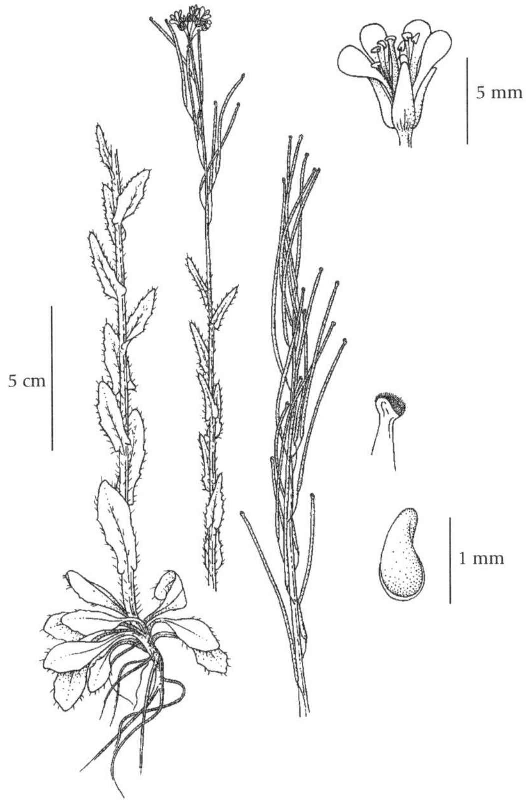

Arabis eschscholtziana

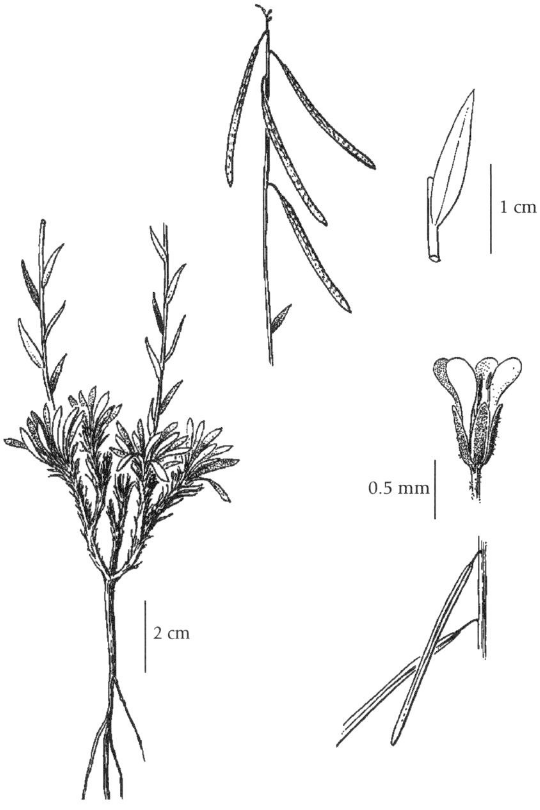

Arabis exilis

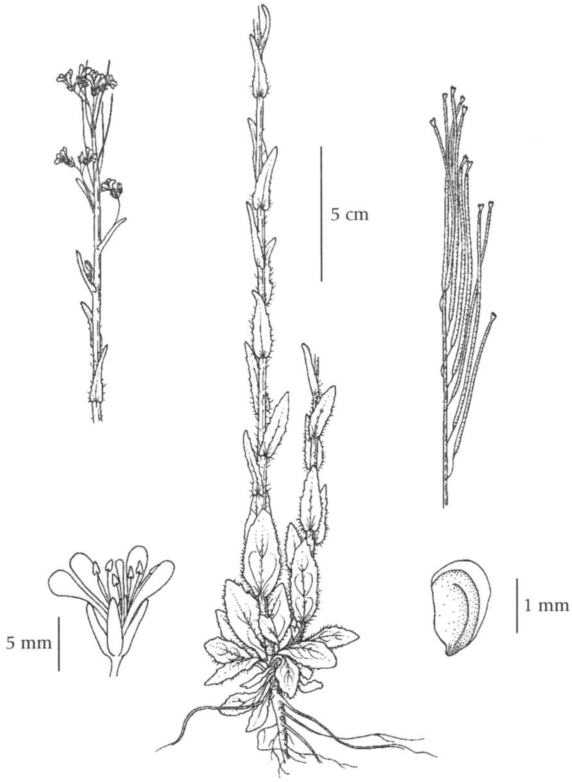

Arabis glabra

***Arabis hirsuta* (L.) Scop.** (*A. hirsuta* var. *glabrata* T.& G. = var. *pycnocarpa*)
hairy rockcress

HABITAT/RANGE:

Moist to mesic meadows, gravel bars, open woods, rock outcrops and waste places in the lowland and montane zones; var. *hirsuta* known in BC only from Glacier National Park; var. *pycnocarpa* common throughout BC east of the Coast-Cascade Mountains; circumpolar, S to CA, NM and GA; Eurasia.

DESCRIPTION:

General: Biennial or short-lived perennial herb from a simple stem-base; stems usually simple, 15-100 cm tall, strongly stiff-hairy at the base, usually glabrous above.

Leaves: Basal leaves oblanceolate, 2-8 cm long, 1-1.5 cm wide, narrowed to a fringed, winged stalk, often purplish beneath, glabrous to hairy, remotely toothed; stem leaves broadly egg-shaped, 2-10 cm long, 1-3 cm wide, broadly unstalked, usually ear-like at the bases and clasping.

Flowers: Racemes few to many flowered; petals 4-9 mm long, white to cream or pale pink; outer sepals bulging on one side at the base.

Fruits: Siliques, erect, tightly appressed to the axis of the inflorescence or angled slightly outward, 4-8 cm long, 0.75-1.5 mm wide; seeds in 1 row, 1.5 mm long, with narrow wings.

Notes: All of our BC material is referable to the var. *pycnocarpa* (Hopkins) Roll., except for a single collection of the European var. *hirsuta*, from Glacier National Park.

1. Plants perennials; siliques beakless .... var. *hirsuta*
1. Plants biennials or short-lived perennials; siliques with a beak 0.5-1.25 mm long .... var. *pycnocarpa* (Hopkins) Roll.

***Arabis holboellii* Hornem.** (*A. holboellii* var. *collinsii* [Fern.] Roll. = var. *secunda; A. pinetorum* Tidestrom = var. *pinetorum*)
Holboell's rockcress

HABITAT/RANGE:

Mesic to dry rocky slopes, gravelly areas, shrublands and grasslands in the steppe to alpine zones; var. *pinetorum* rare in S BC, var. *secunda* infrequent in C and S BC east of the Coast-Cascade Mountains, var. *retrofracta* common throughout BC east of the Coast-Cascade Mountains (probably introduced on the coast), var. *consanguinea* infrequent throughout BC east of the Coast-Cascade Mountains; N to AK, YT and NT, E to PQ and ON and S to ND, SD, WY, UT, NV and CA; Greenland.

DESCRIPTION:

General: Biennial or short-lived perennial herb from a simple or branched stem-base; stems 1-several, simple or branched, 0.1-1.0 m tall, hairy basally, usually glabrous above.

Leaves: Basal leaves in a prominent rosette, oblanceolate, 1-5 cm long, 2-5 mm wide, narrowed to slender stalk, usually densely grey-hairy; stem leaves 1-3 cm long, 2-7 mm wide, unstalked, ear-like at the bases, narrowly oblong to linear, hairy.

Flowers: Racemes many-flowered, greatly elongated in fruit; petals white to purple, 6-10 mm long; sepals not bulging on one side at the base.

Fruits: Siliques, drooping or directed downward due to abruptly bent stalks, 3-6 cm long, 1.5-2 mm wide, straight or slightly curved; seeds in 1 row, about 2 mm long, with very narrow wings less than 0.5 mm wide.

Notes: Mulligan (1995) considers var. *holboellii* to occur in North America only in Greenland. He gives var. *pinetorum* (Tidestrom) Rollins specific status as *A. pinetorum*. Four often poorly-defined varieties occur in BC.

1. Stalks of siliques gradually arched out and down, not abruptly bent downward; siliques usually curved downward .... var. *pinetorum* (Tidestrom) Rollins
1. Stalks of siliques abruptly bent downward; siliques usually straight.
    2. Middle and upper stem leaves mostly with flat margins; inflorescences 1-sided .... var. *secunda* (Howell) Jepson
    2. Middle and upper stem leaves with rolled margins; inflorescences symmetric.
        3. Basal leaves densely grey-hairy; siliques 1.25-1.75 mm wide .... var. *retrofracta* (Grah.) Rydb.
        3. Basal leaves usually not densely grey-hairy; siliques 1.5-2.0 mm wide .... var. *consanguinea* (Greene) Mulligan

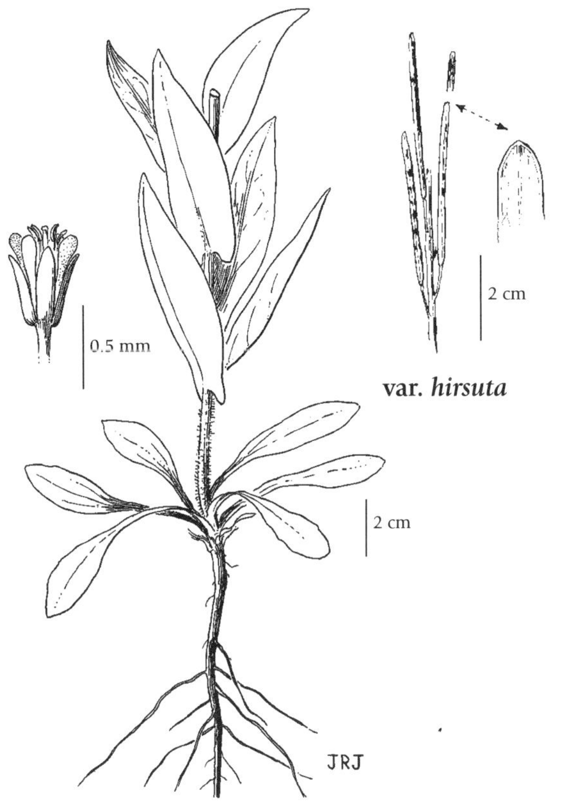

Arabis hirsuta var. *pycnocarpa*

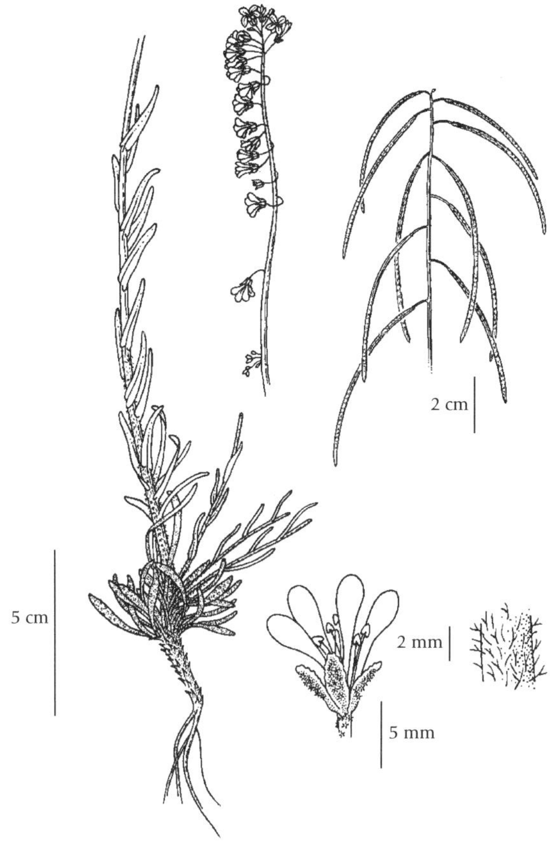

Arabis holboellii var. *pinetorum*

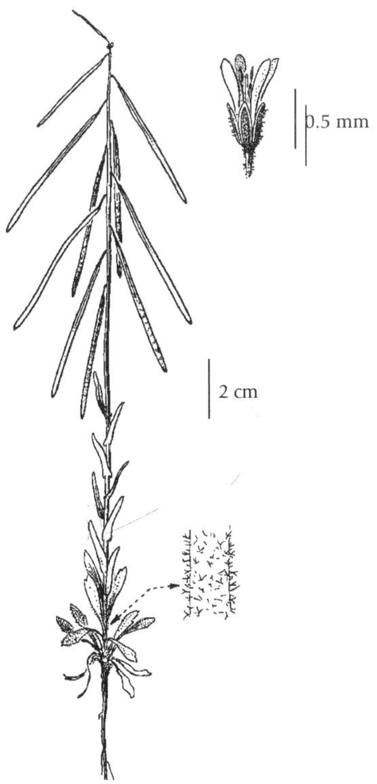

Arabis holboellii var. *secunda*

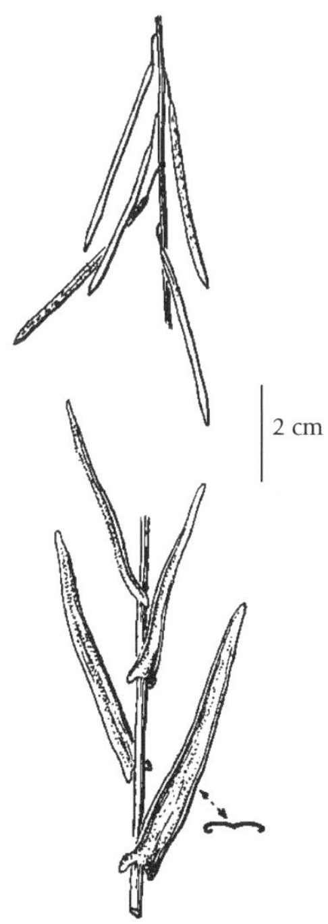

Arabis holboellii var. *retrofracta*

***Arabis kamchatica* (Fisch.) Ledeb.** (*A. lyrata* L. ssp. *kamchatica* [Fisch.] Hult.)

Kamchatka rockcress

HABITAT/RANGE:

Mesic to moist gravel bars, moraines, scree slopes, talus slopes and disturbed areas in the montane to alpine zones; frequent throughout BC; N to AK, E to SK and S to WA; E Asia.

DESCRIPTION:

General: Short-lived perennial herb from a simple or branched stem-base; stems 1-several, simple or branched, 10-30 cm tall, glabrous or sparsely short-hairy basally.

Leaves: Mostly in a basal rosette, broadly egg-shaped, 1.5-6 cm long, 5-17 mm wide, at least some leaves pinnately cleft, others entire or toothed; stem leaves 3-7, oblanceolate, 1-4 cm long, narrowed to a short stalk or unstalked but not ear-like at the bases, entire or toothed.

Flowers: Racemes few-flowered (3 to 20); petals 5-5.5 mm long, white; sepals glabrous, outer two slightly bulged at base.

Fruits: Siliques, glabrous, erect, but not appressed to the axis of the inflorescence, 2-4 cm long, 1-1.5 mm wide, slightly compressed; seeds 1-1.5 mm long, wingless.

***Arabis lemmonii* S. Wats.**

Lemmon's rockcress

HABITAT/RANGE:

Mesic meadows, fellfields, scree slopes and talus slopes in the alpine zone; infrequent in C and S BC; N to YT, E to AB and S to UT, NV and CA.

DESCRIPTION:

General: Perennial herb from a branched stem-base; stems simple, several, 5-20 cm tall, glabrous to densely hairy.

Leaves: Basal leaves lanceolate, 1.5-2 cm long, 3-6 mm wide, entire or remotely toothed, generally densely grey-hairy; stem leaves unstalked, ear-like at the bases, oblong to lanceolate, glabrous to hairy.

Flowers: Racemes few flowered, 1-sided; petals 4-7 mm long, purple; sepals not bulging on one side at the base.

Fruits: Siliques, horizontal, spreading, or slightly arching upward or downward, 3-4 cm long, 1-2 mm wide; seeds in 1 row, about 1.5 mm long, with very narrow wings.

Notes: Two poorly-defined varieties occur in BC.

1. Siliques more than 2 mm wide; stems mostly more than 20 cm tall .... var. *drepanoloba* (Greene) Roll.
1. Siliques 2 mm or less wide; stems mostly less than 20 cm tall .... var. *lemmonii*

***Arabis lignifera* A. Nels.**

Woody-branched rockcress

HABITAT/RANGE:

Dry to mesic gravelly slopes, shrublands and open forests in the steppe and montane zones; infrequent in SC BC; S to ID, CO and NV.

DESCRIPTION:

General: Biennial to short-lived perennial herb from a simple or usually branched stem-base; stems 1-several, simple or branched, 30-100 cm tall, sparsely hairy basally or nearly glabrous throughout.

Leaves: Basal leaves in an indistinct rosette, oblanceolate, 1.5-4.5 cm long, 3-5 mm wide, gradually narrowed to slender stalk, sparsely to densely hairy or glabrous; lower stem leaves similar, becoming unstalked upwards, ear-like and clasping at the bases, lanceolate, 2-4 (7) cm long, 2-5 (15) mm wide, glabrous or hairy.

Flowers: Racemes few-flowered, stalks generally glabrous; petals pink to purple; sepals bulging on one side at the base, 6-9 mm long.

Fruits: Siliques, spreading at right angle to the axis of the inflorescence, or curved downward (not bent), usually glabrous, siliques usually curved downward, 4-6 cm long, 1-1.5 mm wide; seeds in 1 row, about 2 mm long, with wings less than 0.5 mm wide.

Notes: Easily confused with *A. holboellii*.

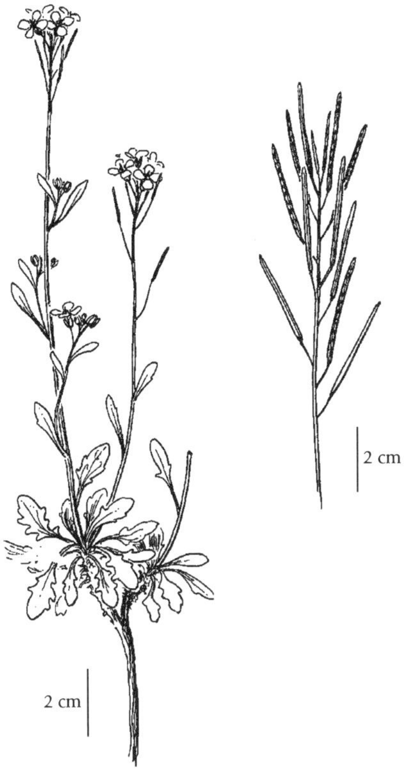

Arabis kamchatica

Arabis lemmonii var. *drepanoloba*

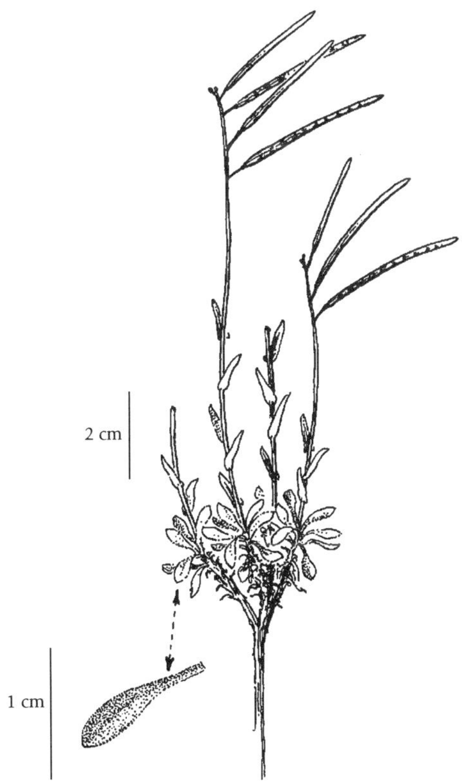

Arabis lemmonii var. *lemmonii*

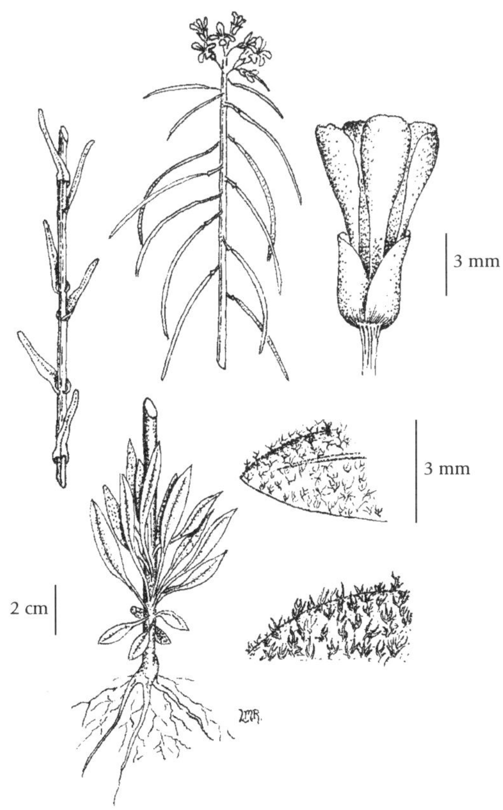

Arabis lignifera

Arabis lyallii* S. Wats. var. *lyallii (*A. codyi* G. Mulligan, *A. depauperata* Nels. & Kennedy, *A. murrayi* G. Mulligan)
Lyall's rockcress
Habitat/Range:
Dry to mesic meadows, rock outcrops, scree slopes and talus slopes in the subalpine to alpine zones; frequent in S BC east of the Coast-Cascade Mountains; N to YT and S to CA and UT.
Description:
General: Perennial herb from a branched stem-base; stems several, 10-25 cm tall, glaucous and glabrous throughout or with a few hairs toward base.

Leaves: Basal leaves in a prominent rosette, lanceolate, entire or few-toothed, 1.5-3 cm long, 0.3-7 mm wide, gradually narrowed to a long, slender stalk, glaucous and glabrous; stem leaves unstalked, ear-like at the bases, 2-3 cm long, 5-8 mm wide, entire or few-toothed, glabrous and glaucous.

Flowers: Racemes few-flowered, symmetric; petals 6-10 mm long, purple; sepals not bulging on one side at the base.

Fruits: Siliques, erect, somewhat spreading from the axis of the inflorescence, 3-6 cm long, 2-3 mm wide; seeds about 2 mm long, with wings less than 0.5 mm long.

Notes: *Arabis depauperata* is tentatively placed here. It has been variously treated as a variety of *A. lemmonii*, as a synonym of *A. microphylla* and *A. lyallii* (Hitchcock *et al.* 1964, Rollins 1993), or as a separate species (Mulligan 1996).

***Arabis lyrata* L.**
lyre-leaved rockcress
Habitat/Range:
Gravel slopes and waste places in the lowland zone; rare in C and SW BC; probably introduced from E North America.
Description:
General: Biennial or short-lived perennial from a simple or branched stem-base; stems 1-several, simple or branched, 10-30 cm tall, glabrous throughout or sparsely stiff-hairy near base.

Leaves: Basal leaves in a distinct rosette, spoon-shaped, 1-4 cm long, 3-5 mm wide, stalks usually not fringed, blades glabrous or sparsely hairy, simple and entire or with a few large teeth or some pinnately cleft; stem leaves few, short-stalked or unstalked, not ear-like at the bases, entire.

Flowers: Racemes few-flowered (5-15); petals white to pale pink, 6-8 mm long; sepals glabrous, outer two slightly bulging on one side at the base.

Fruits: Siliques, erect and angled away from the axis of the inflorescence, 2-4 cm long; seeds in 1 row, 0.75-1.0 mm long, wingless.

***Arabis microphylla* Nutt.** (*A. microphylla* var. *macounii* [S. Wats] Rollins)
littleleaf rockcress
Habitat/Range:
Mesic meadows and rocky slopes in the montane, subalpine and alpine zones; infrequent in S BC east of the Coast-Cascade Mountains; S to UT and CA.
Description:
General: Perennial herb from a branched stem-base; stems several, 10-20 cm tall, slender, glabrous and glaucous.

Leaves: Basal leaves oblanceolate, 6-8 mm long, 1-2 mm wide, glabrous to densely hairy, especially beneath, narrowed to a slender stalk; stem leaves unstalked, ear-like at the bases, lanceolate, 2-3 cm long, 2-4 mm wide, glabrous.

Flowers: Racemes few-flowered; petals 5-8 mm long, pale to dark pink; sepals not bulging on one side at the base.

Fruits: Siliques, erect but angled away from the axis of the inflorescence, 3-4 cm long, 1-2 mm wide; seeds about 1 mm long, with very narrow wings.

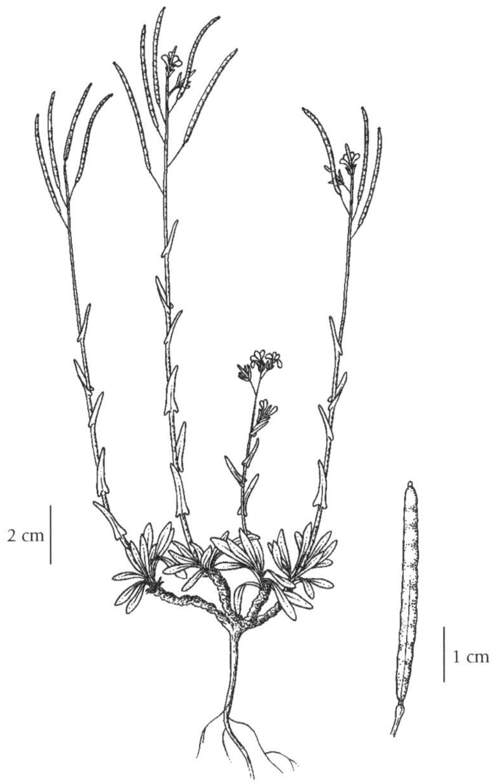

Arabis lyallii var. *lyallii*

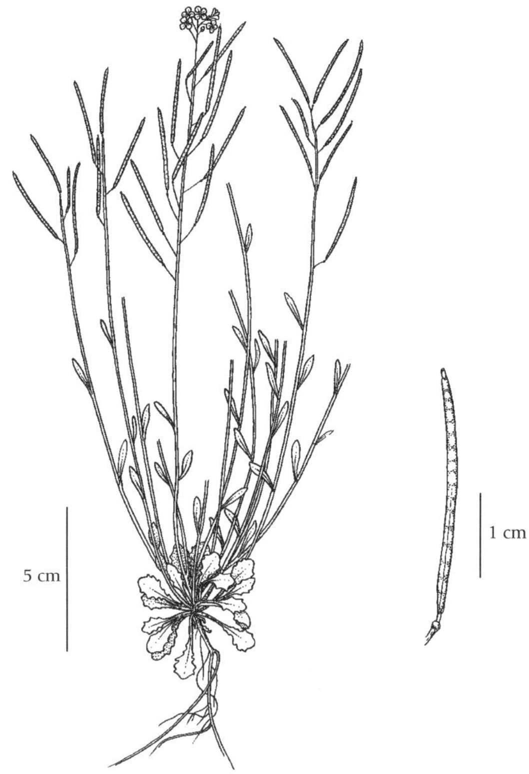

Arabis lyrata

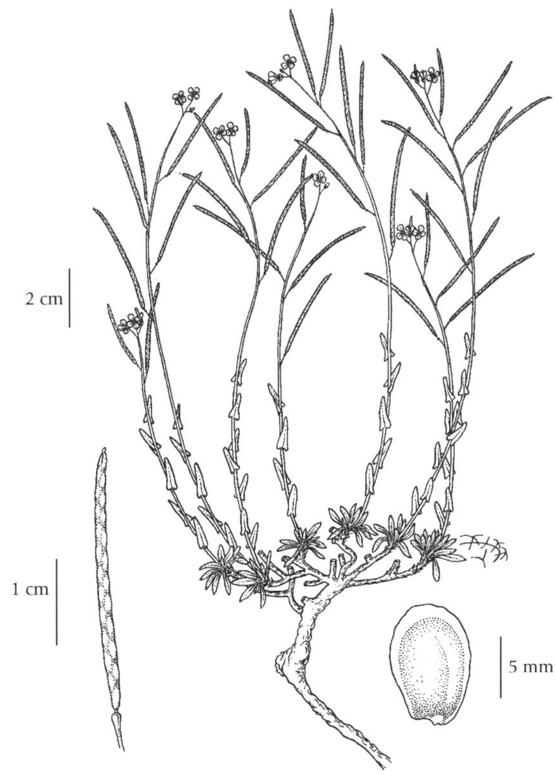

Arabis microphylla

***Arabis nuttallii* Robins.**
Nuttall's rockcress
HABITAT/RANGE:
Moist to mesic meadows, talus slopes, rock outcrops, open forests and roadsides in the montane to alpine zones; rare in SE BC; N to YT, E to SW AB and S to WY, UT and NV.
DESCRIPTION:
General: Perennial herb from a simple stem-base; stems simple, one to several, 10-30 cm tall, with long, simple or 2-branched hairs at the bases, becoming glabrous above.

Leaves: Basal leaves numerous in a rosette, egg-shaped, 1-3 cm long, 3-5 mm wide, drying and remaining prominent, stalks and blades with many long, simple or 2-branched hairs; stem leaves unstalked to short-stalked, not clasping stem, lanceolate to oblanceolate, 8-12 mm long, 2-5 mm wide, long hairs on margins and undersurface.

Flowers: Racemes few-flowered (5-25); petals 5-6 mm long, white to pale lavender; outer two sepals not bulging on one side at the base.

Fruits: Siliques, angled upward, straight to slightly curved, 1-2 cm long, 1-1.5 mm wide; seeds in 1 row, 0.75-1.0 mm long, wingless.

***Arabis sparsiflora* Nutt.** (*A. columbiana* Macoun)
sickle-pod rockcress
HABITAT/RANGE:
Mesic to dry grasslands, gravelly river banks and disturbed areas in the steppe and lower montane zones; rare in SC BC, known recently only from Princeton; S to UT, ID and CA.
DESCRIPTION:
General: Biennial or short-lived perennial herb from a simple or branched stem-base; stems 1-several, simple or branched, 0.3-1.0 m tall, glabrous above, more or less hairy toward base.

Leaves: Basal leaves numerous in a rosette, oblanceolate, entire, 2-3 cm long, 1.5-2.5 mm wide, gradually narrowed to a slender, fringed stalk; stem leaves unstalked, ear-like at the bases, lanceolate, entire or remotely toothed, 1.5-2.0 cm long, 2-3 mm wide, glabrous to sparsely hairy.

Flowers: Racemes many-flowered, symmetric; petals 6-10 mm long, white to purple; sepals slightly bulging on one side at the base.

Fruits: Siliques, directed at right angles from the axis of the inflorescence or slightly upward, but the siliques curved downward, 5-7 cm long, 2 mm wide; seeds 1.5-2 mm long, with very narrow wings.

Notes: Mulligan (1995) separates *A. columbiana* from *A. sparsiflora* based on stem and base leaves with either 2- and 3-branched hairs on the latter or 3- and 4-branched hairs on the former; a rather minor difference.

ARMORACIA

***Armoracia rusticana* Gaertn., Mey. & Scherb.** (*A. lapathifolia* Gilib.)
common horseradish
HABITAT/RANGE:
Moist ditches and disturbed areas in the lowland, steppe and montane zones; rare garden escape in S BC; introduced from Eurasia.
DESCRIPTION:
General: Perennial herb from a deep, swollen, spongy root; stems often branched, 0.6-2.0 m tall, glabrous.

Leaves: Basal leaves long-stalked, 30-50 cm long, widely egg-shaped to oblong, toothed; stem leaves stalked and unstalked; lower stem leaves like basal ones, middle and upper stem leaves often lobed.

Flowers: Inflorescence a flat-topped cluster; petals white, short-clawed, 5-7 mm long.

Fruits: Silicles, about 3 mm wide, flattened perpendicular to the partitions, styles about 0.3 mm long; seeds 4-6 per chamber.

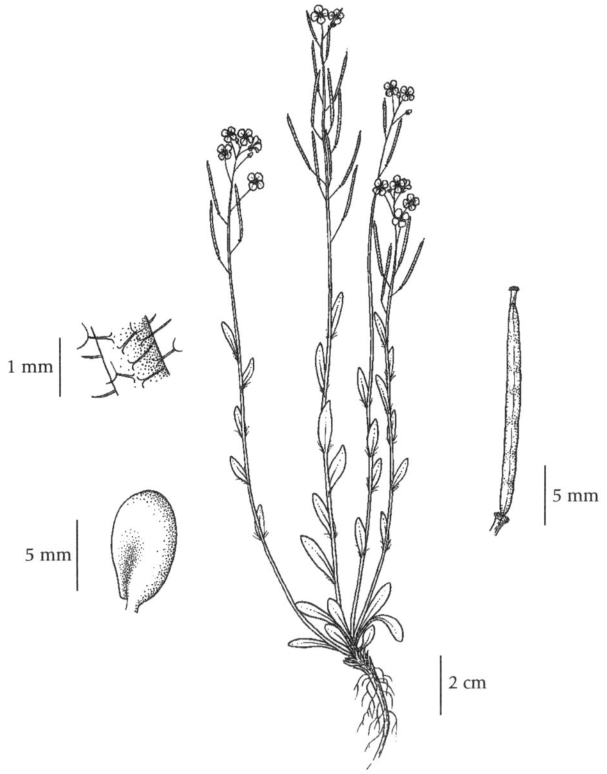

Arabis nuttallii

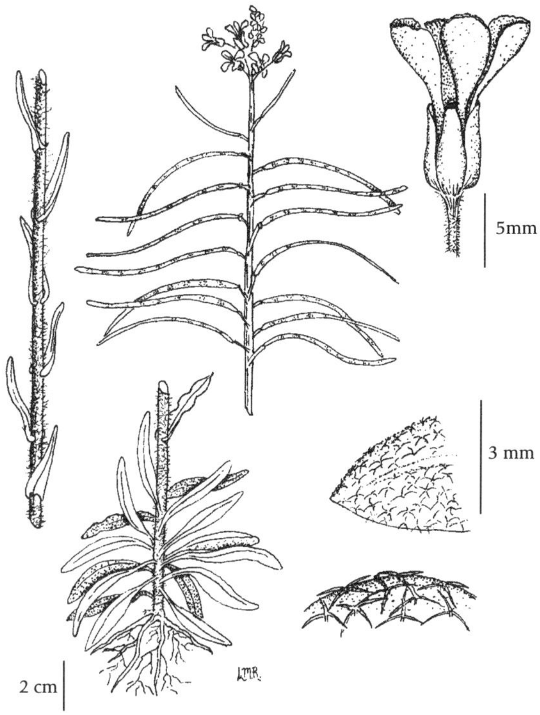

Arabis sparsiflora

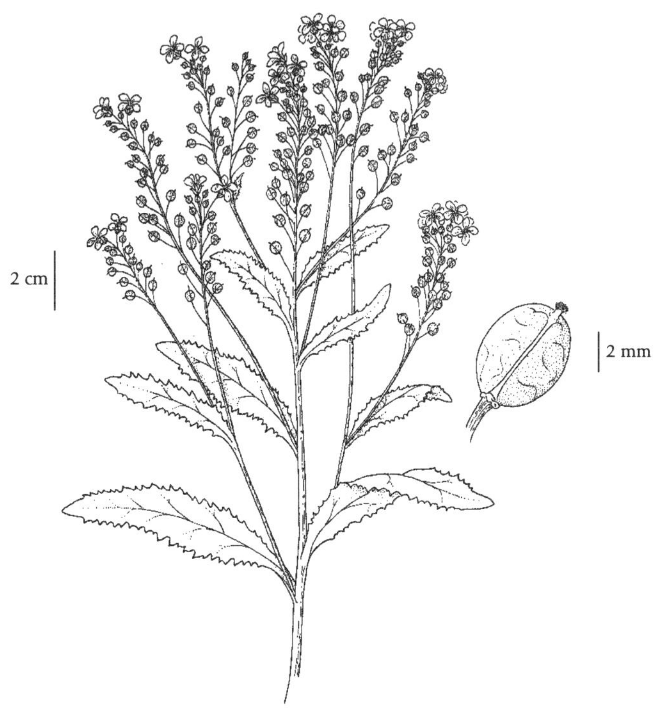

Armoracia rusticana

ATHYSANUS

***Athysanus pusillus* (Hook.) Greene**
common sandweed

HABITAT/RANGE:

Dry (sometimes vernally moist) grassy slopes and rock outcrops in the lowland and steppe zones; infrequent in S BC; S to ID and CA.

DESCRIPTION:

General: Annual herb from a slender taproot; stems usually branched from leafy base, 4-35 cm tall, short-hairy with simple and 2-to 5-branched hairs.

Leaves: Basal leaves oblanceolate to egg-shaped, entire or obscurely toothed, 6-30 mm long, 2-10 mm wide, short-stalked, with coarse, cross-shaped hairs; stem leaves mostly unstalked, reduced upward.

Flowers: Inflorescence of 1-several long, narrow, 1-sided clusters with well-spaced, inconspicuous flowers; petals white, 1-2 mm long, spoon-shaped, sometimes lacking; sepals about 1 mm long, dropping early.

Fruits: Silicles, 2.0-2.5 mm wide, round, flattened parallel to the partition, often with 2 types of hairs, the larger hairs flat and hooked, the smaller hairs branched or lacking; styles 0.2-0.3 mm; stalks bent downward, 1-4 mm long; seeds 1, flat.

BARBAREA

1. Uppermost leaves merely lobed, rarely pinnately cleft, lowermost leaves with 1-2 pairs of lateral lobes; styles 2-3 mm long, distinctly beaklike .... *B. vulgaris*

1. Uppermost leaves pinnately cleft with the terminal lobe the largest, lowermost leaves with 2-10 pairs of lateral lobes; styles 0.5-1.5 mm long, not beaklike.

   2. Siliques 1.5-5 cm long; basal leaves with 2-5 pairs of lateral lobes .... *B. orthoceras*

   2. Siliques 4.5-8 cm long; basal leaves with 4-10 pairs of lateral lobes .... *B. verna*

***Barbarea orthoceras* Ledeb.** (*B. americana* Rydb., *B. orthoceras* var. *dolichocarpa* Fern.)
American winter cress

HABITAT/RANGE:

Moist streambanks, meadows and forests in the lowland, steppe and montane zones; common along the coast, less frequent elsewhere in BC; N to AK, YT and NT, E to NF, and S to MN, NH, CO, AZ and CA.

DESCRIPTION:

General: Biennial (occasionally perennial) from a taproot and a simple stem-base; stems stiff, erect, angled, branched, (10) 20-60 cm tall.

Leaves: Basal leaves pinnately lobed, long-stalked, less than 12 cm long, with 1-5 pairs of lateral lobes, terminal lobe largest, egg-shaped and entire or irregularly toothed; stalks and sometimes leaflets long-fringed or sparsely short-hairy; lowermost stem leaves with 2-10 pairs of lateral lobes, upper pinnately cleft with the terminal lobe the largest, often clasping the stems.

Flowers: In single and terminal racemes or more often in compound, reduced racemes in lower leaf axils; flower stalks not stout, less than 1 mm thick, 2-3 mm long; petals pale yellow, broadly spoon-shaped, 3-5 mm long; sepals pale yellow-green, about 2 mm long.

Fruits: Erect to spreading-ascending siliques, (1.5) 2.5-5 cm long, straight; styles 0.5-1.5 mm long, not beaklike; seeds in 1 row, finely pitted, about 1.5 mm long.

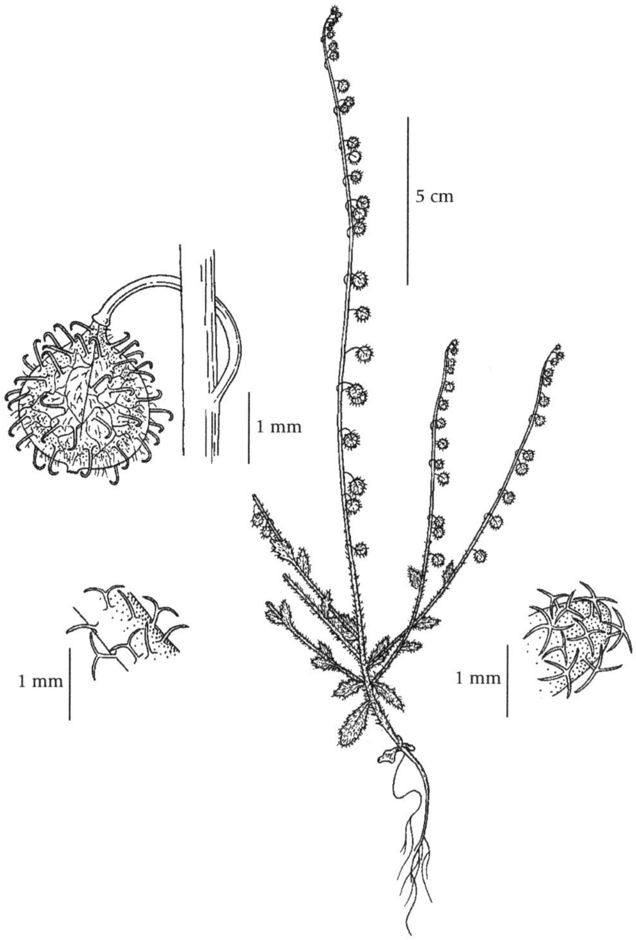

Athysanus pusillus

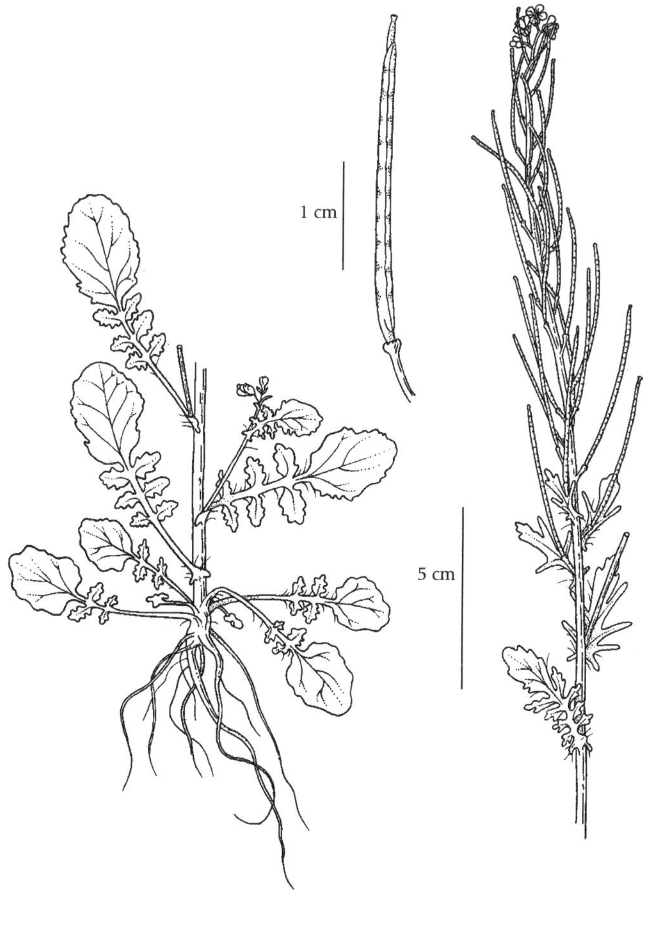

Barbarea orthoceras

Barbarea verna **(P. Mill.) Asch.**
early winter cress
Habitat/Range:
Mesic to dry fields and waste places in the lowland, steppe and montane zones; rare in S BC; introduced from Eurasia.
Description:
General: Biennial (rarely perennial) from a taproot and a simple stem-base; stems (20) 30-80 cm tall, stiff, erect, angled and branched.

Leaves: Basal leaves pinnately lobed, long-stalked, less than 12 cm long, with 4-10 pairs of lateral lobes, terminal lobe largest, egg-shaped and entire or irregularly toothed; stalks glabrous or long, fringed-hairy; lowermost stem leaves with 2-10 pairs of lateral lobes, upper pinnately cleft with the terminal lobe the largest, often clasping the stem.

Flowers: In single and terminal racemes or more often in compound, reduced racemes in lower leaf axils; flower stalks stout, more than 1 mm thick, 2-3 mm long; petals bright yellow, broadly spoon-shaped, 6-8 mm long; sepals pale yellow-green, about 2 mm long.

Fruits: Erect to spreading-ascending siliques, 4.5-8 cm long, straight or nearly straight; styles 0.5-1.5 mm long, not beaklike; seeds 1-1.5 mm long, greyish-brown.

Barbarea vulgaris **R. Br.** (*B. vulgaris* var. *brachycarpa* Rouy & Foucaud)
bitter winter cress; common winter cress; yellow rocket
Habitat/Range:
Moist to wet roadsides, fields and disturbed areas in the steppe and montane zones; frequent in SC BC, rare northward; introduced from Eurasia.
Description:
General: Biennial from a taproot and a simple stem-base; stems 20-80 cm tall, coarse, erect, angled and branched.

Leaves: Basal leaves pinnately lobed, long-stalked, less than 12 cm long, with 1-2 pairs of lateral lobes, terminal lobe largest, egg-shaped and entire or irregularly toothed; stalks and leaflets glabrous; lowermost stem leaves with 1-2 pairs of lateral lobes, upper generally lobed, not pinnately cleft.

Flowers: In single and terminal racemes or more often in compound, reduced racemes in lower leaf axils; flower stalks slender, less than 1 mm wide, 3-5 mm long; petals bright yellow, spoon-shaped to oblanceolate, 6-8 mm long; sepals pale yellow-green, about 2 mm long.

Fruits: Erect to ascending siliques, 1-3 cm long, straight; styles 2-3 mm long, distinctly beaklike; seeds 1-1.5 mm long, oblong.

BERTEROA

Berteroa incana **(L.) DC.**
hoary alyssum
Habitat/Range:
Dry fields and waste places in the steppe and montane zones; frequent in SC and SE BC; introduced from Eurasia.
Description:
General: Biennial herb from a slender taproot; stems 30-110 cm tall, erect, branched, with starlike hairs.

Leaves: Basal leaves oblanceolate, entire, 3-5 cm long, slender-stalked; stem leaves appressed-ascending, short-stalked to unstalked and reduced upward; grey with starlike hairs.

Flowers: In simple to compound racemes; flower stalks slender, erect, 5-10 mm long; petals white, deeply bilobed, 4-6 mm long; sepals hairy, not persistent, 2-3 mm long.

Fruits: Elliptic silicles, erect, 5-7 mm long, about half as broad, moderately inflated; styles persistent, 2-3 mm long; seeds 3-7 per chamber, in 2 rows, lightly honeycombed and narrowly wing-margined.

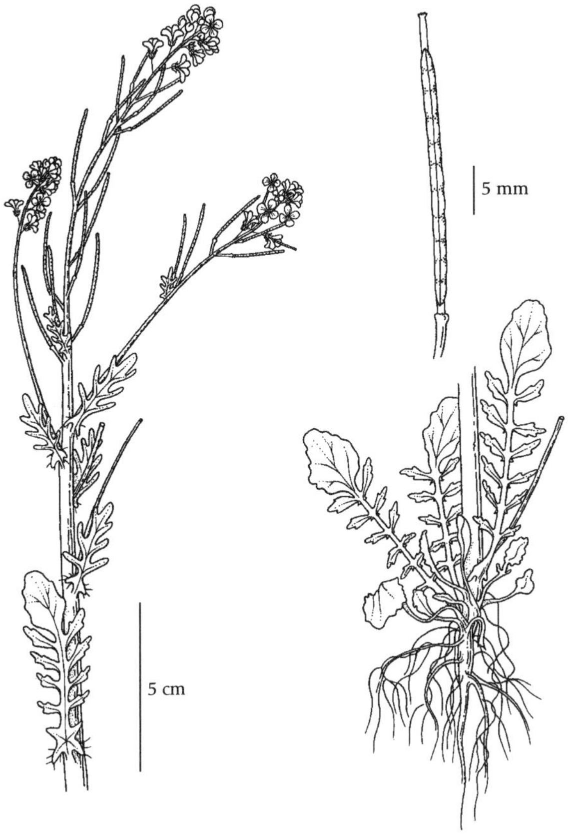

Barbarea verna

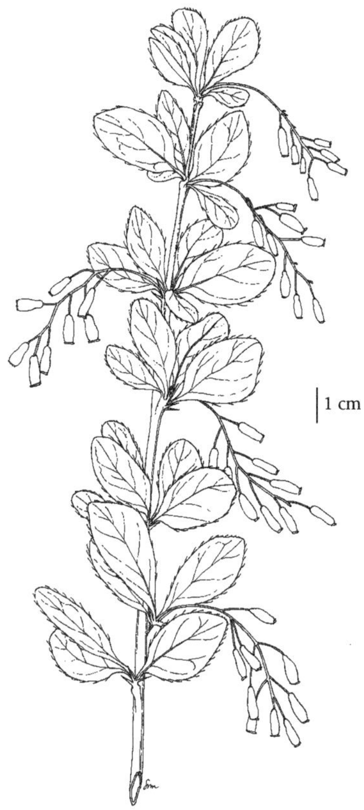

Barbarea vulgaris

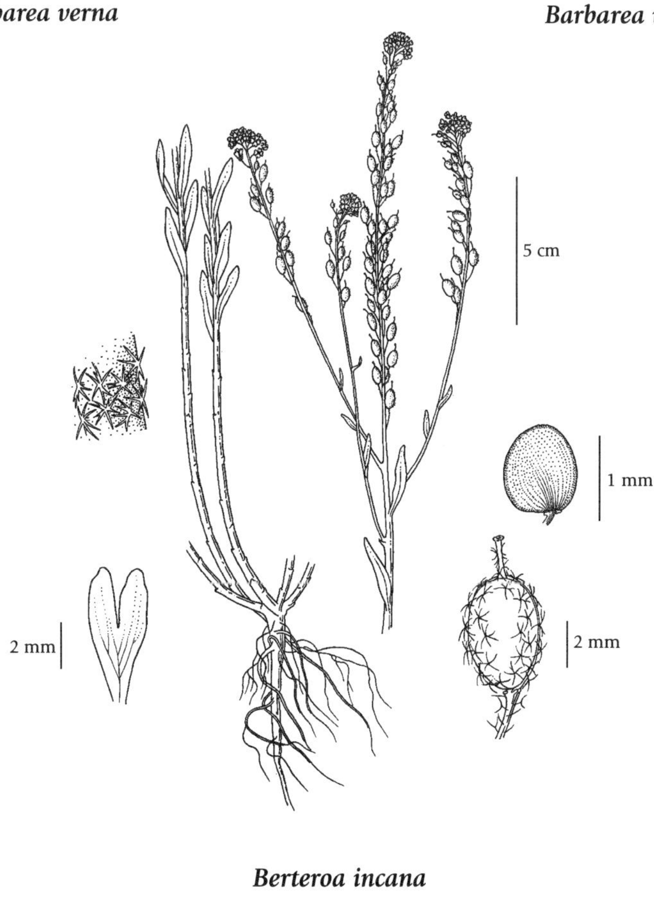

Berteroa incana

BRASSICA

1. Silique beaks conspicuously flattened, 3-nerved; silique segments 3-nerved.
    2. Siliques coarsely short-hairy, 3.5-4.5 mm wide .... *B. hirta*
    2. Siliques usually glabrous, 2-3 mm wide .... *B. kaber*
1. Silique beaks more or less rounded in cross section, usually 1-nerved; silique segments 1-nerved.
    3. Stem leaves unstalked, ear-like at the bases and clasping.
        4. Leaves glabrous to sparsely stiff-hairy; petals 10-14 mm long .... *B. napus*
        4. Leaves bristly; petals 6-8 mm long .... *B. campestris*
    3. Stem leaves unstalked to stalked, not ear-like at the bases or clasping.
        5. Siliques more or less tightly appressed-ascending, 1-2.5 cm long, the midnerve as prominent as the seams of dehiscence .... *B. nigra*
        5. Siliques spreading-ascending, 2-4 cm long, the midnerve less prominent than the seams of dehiscence .... *B. juncea*

***Brassica campestris* L.** (*B. rapa* L. ssp. *campestris* [L.] Clapham)
field mustard; rape mustard; bird-rape

HABITAT/RANGE:

Mesic to dry roadsides, fields and waste places in the lowland, steppe and montane zones; frequent in BC S of 56° N; introduced from Eurasia.

DESCRIPTION:

General: Annual herb from a taproot; stems glabrous or sparsely hairy, glaucous, 20-40 cm tall, simple to freely branched.

Leaves: Basal leaves slender-stalked, pinnately cleft with the terminal lobe the largest, 2-4 lateral lobes, the terminal lobe egg-shaped, wavy-toothed; lowermost stem leaves similar to the basal but ear-like at the bases and clasping or unstalked, upper oblong, narrowly egg-shaped or lanceolate, subentire.

Flowers: In terminal clusters; flower stalks stout, ascending, 7-25 mm long; petals yellow, 6-8 mm long; sepals 3-4 mm long.

Fruits: Siliques, ascending to spreading, 3-7 cm long, 2.5-3.5 mm wide, straight; styles beaked, 10-15 mm long; seeds 1.5 mm long, minutely honeycombed.

***Brassica hirta* Moench** (*Sinapis alba* L.)
white mustard

HABITAT/RANGE:

Mesic to dry roadsides, fields and waste places in the lowland, steppe and montane zones; rare in S BC; introduced from Eurasia.

DESCRIPTION:

General: Annual herb from a taproot; stems 0.2-1.0 m tall, freely branched, coarsely short-hairy to nearly glabrous, hairs sharp and bristlelike.

Leaves: Basal leaves pinnately lobed into 3-7 segments, coarsely-toothed, 8-20 cm long, 5-15 cm wide, the terminal lobe largest; reduced above, often just wavy-margined or lobed.

Flowers: In open terminal clusters, 10-30 cm long; flower stalks 5-10 mm long, spreading; petals pale yellow to whitish, 10-15 mm long; sepals 4-5 mm long.

Fruits: Siliques, spreading to slightly ascending, often curved, 3-4 cm long, 3.5-4.5 mm wide, 3-nerved, coarsely short-hairy, few-seeded, often constricted between seeds; styles beaked, conspicuously flattened; seeds 3-8, minutely honeycombed, about 3 mm long.

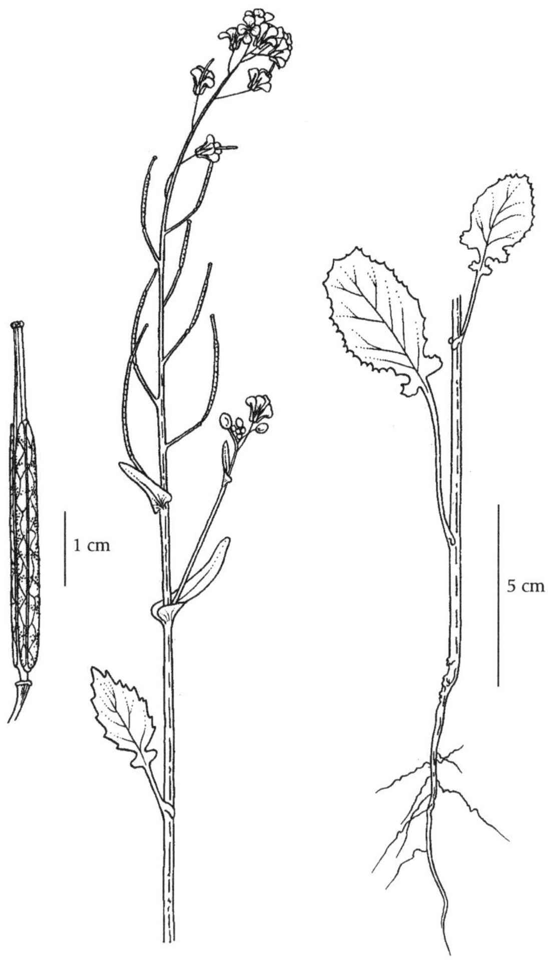

Brassica campestris

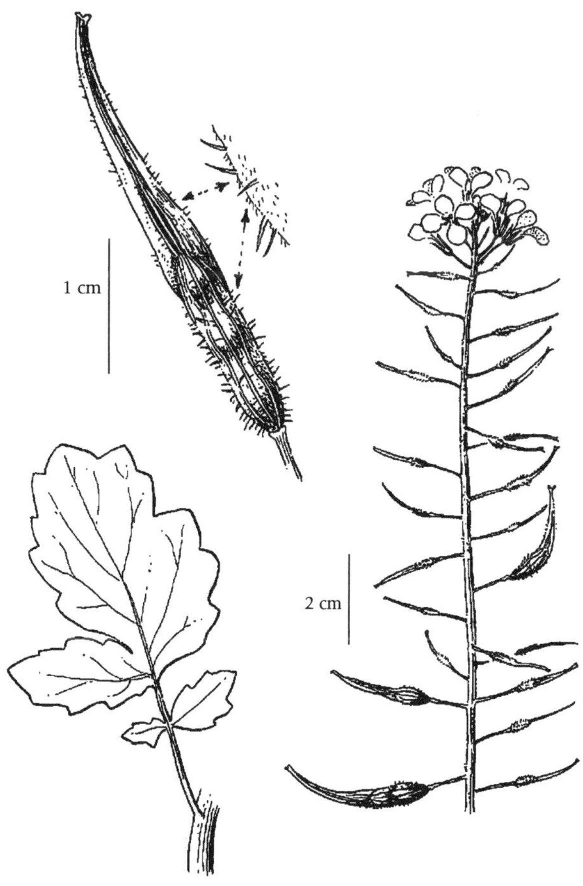

Brassica hirta

Brassica juncea **(L.) Czern.** (*Sinapis juncea* L.)

Indian mustard; brown mustard; leaf mustard; Chinese mustard

HABITAT/RANGE:

Mesic fields and waste places in the lowland and montane zones; rare in SW and NE BC, known only from S Vancouver Island, Vancouver and the Dawson Creek area; introduced from Asia.

DESCRIPTION:

General: Annual herb from a taproot; stems glabrous or somewhat glaucous, 0.1-1.0 m tall, branched.

Leaves: Basal leaves stalked, pinnately cleft with the terminal lobe the largest, 2-3 pairs of lateral lobes or occasionally none, toothed, up to 25 cm long, early deciduous; reduced upward and eventually unstalked, not ear-like at the bases.

Flowers: In terminal clusters, up to 40 cm long; flower stalks 10-15 mm long, spreading-ascending; petals pale yellow, 8-12 mm long; sepals 4-6 mm long.

Fruits: Siliques, spreading-ascending, 2-4 cm long, 2-3 mm wide, the midnerve less prominent than the seams of dehiscence; beak 5-10 mm long, slender; seeds about 2 mm long, finely net-veined and honeycombed.

Brassica kaber **(DC.) L.C. Wheeler** (*Sinapis arvensis* L.)

charlock; wild mustard

HABITAT/RANGE:

Mesic to dry field and waste places in the lowland and steppe zones; rare in SW and SC BC; introduced from Eurasia.

DESCRIPTION:

General: Annual herb from a taproot; stems 3.0-1.0 m tall, simple to freely branched, hairy with awl-like or bristlelike hairs throughout or just on lower stem.

Leaves: Basal leaves variable, from simple and wavy-margined and toothed to pinnately cleft with the terminal lobe the largest, 5-15 cm long; upper stem leaves simple.

Flowers: In compound terminal clusters, to 30 cm long, flower stalks stout, 2-6 mm long, erect or ascending; petals yellow, 8-14 mm long; sepals 4-5 mm long.

Fruits: Siliques, glabrous or somewhat hairy, straight or slightly curved, (3) 4-5 cm long, 2 mm wide, often constricted between seeds; styles beaked, conspicuously flattened, about 2 mm long; seeds 7-12, about 2 mm long, finely honeycombed.

Brassica napus **L.**

turnip; winter rape

HABITAT/RANGE:

Mesic to dry fields and waste places in the lowland, steppe and montane zones; rare in BC south of 55° N; introduced from Eurasia.

DESCRIPTION:

General: Annual or biennial herb from a taproot; stems up to 1.5 m tall, leafy.

Leaves: Glaucous, glabrous or sparsely hairy; basal leaves pinnately lobed, often clasping at the bases; stem leaves somewhat pinnately lobed, unstalked.

Flowers: In elogate racemes; flower stalks slender, spreading to ascending; petals golden to dull yellow, 10-14 mm long.

Fruits: Siliques, ascending, 5-10 cm long; styles beaked, 7-11 mm long; seeds about 1.8 mm long, finely honeycombed.

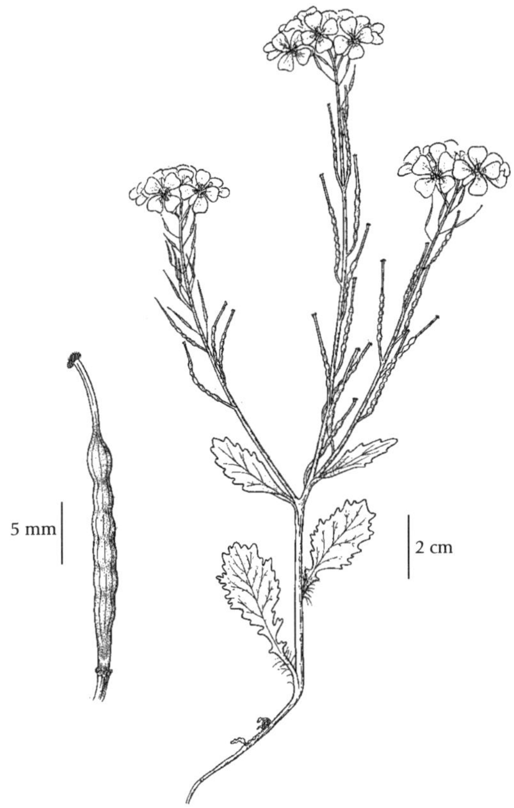

Brassica juncea

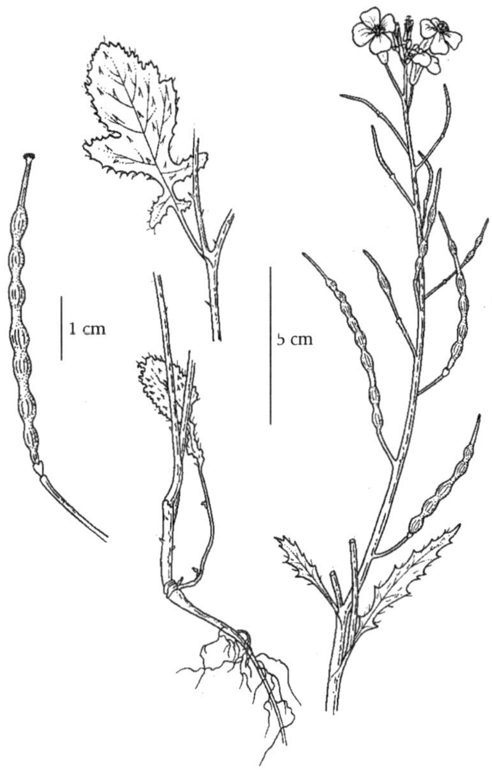

Brassica kaber

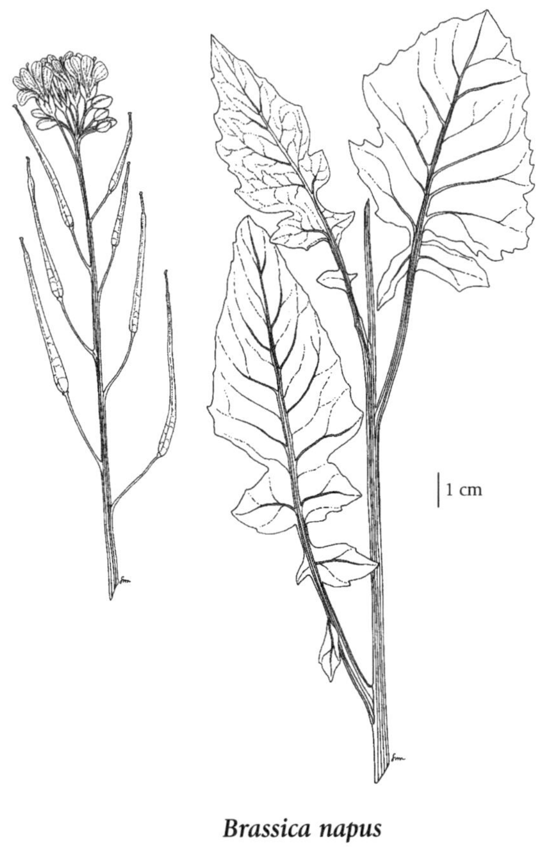

Brassica napus

Brassica nigra **(L.) Koch**
black mustard
HABITAT/RANGE:
Mesic fields and waste places in the lowland zone; rare in SW BC; introduced from Europe.
DESCRIPTION:
General: Annual herb from a taproot; stems sparsely to densely stiff-hairy, 40-150 cm tall, usually branched.

Leaves: Basal leaves stalked, mostly somewhat pinnately cleft, with the terminal lobe the largest, to wavy-margined and lobed or saw-toothed; stem leaves similar but reduced upward, the uppermost elliptic, wavy-margined or saw-toothed, nearly unstalked and not ear-like at the bases.

Flowers: In numerous clusters from the leaf axils; flower stalks stout, erect, 3-6 mm long; petals 7-11 (15) mm long, basal portion slender; sepals 3-4 mm long.

Fruits: Siliques, erect, tightly appressed to stems, 1-2.5 cm long, 1-1.5 mm wide, somewhat quadrangular; beaks smooth; seeds about 2 mm long, minutely honeycombed.

BRAYA

1. Stem leaves present; pods linear, about 1 mm wide .... *B. humilis*
1. Stem leaves lacking; pods plump, 2-3 mm wide .... *B. purpurascens*

Braya humilis **(C.A. Mey.) B.L. Robins.** (*B. richardsonii* [Rydb.] Fern.)
dwarf braya; low braya
HABITAT/RANGE:
Moist to dry forests, river bars, scree slopes and gravelly slopes in the montane to alpine zones; infrequent in N BC, rare southward in the Rocky Mountains; N to AK, YT and NT, E to NF and disjunct in CO.
DESCRIPTION:
General: Perennial or biennial from a taproot or stem-base; stems several, 5-30 cm long, erect or spreading, simple or branched, more or less hairy with forked hairs, purplish.

Leaves: Basal leaves thick, linear-oblanceolate, 1-4 cm long, entire to somewhat toothed or wavy-margined, glabrous or somewhat hairy; stem leaves reduced, unstalked, lanceolate or linear, clasping at the bases.

Flowers: Racemes somewhat headlike in flower, greatly elongated in fruit; flower stalks erect or closely ascending, 2-6 mm long, hairy; petals white or purplish, 3-4 mm long; sepals 1.5-2.5 mm long, hairy, deciduous;

Fruits: Siliques, erect or closely ascending, 1-3 cm long, narrowly cylindric, about 1 mm wide, minutely hairy, somewhat constricted between the seeds; styles slender, 0.5-1 mm long; seeds in 1 row, 0.7-1 mm long.

Notes: An adequate monograph of North American *Braya* (especially those of the far north) has yet to be done, thus our taxonomy of these species is often confusing.

Braya purpurascens **(R. Br.) Bunge *ex* Ledeb.** (*B. americana* [Hook.] Fern., *B. glabella* Rich., *B. henryae* Raup)
purple braya
HABITAT/RANGE:
Dry scree slopes in the alpine zone; rare in N BC; circumpolar, N to AK, YT and NT and E to PQ; Eurasia, Greenland.
DESCRIPTION:
General: Perennial herb from a taproot; stems mostly decumbent-ascending, 2-10 (12) cm tall, sparsely hairy, greenish or purple-tinged; usually leafless.

Leaves: Basal leaves fleshy, often purplish, spoon-shaped or oblong to linear, glabrous or sparsely hairy, 0.7-3.6 cm long, 1-4 mm wide; stem leaves lacking.

Flowers: In a somewhat lax raceme; flower stalks spreading or ascending, 1-7 mm long, hairy; petals white to reddish or purplish at base, 2.5-4 mm long; sepals 1.5-2.5 mm long, glabrous or with a few hairs, tardily deciduous.

Fruits: Siliques, spreading, ascending, or bent downward, oblong to egg-shaped, plump, 5-10 mm long and up to 3 mm wide, glabrous to hairy, often purplish; seeds in 2 rows, about 1 mm long.

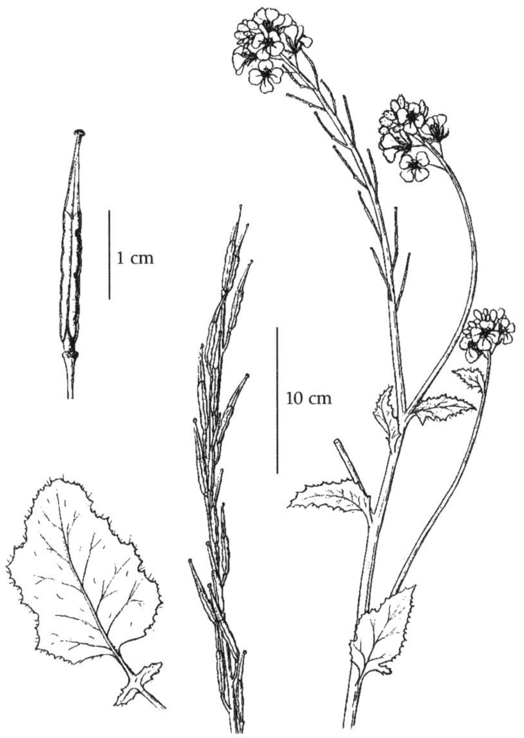

Brassica nigra

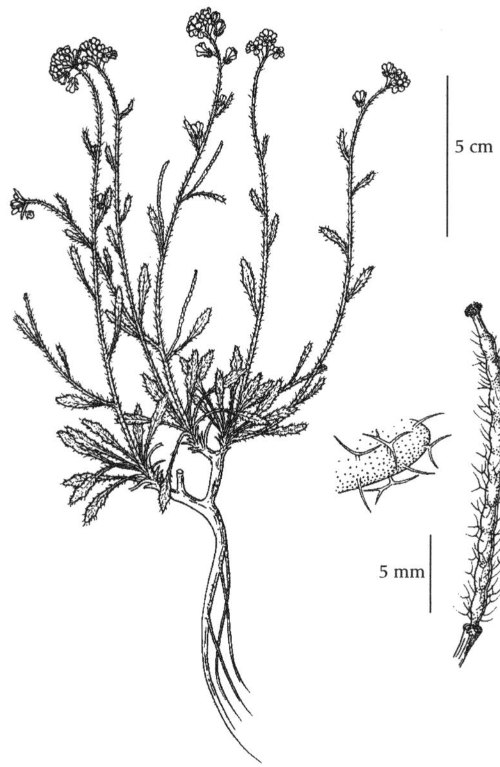

Braya humilis

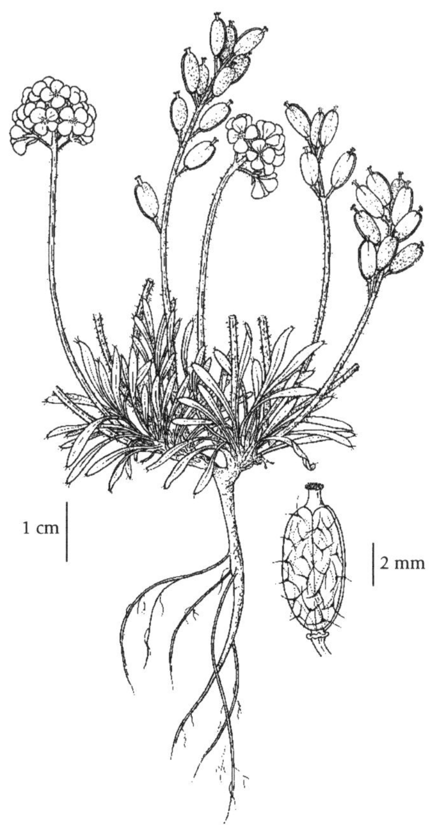

Braya purpurascens

BUNIAS

***Bunias orientalis* L.**

Turkish rocket

Habitat/Range:

Mesic to dry meadows and waste places in the steppe zone; rare in SC BC; introduced from Eurasia.

Description:

General: Biennial to perennial herb from a taproot; stems erect, sparsely glandular, 25-100 cm tall.

Leaves: Lower stem leaves oblong, long-pointed, pinnately cleft to coarsely toothed, often with 2 large, outward-projecting teeth near the base, stalked, to 20 cm long; upper stem leaves unstalked, reduced.

Flowers: Inflorescence 5- to 8-flowered, elongate in fruit; flower stalks 10-15 mm long; petals narrowly egg-shaped, 6 mm wide, constricted toward the base, yellow.

Fruits: Silicles, asymmetricly egg-shaped, 6-10 mm long 1-2 celled, covered with glandular-warty prominences; styles short and almost lateral; seeds 1-2.

CAKILE

1. Silicles slightly to much constricted at the joints; leaves mostly spoon-shaped or egg-shaped, wavy-margined and toothed, sometimes pinnately lobed or nearly entire .... *C. edentula*
1. Silicles usually expanded at the joints into projecting wings; at least the lower leaves usually distinctly pinnately cleft .... *C. maritima*

***Cakile edentula* (Bigel.) Hook.** (*C. edentula* var. *californica* [Heller] Fern.)

American searocket

Habitat/Range:

Moist to wet sandy shorelines in the lowland zone; common in coastal BC; N to AK and S to CA, also along the Great Lakes and the Atlantic Coast.

Description:

General: Annual herb from a taproot; stems freely branched, 10-50 cm tall, glabrous, base of lower stems often decumbent.

Leaves: Basal leaves widely egg- or spoon-shaped, fleshy, deeply toothed to wavy-margined and saw-toothed, with a broad stalk, 2-7 cm long; stem leaves similar.

Flowers: Inflorescence a raceme, 4-15 cm long; flower stalks 3-7 mm long, stout, ascending; petals white to purplish-tinged, 6-8 mm long; sepals sparsely soft-hairy at tip, early deciduous, 3-4 mm long.

Fruits: Siliques, 1.5-2.5 cm long, slightly to considerably constricted at the joints, the upper segment triangular to egg-shaped, constricted to somewhat flattened beak, tip acute to notched.

***Cakile maritima* Scop.**

European searocket

Habitat/Range:

Mesic to moist disturbed areas in the lowland zone; rare in SW BC, known only from the W coast of Vancouver Island; introduced from Europe.

Description:

General: Annual herb from a taproot; stems branched, 15-45 cm long, prostrate or ascending.

Leaves: Basal leaves egg-shaped or oblanceolate to deeply pinnately lobed, fleshy, the lower leaves usually distinctly pinnately lobed, stalked, 3-6 cm long; upper stem leaves less lobed or entire, unstalked.

Flowers: Inflorescence a raceme; flower stalks 3-7 mm long, stout, ascending; petals purple, lilac or white, 6-10 mm long; sepals 3-5 mm long.

Fruits: Siliques, 1.5-3 cm long, usually expanded at the joints into projecting wings, the upper segment twice as long as lower, conic to cylindric, tips acute to blunt.

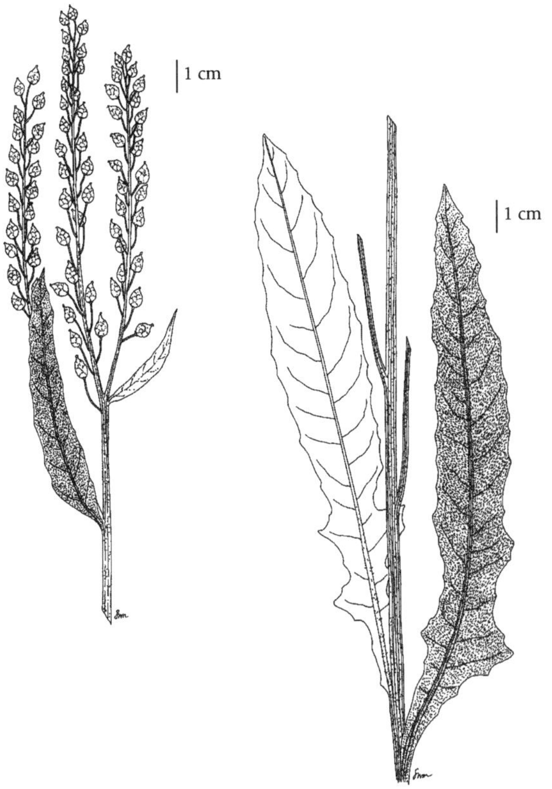

Bunias orientalis

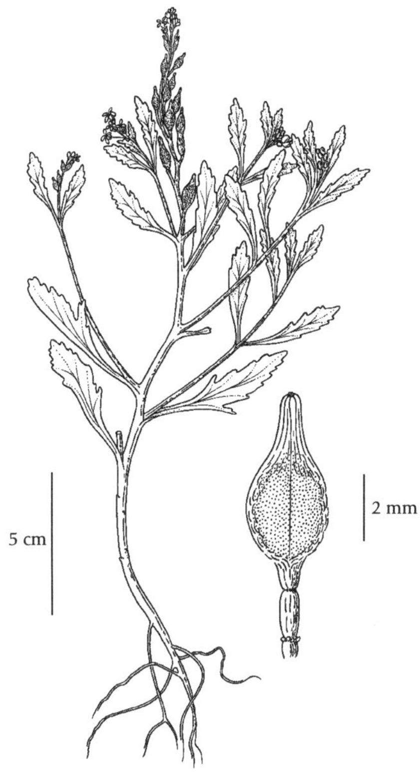

Cakile edentula

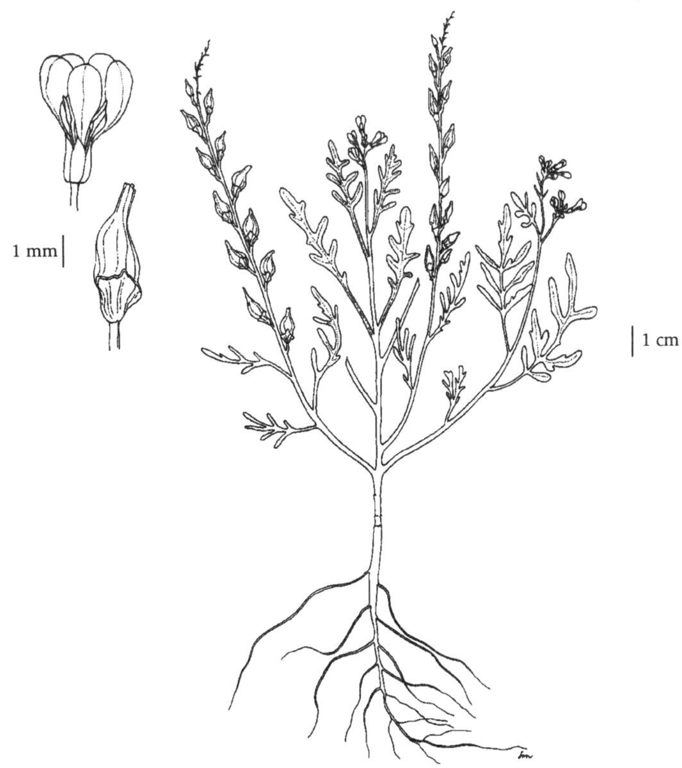

Cakile maritima

CAMELINA

1. Plants hairy at least below, with both long, simple and starlike hairs; silicles 4-7 mm long; seeds mostly less than 1 mm long .... *C. microcarpa*
1. Plants glabrous or, if sparsely hairy, then the simple hairs not exceeding the starlike hairs; silicles 6-9 mm long; seeds 1-2 mm long .... *C. sativa*

Camelina microcarpa **Andrz. *ex* DC.**

littlepod flax; hairy flax

HABITAT/RANGE:

Dry roadsides, fields and waste places in the steppe and montane zones; infrequent in SW, SC and NE BC; introduced from Eurasia.

DESCRIPTION:

General: Annual herb from a taproot; stems 0.3-1.0 m tall, simple or sometimes branched, conspicuously hairy at the base with simple and branched to starlike hairs, little branched and usually glabrous above.

Leaves: Basal leaves few, soon deciduous; stem leaves lanceolate to oblanceolate, nearly entire, 2-8 cm long, short-stalked or unstalked, the tips sharp-pointed to long-pointed, upper leaves arrowhead-shaped with earlike lobes at the base, hairs similar to that of the stems.

Flowers: In elongated, often compound racemes; flower stalks slender, ascending-spreading, 10-15 (17) mm long; petals white to pale yellow, 4-5 mm long; sepals about 3 mm long.

Fruits: Silicles, egg-shaped, 4-7 (8) mm long, 3-5 mm wide, obtuse at tip; styles 2-2.5 mm long; seeds mostly 0.8-1 mm long, slightly longer than wide, plump, finely veined and honeycombed, reddish-brown.

Camelina sativa **(L.) Crantz**

falseflax; gold-of-pleasure

HABITAT/RANGE:

Mesic to dry roadsides, fields and waste places in the lowland and steppe zones; rare in SC BC and S Vancouver Island; introduced from Eurasia.

DESCRIPTION

General: Annual herb from a taproot; stems 0.3-1.0 m tall, simple or sometimes branched, glabrous, or if sparsely hairy then the simple hairs not exceeding the starlike hairs.

Leaves: Basal leaves few, soon deciduous; stem leaves lanceolate to oblanceolate, nearly entire, 2-8 cm long, short-stalked or unstalked, mostly glabrous or only slightly hairy, upper leaves lanceolate or linear-oblong, unstalked, with earlike lobes at the base.

Flowers: In elongated racemes; flower stalks slender, ascending-spreading, 10-25 (30) mm long; petals white to pale yellow, 4-5 mm long; sepals 2-3 mm long.

Fruits: Silicles, 7-9 mm long, 5-6 (7) mm wide, nearly squared-off at tip; styles 2-2.5 mm long; seeds 1-2 mm long, pale yellow-brown.

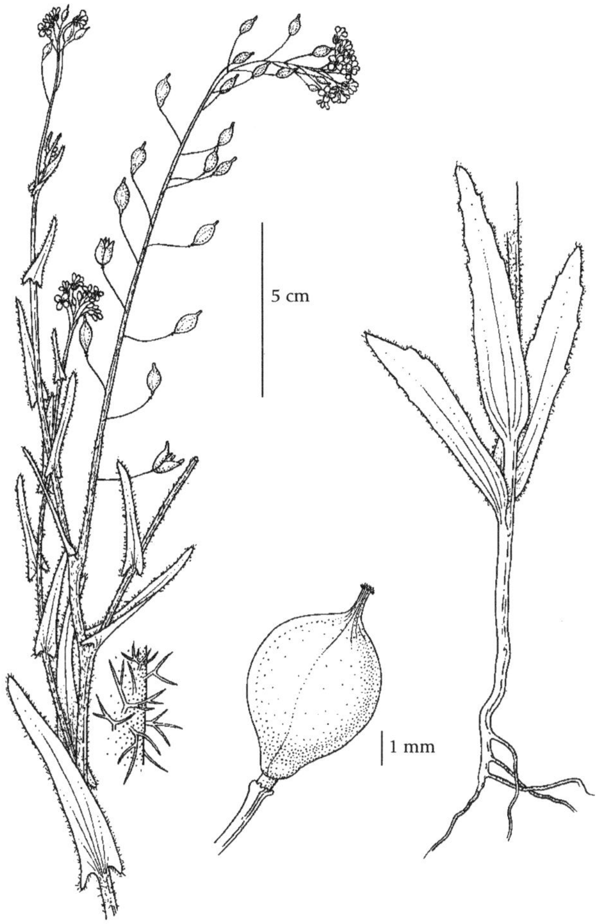

Camelina microcarpa

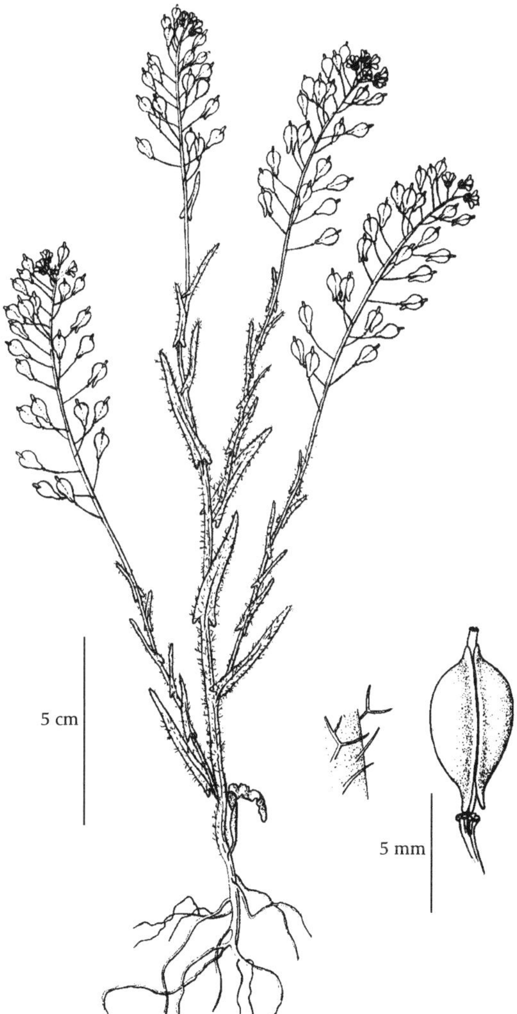

Camelina sativa

CAPSELLA

***Capsella bursa-pastoris* (L.) Medic.**
shepherd's purse
HABITAT/RANGE:
Mesic to dry roadsides, fields and waste places in the lowland, steppe and montane zones; common throughout BC; introduced from Eurasia.
DESCRIPTION:
General: Annual herb from a taproot; stems simple to branched, 10-50 cm long, with simple and starlike hairs.

Leaves: Basal leaves in rosettes, stalked, 3-6 cm long, oblanceolate, more or less entire to pinnately lobed or dissected; stem leaves unstalked and clasping, lanceolate to oblong-oblanceolate, mostly remotely saw-toothed, with earlike lobes at the base.

Flowers: Inflorescene a many-flowers raceme; flower stalks slender, spreading, 7-15 mm long; petals white, with a distinct stalklike base, 1.5-4 mm long; sepals 2 mm long.

Fruits: Silicles, heart-shaped to triangular, 4-8 mm long, 3-5 mm wide, strongly flattened, broad at tip, squared-off to slightly notched; seeds numerous, minutely net-veined.

CARDAMINE

Notes: The complex of similar species including *C. breweri*, *C. hirsuta*, *C. oligosperma*, *C. parviflora*, *C. pensylvanica* and *C. umbellata* is often considered to be one polymorphic species, *C. oligosperma*, and is variously treated in different manuals.

1. Lower leaves mostly simple.
    2. Plants 2-10 cm tall, leaves in a distinct basal rosette; petals 3-5 mm long .... *C. bellidifolia*
    2. Plants 10-60 cm tall, leaves not in a distinct basal rosette; petals 7-12 mm long.
        3. Stem leaves lobed or compound .... *C. nuttallii*
        3. Stem leaves entire and simple .... *C. cordifolia*
1. Lower leaves mostly compound.
    4. Petals 8-15 mm long.
        5. Basal leaves pinnately compound .... *C. pratensis*
        5. Basal leaves with 3-5 leaflets.
            6. Rhizomes tuberous; upper stem leaves with lanceolate, usually entire segments; common in SW BC .... *C. nuttallii*
            6. Rhizomes slender, not tuberous; upper stem leaves egg-shaped, with 3-5 large teeth; rare on the Queen Charlotte Islands and W Vancouver Island .... *C. angulata*
    4. Petals 2-6 mm long.
        7. Plants rhizomatous perennials.
            8. Basal or lower stem leaves, or at least some of them, simple .... *C. breweri*
            8. Basal and lower stem leaves all pinnately compound.
                9. Stem leaves numerous; basal rosettes often not present; plants often partially submerged .... *C. occidentalis*
                9. Stem leaves few (usually 1-3); basal rosettes more prominent than stem leaves; plants not partially submerged .... *C. oligosperma*
        7. Plants annuals or biennials, lacking rhizomes.
            10. Lateral leaflets of stem leaves linear or narrowly oblanceolate, not over 2 mm wide *C. parviflora*
            10. Lateral leaflets of stem leaves wider, oblanceolate to egg-shaped, greater than 2 mm wide.
                11. Stems glabrous at base; leaf blades sparsely stiff-hairy .... *C. hirsuta*
                11. Stems usually stiff-hairy at base; leaf blades glabrous.
                    12. Leaf stalks hairy-fringed; mature siliques mostly 1.2-1.5 mm thick .... *C. oligosperma*
                    12. Leaf stalks lacking fringing hairs; mature siliques mostly 0.7-10 mm thick .... *C. pensylvanica*

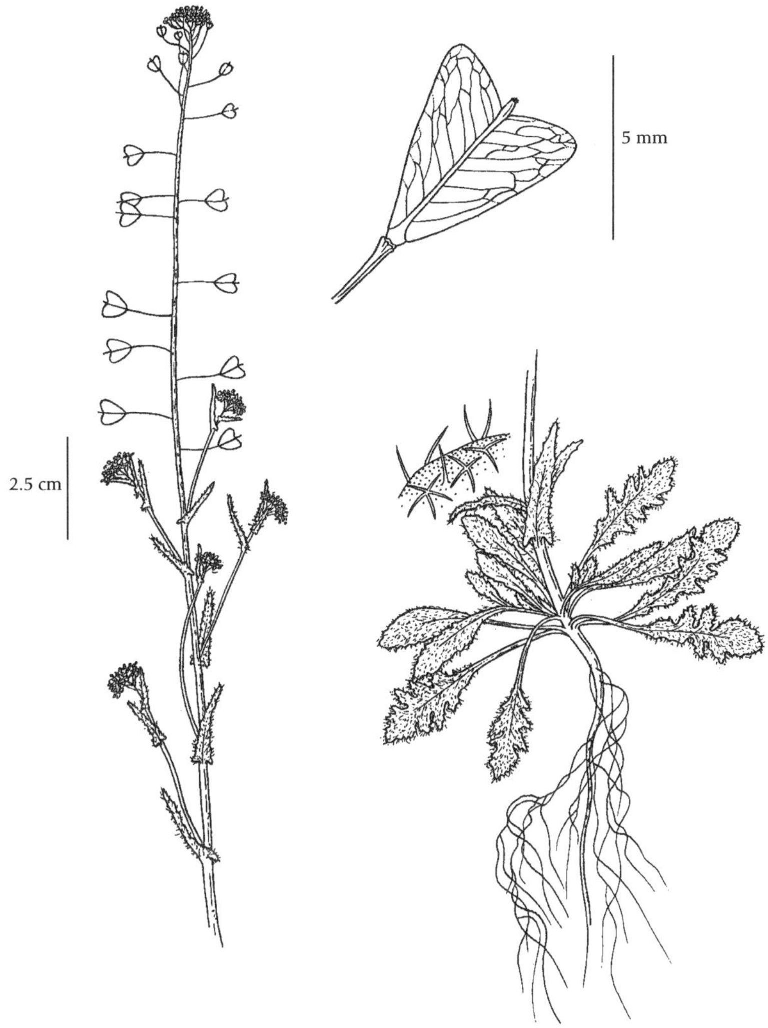

Capsella bursa-pastoris

Cardamine angulata **Hook.**

angled bitter-cress

Habitat/Range:

Moist woods and river banks in the lowland zone; rare on the Queen Charlotte Islands and Vancouver Island; N to S AK and S to N CA.

Description:

General: Perennial herb from slender rhizome, greater than 3 cm long, not tuberous; stems simple, 40-80 cm tall, glabrous or sparsely hairy.

Leaves: With 3-5 leaflets, long-stalked, leaflets narrowly egg-shaped to egg-shaped, widely toothed to lobed, the terminal leaflet 1.5-5 cm long, 1.2-3 cm wide, somewhat larger than the lateral ones, hairy-fringed; upper stem leaves egg-shaped, with 3-5 large teeth.

Flowers: Inflorescence a simple raceme; flower stalks 8-20 mm long, ascending, glabrous; petals white to pinkish, 8-14 mm long; sepals pale yellow-green, about 2 mm long.

Fruits: Siliques, 1.5-3.5 cm long, 1.5-2 mm wide, erect, glabrous; styles 1.5-3 mm long.

Cardamine bellidifolia **L. var.** ***bellidifolia***

alpine bitter-cress

Habitat/Range:

Moist scree slopes, gravelly areas and mossy rocks in the upper montane to alpine zones; frequent in N BC, rare southward and on Vancouver Island; circumpolar, E to NF and S to ME, NH and N CA.

Description:

General: More or less tufted, perennial herb from a taproot and often a branched stem-base with somewhat rhizome-like branches, leaf bases persistent; stems several, few-leaved, 2-10 cm long, glabrous.

Leaves: Basal leaves in rosettes, thick, fleshy, entire or shallowly 1-3 lobed, blades 5-30 mm long, diamond-shaped to elliptic to egg-shaped, narrowed to slender stalks 2-4 times as long; stem leaves 0-3, reduced.

Flowers: Inflorescence a few-flowered, umbel-like raceme; flower stalks 4-10 mm long, ascending; petals white, 3-5 mm long; sepals 1.5-2 mm long.

Fruits: Siliques, erect, 1.5-3.5 cm long, 1-1.5 mm wide; styles stout, 1-3 mm long.

Cardamine breweri **S. Wats.**

Brewer's bitter-cress

Habitat/Range:

Wet to moist streambanks, shallow ponds and lakesides in the lowland and montane zones; infrequent in S BC; S to WY, NV and N CA.

Description:

General: Perennial herb from a rhizome; stems erect to spreading or nearly prostrate, simple to freely branched, 20-60 cm tall, glabrous or sparsely hairy near base.

Leaves: Basal leaves few, simple, blades 1-3 cm long, heart- to kidney-shaped or egg-shaped, shallowly wavy-margined and lobed; stem leaves compound with 2-4 (6) lanceolate to egg-shaped lateral lobes, the terminal lobe heart-shaped to egg-shaped to widely lanceolate, wavy-margined and toothed to shallowly lobed.

Flowers: Inflorescence a few-flowered racemes; flower stalks 5-20 mm long, ascending; petals white, spoon- to egg-shaped, 3-6 mm long; sepals 1.5-2.5 mm long.

Fruits: Siliques, erect, 2-3 cm long, 1-1.5 mm wide; styles 0.5-2 mm long.

Notes: Two varieties occur in BC.

1. Terminal lobes of lower stem leaves heart-shaped at base. .... var. *orbicularis* (Greene) Detl.

1. Terminal lobes of lower stem leaves cuneate or rounded at base .... var. *breweri*

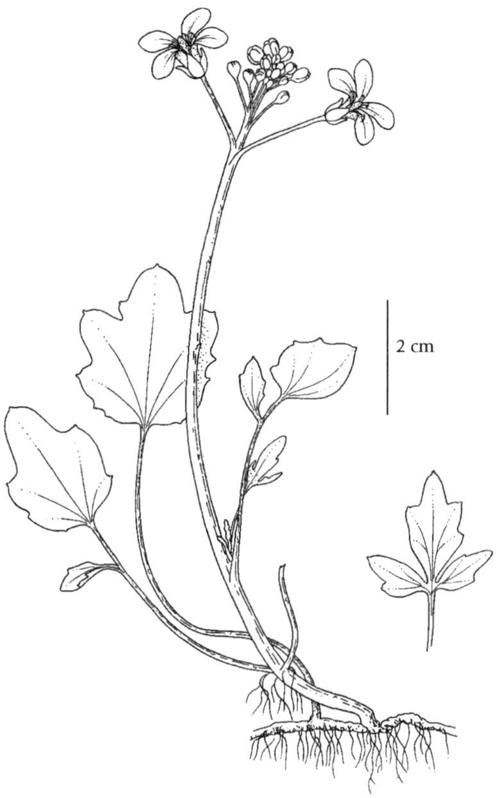

Cardamine angulata

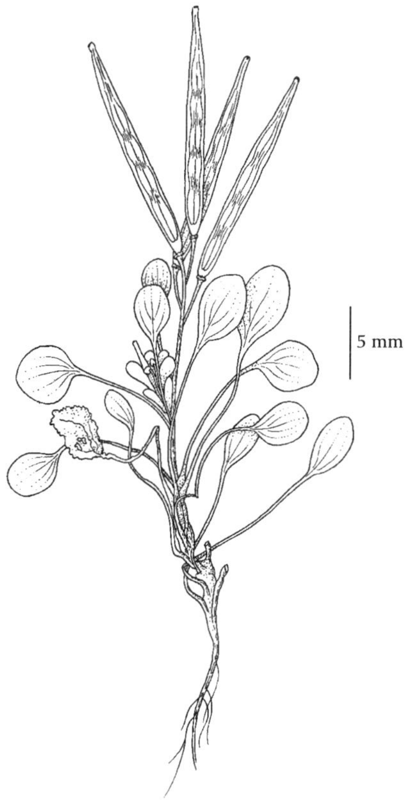

Cardamine bellidifolia var. *bellidifolia*

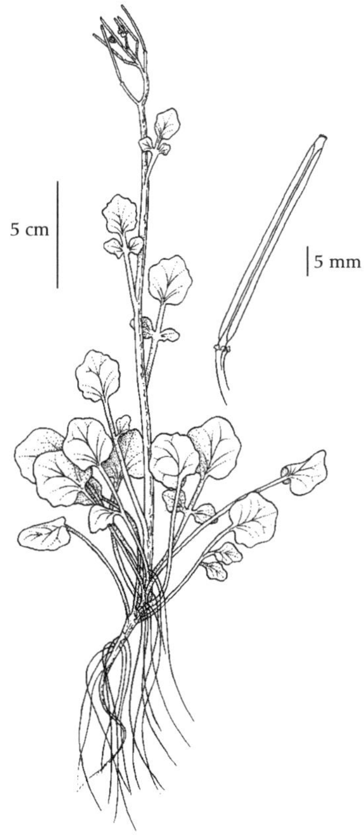

Cardamine breweri

Cardamine cordifolia **A. Gray var.** ***lyallii*** **(S. Wats.) Nels. & Macbr.**
heart-leaved bitter-cress; large mountain bitter-cress

Habitat/Range:

Moist streambanks and meadows in the montane to alpine zones; infrequent in S BC east of the Coast-Cascade Mountains; S to ID, NV and N CA.

Description:

General: Perennial herb from an extensive, slender rhizome; stems erect, simple, 2-2.5 mm thick, (10) 20-60 cm tall, glabrous to densely hairy near base with short, spreading, simple hairs.

Leaves: Basal leaves few, simple, fleshy, glabrous, with slender stalks 2-5 times as long as blade, the blade kidney-shaped, heart-shaped or deltoid, usually wavy-margined and round-toothed, (2) 3-10 cm wide; basal and lower stem leaves usually wider than long; upper stem leaves with stalks less that half as long as those on lower leaves, blades more heart-shaped or deltoid, wavy-margined to lobed, bristly-tipped, 2-6 cm long.

Flowers: Inflorescence a simple raceme, lacking bracts; flower stalks 10-20 mm long, spreading-ascending; petals white, spoon- to egg-shaped, 7-12 mm long; sepals 3-4 mm long, glabrous.

Fruits: Siliques, erect, 2-3.5 cm long, (1) 1.5-4 mm wide; styles 0.5-2 (6) mm long; seeds about 1.5 mm wide, flat, wingless, smooth.

Cardamine hirsuta **L.**
hairy bitter-cress

Habitat/Range:

Mesic to dry waste places, roadsides and open woods in the lowland zone; frequent in SW BC; introduced from Eurasia.

Description:

General: Annual herb from slender taproot; stems usually solitary, stiff, 5-30 cm tall, usually glabrous.

Leaves: Basal leaves in compact rosettes, compound, stalked, hairy-fringed, pinnate with 2-3 or more pairs of egg-shaped to orbicular leaflets and a larger, somewhat kidney-shaped terminal leaflet; stem leaves few, almost unstalked, with smaller, narrower leaflets; all leaflets wavy-margined to shallowly few-lobed, sparsely stiff-hairy above and on the margins.

Flowers: Inflorescence a simple raceme; flower stalks ascending, 5-10 mm long; petals white, narrow, 1.5-2 mm long; sepals half as long as petals, greenish-violet with narrow white margins; stamens 4.

Fruits: Siliques, erect, 1.8-2.5 cm long, about 1 mm wide; styles 0.5-2 (6) mm long; seeds 22-36.

Cardamine nuttallii **Greene var.** ***nuttallii*** (*C. pulcherrima* Greene var. *tenella* [Pursh] C.L. Hitchc.)
slender toothwort; beautiful bitter-cress

Habitat/Range:

Moist woods and bottom lands in the lowland zone; common in SW BC, known from S Vancouver Island, the Gulf Islands and adjacent mainland of BC; S to N CA.

Description:

General: Perennial herb from a short, slender, fleshy rhizome, less than 3 cm long and 1-3 (4) mm thick; stems erect, 20-30 cm tall, glabrous.

Leaves: Basal leaves long-stalked, blades simple, orbicular to heart-shaped, shallowly round-toothed to lobed, sometimes deeply 3- or 5-lobed or compound with 3-5 (7) leaflets; stem leaves 1-3, grouped mostly above middle of stem, lobed, or compound with 3 or 5 lanceolate to egg-shaped-lanceolate, entire leaflets, 1-4 cm long.

Flowers: Inflorescence a few-flowered racemes, without bracts; flowers stalks erect to ascending, 10-20 mm long; petals pink to reddish or purplish, 7-12 mm long; sepals 3-5 mm long.

Fruits: Siliques, erect, 1.5-5 cm long, about 1.5 mm wide; styles slender, 3-6 mm long.

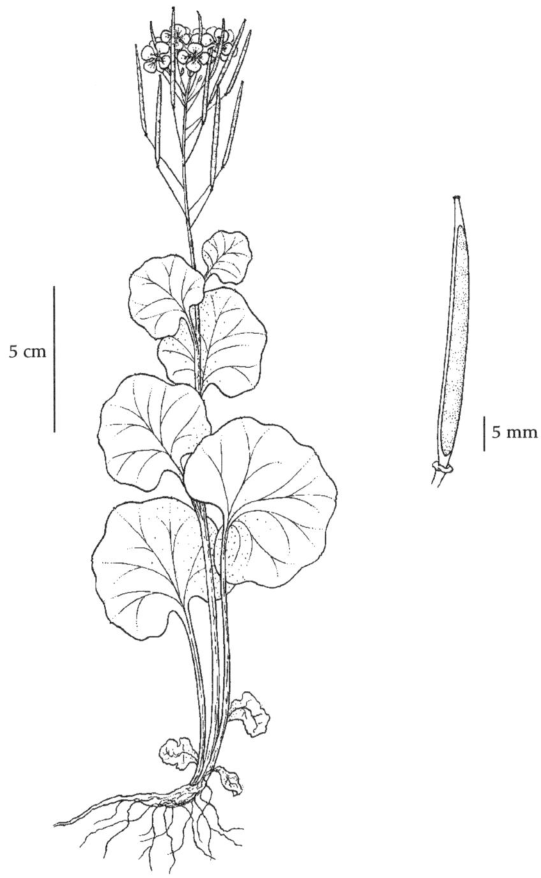

Cardamine cordifolia var. *lyallii*

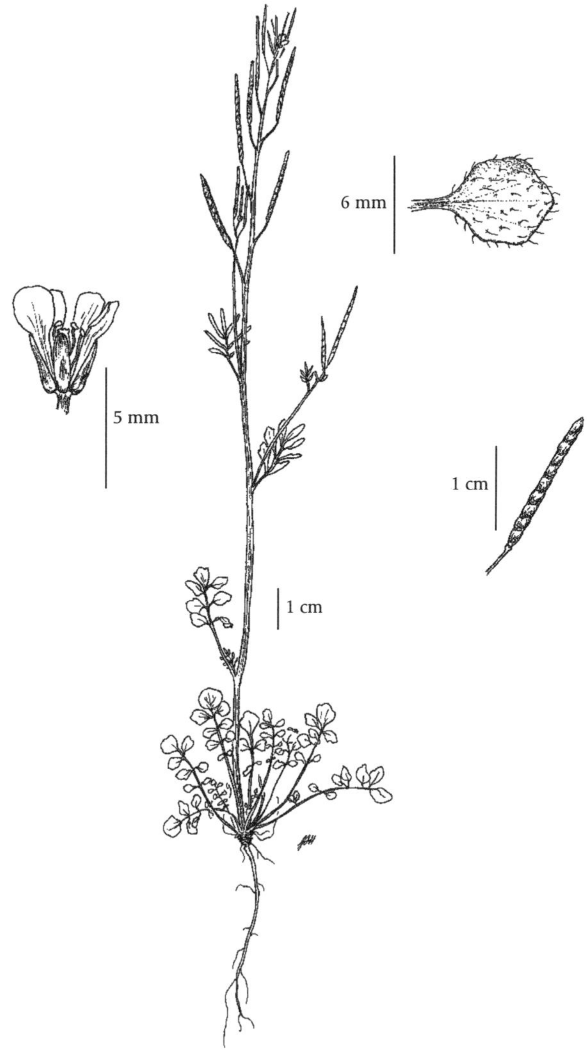

Cardamine hirsuta

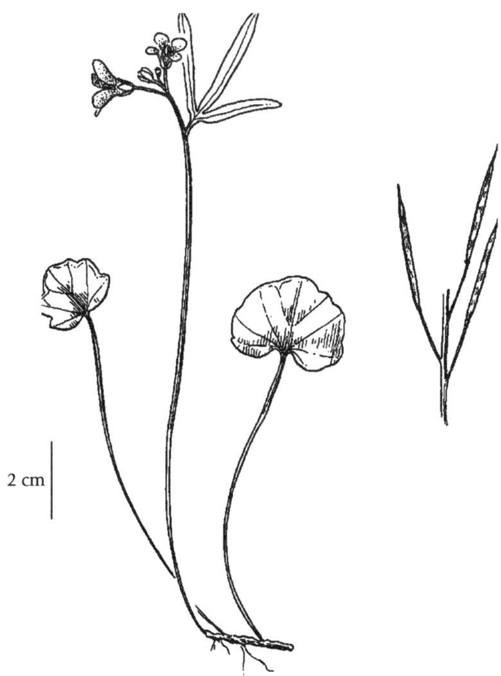

Cardamine nuttallii var. *nuttallii*

***Cardamine occidentalis* (S. Wats. *ex* B.L. Robins.) Howell**
western bitter-cress

HABITAT/RANGE:

Wet to moist streambanks and lakeshores in the lowland zone; infrequent in SW BC; N to AK and S to CA.

DESCRIPTION:

General: Perennial herb from a short, slender rhizome; stems 1-1.5 mm thick, enlarged and tuberous near base of leafy stems, 20-40 cm tall, simple to freely branched, erect to decumbent and rooting at nodes, glabrous to sparsely hairy.

Leaves: Basal leaves few, slender-stalked, pinnate, lateral lobes (2) 4-6, egg-shaped to heart-shaped, short-stalked to unstalked, 3-8 mm long, generally entire, terminal lobe 1-2 cm long, shallowly wavy-margined; stem leaves several, upper ones unstalked to nearly unstalked, fewer lateral lobes, narrowly egg-shaped to linear, mostly 1-2 cm long, entire or slightly toothed, terminal leaflet egg- to wedge-shaped, 1.5-3 cm long.

Flowers: Inflorescence an elongate raceme, without bracts; flower stalks ascending, 10-15 mm long; petals white, 3.5-5 mm long; sepals 1-2 (2.5) mm long.

Fruits: Siliques, erect, 2-3 cm long, 1-1.5 mm wide; styles about 1 mm long.

***Cardamine oligosperma* Nutt.** (*C. umbellata* Greene = var. *kamtschatica*)
little western bitter-cress; few-seeded bitter-cress (var. *oligosperma*), **Siberian bitter-cress**; umbellate bitter-cress (var. *kamtschatica*)

HABITAT/RANGE:

Wet to mesic streamsides, ditches, waste places, roadsides, meadows, gravelly slopes and open woods in the lowland to alpine zones; var. oligosperma – common in S BC, less frequent northward, var. *kamtschatica* – frequent in coastal and N BC; var. *oligosperma* – S to MT and CA, var. *kamtschatica* – amphiberingian, N to AK, YT and NT, E to AB and S to OR; E. Asia.

DESCRIPTION:

General: Annual or biennial (var. *oligosperma*) or perennial (var. *kamtschatica*) herb from a taproot; stems erect to ascending, one or more, freely branched, 10-50 cm tall, hairs short or lacking.

Leaves: Basal leaves forming a rosette; stem leaves similar, pinnate, sparsely stiff-hairy to hairy-fringed to almost glabrous, lateral lobes (2) 3-10, shortly-stalked to stalked, oblanceolate to egg-shaped, entire or shallowly 2 to several times round-toothed or lobed, 3-20 mm long, the terminal leaflet usually larger, to 3 cm long.

Flowers: Inflorescence an elongate racemes, more or less umbel-like, 3-10 cm long, generally lacking bracts; flower stalks erect to ascending, 5-20 mm long; petals white, 2-4 mm long; sepals 1-2 mm long.

Fruits: Siliques, erect, 1.5-2.5 cm long, 1-1.5 mm wide, glabrous to sparsely hairy; styles less than 0.5 mm long; seeds 15-22 (24), 1.5-2 mm long, oblong-oval, narrowly wing-margined.

Notes: Two rather weak varieties occur in BC.

1. Racemes more or less umbel-like, the central axis 1-2 cm long; plants of the upper montane to alpine zones .... var. *kamtschatica* (Regel) Detling

1. Racemes more elongate, the central axis greater than 3 cm long; plants of the lowland and lower montane zones .... var. *oligosperma*

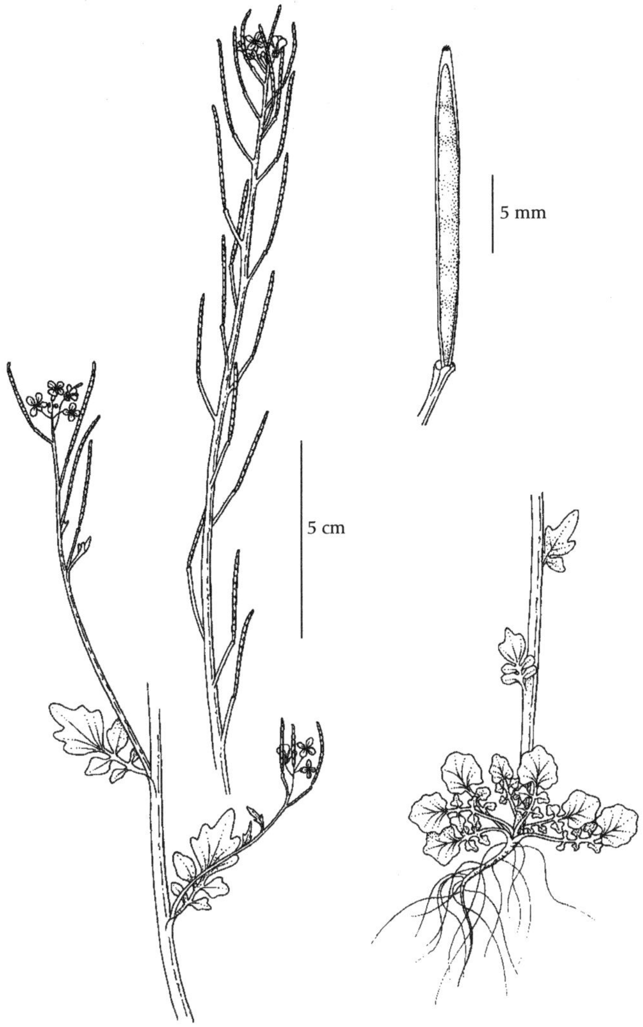

Cardamine occidentalis

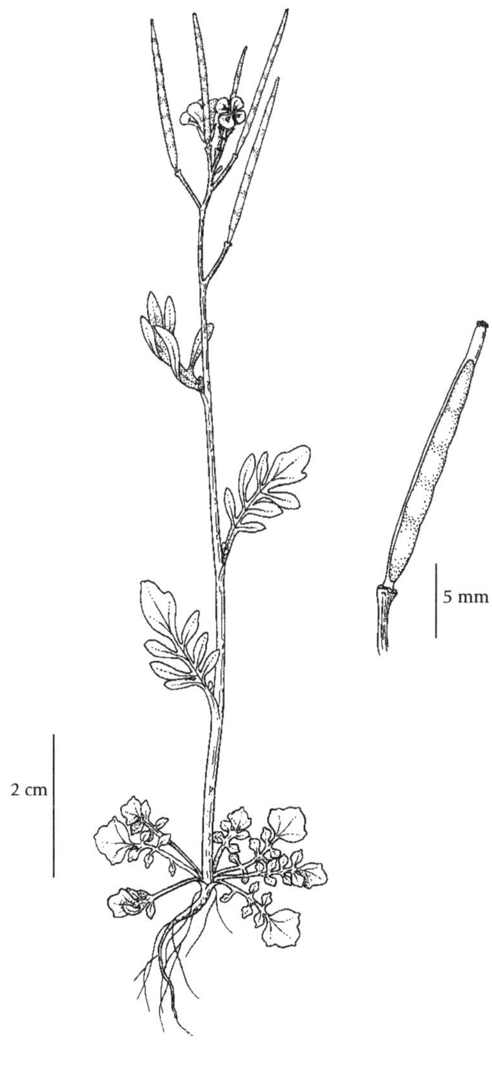

Cardamine oligosperma

Cardamine parviflora **L. var.** *arenicola* **(Britt.) O.E. Schulz**
small-flowered bitter-cress
Habitat/Range:
Dry sandy or rocky places in the lowland and montane zones; rare, scattered throughout BC; circumpolar, N to NT, E to NF, S to TX, FL and OR; Eurasia.
Description:
General: Annual or biennial herb from a taproot; stems usually solitary, 10-30 cm tall, simple or branched, glabrous.
Leaves: Basal leaves pinnate, terminal leaflet linear-oblong to oblanceolate; stem leaves with 3-6 pairs of leaflets, linear or narrowly oblanceolate, not over 2 mm wide, entire.
Flowers: Inflorescence an elongated raceme; flowers about 4 mm wide, petals 1.5-3 mm long.
Fruits: Siliques, ascending, 1-3 cm long; styles 0.5-1.5 mm long.

Cardamine pensylvanica **Muhl.** *ex* **Willd.**
Pennsylvanian bitter-cress
Habitat/Range:
Moist streamsides, ditches, open woods and waste places in the lowland and montane zones; frequent throughout BC; E to NF and S to TX, FL and N CA.
Description:
General: Biennial or annual herb from a taproot, also some fibrous roots from the lower nodes; stems erect to ascending, one or more, freely branched, 10-40 cm tall, usually stiff-hairy at the base.
Leaves: Basal and stem leaves pinnate, generally glabrous, lateral lobes (2) 4-10, shortly-stalked to stalked; basal leaves not rosetted, oblanceolate to egg-shaped, entire or shallowly lobed, 3-15 mm long, the terminal leaflet usually largest, up to 2 cm long; stem leaves linear or lanceolate to linear-oblanceolate.
Flowers: Inflorescence an elongated racemes, more or less umbel-like, 3-10 cm long, usually lacking bracts; flower stalks ascending, 5-10 mm long; petals white, 2-4 mm long; sepals 1.2-1.8 mm long.
Fruits: Siliques, erect, 1.5-2.5 cm long, 0.7-1 (1.5) mm wide, glabrous; styles 0.4-0.8 mm long; seeds (20) 24-40, about 1 mm long, usually wingless.

Cardamine pratensis **L. var.** *angustifolia* **Hook.**
cuckoo bitter-cress; cuckoo-flower; lady's-smock
Habitat/Range:
Wet to moist streamsides, meadows, floodplains and open woods in the montane zone; frequent in SW BC (lower Fraser Valley, where it may be introduced), infrequent in N BC; circumpolar, N to AK, YT and NT, E to NF, S to OH and MS; Eurasia.
Description:
General: Perennial herb from a rhizome; stems 8-45 cm tall, glabrous.
Leaves: Basal leaves with 9-17 orbicular to lanceolate leaflets, 0.2-0.9 cm long, glabrous; stem leaves with 9-15 linear to narrowly oblong leaflets, linear or lanceolate to linear-oblanceolate.
Flowers: Inflorescence a simple raceme, without bracts; flower stalks spreading-ascending, 5-20 mm long, glabrous; petals pink, pink-purple, or white, 8-15 mm long; sepals 3-4.5 mm long.
Fruits: Siliques, erect, 1.4-3 cm long, 1-1.5 mm wide, glabrous; styles 1-1.5 mm long.

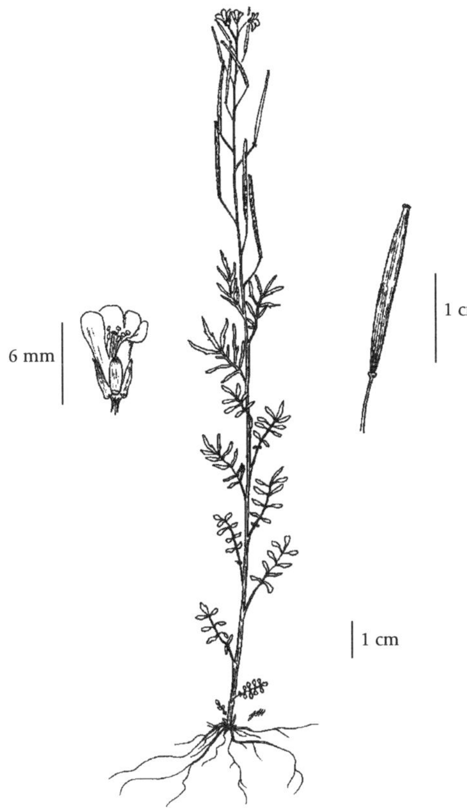

Cardamine parviflora

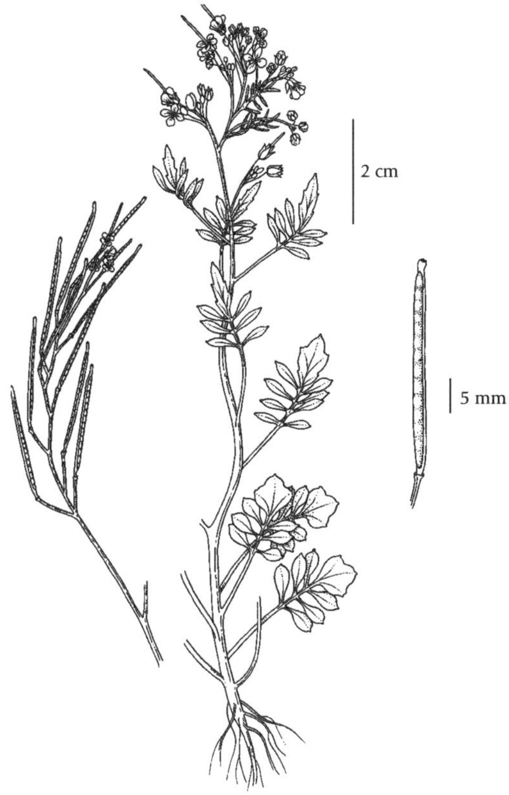

Cardamine pensylvanica

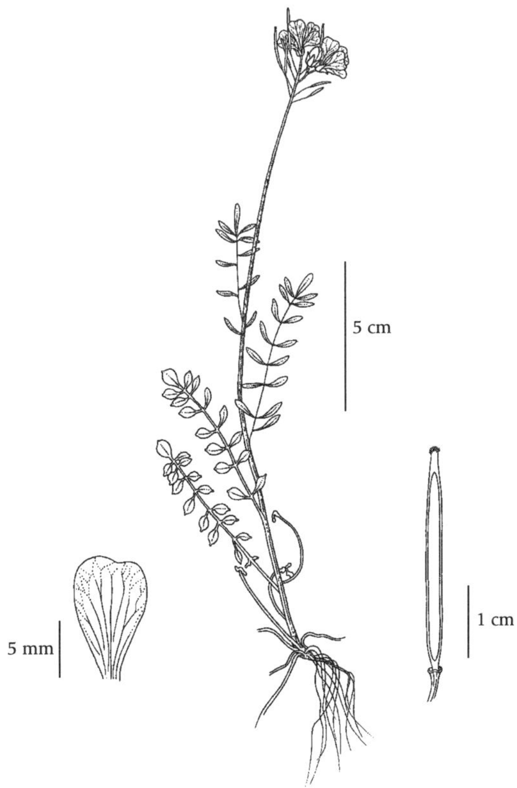

Cardamine pratensis* var. *angustifolia

CARDARIA

1. Silicles and sepals with short, simple hairs .... *C. pubescens*
1. Silicles and sepals glabrous .... *C. draba*

Cardaria draba **(L.) Desv.** (*Cardaria chalepensis* (L.) Handel-Mazzetti = ssp. *chalepensis*)
Chalapa hoary-cress (ssp. *chalepensis*), **heart-podded hoary-cress** (ssp. *draba*)

Habitat/Range:

Dry roadsides, fields and waste places in the lowland, steppe and montane zones; infrequent in S BC (ssp. *draba*) or rare in SC BC (ssp. *chalepensis*); introduced from Eurasia.

Description:

General: Perennial herb from a large, white rhizome; stems stout, somewhat white-hairy, 10-50 cm tall, branched above.

Leaves: Basal leaves simple, toothed to entire, widely oblanceolate to egg-shaped, short-hairy, short-stalked; stem leaves similar, unstalked, clasping.

Flowers: Racemes in a flat-topped group, bractless with many small flowers; petals white, 3-4 mm long; sepals early deciduous, glabrous, margins white.

Fruits: Silicles, glabrous, 2.5-6.0 mm long, 3-5 mm wide, round to widely egg-shaped or heart-shaped, inflated, somewhat compressed.

Notes: Two subspecies occur in BC.

1. Silicles heart-shaped .... ssp. *draba*
1. Silicles not heart-shaped .... ssp. *chalepensis* (L.) O.E. Schulz

Cardaria pubescens **(C.A. Mey.) Jarm.**
globe-pod hoary-cress

Habitat/Range:

Dry roadsides, fields and waste places in the steppe and montane zones; infrequent in SC BC, rare northward to NE BC (Dawson Creek); introduced from Eurasia.

Description:

General: Perennial herb from a large white rhizome; stems stout, somewhat white-hairy, 10-50 cm tall, branched above.

Leaves: Basal leaves simple, toothed to entire, short-hairy, oblong-linear to lanceolate, short-stalked; stem leaves similar, reduced, unstalked, clasping.

Flowers: Racemes corymbose, bractless with inconspicuous flowers; petals white, 2-3.5 mm; sepals early deciduous, hairy, margins white.

Fruits: Silicles, hairy, not heart-shaped at base, 3-4.5 mm wide, inflated, somewhat obcompressed.

CHORISPORA

Chorispora tenella **(Pall.) DC.**
blue mustard

Habitat/Range:

Dry roadsides and waste places in the steppe zone; frequent in SC BC; introduced from Eurasia.

Description:

General: Annual herb from a taproot; stems 10-50 cm tall, branched from near base, stalked-glandular and often sparsely stiff-hairy.

Leaves: Basal and lower stem leaves stalked, wavy-toothed to pinnately lobed, 3-8 cm long, elliptic-oblong to lanceolate or oblanceolate; upper stem leaves similar, reduced, unstalked, toothed to entire.

Flowers: Racemes elongate, the lower flowers from the axils of slightly reduced leaves, the raceme mostly bractless; flower stalks ascending, 2-4 mm long, stout; petals magenta, with narrow, stalklike bases that slightly exceed the sepals, the narrowly oblong blades spreading, 5 mm long; sepals erect, 6-8 mm long, free but forming a tube.

Fruits: Siliques, spreading to ascending, 35-45 mm long, often upcurved; beak 7-20 mm long.

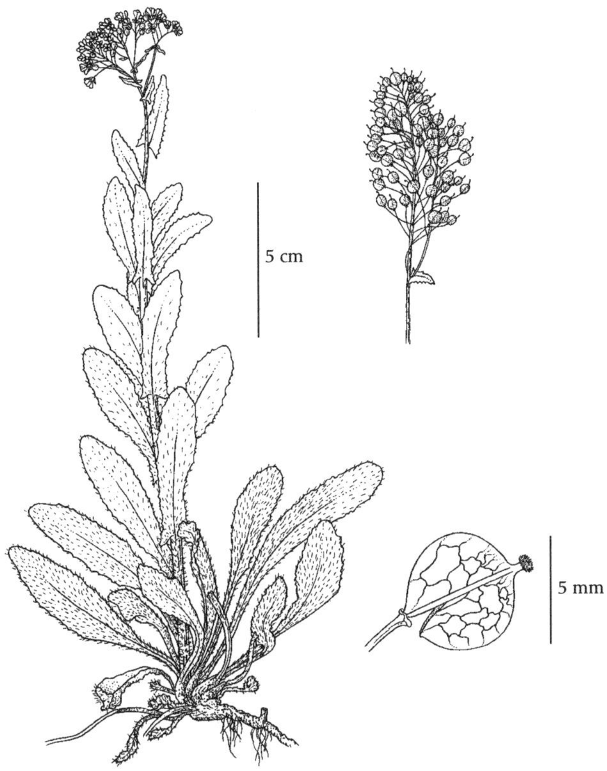

Cardaria draba ssp. *chalepensis*

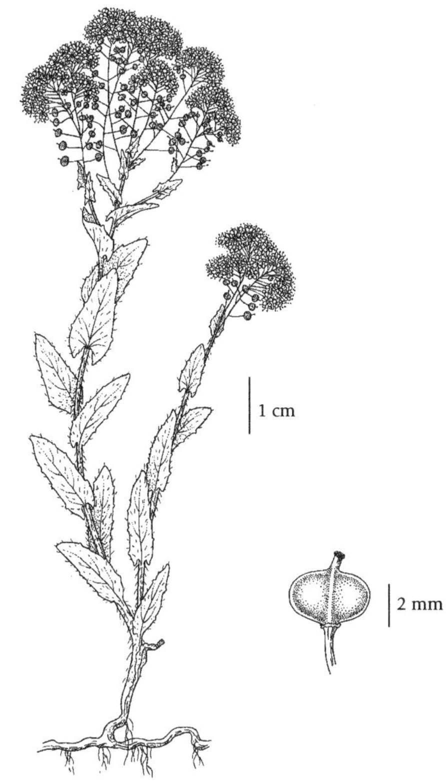

Cardaria draba ssp. *draba*

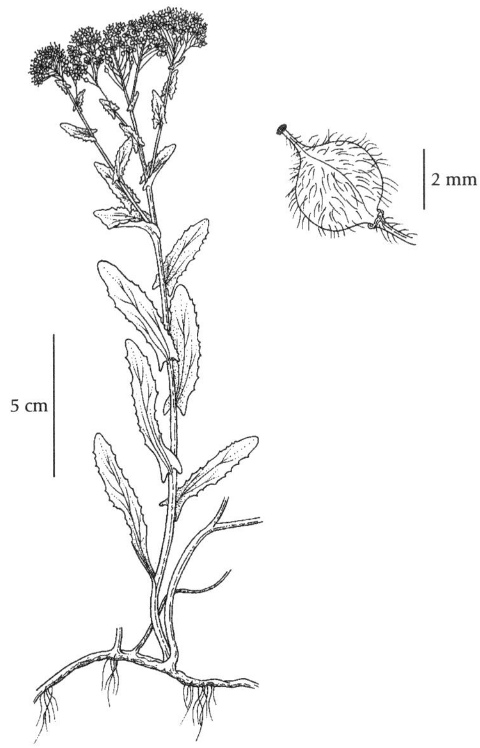

Cardaria pubescens

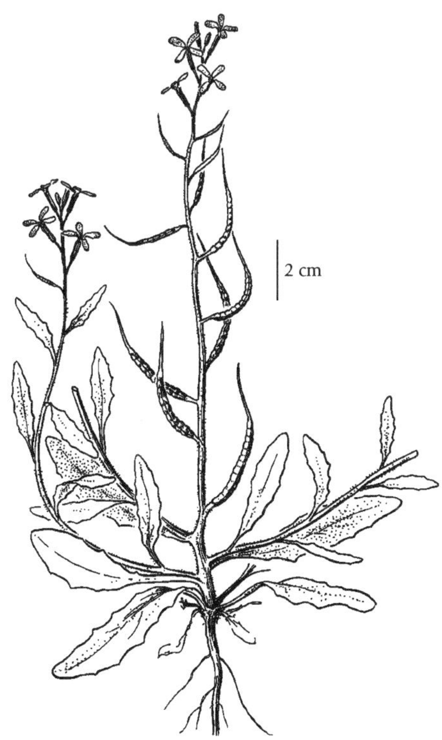

Chorispora tenella

COCHLEARIA

***Cochlearia groenlandica* L.** (*C. officinalis* L. ssp. *oblongifolia* [DC.] Hult.)

scurvy-grass; spoonwort

HABITAT/RANGE:

Moist sandy seashores, tidal marshes and mudflats in the lowland zone; frequent in coastal BC; circumboreal, N to AK, YT and NT and S to WA; Eurasia.

DESCRIPTION:

General: Biennial or perennial herb from a short, stout taproot; stems spreading to erect, (5) 10-30 (35) cm tall; glabrous.

Leaves: Basal leaves in a rosette, stalks slender and several times longer than the blade, the blades (5) 10-20 mm long, kidney-shaped to heart-shaped or egg-shaped, entire to wavy-margined; stem leaves usually with larger blades, unstalked or broadly short-stalked, generally toothed.

Flowers: Racemes elongate; flower stalks ascending, 5-15 mm long; petals white, 3-5 mm long; sepals pale yellow-green, about 2 mm long.

Fruits: Silicles, oval to elliptic, (3) 4-7 mm long; styles 0.2-0.5 mm long.

CONRINGIA

***Conringia orientalis* (L.) Andrz.**

hare's-ear mustard

HABITAT/RANGE:

Mesic to dry roadsides, fields and waste places in the steppe and montane zones; rare in SW and SC BC; introduced from Eurasia.

DESCRIPTION:

General: Annual herb from a taproot; stems 30-70 cm tall, mostly simple, glabrous and glaucous.

Leaves: Basal leaves 5-9 cm long, oblanceolate to egg-shaped, narrowed to base, more or less entire; stem leaves oblong-lanceolate, unstalked and heart-shaped, clasping at the bases.

Flowers: Racemes few-flowered, flat-topped; flower stalks ascending, 10-15 mm long; petals cream to lemon, with a slender stalklike base and a narrowly egg-shaped blade, 7-12 mm long; sepals sharp-pointed, erect, 6-8 mm long.

Fruits: Siliques, 8-13 cm long, 1.5-2 mm thick, slightly constricted between the seeds, tapered to a slender tip; thick styles about 1 mm long.

CORONOPUS

***Coronopus didymus* (L.) Sm.**

lesser swine-cress

HABITAT/RANGE:

Mesic fields and waste places in the lowland zone; infrequent in SW BC, known only from SE Vancouver Island and Vancouver; introduced from S America.

DESCRIPTION:

General: Annual herb from a taproot; stems many-branched, prostrate to decumbent, 20-50 cm long; glabrous to scurfy-hairy or lightly stiff-hairy.

Leaves: Basal leaves few; stem leaves many, egg-shaped-oblong, very deeply pinnately lobed, 1.5-3 cm long, the lobes narrow, entire to toothed or dissected, stalked below, becoming shortly-stalked above.

Flowers: In numerous axillary racemes; flower stalks slender, 1.5-2.5 mm long; petals white, linear, 0.5 mm long; sepals deciduous, about 0.5 mm long.

Fruits: Silicles, inflated, wrinkled, about 2 mm wide, 1.5 mm long.

Cochlearia groenlandica

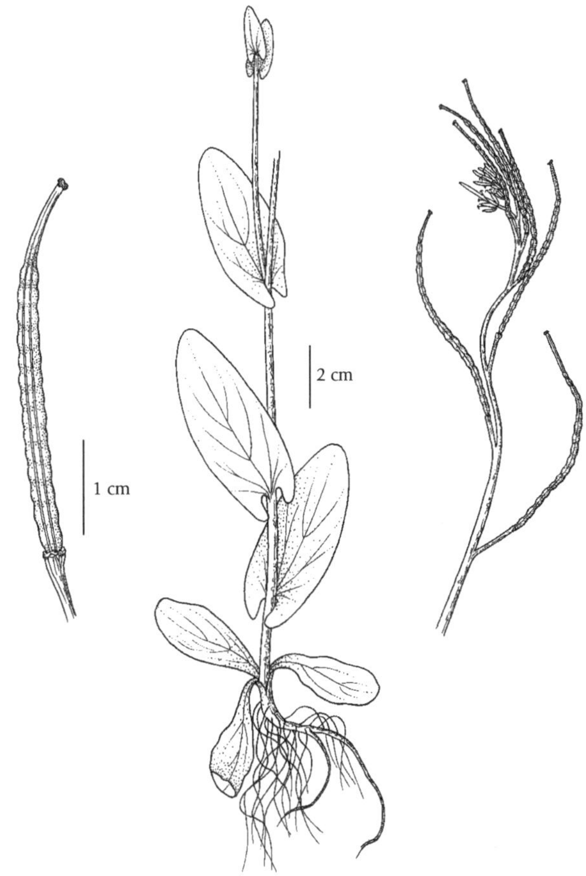

Conringia orientalis

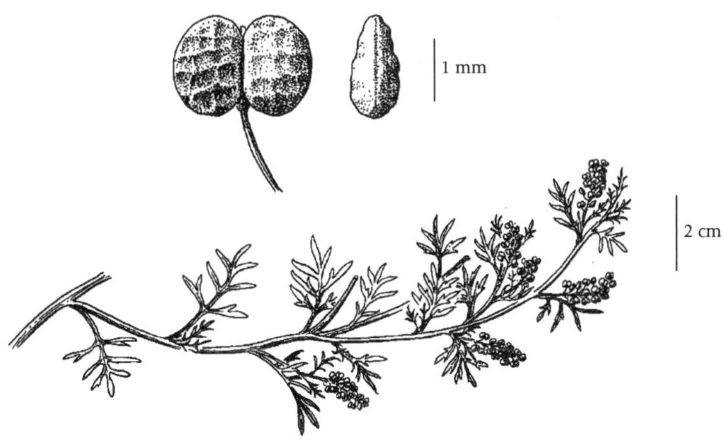

Coronopus didymus

DESCURAINIA

1. Siliques, 7-12 (rarely 15) mm long; stalks more than half the length of the siliques, seeds in 1 or 2 rows; leaves generally pinnate.
   2. Siliques somewhat club-shaped, rounded above, about as long as or shorter than the stalks; seeds in 2 rows, at least in part .... *D. pinnata*
   2. Siliques linear, pointed above, usually longer than the stalks; seeds in a single row .... *D. incana*
1. Siliques 11-33 mm long; stalks less than half the length of the siliques; seeds in 1 row, leaves, or at least some of them twice to thrice pinnate.
   3. Flower stalks 1-9 mm long; stems sparsely to densely stalked-glandular; racemes clustered at the top .... *D. sophioides*
   3. Flower stalks 7-14 mm long; stems not stalked-glandular; racemes elongated .... *D. sophia*

***Descurainia incana* (Bernh. *ex* Fisch. & C.A. Mey.) Dorn** (*D. richardsonii* O.E. Schulz in Engl. = ssp. *incana*; *D. richardsonii* var. *sonnei* [B.L. Robins.] C.L. Hitchc. = ssp. *incisa*; *D. richardsonii* var. *viscosa* [Rydb.] M.E. Peck = ssp. *viscosa*)

Richardson's tansy mustard

HABITAT/RANGE:

Mesic to dry roadsides, disturbed areas and waste places in the montane and steppe zones; rare to infrequent in S BC, rare (and possibly introduced) northward; N to AK, YT and NT, E to PQ and S to KS, MN, CO, NM, and CA.

DESCRIPTION:

General: Annual or biennial herb from taproot; stems 20-100 cm tall, usually branched, at least above, nearly glabrous and greenish or the hairs whitish-starlike, often stalked-glandular above.

Leaves: Basal and lower stem leaves once or twice pinnately lobed or 1-2 compound, widely lanceolate or oblanceolate to egg-shaped, 1.5-10 cm long; upper stem leaves simple to 1-pinnate, entire.

Flowers: Racemes densely flowered; flower stalks slender, spreading to strongly ascending, 4-8 mm long; petals pale to deep yellow, 1.5-3.5 mm long; sepals 1-2.5 mm long.

Fruits: Siliques, straight or somewhat curved, 5-17 mm long, 1-1.5 mm wide, usually longer than the stalks, strongly constricted between the seeds, tapered and sharp-pointed at each end but more at tip; styles 0.1-0.7 mm long; seeds in one row, usually 4-10 per chamber.

Notes: Three subspecies may be recognized in BC.

1. Stalks and siliques erect and closely appressed to the axis of the inflorescence; plants with greyish, starlike hairs, infrequent in SE BC, rare northwards .... ssp. *incana*
1. Stalks and siliques ascending to spreading; plants with or without glands, infrequent in SW and SC BC.
   2. Plants with stalked glands, infrequent in SW and SC BC .... ssp. *viscosa* (Rydb.) Kartesz & Gandhi
   2. Plants without glands, rare in SC BC .... ssp. *incisa* (Engelm.) Kartesz & Gandhi

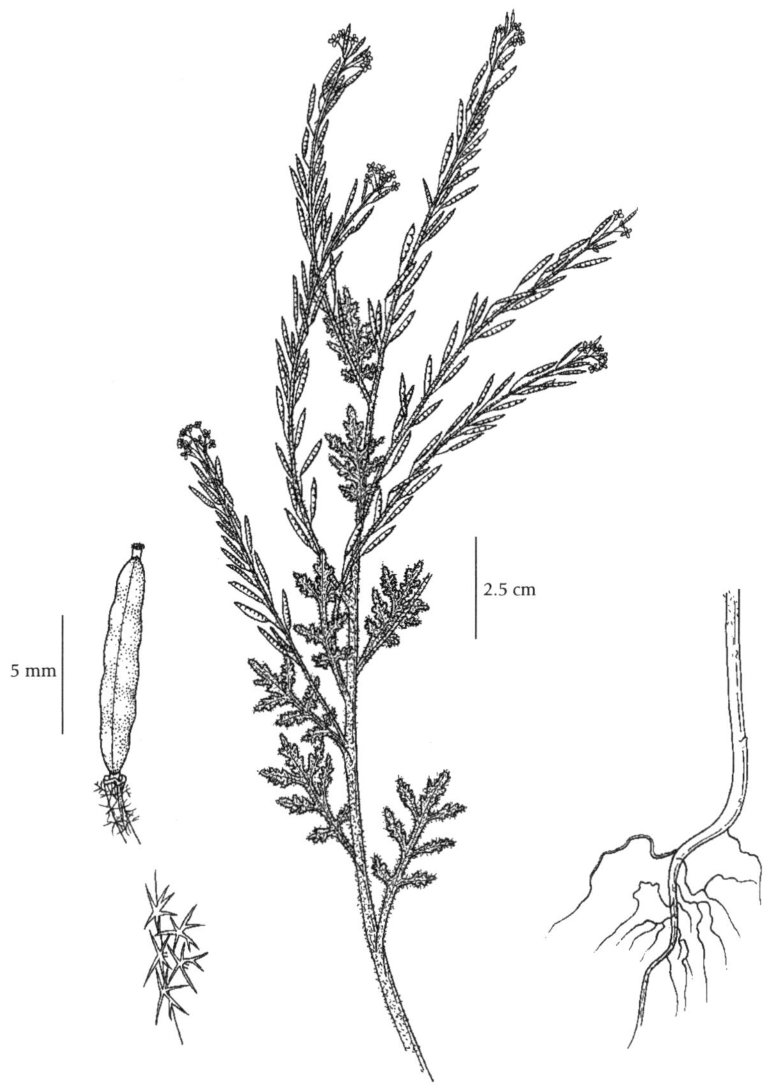

Descurainia incana

***Descurainia pinnata* (Walt.) Britt.** (*D. incisa* [A. Gray] Britt. ssp. *filipes* [A. Gray] Rollins, *D. pinnata* var. *filipes* [A. Gray] Peck = ssp. *filipes*)
western tansy mustard
HABITAT/RANGE:
Dry shrublands and waste places in the montane and steppe zones; common in S BC, S to E WA, CO, NV and E CA.
DESCRIPTION:
General: Annual herb from a taproot; stems 10-60 cm tall, simple or branched above, sparsely hairy, multibranched, often stalked-glandular above.

Leaves: Basal and lower stem leaves once or twice pinnately lobed, linear to egg-shaped, hairy, 3-10 cm long; upper stem leaves once pinnate to almost entire, generally glabrous, rarely glandular, 1-10 cm long.

Flowers: Racemes elongate; flower stalks 6-12 mm long, at about 45 degrees to axis of inflorescence; petals yellow, erect, 2-3 mm long, exceeding sepals by about 0.5 mm; sepals 1.5-2.5 mm long.

Fruits: Siliques, somewhat thicker towards the tips, rounded above, about as long as or shorter than the stalks, 7-12 (15) mm long; styles short, not beaklike; seeds 1-20 per chamber, partially in 2 rows, about 0.8 mm long.

Notes: Two highly variable, and often intergrading, varieties occur in our region.

1. Silicles 4-12 mm long, usually nearly equal to or longer than the stalks .... ssp. *intermedia* (Rydb.) Detl.
1. Silicles 10-20 mm long, usually shorter than the stalks .... ssp. *filipes* (A. Gray) Detl.

***Descurainia sophia* (L.) Webb *ex* Prantl**
flixweed
HABITAT/RANGE:
Mesic to dry fields and waste places in the lowland, steppe and montane zones; frequent throughout BC except Queen Charlotte Islands and adjacent coast; introduced from Eurasia.
DESCRIPTION:
General: Annual or biennial herb from a taproot; stems 30-100 cm tall, usually branched above, occasionally from the base; finely starlike-hairy, often greyish, sometimes with some simple hairs.

Leaves: Basal leaves few; stem leaves very deeply twice or thrice pinnately lobed to twice or thrice compound, oblanceolate to widely egg-shaped, short-stalked, 2-10 cm long, ultimate lobes of leaves generally linear.

Flowers: Racemes elongated; flower stalks slender, ascending, 7-14 mm long, less than half the length of the siliques; petals pale yellow, needle-like, spoon-shaped, about length of sepals; sepals 2-2.5 mm long, greenish or yellowish.

Fruits: Siliques, nearly erect, linear, 11-33 mm long, 1-1.5 mm wide, inconspicuously constricted between the seeds, rounded at tip; seeds (12) 15-25 per chamber in one row, 0.7-1.5 mm long, oblong-elliptic.

***Descurainia sophioides* (Fisch. *ex* Hook.) O.E. Schulz**
northern tansy mustard
HABITAT/RANGE:
Moist to mesic roadsides, disturbed areas and waste places in the montane zone; rare in N BC; amphiberingian, N to AK, YT and NT and E to N to MB; E. Asia.
DESCRIPTION:
General: Annual or sometimes biennial herb from a taproot; stems 10-150 cm tall; sparsely to densely hairy with starlike and stalked-glandular hairs.

Leaves: Basal leaves few; stem leaves deeply once to thrice compound below, becoming once or twice pinnately compound above, 2-12 cm long, segments oblong to oblanceolate, 1-2 mm wide.

Flowers: Racemes clustered at the top; flowers stalks spreading-ascending, often curved, 1-9 mm long, less than half the length of the siliques; petals yellowish, about length of sepals; sepals 1.5-2 mm long, yellowish.

Fruits: Siliques, spreading-ascending, 16-33 mm long, 0.7-1 mm wide; style to 0.2 mm long; seeds in one row.

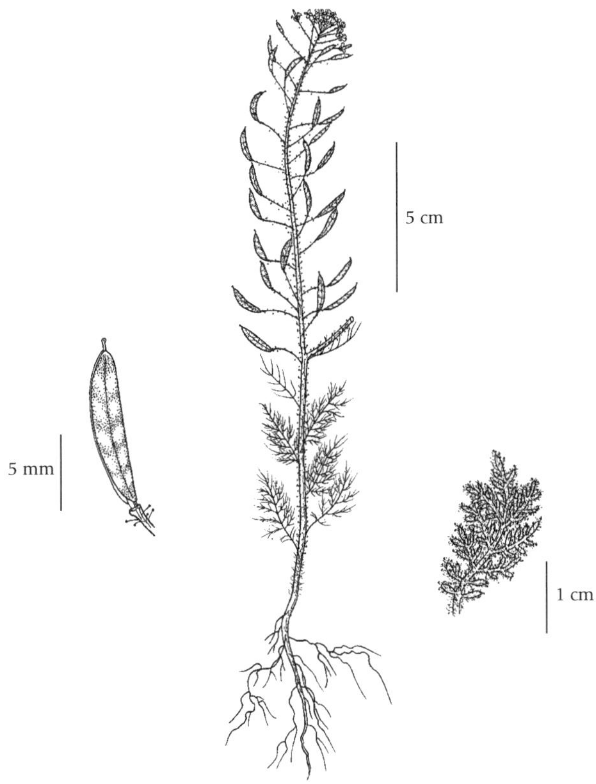

Descurainia pinnata **ssp. *intermedia***

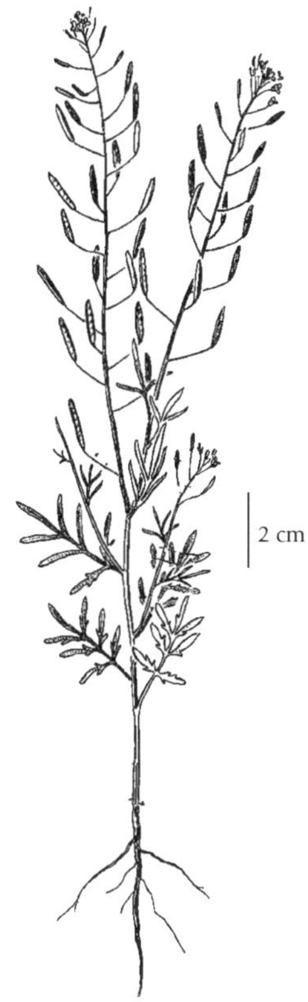

Descurainia pinnata **ssp. *filipes***

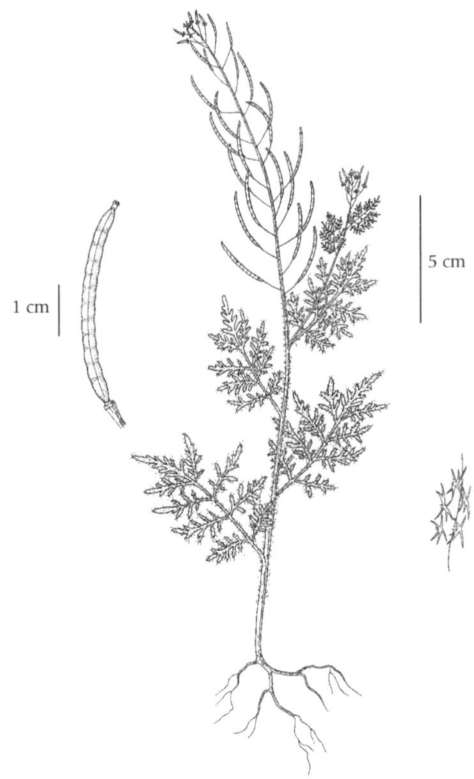

Descurainia sophia

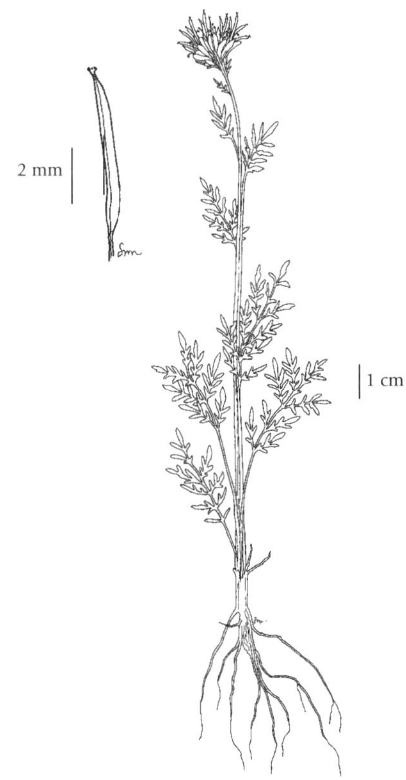

Descurainia sophioides

DRABA [5, 6]

1. Large succulent plants with leaves up to 30 cm long and silicles 18-22 mm long; seeds 1.5 mm long, black; seacoast plants from Vancouver Island and northward; often on highly nitrified substrate of bird-nesting sites .... *D. hyperborea*
1. Small non-succulent plants with leaves shorter than 15 cm and silicles less than 18 mm long; seeds less than 0.5 mm long, brown; range various.
    2. Styles less than 0.15 mm long.
        3. Annual or biennial plants of the lowland, steppe and montane zones.
            4. Petals yellow; stem leaves several on the lower stems .... *D. nemerosa*
            4. Petals white; stem leaves lacking or 1-2 near the base.
                5. Petals deeply bilobed; stem leaves absent; siliques elliptic to egg-shaped .... *D. verna*
                5. Petals slightly notched at the tips; stem leaves present; siliques linear .... *D. reptans*
        3. Perennial plants of the subalpine and alpine zones (sometimes montane in *D. albertina*).
            6. Lower leaf surfaces glabrous or with simple to many-branched hairs; crosslike to starlike hairs absent .... *D. crassifolia*
            6. Lower leaf surfaces mainly with crosslike to starlike hairs. ....
                7. Lower leaf surfaces mainly with starlike hairs .... *D. praealta*
                7. Lower leaf surfaces mainly with crosslike hairs.
                    8. Upper leaf surfaces mainly with 2-branched, 3-branched and crosslike hairs .... *D. stenoloba*
                    8. Upper leaf surfaces mainly with simple or 2-branched hairs; stems mainly with simple hairs .... *D. albertina*
    2. Styles more than 0.15 mm long (rarely absent in *D. fladnizensis*).
        9. Silicles mainly with short-stalked crosslike or starlike hairs.
            10. Stem leaves (3) 5-8 (12); usually some flowers or silicles in leaf axils; fruiting racemes usually occupying top half of stems; silicles frequently twisted; stalks usually appressed to axis of the inflorescence .... *D. cana*
            10. Stem leaves (0) 1-4 (5); inflorescence leafless; fruiting racemes occupying top third of stems; silicles not twisted; stalks spreading to semi-erect .... *D. cinerea*
        9. Silicles glabrous or with simple to branched hairs.
            11. Lower leaf surfaces with predominantly simple to many-branched hairs.
                12. Lower leaf surfaces mainly with 4- to many-branched hairs.
                    13. Branched hairs less than 0.25 mm wide, concentrated towards apex of leaves, mainly 7- to many-branched .... *D. lactea*
                    13. Branched hairs more than 0.25 mm wide, uniformly distributed on lower surfaces of leaves, mainly less than 7-branched .... *D. alpina*
                12. Lower leaf surfaces mainly with simple to 3-branched hairs.
                    14. Stems glabrous .... *D. fladnizensis*

---

[5]Key adapted from Mulligan (1976).

[6]Strong magnification may be required when examining hairs of *Draba*.

14. Stems hairy.

15. Petals narrower than sepals; stigmas headlike .... *D. stenopetala*

15. Petals wider than sepals; stigmas bilobed.

16. Styles over 0.5 mm long .... *D. densifolia*

16. Styles less than 0.5 mm long.

17. Lower leaf surfaces with some simple hairs; silicles usually glabrous .... *D. alpina*

17. Lower leaf surfaces lacking simple hairs; silicles hairy *D. corymbosa*

11. Lower leaf surfaces with predominently crosslike to starlike hairs.

18. Lower leaf surfaces mainly with crosslike hairs.

19. Stem leaves (2) 3-6 (10); fruiting stems usually more than l5 cm high .... *D. borealis*

19. Stem leaves 0-2 (3); fruiting stems usually less than l5 cm high.

20. Lowest stalks shorter than silicles; basal leaves compacted.

21. Petals white; stems leafless or with one leaf; upper leaf surfaces with some starlike hairs .... *D. porsildii*

21. Petals yellow, drying white; stems leafless; upper leaf surfaces lacking starlike hairs .... *D. macounii*

20. Lowest stalks usually about same to twice length of silicles; basal leaves loosely tufted .... *D. longipes*

18. Lower leaf surfaces mainly with starlike hairs.

22. Lower leaf surfaces mainly with long-stalked starlike hairs.

23. Upper leaf surfaces with long-stalked starlike hairs .... *D. ventosa*

23. Upper leaf surfaces with long, simple or once- or twice-branched hairs.

24. Stems and stalks with simple, rarely branched hairs; leaves greater than 2 mm wide .... *D. ruaxes*

24. Stems and stalks with starlike and branched, rarely simple, hairs; leaves less than 2 mm wide .... *D. paysonii*

22. Lower leaf surfaces with predominently short-stalked or unstalked hairs.

25. Starlike hairs on lower leaf surfaces mainly with 9 or more rays.

26. Starlike hairs on lower leaf surfaces unstalked .... *D. oligosperma*

26. Starlike hairs on lower leaf surfaces stalked.

27. Starlike hairs less than 0.2 mm in diameter.

28. Both leaf surfaces white-hairy with starlike hairs; stems and stalks white-hairy with starlike hairs; rays of starlike hairs mainly parallel to leaf surfaces .... *D. nivalis*

28. Starlike hairs mostly on underside of green leaves, especially towards apex; stems and stalks usually glabrous; rays of starlike hairs pointing in many directions .... *D. lactea*

27. Starlike hairs more than 0.2 mm in diameter.

29. Stem leaves 2-10, rarely fewer; stems over 10 cm high .... *D. glabella*

29. Stem leaves 0-1; stems less than 10 cm high (occasionally taller in *D. incerta*).

30. Starlike hairs mostly with one longer central stalk; lowest flower stalks usually twice length of silicle; petals yellow .... *D. incerta*

30. Starlike hairs without a longer central axis; lowest flower stalk less than twice length of silicle; petals white.

31. Petals over 4.0 mm long; styles over 0.5 mm long; silicles narrowly elliptic .... *D. palanderiana*

31. Petals 4.0 mm long or shorter; styles less than 0.5 mm long; silicles narrowly oblong .... *D. lonchocarpa*

25. Starlike hairs on lower surfaces of leaves mainly with 8 or fewer rays.

32. Stems with 0-2 leaves.

33. Stems with 1-2 leaves, usually more than 2 mm wide; silicles appressed to axis of inflorescence; stem and stalks frequently hairy. .... *D. lonchocarpa (var. vestita)*

33. Stems leafless or rarely with 1 leaf, less than 2 mm wide; silicles spreading; stem and stalks usually glabrous .... *D. porsildii*

32. Stems with 4 or more leaves.

34. Stems with more than 10 leaves; stalks appressed to stem .... *D. aurea*

34. Stems with 4-10 leaves; stalks spreading.

35. Lowest flower stalks about same length as or twice length of silicles; some crosslike and starlike hairs long-stalked .... *D. praealta*

35. Lowest flower stalks shorter than silicles; all of the crosslike and starlike hairs short-stalked.

36. Styles more than 0.5 mm long .... *D. borealis*

36. Styles less than 0.5 mm long .... *D. glabella*

Draba albertina Greene (*D. stenoloba* var. *nana* [O.E. Schulz] C.L. Hitchc.)
slender draba; slender whitlow-grass

HABITAT/RANGE:

Mesic to dry forest openings and meadows from the montane to alpine zones; infrequent in SC and SE BC, rare northward to NW BC (Atlin) and westward to central Vancouver Island; N to YT and NT, E to SK and S to CO and CA.

DESCRIPTION:

General: Biennial or short-lived perennial herb from a taproot and a simple stem-base; stems one to several, 4-30 cm long, often branched from the leaf axils, basal hairs coarse, mainly simple hairs, some 2-branched.

Leaves: Basal leaves oblanceolate to egg-shaped, entire to wavy-margined and minutely toothed, 0.3-3.5 cm long, 1.5-7 mm wide; stem leaves lacking or 1-3 (4), generally lanceolate (narrowly elliptic to widely lanceolate), 0.4-3 cm long, 2-18 mm wide; lower surface hairs short-stalked, 2-branched, 3-branched and crosslike; upper surface hairs none or simple to 2-branched.

Flowers: Racemes less than 35-flowered, open, generally glabrous; petals yellow, 2.5-3 mm long; sepals 1.5-2 mm long, glabrous or with some simple hairs.

Fruits: Silicles, 5-15 mm long, almost linear to narrowly egg-shaped, 1-2 mm wide, glabrous, rarely with short hairs; styles less than 0.15 mm long; seeds up to 32, oval, light brown, about 1 mm long.

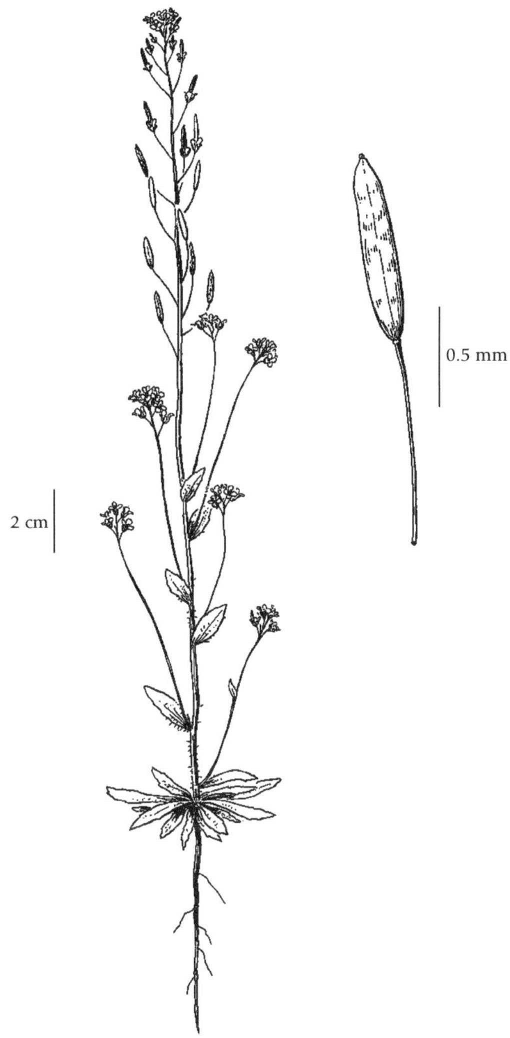

Draba albertina

***Draba alpina* L.** (*D. alpina* var. *pilosa* [Adams] Regel, *D. pilosa* Adams *ex* DC.)
alpine draba; alpine rock-cress

HABITAT/RANGE:

Moist to mesic meadows, rocky slopes, and snowbeds in the subalpine and alpine zones; rare in N BC; N to AK, YT and NT, and E to PQ.

DESCRIPTION:

General: Tufted perennial herb from a taproot and a branched stem-base with many persistent leaves; stems branched, 2.5-13 cm long, sometimes glabrous, more often with simple, branched, and starlike hairs.

Leaves: Basal leaves (or 1 stem leaf), oblanceolate to oblong or elliptic, 0.5-2.5 cm long, entire, lower surface with simple to many-branched hairs, mainly less than 7-branched, uniformly distributed, hairs more than 0.25 mm wide; midribs often conspicuous in age and persistent; stem leaves 1 or lacking.

Flowers: Racemes 2- to 10-flowered; flower stalks 2-10 mm long, hairy (like the stems) or glabrous; petals yellow, notched at the tips, 4-5 mm long, wider than sepals; sepals 2-3.5 mm long, sparsely soft-hairy with simple or, rarely branched, hairs.

Fruits: Silicles, 5-11 mm long, 2-3 (4) mm wide, lance egg-shaped to elliptic, plane, hairy with simple to branched hairs or glabrous; styles 0.3-0.7 mm long; seeds 2-10 (or more).

***Draba aurea* Vahl** (*D. aurea* var. *leiocarpa* [Pays. & St. John] C.L. Hitchc.)
golden draba; golden whitlow-grass

HABITAT/RANGE:

Mesic to dry forest openings, meadows, rocky slopes and talus slopes from the montane to alpine zones; frequent throughout BC east of the Coast-Cascade Mountains; N to AK, YT and NT, E to PQ and S to N WA, ID, MT, NM and AZ.

DESCRIPTION:

General: Perennial herb from a taproot; stems simple or branched, erect or decumbent at base, 10-50 cm tall, hairs long and simple or often branched.

Leaves: Basal leaves oblanceolate, 1-4 cm long, 2-13 mm wide, entire or denticulate; stem leaves mostly 10-20, egg-shaped to lanceolate or oblanceolate, 0.5-3 cm long, 0.3-1.2 cm wide, entire to toothed; densely hairy, lower leaf surfaces with predominently short-stalked or unstalked hairs, the hairs starlike mainly with 8 or fewer rays.

Flowers: Racemes several- to many-flowered, occasionally with solitary flowers or lateral racemes in the axils of upper leaves; flower stalks 3-20 mm long, appressed to axis of the inflorescence, hairy; petals pale to deep yellow, rounded, 4.5-6 mm long; sepals 2-3.5 mm long, soft-hairy.

Fruits: Silicles, 7-20 mm long, 2-4 mm wide, lance-elliptic, plane or twisted, soft-hairy or glabrous; styles 0.3-1.5 mm long; seeds 20-50, about 1 mm long.

***Draba borealis* DC.** (*D. maxima* Hult., *D. mccallae* Rydb.)
northern draba; northern whitlow-grass

HABITAT/RANGE:

Moist to mesic meadows and gravelly floodplains in the subalpine and alpine zones; frequent in N BC; amphiberingian, N to AK, YT and NT and E to SW AB; E. Asia.

DESCRIPTION:

General: Loosely tufted perennial herb from a taproot and a simple or branched stem-base; stems branched, erect or decumbent at base, 7-55 cm tall, hairy with some long, simple or occasionally branched hairs.

Leaves: Basal leaves oblanceolate, 0.6-5 cm long, 3-18 mm wide, toothed or entire; stem leaves (2) 3-6 (10), egg-shaped to lanceolate, elliptic or oblong, 0.4-4 cm long, 0.2-1.8 cm wide, toothed or entire, hairy with 4- to 6-rayed, starlike hairs and often straight, simple or branched hairs.

Flowers: Racemes several- to many-flowered, occasionally with solitary flowers in axils of upper leaves; flower stalks 3-16 mm long, ascending, straight, soft-hairy; petals white to cream, 4-6 mm long; sepals 2-3 mm long, soft-hairy.

Fruits: Silicles, 7-14 mm long, 2-4 mm wide, lanceolate, twisted or plane; hairy with simple or branched hairs; styles 0.3-1 mm long.

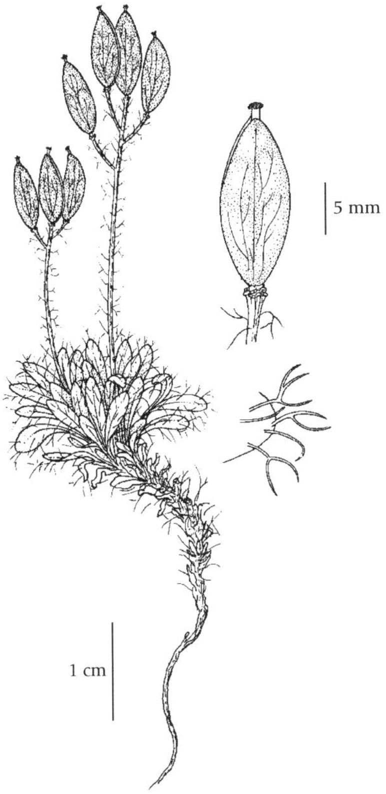

Draba alpina

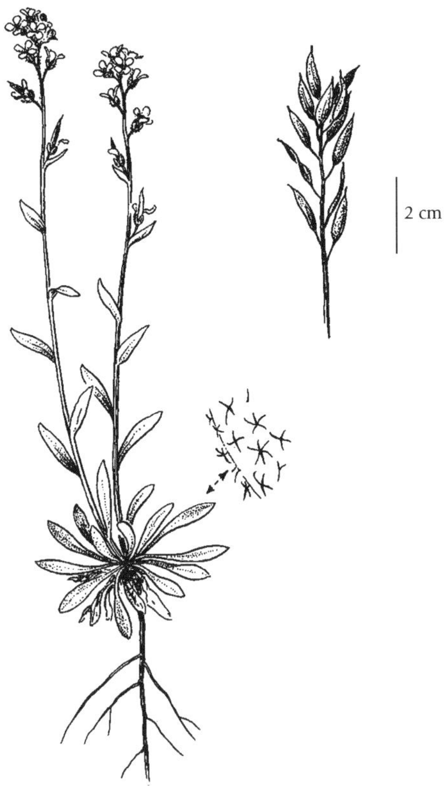

Draba aurea

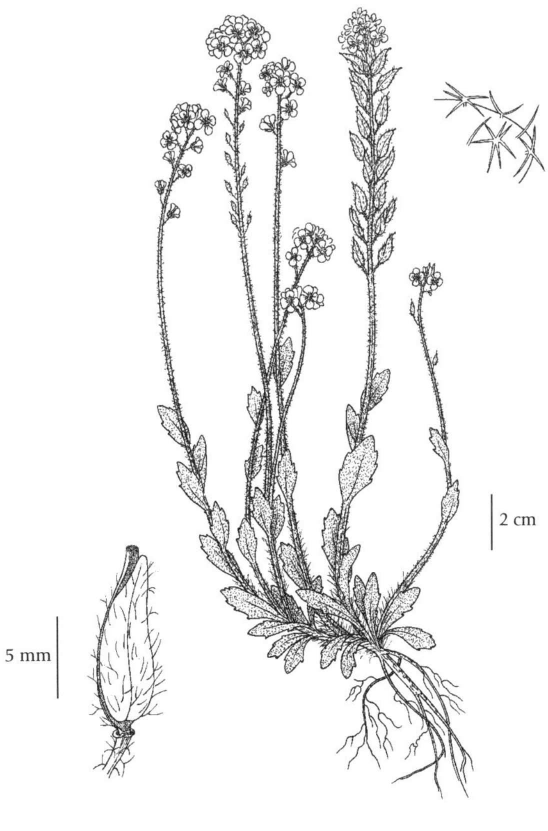

Draba borealis

Draba cana **Rydb.** (*D. lanceolata auct. non* Royle)

lance-leaved draba; hoary draba; lance-leaved whitlow-grass

HABITAT/RANGE:

Mesic to dry meadows, cliffs, rocky slopes and river terraces in the subalpine and alpine zones; frequent in N BC, less frequent southward E of the Coast-Cascade Mountains; circumpolar, N to AK, YT and NT, E to PQ and S to CO, VT and NV; Eurasia, Greenland.

DESCRIPTION:

General: Loosely tufted perennial herb from a taproot and a simple or branched stem-base; stems branched, erect or decumbent at base, 5-25 cm tall, densely hairy with simple and starlike hairs.

Leaves: Basal leaves oblanceolate, 1-4 cm long, 1-4 mm wide, usually toothed; stem leaves (3) 5-8 (12), lanceolate, commonly toothed; densely hairy with simple or branched hairs, generally starlike with branches again branched.

Flowers: Racemes several- to many-flowered, less than 60, usually some in leaf axils; flower stalks 2-9 mm long, appressed, erect, hairy, lower ones often bracted and less than 3 mm; petals white, 3-5 mm long; sepals 1.5-2 mm long, sparsely soft-hairy.

Fruits: Silicles, 4-12 mm long, 1.5-2.5 mm wide, lanceolate to oblong, commonly twisted; usually densely soft-hairy, the hairs less than 3-branched and also starlike; styles 0.2-0.8 mm long; fruiting racemes usually occupying the top half of the stems.

Draba cinerea **J.E. Adams**

gray-leaved draba; gray-leaved whitlow-grass

HABITAT/RANGE:

Mesic to dry meadows and cliffs in the alpine zone; rare in N BC; circumpolar, N to AK, YT and NT and E to N PQ; Eurasia.

DESCRIPTION:

General: Tufted perennial herb from a taproot and a simple or branched stem-base; stems branched, erect, 6-20 cm tall, starlike-hairy, mixed with simple or branched hairs.

Leaves: Basal leaves oblanceolate, 0.5-3 cm long, 2-6 mm wide, entire or toothed; stem leaves lacking or 1-4 (5), lanceolate to egg-shaped, entire or toothed, 0.3-1.5 cm long, 0.1-0.7 cm wide; hairy with overlapping starlike hairs or nearly glabrous.

Flowers: Racemes several- to many-flowered, sometimes with solitary flowers in upper leaf axils; stalks 3-8 mm long (or more), spreading to semi-erect, hairy; petals white to cream, notched at the tips, 3-4 mm long; sepals 1.8-2.2 mm long, hairy.

Fruits: Silicles, 6-8 mm long, 2-2.5 mm wide, lanceolate to oblong, narrowed equally at base and apex, flat; often densely hairy, mainly with short-stalked crosslike or starlike hairs; fruiting racemes occupying the top third of the stems; styles 0.3-0.8 mm long.

Draba corymbosa **R. Br. *ex* DC.** (*D. bellii* Holm, *D. macrocarpa* J.E. Adams)

Baffin Bay draba; Baffin Bay whitlow-grass

HABITAT/RANGE:

Mesic to dry meadows in the alpine zone; rare in NW BC; N to AK, YT and NT and E to PQ.

DESCRIPTION:

General: Densely tufted perennial from a taproot, forming compact tussocks; stems 1-5 cm tall, hairy with simple and branched hairs.

Leaves: All basal leaves withering but persistent, in a densely crowded rosette, long-hairy on upper surface with simple and branched hairs, hairy with branched, crosslike and starlike hairs on lower surface; stem leaves lacking.

Flowers: Racemes 2- to 8-flowered; petals pale yellow.

Fruits: Silicles, elliptic-oblong, rarely twice as long as wide, hairy with short, simple and branched hairs; styles 0.3-0.7 mm long.

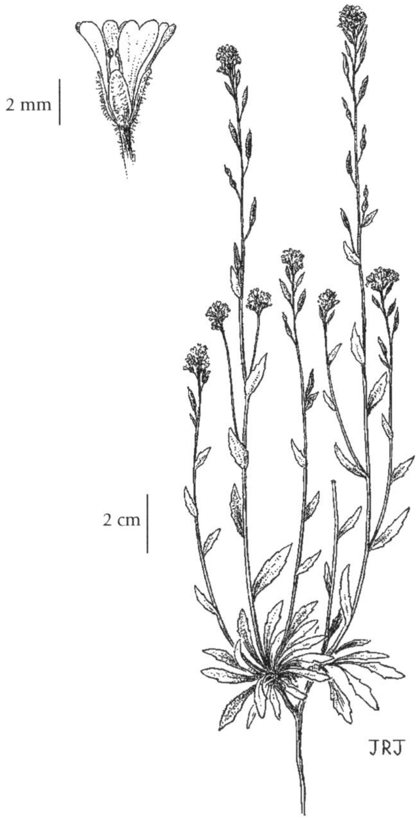

Draba cana

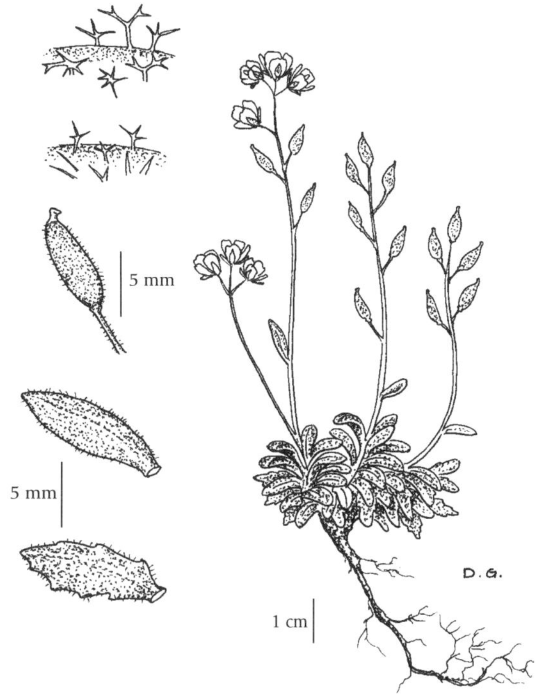

Draba cinerea

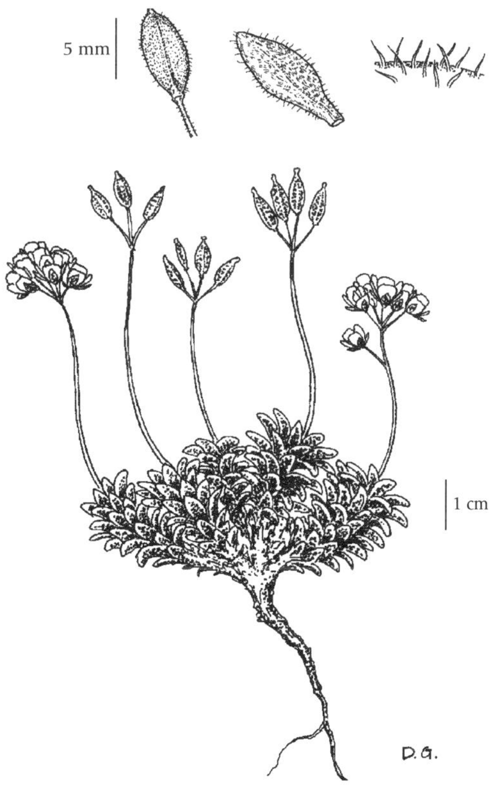

Draba corymbosa

***Draba crassifolia* Grah.**

Rocky Mountain draba; Rocky Mountain whitlow-grass

HABITAT/RANGE:

Mesic to dry meadows and cliffs in the subalpine and alpine zones; infrequent throughout BC E of the Coast-Cascade Mountains; N to AK, YT, and NT, E to PQ and S to CO and AZ.

DESCRIPTION:

General: Biennial or perennial herb from a taproot and a simple or branched stem-base; stems branched, erect, 2-12 cm tall, usually glabrous.

Leaves: all basal leaves egg-shaped to oblanceolate, entire to toothed, 0.2-2.5 cm long, 1-6 mm wide, usually glabrous or with a few simple and 2-branched hairs, margins glabrous or hairy-fringed with simple to 2-branched hairs; stem leaves lacking, rarely 1.

Flowers: Racemes 2- to 15-flowered, open; petals yellow, 1.5-2 mm long; sepals 1-1.5 mm long, glabrous.

Fruits: Silicles, 4-10 mm long, 1.5-2.5 mm wide, narrowly elliptic to narrowly egg-shaped, plane, glabrous; fruiting stalks 2-6 mm long, generally shorter than silicles, glabrous;styles lacking or up to 0.15 mm long; seeds 10-24, oval, light brown, about 1 mm long.

***Draba densifolia* Nutt. in T.& G.** (*D. stenopetala auct. non* Trautv.)

Nuttall's draba; Nuttall's whitlow-grass

HABITAT/RANGE:

Mesic to dry forest openings, meadows, rocky slopes and cliffs from the upper montane to alpine zones; rare in SC BC; E to SW AB, S to WY, VT and CA, disjunct in AK.

DESCRIPTION:

General: Tufted perennial herb from a taproot and a usually branched stem-base; stems erect, 0.5-15 cm tall, glabrous to hairy with simple and branched hairs.

Leaves: All basal, linear to linear-oblanceolate, 0.2-1.2 (1.9) cm long, 0.5-3 mm wide, upper surface usually glabrous, lower leaf surfaces mainly with simple to 3-branched hairs, margins hairy-fringed with stiff simple hairs, 0.5-1 mm long; stem leaves lacking.

Flowers: Racemes 3- to 15-flowered; flower stalks 2-5 mm long, hairy; petals yellow, 2-6 mm long; sepals 2-2.5 mm long, soft-hairy.

Fruits: Silicles, 2-7 mm long, 2-3.5 mm wide, egg-shaped to elliptic, plane; hairy or glabrous; styles 0.5-1 mm long; seeds 2-12, about 2 mm long.

***Draba fladnizensis* Wulf.**

Austrian draba; Austrian whitlow-grass

HABITAT/RANGE:

Mesic to dry meadows, cliffs and talus slopes in the subalpine and alpine zones, rare in N BC; circumpolar, N to AK, YT and NT, E to AB and S to CO and UT; Eurasia.

DESCRIPTION:

General: Tufted perennial herb from a taproot and a simple or branched stem-base; stems erect, 1.5-12 cm tall, glabrous.

Leaves: Usually all basal, oblanceolate, 0.3-1 cm long, 1-2 mm wide, somewhat hairy with simple and some branched hairs; margins strongly hairy-fringed with long, simple hairs; stem leaves lacking or sometimes 1.

Flowers: Racemes few-flowered; flower stalks 2-10 mm long, glabrous; petals white, 1.5-3 mm long; sepals 1.0-1.5 mm long, sparsely long-hairy.

Fruits: Silicles, 3-6 mm long, 1.2-2 mm wide, oblong-egg-shaped, glabrous or short-hairy; styles up to 0.2-0.5 mm long, rarely lacking; seeds 10-20, 0.7-0.9 mm long.

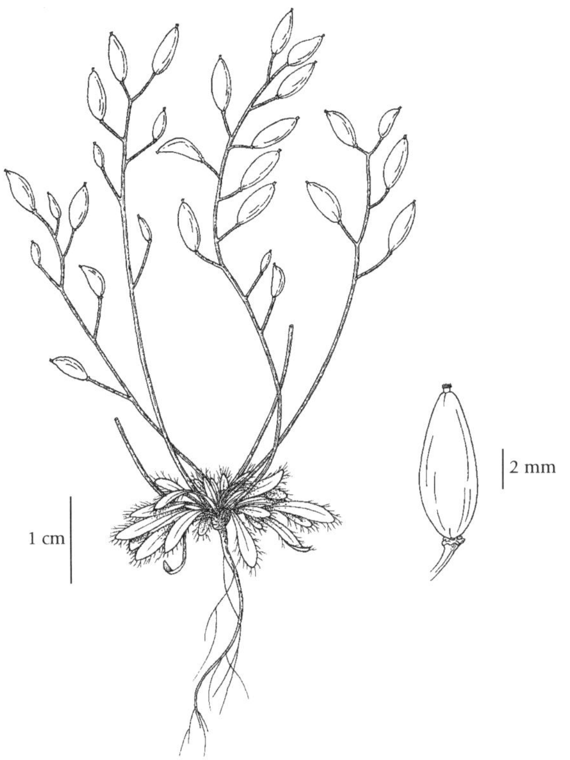

Draba crassifolia

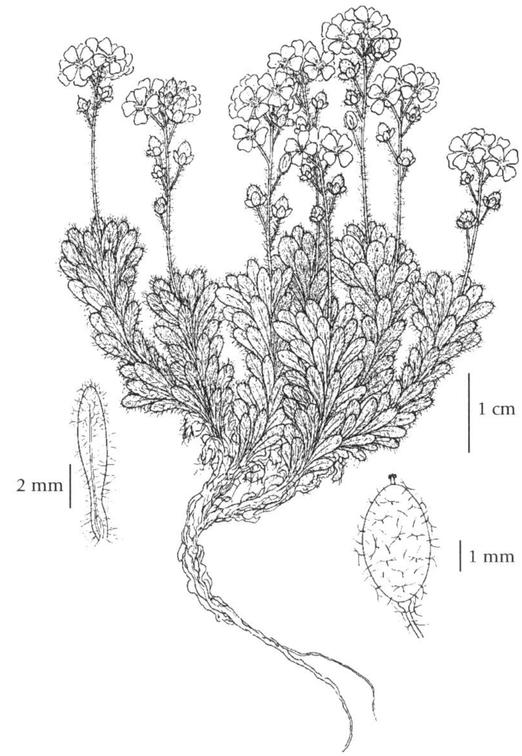

Draba densifolia

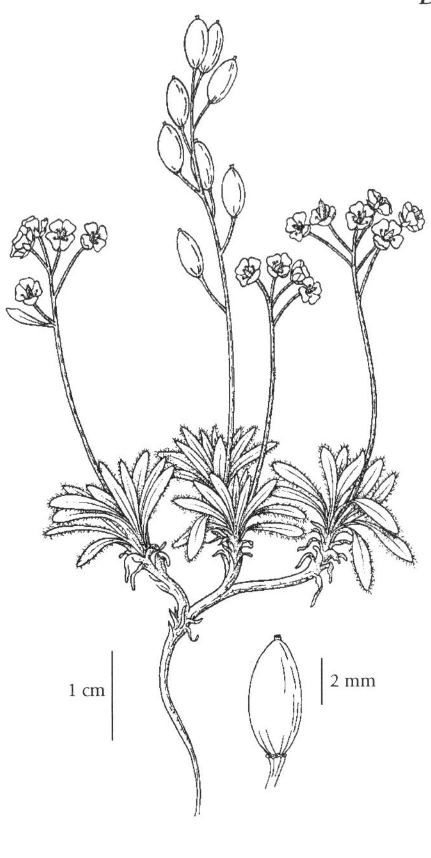

Draba fladnizensis

Draba glabella **Pursh var. *glabella*** (*D. hirta auct. non* L.)

smooth draba; smooth whitlow-grass

HABITAT/RANGE:

Moist to mesic meadows and cliffs in the subalpine and alpine zones; rare in N BC; circumpolar, N to AK, YT and NT and E to NF.

DESCRIPTION:

General: Loosely tufted perennial herb from a taproot and a simple or branched stem-base; stems erect, 10-40 cm tall, starlike-hairy, rarely with a few simple hairs below.

Leaves: Basal leaves oblanceolate, 0.5-4 cm long, 2-10 mm wide, entire or toothed; stem leaves 2-10, egg-shaped to lanceolate, 0.8-3 cm long, 3-12 mm wide, toothed or entire; hairy with stalked starlike hairs, 4- to 5-rayed with the longer 2 rays usually paralleling the leaf axis and again branched; margins hairy-fringed with mostly simple or branched hairs, generally starlike with branches again branched.

Flowers: Racemes several- to many-flowered, occasionally with solitary flowers in upper leaf axils; flower stalks 2-15 mm long, ascending, straight, sparsely soft-hairy to glabrous; petals white, occasionally cream to yellow, notched at the tips, 3.5-5 mm long; sepals 2-3 mm long, sparsely soft-hairy.

Fruits: Silicles, 6-15 mm long, 1.5-3 mm wide, lanceolate to oblong, plane or rarely twisted, glabrous or short-hairy; styles less than 0.25 mm long.

Draba hyperborea **(L.) Desv.**

North Pacific draba; North Pacific whitlow-grass

HABITAT/RANGE:

Moist to mesic, often nitrified, rocky coastal cliffs, bluffs and beaches in the lowland zone; locally frequent from N Vancouver Island northward in coastal BC; amphiberingian, N to AK; E Asia.

DESCRIPTION:

General: Tufted perennial herb from a thick taproot and a stem-base with persistent peglike leaf bases, sometimes elongated and rhizomatous; stems erect, 2-8 cm tall, glabrous.

Leaves: Basal leaves widely wedge-shaped to oblong spoon-shaped, usually toothed, fleshy, 1-30 cm long, 3-40 mm wide, stalks widely winged; stem leaves 3-10, oblanceolate to elliptic or lanceolate, toothed or entire, fleshy, 0.9-9.5 cm long, 3-20 mm wide, hairy with simple hairs or sometimes with branched or starlike hairs.

Flowers: Racemes few- to many-flowered; flower stalks 3-25 mm long, ascending, hairy to glabrous; petals yellow, 4-6 mm long; sepals 3-4 mm long.

Fruits: Silicles, 6-22 mm long, 4-7 mm wide, oval to lanceolate, glabrous, plane, often inflated; styles 0.5-1.5 mm long; seeds 1.5 mm long, black.

Draba incerta **Pays.**

Yellowstone draba; Yellowstone whitlow-grass

HABITAT/RANGE:

Dry meadows, cliffs, fellfields and scree slopes in the upper montane, subalpine, and alpine zones; common throughout BC E of the Coast-Cascade Mountains; N to AK and YT, E to AB and S to WA, ID, MT and WY, disjunct in PQ.

DESCRIPTION:

General: Tufted perennial herb from a taproot and a usually much branched stem-base; stems erect, 2-20 cm tall, starlike-hairy.

Leaves: All basal, narrowly oblanceolate, 0.7-1.5 cm long, 1-4 mm wide, hairy with stalked, 4-rayed, starlike hairs, the hairs again branched; margins hairy-fringed with branched or starlike hairs; midribs somewhat persistent as threadlike processes; stem leaves absent except on new shoots.

Flowers: Racemes 3- to 15-flowered; petals yellow, 4-5.5 mm long; sepals 2.5-3.5 mm long, sparsely soft-hairy.

Fruits: Silicles, 6-10 mm long, 2.5-3.5 mm wide, lanceolate to egg-shaped, plane, hairy to nearly glabrous or glabrous; lowest fruiting stalks 4-18 mm long, ususally twice the length of the silicles, ascending, starlike-hairy; styles 0.4-1 mm long; seeds 8-14, 1.2-1.5 mm long.

Draba glabella var. *glabella*

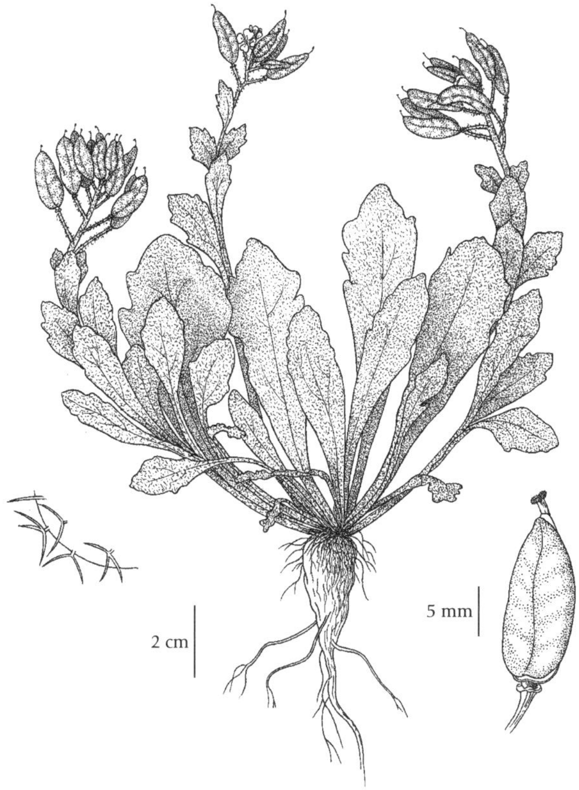

Draba hyperborea

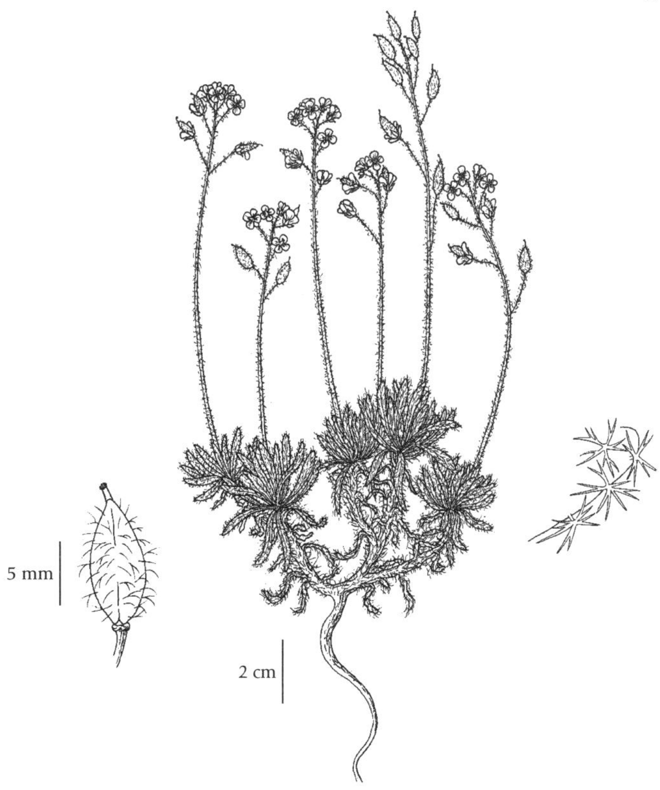

Draba incerta

Draba lactea **Adams** (*D. allenii* Fern.)
milky draba; milky whitlow-grass
HABITAT/RANGE:
Mesic to dry meadows and cliffs in the subalpine and alpine zones; rare in N BC; circumboreal, N to AK, YT, and NT; Eurasia.
DESCRIPTION:
General: Tufted perennial herb from a taproot and a usually branched stem-base; stems erect, 5-10 cm tall, glabrous.

Leaves: Usually all basal, linear-oblong, 0.5-2 cm long, 1-5 mm wide, hairy with at least some branched or starlike hairs concentrated toward apex; lower leaf surfaces mainly with 7- to many-branched hairs less than 0.25 mm wide or stalked, starlike hairs less than 0.2 mm in diameter with 9 or more rays; margins hairy-fringed with simple hairs and some branched or starlike hairs near apex, midribs persistent; stem leaves lacking or 1.

Flowers: Racemes 3- to 7-flowered; flower stalks 1-8 mm long, glabrous or sometimes starlike short-hairy; petals creamy white, 2.5-4 mm long; sepals 1.5-2.2 mm long, sparsely soft-hairy.

Fruits: Silicles, 6-8 mm long, 2-3 mm wide, lanceolate to egg-shaped, plane, glabrous; styles to 0.5 mm long.

Draba lonchocarpa **Rydb.** (*D. lonchocarpa* var. *exigua* O.E. Schulz, *D. nivalis* Lilj. var. *elongata* S. Wats. = var. *lonchocarpa; D. lonchocarpa* var. *denudata* O.E. Schulz, *D. lonchocarpa* ssp. *kamtschatica* [Ledeb.] Calder & Tayl., *D. nivalis* var. *denudata* [O.E. Schulz] C.L. Hitchc. = var. *vestita; D. nivalis* var. *thompsonii* C.L. Hitchc. = var. *thompsonii*)
lance-fruited draba; spear-fruited draba; lance-fruited whitlow-grass
HABITAT/RANGE:
Mesic to dry meadows, cliffs and scree slopes in the subalpine and alpine zones; var. *lonchocarpa* – common throughout BC E of the Coast-Cascade Mountains, var. *thompsonii* – rare in and E of the Coast-Cascade Mountains south of 57° N, var. *vestita* – rare in coastal and west-central BC; var. *lonchocarpa* – N to AK, YT and NT, E to AB, and S to OR, MT and WY, var. *thompsonii* – S to N WA and var. *vestita* – N to SC AK.
DESCRIPTION:
General: Tufted perennial herb from a taproot; stems tufted, 1-12 cm tall, glabrous to finely feltlike and woolly with starlike hairs.

Leaves: Usually all basal, linear to oblanceolate or egg-shaped, 0.5-1.5 cm long, 1-5 mm wide, densely feltlike and woolly with starlike hairs, toothed, midribs firm and withering but persistent; stem leaves lacking or sometimes 1-2.

Flowers: Racemes 3- to 15-flowered; petals white, 2.5-4 mm long.

Fruits: Silicles, (3) 7-12 (20) mm long, 1-2 (3.5) mm wide, linear to elliptic or narrowly oblong-oblanceolate, somewhat contorted, glabrous to starlike-hairy; fruiting stalks shorter than silicles; styles 0.2-0.5 mm long; seeds 8-30, about 1 mm long.

Notes: Three varieties occur in BC.

1. Stems with 1 or 2 leaves, usually more than 2 mm wide; silicles usually appressed to axis of the inflorescence; stems and stalks frequently hairy .... var. *vestita* O.E. Schulz

1. Stems leafless or rarely with 1 leaf, less than 2 mm wide; silicles usually spreading; upper stems and stalks usually glabrous.

   2. Silicles more than 2 mm wide .... var. *thompsonii* (C.L. Hitchc.) Rollins

   2. Silicles less than 2 mm wide .... var. *lonchocarpa*

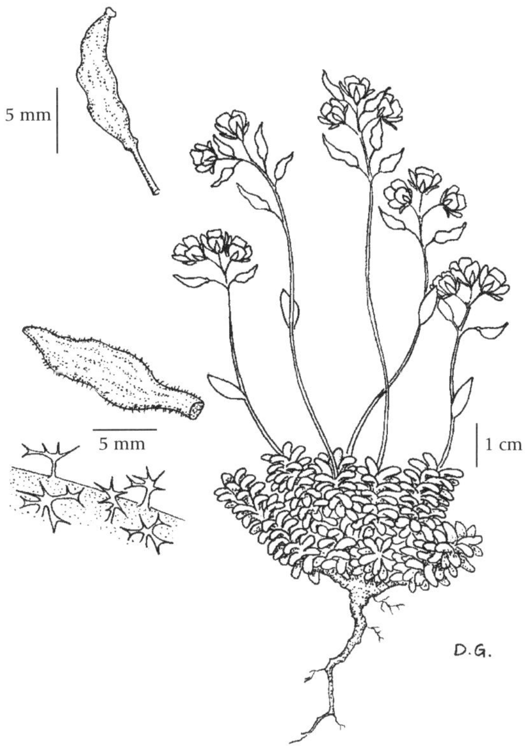

Draba lactea

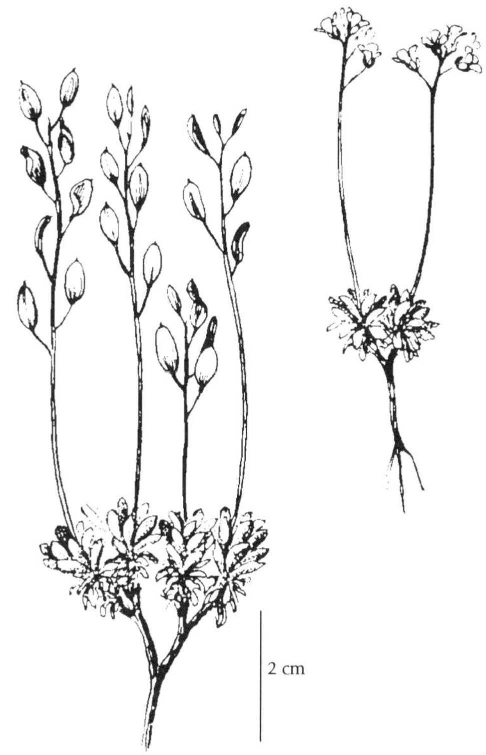

Draba lonchocarpa var. *thompsonii*

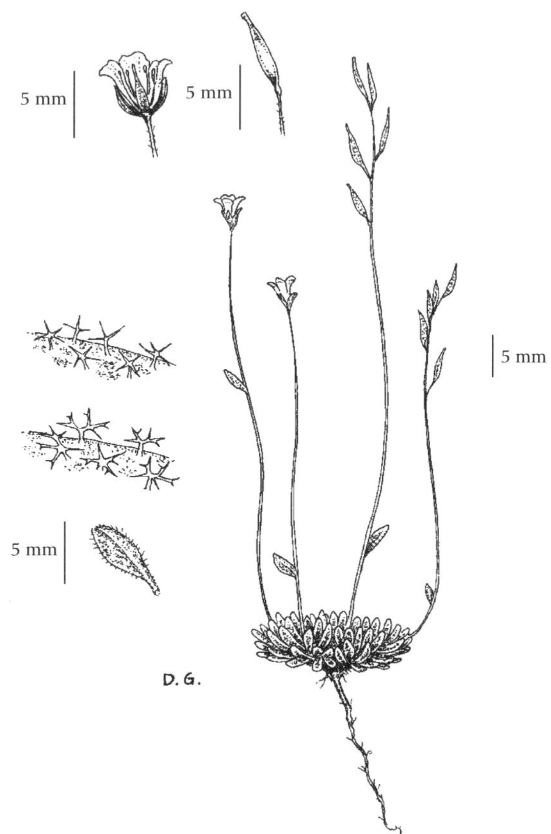

Draba lonchocarpa var. *vestita*

Draba longipes **Raup**

long-stalked draba; long-stalked whitlow-grass

Habitat/Range:

Moist meadows, rocky slopes and cliffs in the subalpine and alpine zones; infrequent in N BC; N to AK, YT and W NT and E to AB.

Description:

General: Loosely tufted perennial herb from a taproot; stems decumbent at base, 7-20 (35) cm tall, hairy with branched or starlike hairs, sometimes with a few long, simple hairs near base.

Leaves: Basal leaves loosely tufted, oblanceolate, 0.5-3 cm long, 2-12 mm wide, hairy with unequally 4-rayed, starlike hairs, occasionally mixed with 5-rayed or crosslike hairs, hairy-fringed with simple, branched or starlike hairs, entire to sometimes obscurely toothed; stem leaves lacking or 2, sometimes 3, lanceolate to egg-shaped-lanceolate.

Flowers: Racemes 2- to -15 flowered; petals white, cream or yellow, 3-5 mm long; sepals 1.5-2.5 mm long, sparsely soft-hairy.

Fruits: Silicles, 6-12 mm long, 1-2.5 mm wide, lanceolate to oblong, plane, glabrous or nearly so; fruiting stalks 3-15 mm long, lowest often about the same length as to twice the length of silicles, hairy; styles 0.5-1 mm long.

Draba macounii **O.E. Schulz**

Macoun's draba; Macoun's whitlow-grass

Habitat/Range:

Mesic to dry meadows, rocky slopes and scree slopes in the subalpine and alpine zones; rare in N BC; N to AK, S YT and SW NT and E to AB.

Description:

General: Tufted perennial herb from a taproot and a branching stem-base; stems 2-3 cm tall, sparsely hairy with simple and branched hairs.

Leaves: All basal, compacted, oblanceolate, 0.5-1 cm long, 2.5-5 mm wide, hairy with tangled, simple and branched hairs, margins hairy-fringed; stem leaves lacking.

Flowers: Short racemes 4- to 8-flowered; petals pale yellow, drying white, 3.5-4.5 mm long.

Fruits: Silicles, 5-7 mm long, about half as wide, oblong-elliptic, glabrous, dark green; fruiting stalks shorter than silicles, spreading; styles less than 1 mm long.

Draba nemorosa **L.**

woods draba; woods whitlow-grass

Habitat/Range:

Moist to dry open forests, grasslands and shrublands in the lowland, steppe and montane zones; common in S BC, rare northward; circumpolar, N to AK, YT and NT, E to PQ and S to CO and N CA; Eurasia.

Description:

General: Annual herb from slender taproot; stems simple or branched, 5-25 (35) cm tall, hairy with both branched and starlike hairs, sometimes with some simple hairs or nearly glabrous.

Leaves: Basal leaves few; stem leaves on lower third of stem, narrowly egg- or spoon-shaped; 1-3 cm long, 0.1-2 cm wide, minutely to regularly toothed; upper surface of leaves hairy with simple hairs and some long-stalked, 2-branched, 3-branched, and crosslike hairs; lower leaf surfaces with mostly crosslike hairs, but also some long-stalked, 2-branched and 3-branched hairs.

Flowers: Racemes 10- to 50-flowered, long and lax; petals yellowish, 2.5-4 mm long; sepals 2-2.5 mm long, sparsely soft-hairy to glabrous.

Fruits: Silicles, 3-10 mm long, 1-2 mm wide, elliptic to elliptic-oblanceolate, glabrous or hairy; fruiting stalks to 35 mm long, 1-5 times as long as silicles, spreading-ascending, glabrous; styles lacking; seeds 25-60, oval, light brown, about 0.7 mm long.

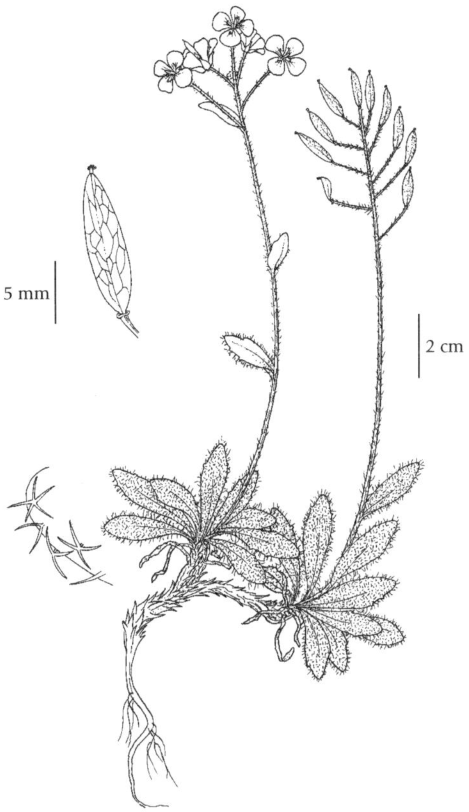

Draba longipes

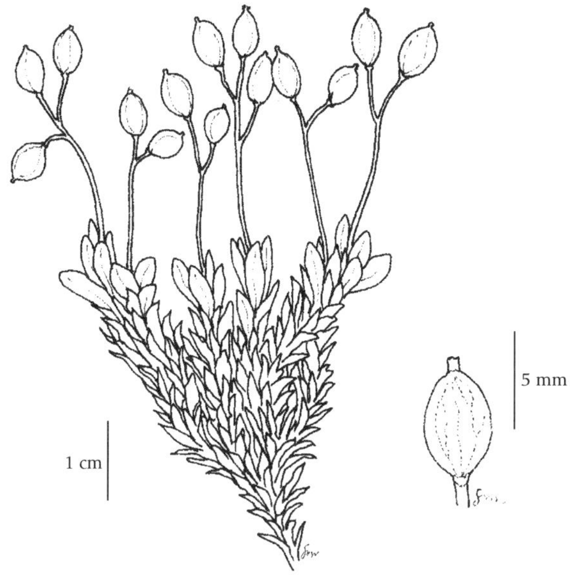

Draba macounii

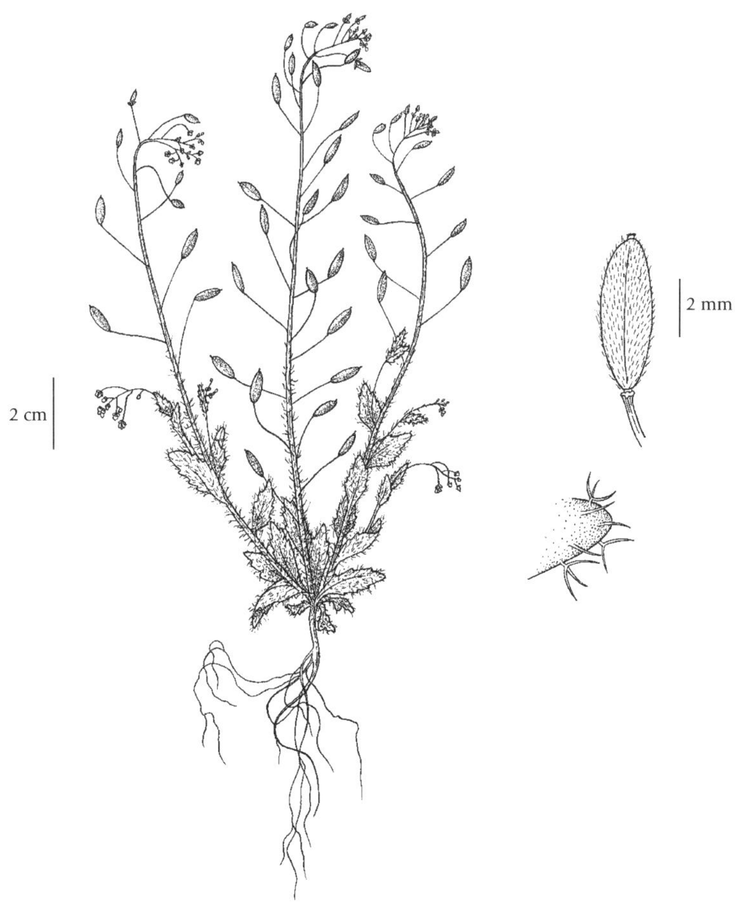

Draba nemorosa

***Draba nivalis* Lilj.**
snow draba; snow whitlow-grass
Habitat/Range:
Mesic to dry meadows, rocky slopes and scree slopes in the subalpine and alpine zones; infrequent throughout BC; N to AK, YT and NT and E to NF.
Description:
General: Tufted perennial herb from a taproot; stems branched, 3-12 (20) cm tall, starlike-hairy.

Leaves: Usually all basal, oblong-egg-shaped, 0.5-1.5 cm long, 1-4 mm wide, greyish-hairy with overlapping starlike hairs, margins with all starlike hairs or with only simple hairs near base, midribs prominent and persistent; stem leaves absent or sometimes 1.

Flowers: Racemes 3- to 12-flowered; flower stalks 1-6 (11) mm long, ascending to erect, grey-hairy with starlike hairs; petals white, 2.5-4 mm long; sepals 1.5-2 mm long, glabrous or hairy.

Fruits: Silicles, 5-12 mm long, 1-2 mm wide, lance-linear to oblong or lance-oblong, plane to contorted, glabrous; styles 0.2-0.5 mm long.

***Draba oligosperma* Hook.**
few-seeded draba; few-seeded whitlow-grass
Habitat/Range:
Dry rocky slopes, cliffs, meadows, fellfields and scree slopes in the steppe to alpine zones; infrequent in S BC, rare northward; N to AK, YT and NT, E to SW AB and S to WY, CO, NV and CA.
Description:
General: Densely tufted perennial herb from a taproot; stems, branched, 1-10 cm tall, glabrous or starlike-hairy near base.

Leaves: All basal, narrowly oblanceolate to oblong, 3-12 mm long, 0.7-1.7 mm wide, hairy with 4-rayed, starlike hairs with the rays branched, margins hairy with starlike hairs, starlike hairs on lower leaf surfaces unstalked with mostly 9 or more rays, slightly V-shaped in cross section, midribs prominent and often persistent; stem leaves lacking.

Flowers: Racemes 3- to 15-flowered; flower stalks 2-10 mm long, ascending, glabrous; petals yellow, 3-5 mm long; sepals 2-2.5 mm long, sparsely soft-hairy.

Fruits: Silicles, 3-8 mm long, 2-4 mm wide, egg-shaped to lanceolate or oval, plane, glabrous or hairy; styles 0.2-1 mm long; seeds 2-10, 1.4-1.8 mm long.

***Draba palanderiana* Kjellm.** (*D. caesia auct. non* M.F. Adams)
Palander's draba; Palander's whitlow-grass
Habitat/Range:
Mesic to dry meadows and cliffs in the subalpine and alpine zones; rare in NW BC, known only from the Cassiar area; N to AK, YT and W NT.
Description:
General: Tufted perennial herb from a taproot; stems branched, 2-12 cm tall, soft starlike-hairy or nearly glabrous.

Leaves: All basal, oblanceolate to egg-shaped, 6-10 mm long, 2-2.5 mm wide, greyish-hairy with overlapping starlike hairs or only sparsely hairy on upper leaf surfaces, midribs persistent; stem leaves lacking.

Flowers: Racemes 5- to 15-flowered; flower stalks 6-14 mm long, hairy to nearly glabrous; petals white or pale yellow, often drying to a darker yellow, 4-5.5 mm long; sepals 2-3 mm long, hairy to nearly glabrous.

Fruits: Silicles, 5-12 mm long, linear-lanceolate, plane, glabrous or hairy when young; styles 0.5-0.8 mm long.

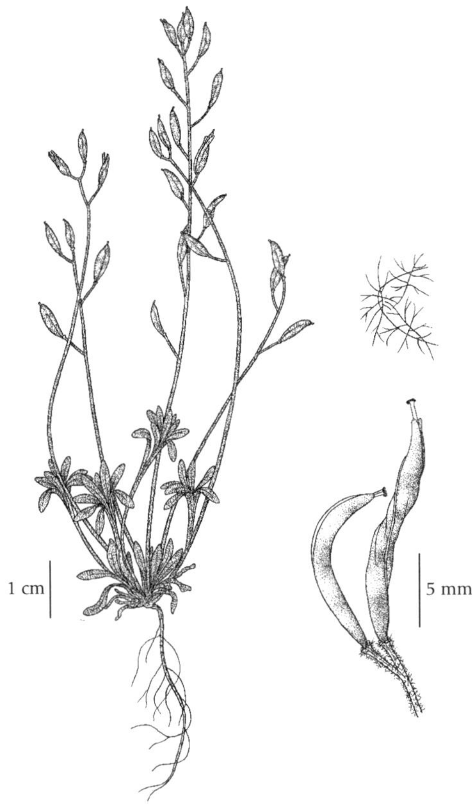

Draba nivalis

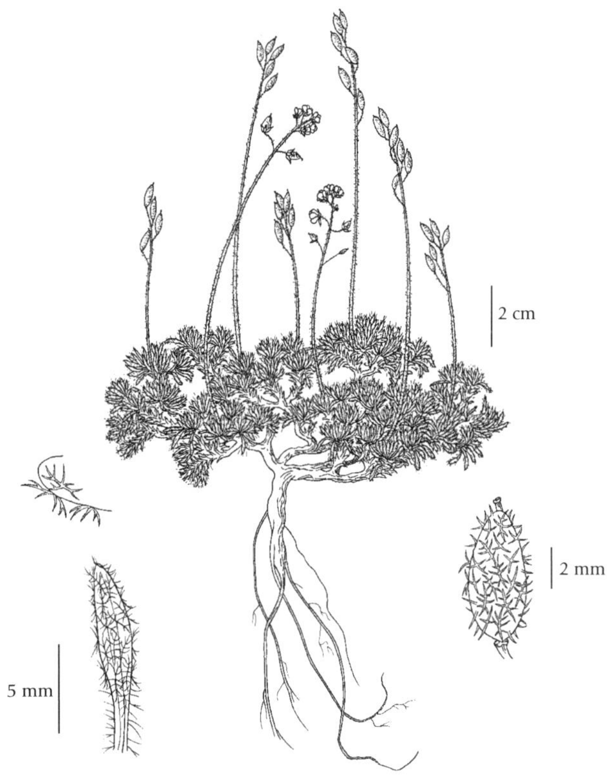

Draba oligosperma

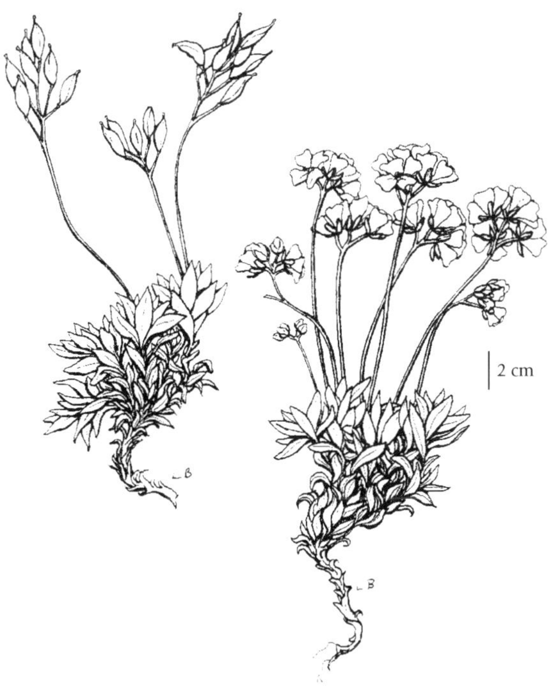

Draba palanderiana

***Draba paysonii* Macbr.** (*D. paysonii* var. *treleasii* [O.E. Schulz] C.L. Hitchc.)

Payson's draba; Payson's whitlow-grass

Habitat/Range:

Dry meadows, cliffs, rocky slopes, fellfields and scree slopes in the subalpine and alpine zones; infrequent in S BC; N to AK, E to SW AB and S to WY, VT, NV and CA.

Description:

General: Tufted perennial herb from a taproot; stems 0.5-3 (6) cm tall, stiff-hairy with long freely branched hairs.

Leaves: All basal, linear to narrowly spoon-shaped, 4-14 mm long, 0.7-1.5 mm wide, hairy, lower leaf surfaces with simple, 2-branched, and long-stalked starlike hairs, upper leaf surfaces with simple and once or twice branched hairs, margins with conspicuous simple and branched hairs about 1 mm long, leaves usually crisped or curled when dried, midribs persistent; stem leaves lacking.

Flowers: Racemes 3- to 10-flowered; petals yellow, 2-4.5 mm long.

Fruits: Silicles, 3-8 mm long, 2.5-4.5 mm wide, egg-shaped to narrowly egg-shaped, inflated, hairy with short stiff simple or branched hairs, occasionally glabrous; fruiting stalks to 7 mm long, shorter than silicles, with starlike and branched hairs; styles 0.5-0.75 mm long; seeds 3-10, 1.3-1.8 mm long.

***Draba porsildii* Mulligan**

Porsild's draba; Porsild's whitlow-grass

Habitat/Range:

Mesic to dry talus slopes, scree slopes and cliffs in the subalpine and alpine zones; rare in NE and SE BC; N to AK, S YT and W NT and E to SW AB.

Description:

General: Tufted perennial herb from a taproot; stems 2-6.5 cm tall, glabrous to sparsely hairy.

Leaves: Usually all basal, oblanceolate, 0.3-1.2 cm long, 1.5-2 mm wide, hairy with simple, branched, crosslike, and starlike (5-8 rays) hairs, entire; stem leaves lacking or sometimes 1.

Flowers: Racemes 3- to 6-flowered; petals white, 2-3 mm long; sepals 1.5-2 mm long.

Fruits: Silicles, 4-8 mm long, 2-3 mm wide, egg-shaped, glabrous; fruiting stalks shorter than silicles; styles 0.25 mm long; seeds 5-9, about 1 mm long.

***Draba praealta* Greene**

tall draba; tall whitlow-grass

Habitat/Range:

Moist to mesic forests, meadows and rocky slopes in the montane and subalpine zones; infrequent in S BC east of the Coast-Cascade Mountains, rare northward; N to AK, YT and NT, E to AB and S to WY, NV and OR.

Description:

General: Biennial or short-lived perennial or occasionally annual herb from a slender taproot; stems simple or branched, 10-30 cm tall, hairy with simple, branched, or starlike hairs.

Leaves: Mostly basal, in a rosette, oblanceolate, 5-30 mm long, 2-6 mm wide; stem leaves 1-6, lanceolate to egg-shaped, 0.5-3 cm long, 2-14 mm wide, entire to few toothed, hairy with 4- to 7-rayed hairs, usually mixed with branched or simple hairs.

Flowers: Racemes 5- to 30-flowered, elongate in fruit; flower stalks 3-10 mm long, ascending, with starlike or branched hairs; petals white, sometimes creamy yellow, 2-3.5 mm long; sepals 1.5-2 mm long, soft-hairy.

Fruits: Silicles, 7-14 mm long, 1.5-2.5 mm wide, narrowly-lanceolate, plane, soft-hairy with simple and branched hairs; styles 0.1-0.2 mm long; seeds 20-70, about 0.8 mm long.

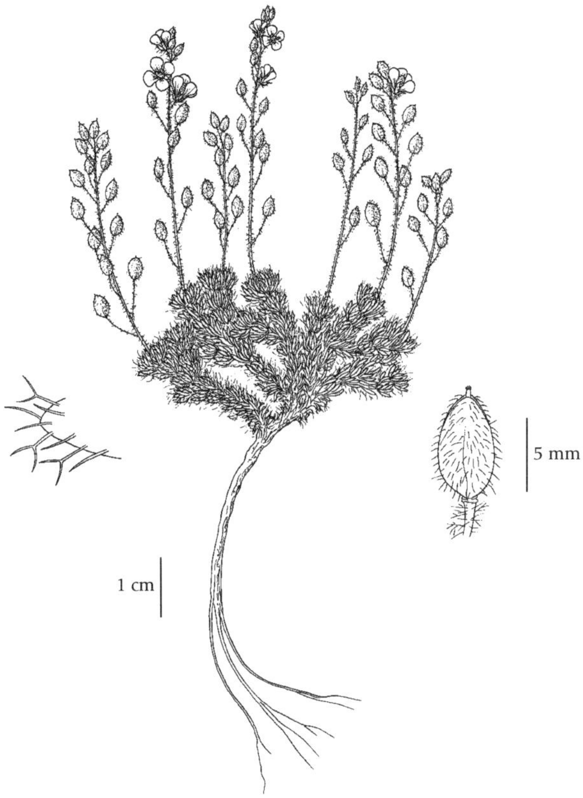

Draba paysonii

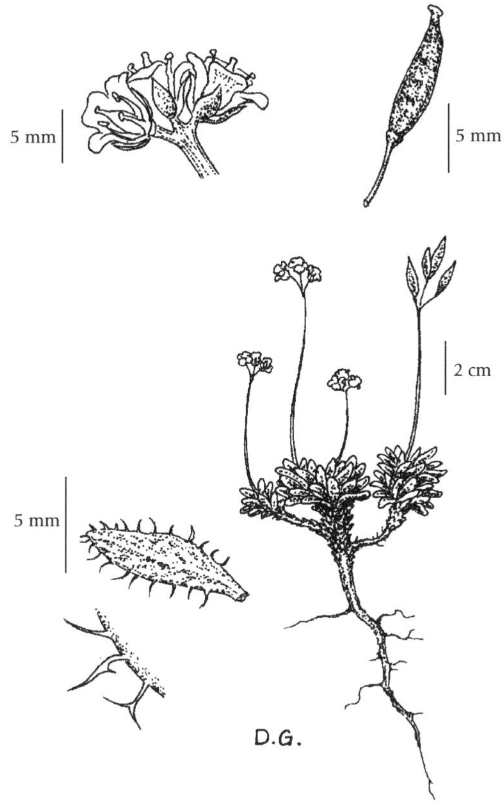

Draba porsildii

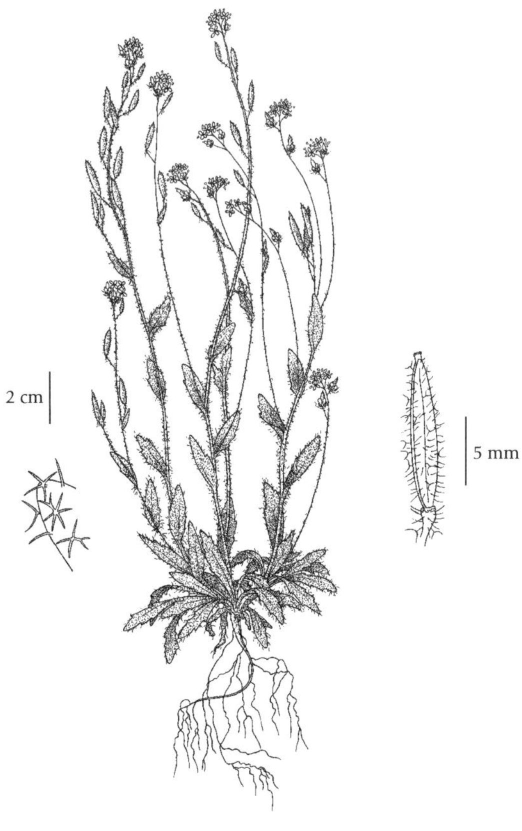

Draba praealta

***Draba reptans* (Lam.) Fern.**

Carolina draba; Carolina whitlow-grass

HABITAT/RANGE:

Dry, often calcareous, cliffs, rocky slopes and forest openings in the steppe and lower montane zones; rare in SC BC, known only from Doc English Bluff (Williams Lake); E to AB and S to TX, GA, CO, NM, AZ and CA.

DESCRIPTION:

General: Annual herb from a taproot; stems simple or branched, 5-20 cm tall, hairy near base with simple, branched and starlike hairs.

Leaves: Mostly basal, in a rosette, oblanceolate to egg-shaped, 5-20 mm long, lower surface hairs branched and starlike, upper surface hairs simple and branched, hairy-fringed with stiff hairs, entire; stem leaves 1-5.

Flowers: Racemes up to 30-flowered; flower stalks 3-10 mm long, erect, with starlike or branched hairs; petals white, slightly notched at the tips, 3-5 mm long; sepals about 2 mm long, sometimes lacking.

Fruits: Silicles, 5-20 mm long, 1-2 mm wide, linear, glabrous to finely stiff-hairy; styles less than 0.15 mm long; seeds 18-80, 0.5-0.7 mm long.

***Draba ruaxes* Pays. & St. John** (*D. ventosa* A. Gray var. *ruaxes* [Pays. & St. John] C.L. Hitchc.)

coast mountain draba; coast mountain whitlow-grass

HABITAT/RANGE:

Dry meadows, cliffs, rocky slopes and scree slopes in the subalpine and alpine zones; rare in the Coast-Cascade Mountains; N to AK and YT and S to N WA.

DESCRIPTION:

General: Tufted perennial herb from a taproot and a branched stem-base, often clothed with dried remains of old leaves; stems 2-10 cm tall, hairy with simple and branched hairs.

Leaves: All basal, egg-shaped to elliptic or oblanceolate, 5-12 mm long, 2-4 mm wide, hairy with long-stalked, starlike hairs on lower surface, long, simple and few-branched hairs on upper; stem leaves lacking.

Flowers: Racemes short, 3- to 10-flowered; petals yellow, 4-5 mm long.

Fruits: Silicles, 5-8 mm long, 3.5-5.5 mm wide, oval to egg-shaped, plane, glabrous to densely soft-hairy with simple and branched hairs; fruiting stalks almost as long as silicles; styles 0.6-1.2 mm long; seeds 10-16, 1.5-2 mm long.

***Draba stenoloba* Ledeb.**

Alaska draba; Alaska whitlow-grass

HABITAT/RANGE:

Moist to dry meadows and cliffs in the subalpine and alpine zones; infrequent throughout BC; N to AK, YT, W NT and E to SW AB.

DESCRIPTION:

General: Biennial or short-lived perennial herb from a slender taproot and a usually simple stem-base; stems simple or branched, 3-32 cm tall, hairy below mainly with crosslike hairs, commonly glabrous above.

Leaves: Basal leaves 10-25, 0.4-3 cm long, 1.5-9 mm wide, in a rosette, egg-shaped to oblanceolate, slightly minutely toothed or entire; stem leaves absent or 1-2 or sometimes as many as 4, elliptic to widely lanceolate, toothed or entire, 0.3-1.6 cm long, 2-10 mm wide; upper leaf surfaces with simple and short-stalked 2-branched, 3-branched, and crosslike hairs (mostly 2-branched or crosslike); undersurface of leaves with short-stalked 2-branched and crosslike hairs.

Flowers: Racemes 2- to 15-flowered, open, often compound; flower stalks 3-15 mm long, commonly glabrous, seldom hairy, ascending; petals 3-3.5 mm long, yellow; sepals 1.5-2.5 mm long, soft-hairy or glabrous.

Fruits: Silicles 10-20 mm long, 1.5-2 mm wide; narrowly elliptic to narrowly egg-shaped, plane, glabrous or rarely hairy; style to 0.1 mm long; seeds 16-40, oval, light brown, 1 mm long.

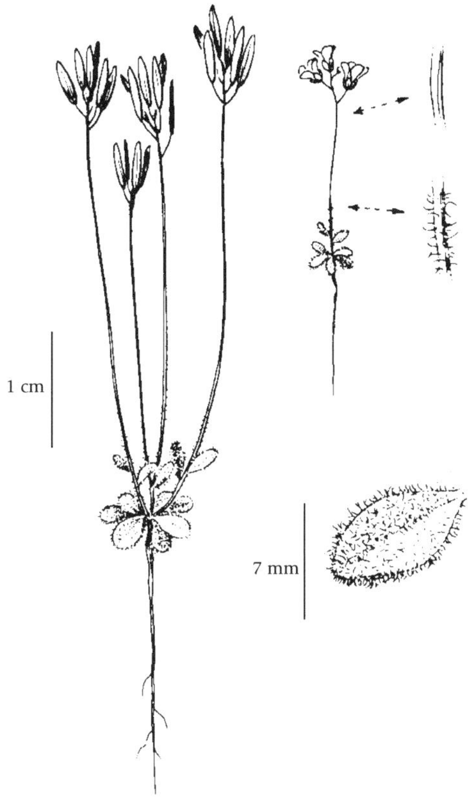

Draba reptans

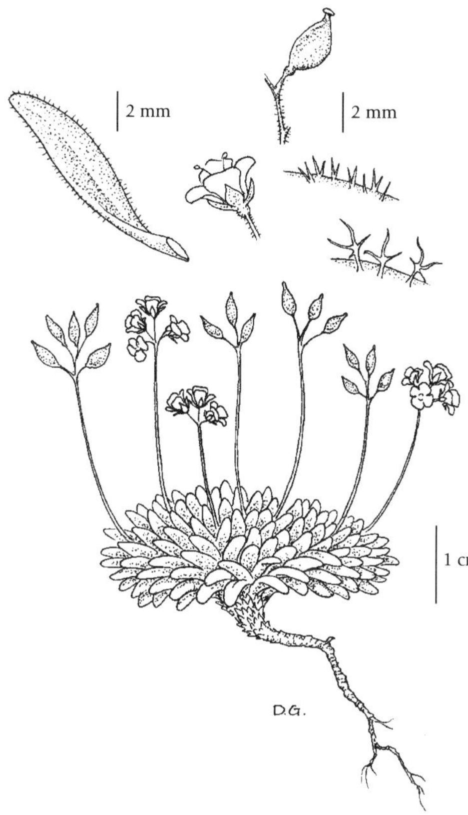

Draba ruaxes

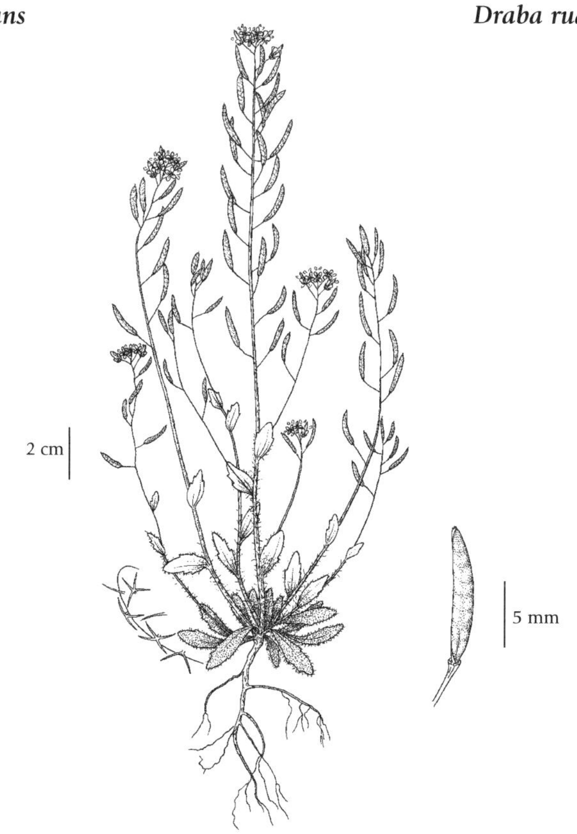

Draba stenoloba

Draba stenopetala **Trautv.**

star-flowered draba; star-flowered whitlow-grass

HABITAT/RANGE:

Mesic to dry meadows, cliffs and scree slopes in the subalpine and alpine zones; rare in NW BC, known only from Mt. Edziza; amphiberingian, N to AK and YT; E. Asia.

DESCRIPTION:

General: Tufted perennial herb from a taproot and a branched stem-base densely clothed with persistent leaves; stems 1-15 cm tall, with simple to 3-branched hairs.

Leaves: All basal, 4-7 mm long, 1-2 mm wide, oblanceolate to oblong or elliptic, with simple and branched hairs or sometimes glabrous; stem leaves lacking.

Flowers: Few-flowered racemes; flower stalks 2-9 mm long, with branched or starlike hairs; petals 2-5 mm long, yellow or purplish, oblong to narrowly oblanceolate, narrower than sepals, entire; sepals 2.5-3.5 mm long, hairy with simple and branched hairs.

Fruits: Silicles 3-4 (5) mm long, 1-2 mm wide, narrowly egg-shaped to elliptic, hairy or glabrous; styles 0.2-0.3 mm long; seeds 4-6.

Draba ventosa **A. Gray**

Wind River draba; Wind River whitlow-grass

HABITAT/RANGE:

Dry meadows, cliffs, rocky slopes and scree slopes in the subalpine and alpine zones; rare in SE BC; N to YT, E to SW AB and S to WY and UT.

DESCRIPTION:

General: Low, tufted perennial herb from a taproot and a branched stem-base, usually covered with withered remains of older leaves; stems leafless, 2-4 cm tall, densely covered with long-stalked starlike hairs.

Leaves: Numerous, egg-shaped-elliptic to oblanceolate, 5-12 mm long, 2-4 mm wide; densely hairy mainly with long-stalked starlike hairs on both leaf surfaces, also with simple and branched hairs; older leaves persistent, withered; stem leaves lacking.

Flowers: Racemes compressed, 3- to 20-flowered; petals yellow, 4-5 mm long.

Fruits: Silicles 5-8 mm long, 3.5-5.5 mm wide, oval to egg-shaped, plane; segments thick, often somewhat corrugated, densely and softly hairy with a mixture of simple and branched (almost crosslike) hairs; fruiting stalks about length of silicles; styles 0.6-1.2 mm long; seeds 10-16, 1.5-2 mm long.

Draba verna **L.** (*Erophila verna* [L.] Chev.)

common draba; common whitlow-grass

HABITAT/RANGE:

Dry disturbed areas and waste areas in the lowland, montane and steppe zones; locally frequent in SW BC, known from SE Vancouver Island, the Gulf Islands and the adjacent lower mainland, frequent in WC and SC BC; introduced from Eurasia.

DESCRIPTION:

General: Annual herb from a slender taproot; stems leafless, 5-20 (25) cm tall, glabrous or hairy with simple and branched hairs near base.

Leaves: Basal leaves in a rosette, spoon-shaped to oblanceolate, entire to toothed, 1-2.5 cm long, hairs simple, branched or starlike; stem leaves lacking.

Flowers: Racemes 3- to 20-flowered; petals white, 3-4 mm long, deeply bilobed.

Fruits: Silicles 3-10 mm long, 1.5-4 mm wide, elliptic to egg-shaped, glabrous; fruiting stalks 2-4 times as long as the silicles; style about 0.1 mm long; seeds 30-60, 0.3-0.5 mm long.

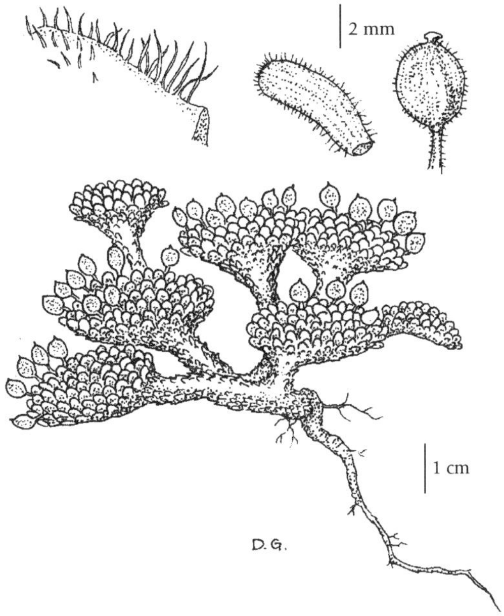

Draba stenopetala

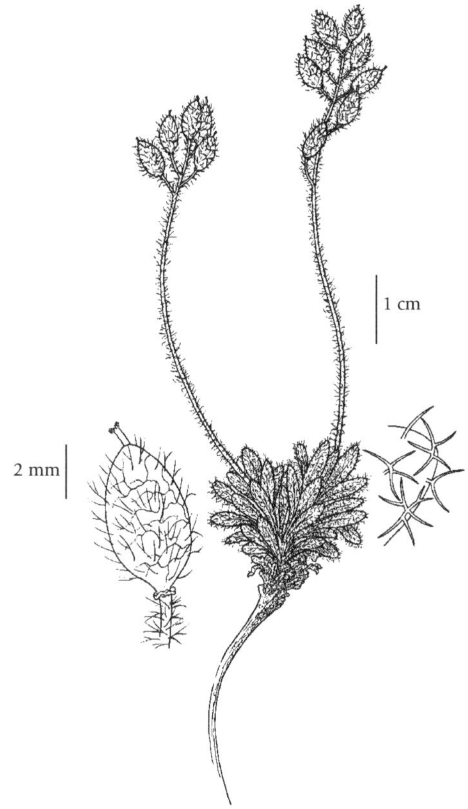

Draba ventosa

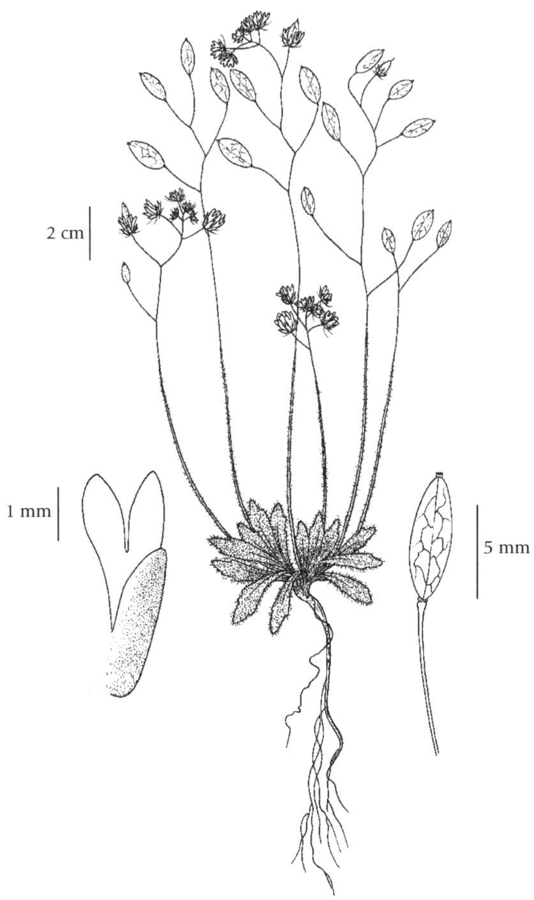

Draba verna

ERUCA

***Eruca vesicaria* (L.) Cav. ssp. *sativa* [Miller] Thell.** (*E. sativa* P. Mill.)
garden-rocket

HABITAT/RANGE:

Mesic to dry roadsides, fields and waste places in the lowland zone; rare in SW BC, known from the lower Fraser River valley; E to PQ, and S to MX; introduced from Eurasia.

DESCRIPTION:

General: Annual herb from a slender taproot; stems usually basally branched, 0.2-1.0 m tall, more or less soft-hairy, especially near the base, hairs often reflexed.

Leaves: Mostly basal and clustered, widely oblanceolate, wavy pinnately cut, 5-15 cm long; stem leaves reduced upward; glabrous on upper surface, few simple hairs on lower surface along midrib.

Flowers: Dense, many-flowered; flower stalks 2-5 mm long, ascending; petals white to yellowish, with reddish-purplish veins, 15-20 mm long; sepals 7-10 mm long.

Fruits: Siliques, lance-oblong to narrowly egg-shaped, 1.5-2.5 cm long, 3-5 mm wide, round in cross section; beaks broad, flat, tapering.

ERUCASTRUM

***Erucastrum gallicum* (Willd.) O.E. Schulz**
dog mustard

HABITAT/RANGE:

Mesic to dry fields and waste places in the montane zone; rare in S BC; introduced from Eurasia.

DESCRIPTION:

General: Annual or sometimes biennial herb from a taproot; stems simple to freely branched, 15-80 cm tall more or less stiff, long-hairy, the hairs pointed down the stem.

Leaves: Deeply pinnately lobed, margins often wavy, 3-20 cm long.

Flowers: Racemes open and elongate, bracteate, the lower leaf-like; flower stalks less than 1 cm long, slender, ascending; petals pale yellow, 4-7 mm long.

Fruits: Siliques, 1-4.5 cm long, about 1.5 mm wide; beaks 1.5-3 mm long.

ERYSIMUM

1. Petals pink to pinkish-purple .... *E. pallasii*
1. Petals yellow to deep orange.
   2. Petals 3-12 mm long.
      3. Petals 3-5 mm long; siliques 1.5-3 cm long .... *E. cheiranthoides*
      3. Petals 5-10 mm long; siliques 2.5-5 cm (rarely 1.5 cm) long .... *E. inconspicuum*
   2. Petals 15-25 mm long.
      4. Plants usually greenish; siliques flattened; plants alpine .... *E. arenicola*
      4. Plants usually greyish; siliques not markedly flattened; plants lowland or montane.
         5. Seeds in 1 row; nectary glands present at base of filaments .... *E. asperum*
         5. Seeds more or less in 2 rows; nectary glands lacking .... *E. cheiri*

Eruca vesicaria ssp. *sativa*

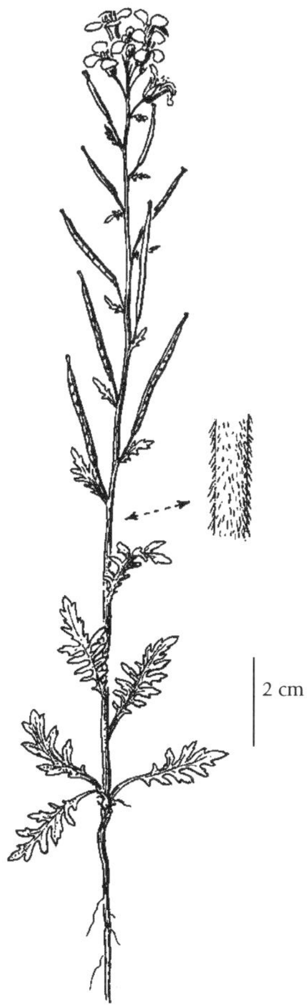

Erucastrum gallicum

Erysimum arenicola **S. Wats. var.** ***torulosum*** **(Piper) C.L. Hitchc.** (*E. torulosum* Piper)
sand-dwelling wallflower

Habitat/Range:

Moist to mesic rocky slopes or talus slopes in the alpine zone; infrequent in SW BC, known only from Vancouver Island; S to OR.

Description:

General: Perennial herb from a taproot; simple or several-stemmed, 10-25 (50) cm tall, sparsely grey-hairy; stem-base often covered with old persistent leaf bases.

Leaves: Basal leaves tufted, numerous, spoon-shaped to oblanceolate, stalked, entire to dentate, obtuse to sharp-pointed, 2.5-6 (7) cm long, 1.5-8 mm wide, hairy with 2- to 3-branched hairs; stem leaves variable, sometimes reduced but often larger than the basal, usually wavy-toothed.

Flowers: Inflorescence slender, 3-10 mm long, somewhat ascending; petals lemon yellow, the claw 10-15 mm long, the blade narrowly to broadly egg-shaped, 5-9 mm long; sepals erect, hairy.

Fruits: Siliques, ascending to erect, 3-12 cm long, 2-2.5 mm wide, compressed parallel to the plane of the partitions, alternately contracted and expanded, glabrous to hairy with 2- to 4-branched hairs; beaks 1-4 mm long; seeds 2-3 mm long, wingless or with a slight apical wing margin.

Erysimum asperum **(Nutt.) DC.** (*E. capitatum* [Dougl.] Greene)
prairie rocket; western wallflower

Habitat/Range:

Dry roadsides and rocky slopes in the montane zone; rare in SE BC; E to S MB and S to KS, MN, NM and N CA.

Description:

General: Biennial herb from a simple or sometimes branched stem-base; stems erect, stout, 5-45 cm tall, simple or branched above, ash grey with a close hairiness.

Leaves: Basal leaves thick, firm, numerous, more or less in a rosette, oblanceolate, wavy-toothed, stalked, 3-10 cm long, (2) 4-13 (25) mm wide, sharp-pointed at the tips, densely grey-hairy mainly with 2-rayed hairs; stem leaves narrowly lanceolate, entire to somewhat toothed.

Flowers: Racemes bractless, 4-40 cm long; flower stalks 5-15 (20) mm long, stiffly erect or spreading; petals bright yellow to deep orange or somewhat reddish, 13-32 mm long, blades 4-8 (9) mm wide; sepals (6) 8-13 mm long, outer ones with a pouch at the base.

Fruits: Siliques, (3) 5-11 (14) cm long, 1.3-2.7 mm wide, round in cross section to flattened; beaks 1-5 mm long, beak-like; seeds 1.5-2.3 mm long, wingless or slightly wing-margined at tip.

Erysimum cheiranthoides **L.** (*E. cheiranthoides* ssp. *altum* Ahti)
wormseed mustard

Habitat/Range:

Moist to mesic roadsides, fields and waste places in the lowland, steppe and montane zones; frequent throughout BC except the Queen Charlotte Islands and adjacent coast; circumboreal, N to AK, YT and NT, E to NF and S to NC, MI, UT and CA; Eurasia, native status uncertain.

Description:

General: Annual or biennial herb from a taproot; stems simple or branched, leafy, 0.2-1.0 m tall, sparsely hairy with 2-pronged hairs.

Leaves: Basal leaves lacking; stem leaves linear to lanceolate or oblanceolate, entire or minutely toothed, 2-8 cm long, 2-18 mm wide, finely hairy with 3-pronged hairs.

Flowers: Racemes bractless; flower stalks 4-15 mm long, spreading or spreading-ascending; petals pale yellow, 3.5-5 mm long; sepals yellowish or greenish, 2-3 mm long.

Fruits: Siliques, ascending to erect, 1.5-2 (3) cm long, about 1 mm wide, round in cross section, finely appressed-hairy; beaks 0.5-1 mm long; seeds 1-1.2 mm long, wingless.

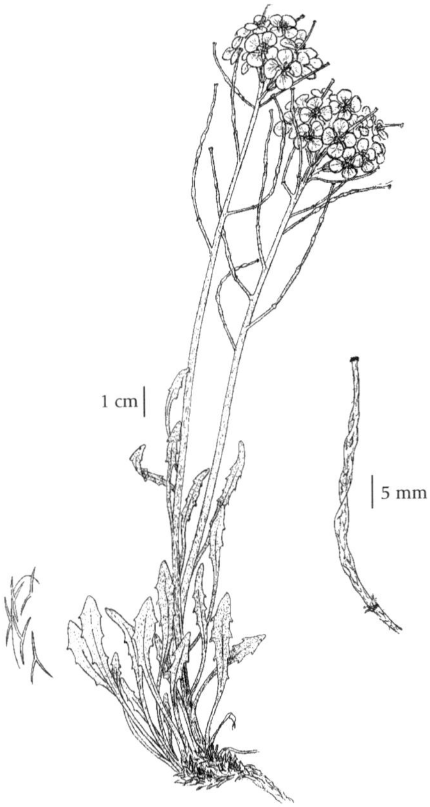

Erysimum arenicola var. *torulosum*

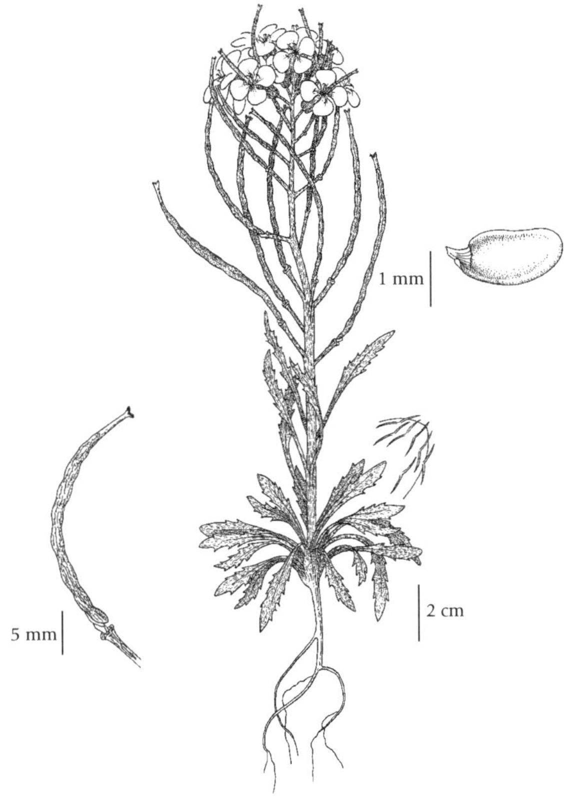

Erysimum asperum

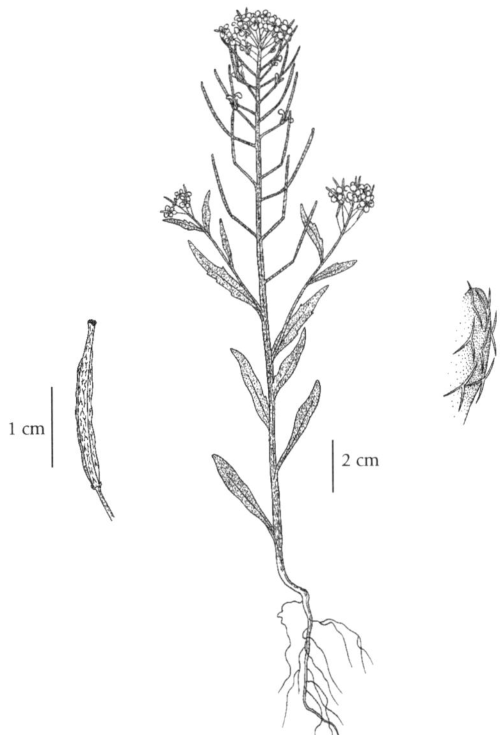

Erysimum cheiranthoides

***Erysimum cheiri* (L.) Crantz** (*Cheiranthus cheiri* L.)
common wallflower
Habitat/Range:
Mesic to dry rock outcrops and waste places in the lowland zone; infrequent garden escape on SE Vancouver Island; introduced from Europe.
Description:
General: Perennial herb from slender taproot; stems erect or ascending, much branched, angled, leafy, 15-80 cm tall, hairy with appressed, branched hairs.

Leaves: Basal and stem leaves oblong-lanceolate, more or less entire, usually greyish-hairy with branched hairs, particularly beneath; basal leaves in rosette, short-stalked, 4-20 cm long; stem leaves crowded, short-stalked.

Flowers: Inflorescence 4- to 6-flowered; flower stalks erect, 5-15 mm long; petals yellow to red or brown, 15-25 mm long; sepals purplish, half the length of petals.

Fruits: Siliques, 3-10 cm long, 3-4 mm wide, densely white-hairy; beaks 0.5-4 mm long; seeds irregularly 2-rowed, more or less spherical, 2-4 mm long, winged at apex.

***Erysimum inconspicuum* (S. Wats.) MacMill.**
small wallflower; small-flowered rocket
Habitat/Range:
Dry open sites in the steppe and montane zones; infrequent throughout BC east of the Coast-Cascade Mountains; N to AK, YT and NT, E to PQ and S to KS, NC, VT, NM and CA.
Description:
General: Biennial or short-lived perennial herb from a short taproot and a usually unbranched stem-base; stems erect, 0.2-1.0 m tall, simple or branched, hairy throughout with appressed 2-pronged hairs.

Leaves: Basal leaves linear to linear-oblanceolate, entire or sparsely toothed, stalked, 1.5-8 cm long, 2-6 mm wide; stem leaves becoming reduced and unstalked upwards.

Flowers: Racemes bractless; flower stalks 3-8 mm long, ascending, stout; petals pale to bright yellow, 5-10 mm long, long narrow, stalklike base, broadly oblong to nearly orbicular blades 2-3 mm across; sepals erect, narrow, often purplish, 4-6 mm long, densely hairy.

Fruits: Siliques, 1.5-5 cm long, 1-2 mm wide, round in cross section, ascending to erect, densely hairy with 2-pronged hairs; beaks thick, 1-2 mm long; seeds oblong, 1.5-2 mm long, wingless.

***Erysimum pallasii* (Pursh) Fern.**
Pallas' wallflower
Habitat/Range:
Dry talus slopes in the alpine zone; rare in NW BC, known only from Mt. Edziza; circumpolar, N to AK, YT and NT, Eurasia.
Description:
General: Perennial or biennial herb from taproot and branched or unbranched stem-base; stems simple, 3-35 cm tall, hairy.

Leaves: Basal leaves in a rosette, linear-lanceolate, 5-7 cm long, blades entire or shallowly toothed, 1-6 cm long, 2-4 mm wide; stalks narrow.

Flowers: Inflorescence racemose, flowers numerous, often with 1 to few bracts, each subtending one of the lower flowers, very short when flowering, greatly elongating in fruit; flower stalks 5-28 mm long, ascending to erect; petals pink to pink-purple, 10-16 mm long, 2-3 mm wide, claw about 5-8 mm long; sepals papery-margined, purplish, 5-8 mm long.

Fruits: Siliques, stiffly erect, dark purple, 3-10 cm long, 1.5-2 mm wide, somewhat curved, distinctly compressed parallel to the septum; beaks (1) 1.5-3 mm long; seeds 2 mm long.

Erysimum cheiri

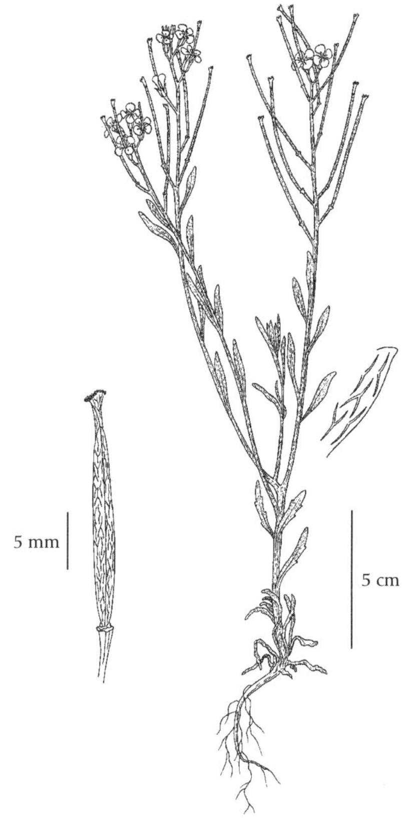

Erysimum inconspicuum

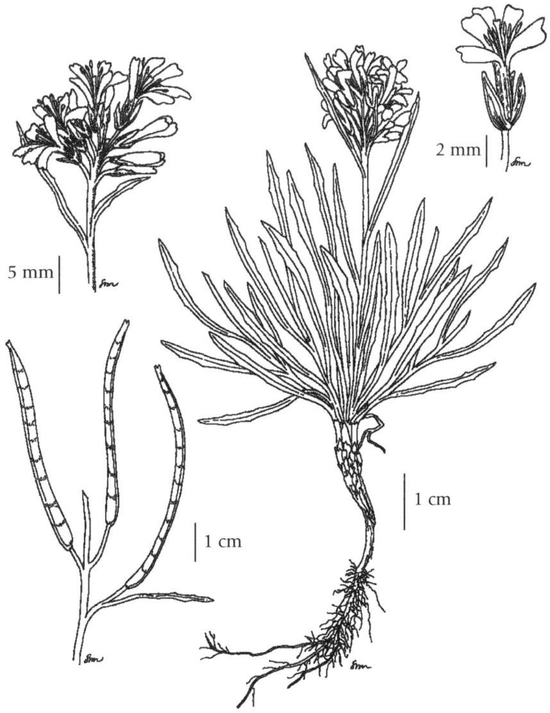

Erysimum pallasii

EUTREMA

Eutrema edwardsii **R. Br.**
Edward's wallflower

HABITAT/RANGE:

Dry talus slopes in the subalpine and alpine zones; rare in N BC; circumboreal, N to AK, YT and NT; Eurasia.

DESCRIPTION:

General: Perennial herb from a slender taproot usually 3-6 mm thick and a simple or branching stem-base; stems 1-20, decumbent-ascending to erect, 10-45 cm tall, glabrous.

Leaves: Basal leaves long-stalked, 2-15 cm long, blades narrowly egg-shaped or elliptic, 0.6-4 cm long, 0.3-3 cm wide, entire; stem leaves 3-7, reduced upwards, lance-oblong to lanceolate, becoming unstalked or short-stalked, entire.

Flowers: Racemes 3- to 20-flowered, short at first but elongating to 5-10 cm; flower stalks 3-14 mm long, spreading-ascending; petals white or cream, 3-4.5 mm long, egg-shaped to broadly spoon-shaped; sepals purplish, 2-2.5 mm long.

Fruits: Siliques, purplish, 9-20 mm long, 1.5-3 mm wide, ascending to erect, oblong to linear, tapering above and below, round in cross section, glabrous, partition perforated; beaks to 0.5 mm long; seeds oblong, plump, about 2.5 mm long, light brown.

HALIMOLOBOS

Halimolobos whitedii **(Piper) Rollins**
Whited's halimolobos

HABITAT/RANGE:

Dry grasslands in the steppe zone; rare in SC BC; S to N WA.

DESCRIPTION:

General: Short-lived perennial herb from a short taproot and usually a branched stem-base; stems usually several, simple to freely branched, 20-50 cm tall, hairy.

Leaves: Basal leaves narrowly spoon-shaped to linear-oblanceolate, slender-stalked, to 10 cm long and 2 cm wide, entire to remotely toothed, densely hairy with treelike hairs; stem leaves gradually reduced upwards, mostly entire, 2-5 cm long, the upper ones unstalked but not with earlike lobes at the base, densely hairy.

Flowers: Racemes open and elongate in fruit; flower stalks slender, ascending, 8-20 mm long, densely hairy; petals white or strongly pinkish-veined, 4-7 mm long, spoon-shaped; sepals oblong to nearly egg-shaped, densely hairy, outer sepals somewhat pouched at base.

Fruits: Siliques ascending-erect, 1-2.2 cm long, 1-1.3 mm wide, with fine, starlike hairs, very slightly alternately contracted or exposed, somewhat compressed; beaks less than 0.5 mm long, stout; seeds in 1 row, oblong, about 1 mm long, wingless.

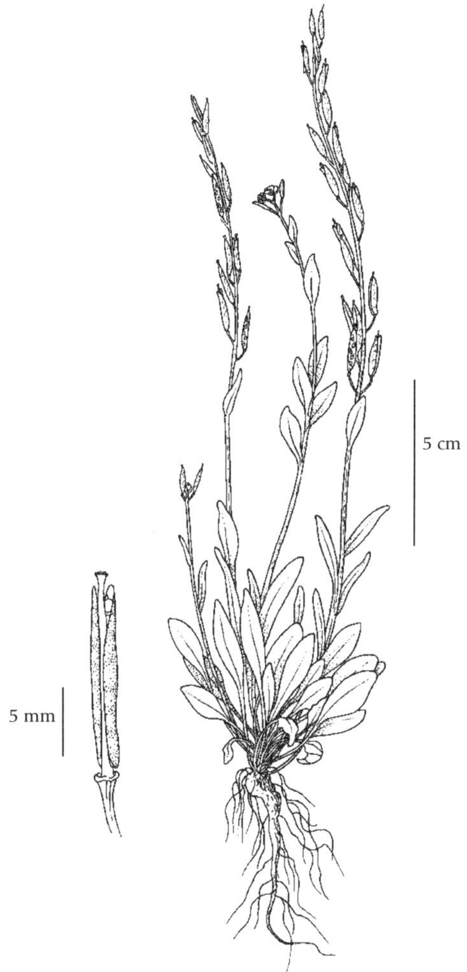

Eutrema edwardsii

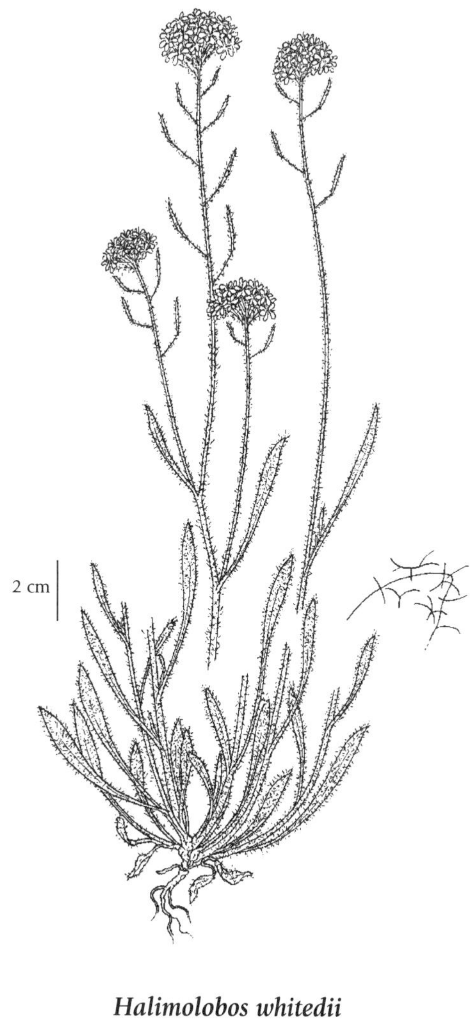

Halimolobos whitedii

HESPERIS

Hesperis matronalis **L.**

sweet rocket; dames'-violet

HABITAT/RANGE:

Mesic to dry roadsides, fields and disturbed areas in the lowland, steppe and montane zones; frequent in SW BC, known from Vancouver Island and the adjacent mainland, rare in SC BC, locally frequent in WC and SE BC; introduced from Eurasia.

DESCRIPTION:

General: Perennial or biennial herb from a taproot; stems 1 to sometimes several, simple or sparingly branched, 0.5-1.3 m tall, leafy, hairy with coarse, spreading, simple or branched hairs.

Leaves: Basal leaves soon deciduous; stem leaves lanceolate to narrowly egg-shaped, saw-toothed, 1.5-20 cm long, 0.5-4 cm wide, lower long-stalked, smaller and unstalked upwards, hairy with simple and branched hairs.

Flowers: Fragrant; inflorescence compound, somewhat corymbiform racemes; flower stalks 3-15 mm long, ascending to spreading; petals white to rose or purple, 15-25 mm long; sepals 5-7 mm long, soft-hairy.

Fruits: Siliques, 4-10 cm long, 1-2 mm wide, round in cross section, usually somewhat alternately contracted and expanded; seeds 3-4 mm long.

HUTCHINSIA

Hutchinsia procumbens **(L.) Desv.** (*Hymenolobus procumbens* [L.] Nutt. *ex* T.& G.)

hutchinsia

HABITAT/RANGE:

Moist saline or alkaline sites in the steppe zone; rare in SC BC; cosmopolitan, S to WA, E to NF and S to CO and CA; Chile, Eurasia, N Africa and Australia.

DESCRIPTION:

General: Annual herb from a slender taproot; stems usually freely branched, often somewhat decumbent, (3) 5-10 (15) cm tall, glabrous or sparsely hairy.

Leaves: Mostly basal, egg-shaped, entire, with stalks equal to the blades, or pinnately-cut with a large teminal lobe, 5-20 mm long; stem leaves gradually reduced upwards to linear bracts, without lobes at the base of the bracts.

Flowers: Racemes lax, elongate, 2-6 times as long as the stem proper; flower stalks slender, spreading, often slightly S-shaped, 3-10 mm long; petals white, egg-shaped, about equal to sepals in length; sepals scarcely 1 mm long.

Fruits: Silicles, elliptic to more or less egg-shaped, 2.5-3.5 mm long; beaks less than 0.2 mm long.

IBERIS

Iberis umbellata **L.**

candytuft

HABITAT/RANGE:

Dry roadsides and waste places in the lowland zone; rare in SW BC, known only from Nanoose Bay area; introduced from Europe.

DESCRIPTION:

General: Annual or biennial herb from a taproot; stems erect, leafy, 10-30 cm tall, essentially glabrous.

Leaves: Leaves mainly of the stems, narrowly elliptic, more or less entire, sharp-pointed, essentially glabrous.

Flowers: Inflorescence short, forming a dense rounded cluster; petals unlike, the lower or outer ones much larger than the 2 upper ones, mauve to purple, sometimes whitish.

Fruits: Silicles, strongly compressed at right angles to the inner partitions, beaked, 1- to 2-seeded; segments keeled, wings prolonged upwards into triangular, long-pointed lobes equal to or exceeding the segments.

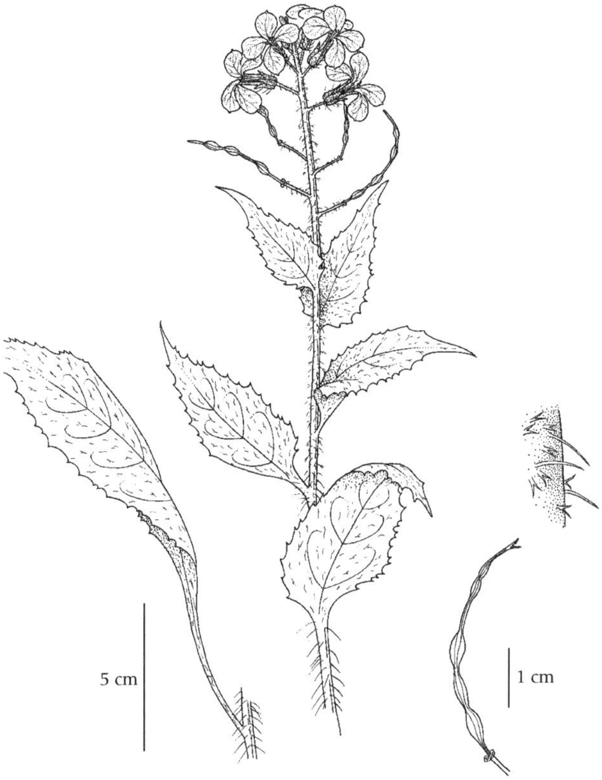

Hesperis matronalis

Hutchinsia procumbens

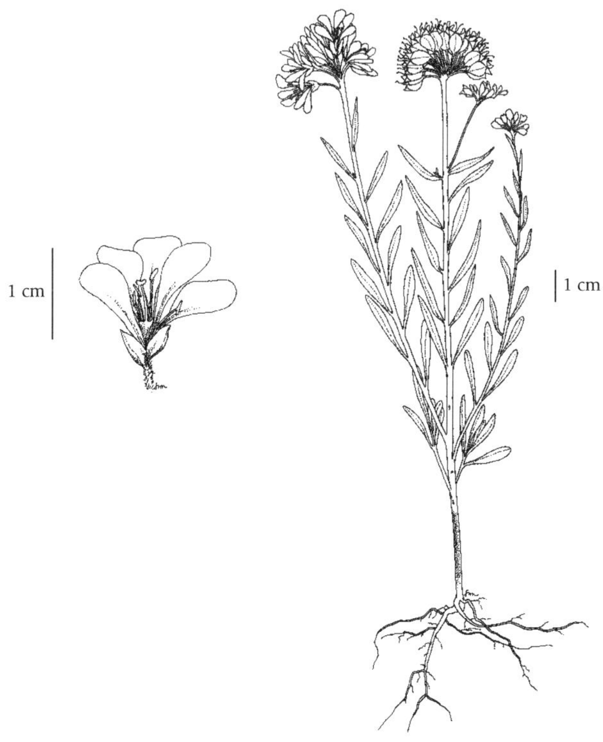

Iberis umbellata

IDAHOA

Idahoa scapigera **(Hook.) A. Nels. & J.F. MacBr.**
scalepod
HABITAT/RANGE:
Moist seepages to dry rocky slopes in the lowland, steppe and montane zones; rare in SW BC, known only from S Vancouver Island and the Gulf Islands, rare eastward in S BC; S to ID and CA.
DESCRIPTION:
General: Annual herb from a slender taproot; stems several, scapose, glabrous.

Leaves: Basal leaves numerous in a rosette; flower stalks slender, usually 1-3 times as long as the blades, blades egg-shaped, entire to deeply lobed, 1-3 cm long; stem leaves lacking.

Flowers: Solitary on leafless stems, (2) 3-13 cm tall; petals white, about equalling the sepals in length; sepals usually red or purplish, 1.4-2 mm long.

Fruits: Silicles, oblong-oval to suborbicular, 6-12 mm long; seeds 6-12, about 5 mm long, 3.5-5 mm wide including about 1 mm wide wing.

ISATIS

Isatis tinctoria **L.**
dyer's woad; asp-of-Jerusalem
HABITAT/RANGE:
Mesic to dry roadsides and waste places in the lowland and montane zones; rare in SW and SE BC, known only from the Duncan and Nelson areas; introduced from Europe.
DESCRIPTION:
General: Biennial or perennial from a stout taproot; stems erect, 0.4-1.2 m tall, simple below, branched above, glabrous, somewhat glaucous.

Leaves: Basal leaves several, clustered, oblanceolate to elliptic-oblanceolate, long-stalked, toothed or wavy-margined, to 18 cm long, soft-hairy on the margins to sparsely soft-hairy with simple hairs; stem leaves lanceolate to elliptic, unstalked, with ear-like basal lobes, mostly entire, glabrous.

Flowers: Racemes compound, numerous; flower stalks slender, conspicuously enlarged below flowers, usually reflexed; petals yellow, about 3.5 mm long, spoon-shaped.

Fruits: Oblong to oblong-oblanceolate, 8-18 mm long, 5-7 mm wide, glabrous, blackish; fruiting stalks somewhat shorter than the silicles; seeds yellowish, ellipsoidal, about 3 mm long.

LEPIDIUM [22]

1. Middle and upper leaves suborbicular, deeply heart-shaped and clasping with slightly overlapping lobes, thus appearing as if the stem passes through the leaf .... *L. perfoliatum*
1. Middle and upper leaves narrower, linear to broadly lanceolate, if clasping then not appearing as if the stem passes through the leaf.
    2. Silicles 5-6 mm long.
        3. Middle and upper leaves not clasping the stems; silicles on strongly ascending to appressed flower stalks .... *L. sativum*
        3. Middle and upper leaves clasping the stems; silicles on spreading flower stalks.
            4. Plants annuals or biennials, usually with a single erect stem; anthers yellow; silicles covered with small white blisters, beaks included to slightly exserted from the shallowly notched tips .... *L. campestre*
            4. Plants perennials, with often decumbent stems; anthers violet; silicles with few or no blisters, beaks mostly exserted from the shallowly notched tips .... *L. heterophyllum*

---

[22] Key adapted from Mulligan (1961).

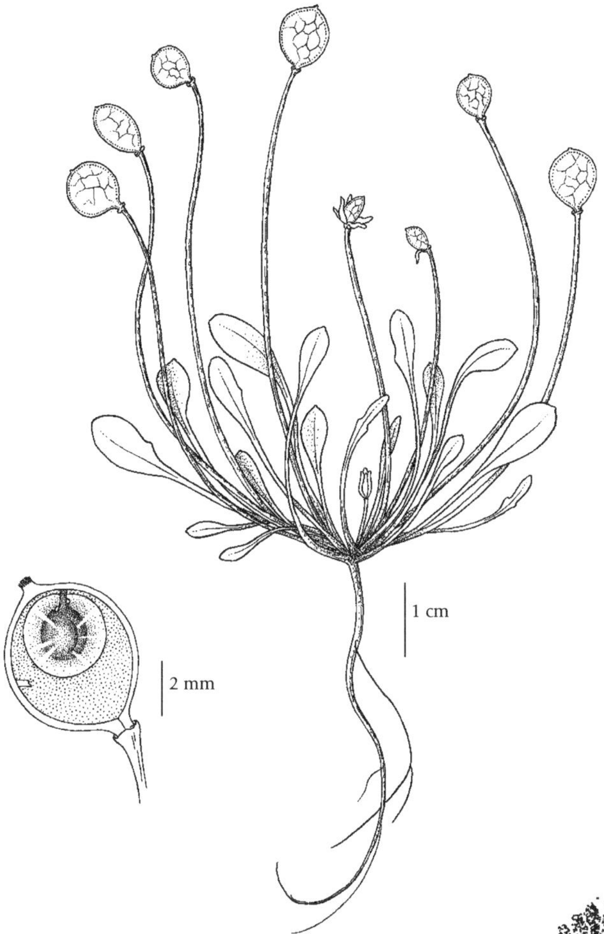

Idahoa scapigera

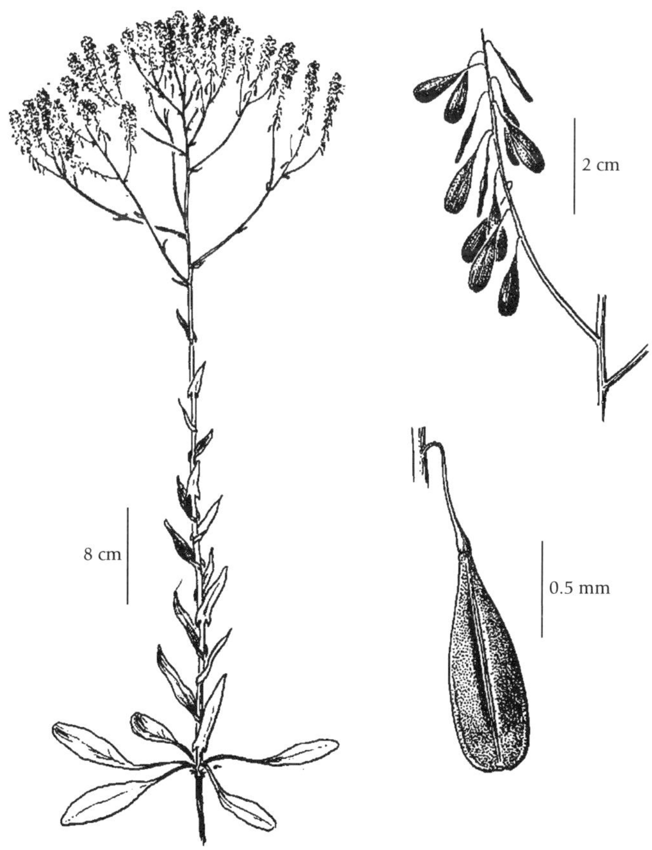

Isatis tinctoria

2. Silicles 2.5-3.5 mm long.
    5. Silicles short-hairy, at least on the margins.
        6. Plants 0.5-2.0 m tall; inflorescence of dense clusters of flowers .... *L. latifolium*
        6. Plants 0.1-0.6 m tall; inflorescence of elongate racemes.
            7. Silicles 2.5-3 by 1.5-2 mm long, nearly elliptic; inflorescence congested into numerous axillary racemes as well as terminal ones .... *L. ramosissimum*
            7. Silicles 3-3.5 by 2.5-3 mm long, orbicular; inflorescence a single raceme or of sparsely branched racemes .... *L. densiflorum*
    5. Silicles glabrous.
        8. Silicles orbicular to nearly round; petals conspicuous, as long or slightly longer than the sepals .... *L. virginicum*
        8. Silicles egg-shaped to round, petals shorter than sepals or lacking.
            9. Silicles egg-shaped, narrowed into sharp-pointed teeth at the tips .... *L. bourgeauanum*
            9. Silicles round to short egg-shaped, abruptly curved into obtuse teeth at the tips .... *L. densiflorum*

Lepidium bourgeauanum **Thell.** (*Lepidium ramosissimum* A. Nels. var. *bourgeauanum* [Thell.] Rollins)
branched pepper-grass; Bourgeau's pepper-grass

HABITAT/RANGE:

Mesic to dry disturbed areas in the montane zone; rare in C and SE BC; N to YT and NT and E to MB, also introduced E to NF.

DESCRIPTION:

General: Biennial herb from taproot; stems erect, 15-60 cm tall, branches ascending to somewhat appressed, sparsely to densely short-hairy.

Leaves: Stem leaves linear to broadly lanceolate, sparsely to densely short-hairy, lower ones incised, middle ones incised or slightly toothed, upper ones entire or slightly toothed.

Flowers: Inflorescence of many racemes, terminal on the branches; flower stalks spreading to ascending, slightly flattened; petals white, lacking or if present then up to 1/2 length of sepals.

Fruits: Silicles, egg-shaped, 2.5-3.5 mm long, 1.5-2 mm wide, narrowed into sharp-pointed teeth at the tips, glabrous; beaks lacking or very short.

Lepidium campestre **(L.) R. Br.**
field pepper-grass

HABITAT/RANGE:

Dry fields, roadsides and waste places in the steppe and montane zones; infrequent in SE BC; introduced from Eurasia.

DESCRIPTION:

General: Annual or biennial herb from a taproot; stems mostly single, erect, branched above, leafy to the inflorescenses, 20-50 cm tall, spreading short-hairy.

Leaves: Basal leaves in a rosette, oblanceolate, stalked, entire to pinnately cut with a large terminal lobe, 5-7 cm long, to 12 mm wide; stem leaves numerous, arrowhead-shaped, minutely toothed, unstalked, upper strongly ear-like and clasping at the base.

Flowers: Inflorescence of often compound racemes; flower stalks widely spreading, 4-8 mm long, slightly flattened; petals white, about 2 mm long, narrowly spoon-shaped; sepals often pinkish, about 1.5 mm long, nearly glabrous to long-hairy; stamens 6; anthers yellow.

Fruits: Silicles, oblong egg-shaped, 5-6 mm long, 4 mm wide, covered with white blisters, otherwise glabrous to short-hairy, margins and tips broadly winged, the tips slightly notched; beaks 0.2-0.6 mm long, included to slightly exserted at the shallowly notched tips; seeds 2.5-3 mm long.

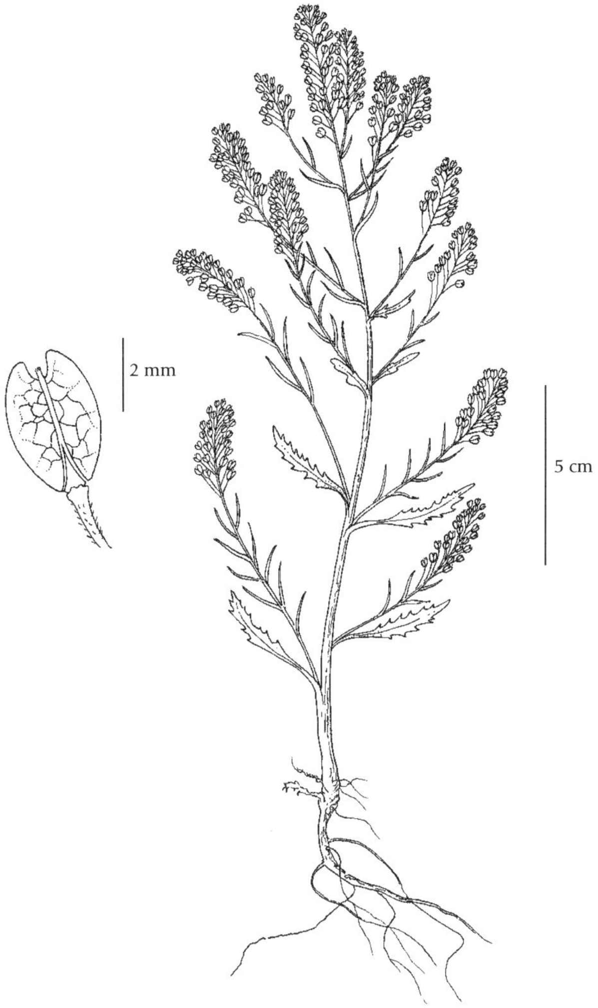

Lepidium bourgeauanum

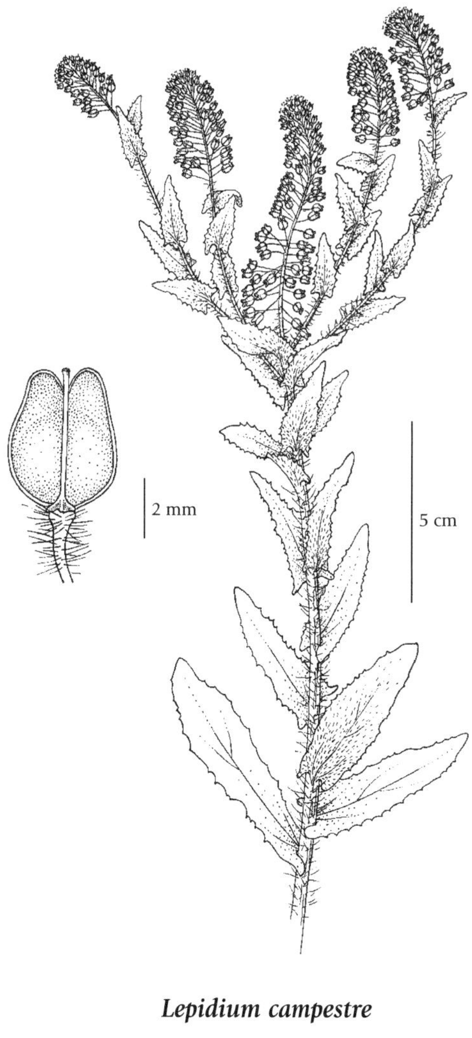

Lepidium campestre

***Lepidium densiflorum* Schrad.**

prairie pepper-grass

Habitat/Range:

Dry disturbed areas and waste places in the steppe and montane zones; var. *densiflorum* – frequent throughout BC, var. *elongatum* – frequent in S and NW BC, var. *macrocarpum* – frequent in S and E BC, and var. *pubicarpum* – rare in SC BC; var. *densiflorum* – native from SE BC to MB, introduced elsewhere, var. *elongatum* – ranges N to YT and NT, var. *macrocarpum* – ranges E to SK, and var. *pubicarpum* – ranges S to VT and CA.

Description:

General: Annual herb from a slender taproot; stems usually single, erect, 10-50 cm tall, branched above, short-hairy.

Leaves: Basal leaves soon deciduous; stem leaves 3-10 cm long, 1-2 cm wide, entire to saw-toothed or pinnately cut with toothed segments, reduced upwards, entire to lobed, not clasping, lower sometimes shallowly pinnately-cut, upper unstalked.

Flowers: Inflorescence ofnumerous, elongate racemes; flower stalks slightly ascending, 2-4 mm long, slightly flattened on lower surfaces, usually hairy on upper surfaces; petals lacking, rarely vestigial, white, to 1 mm; sepals about 1 mm long.

Fruits: Silicles, round to short egg-shaped, 3-3.5 by 2.5-3 mm long, tips slightly winged, abruptly curved into obtuse tee at the tips, shallowly slightly notched, glabrous to short-hairy; beaks lacking.

Notes: Four, sometimes difficult to distinguish, varieties occur in BC.

1. Silicles averaging 2.5 mm long, glabrous; flower stalks slightly flattened .... var. *densiflorum*
1. Silicles averaging 3 mm long, glabrous or short-hairy; flower stalks conspicuously flattened.
    2. Silicles glabrous .... var. *macrocarpum* Mulligan
    2. Silicles short-hairy.
        3. Silicles short-hairy on both surfaces .... var. *pubicarpum* (A. Nelson) Thell.
        3. Silicles short-hairy only on the margins .... var. *elongatum* (Rydb.) Thell.

***Lepidium heterophyllum* (DC.) Benth.** (*L. smithii* Hook.)

Smith's pepper-grass

Habitat/Range:

Mesic to dry fields, roadsides and waste places in the lowland zone; infrequent in SW BC, known from S Vancouver Island and the Vancouver area; introduced from Europe.

Description:

General: Perennial herb from a taproot; stems several, usually ascending, 15-45 cm tall, simple or branched above, stiff-hairy with spreading hairs.

Leaves: Basal leaves oblanceolate to elliptic, stalked, entire to sparsely toothed; stem leaves oblong to narrowly triangular egg-shaped, toothed with narrow teeth, unstalked, clasping, glabrous to hairy.

Flowers: Inflorescence of spreading stalks, 3-6 mm long, densely stiff-hairy, slightly flattened; petals white, 2.5-3 mm long, spoon-shaped, claws narrow; sepals about 2 mm long; anthers violet.

Fruits: Silicles, strongly spreading, oblong, 4-7 mm long, 3-6 mm wide, glabrous or with a few hairs, winged for about 1/3 length, usually with a notch at the tip; segments with few or no vesicles; beaks about 1 mm long, mostly exserted from the shallowly notched tip.

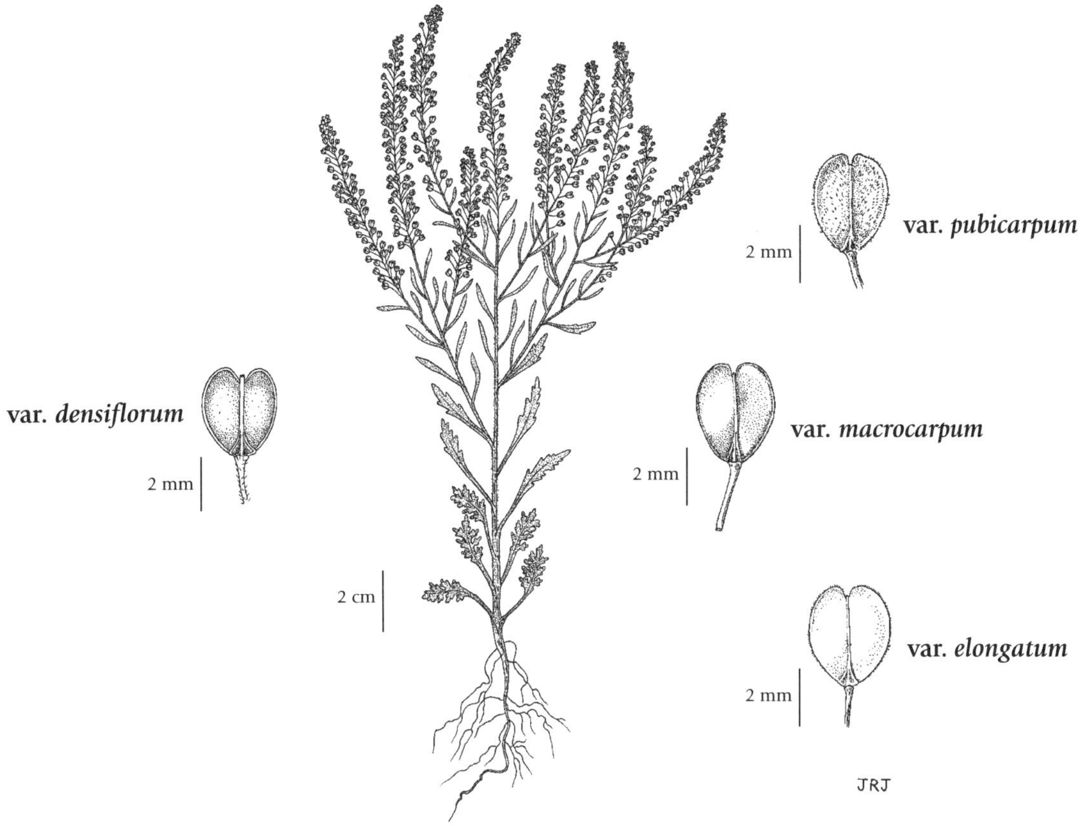

Lepidium densiflorum

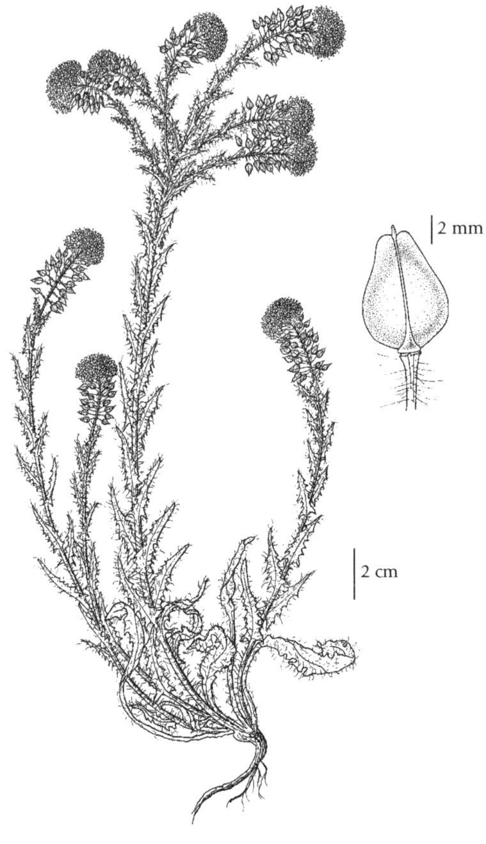

Lepidium heterophyllum

***Lepidium latifolium* L.**
broad-leaved pepper-grass; dittander
HABITAT/RANGE:
Wet to mesic beaches, tidal shores, fields, roadsides and waste places; rare in SW and SE BC, known from Vancover and Windermere; introduced from Europe.
DESCRIPTION:
General: Perennial from widely spreading underground root system; stems numerous, ascending, 0.5-2.0 m tall; glabrous (or almost).

Leaves: Basal leaves entire to evenly toothed, up to 30 cm long, 6-8 cm wide, stalks nearly equaling blades; stem leaves reduced, narrower (1-4 cm wide), lower stalked, upper unstalked.

Flowers: Inflorescence of dense clusters in a large, dense pyramidal panicle; flowers minute; petals white, to twice length of sepals; sepals less than 1 mm long, oval to egg-shaped, widely white-margined.

Fruits: Silicles, orbicular, about 2 mm long, sparsely soft-hairy, unnotched or slight notched at the tips; stalks longer than silicles, round in cross section, glabrous or sparsely hairy; beaks nearly obsolete.

***Lepidium perfoliatum* L.**
clasping-leaved pepper-grass
HABITAT/RANGE:
Dry roadsides and waste places; infrequent in SC BC; introduced from Eurasia.
DESCRIPTION:
General: Annual or biennial herb from a taproot; stems simple to freely branched, (10) 20-60 cm tall, usually somewhat short-hairy or hairy below, glabrous and glaucous above.

Leaves: Leave of 2 distinct types; basal leaves 2- to 3- times pinnately cut into linear segments, to 10 cm long, long-stalked; lower stem leaves usually similar; middle and upper leaves suborbicular, deeply heart-shaped and clasping with slightly overlapping lobes, thus appearing as if the stem passes through the leaf.

Flowers: Flower stalks slender, round in cross section, usually somewhat longer than the silicles; petals pale yellow, about 1.5 mm long (half length of sepals), narrowly spoon-shaped.

Fruits: Silicles, spreading, suborbicular, 3-4 mm long and wide, glabrous or rarely slightly hairy, barely winged at the tips, slightly notched; sinuses about 0.2 mm deep; beaks about equalling the sinus.

***Lepidium ramosissimum* A. Nels.**
branched pepper-grass
HABITAT/RANGE:
Dry roadsides and disturbed area in the lowland and steppe zones; rare in E BC and on the Gulf Islands; E to ON and S to NM, native status uncertain in BC.
DESCRIPTION:
General: Winter annual or biennial herb from a taproot; stems freely branched, leafy, 10-60 cm tall, short-hairy with fine hairs, the hairs often pointing toward the stem-base.

Leaves: Basal leaves usually deciduous by flowering time, pinnately cut, the lobes more or less toothed; upper stem leaves mostly linear, entire to sparsely lobed.

Flowers: Inflorescences in terminal, open, few- to many-flowered racemes, usually with numerous smaller racemes from the upper leaf bases; flower stalks about the length of silicles, nearly round in cross section, very slightly laterally winged, densely hairy; petals white, more or less vestigial, linear, shorter than the sepals or lacking; sepals about 1 mm long.

Fruits: Silicles, nearly elliptic, 2.5-3.0 mm long, glabrous, or hairy near the margins and minutely hairy on the margins, slightly winged at the tip, shallowly slightly notched; sinuses about 0.2 mm deep; beaks lacking or nearly so.

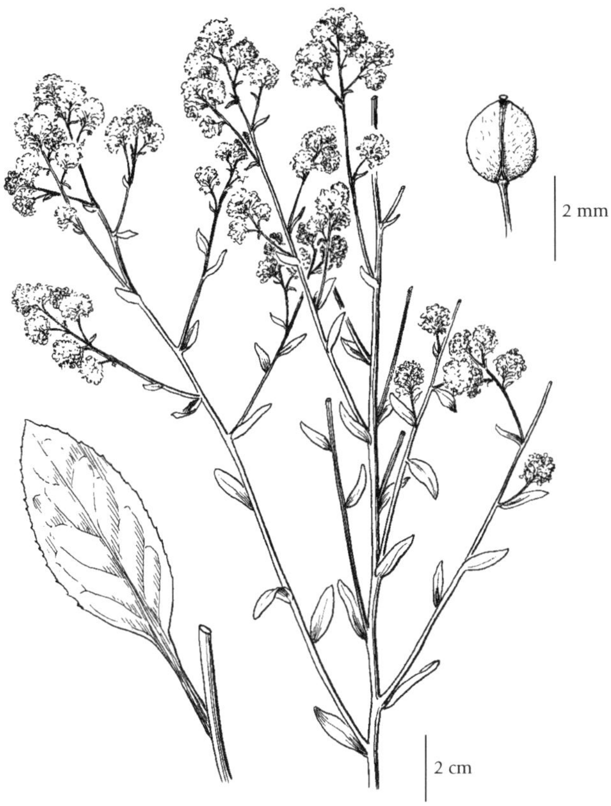

Lepidium latifolium

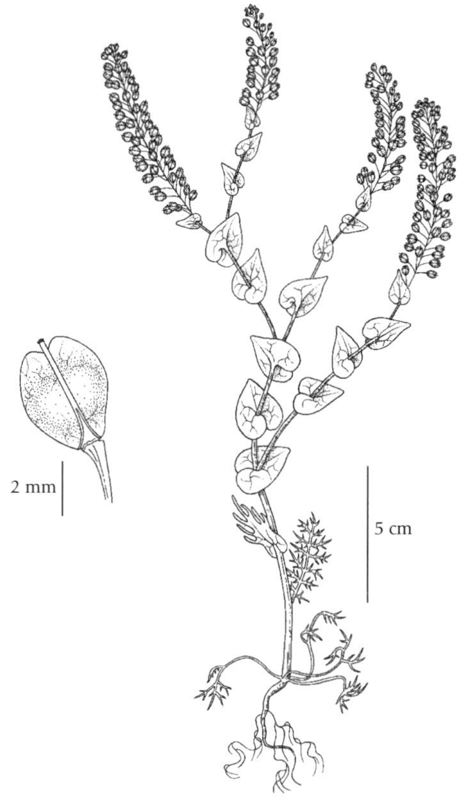

Lepidium perfoliatum

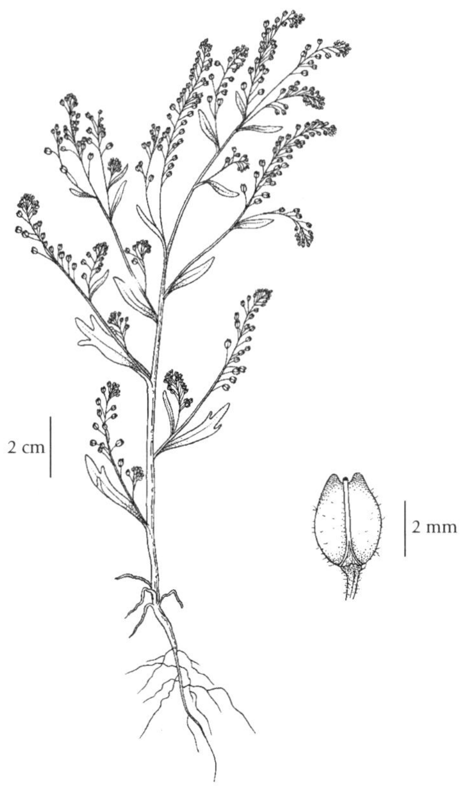

Lepidium ramosissimum

***Lepidium sativum* L.**

garden cress

Habitat/Range:

Dry to mesic roadsides and waste places; rare in SE BC and on S Vancouver Island; introduced from Eurasia.

Description:

General: Annual herb from a slender taproot; stems erect, simple or branched above, 25-70 cm tall, sometimes hairy below, usually glabrous above.

Leaves: Basal leaves commonly deciduous by flowering time; stem leaves pinnately cut and mostly somewhat dissected, the middle and upper ones not clasping.

Flowers: Racemes elongate, spikelike; petals white, about 3 mm long.

Fruits: Silicles, oblong egg-shaped, 5-6 mm long, about 4 mm wide, glabrous, margins somewhat winged and upturned, slightly notched on strongly ascending to appressed stalks; stalks shorter than silicles, nearly round in cross section to slightly flattened; sinuses 0.2-0.4 mm deep; beaks slightly shorter than the sinus margins.

***Lepidium virginicum* L.**

tall pepper-grass

Habitat/Range:

Mesic to dry disturbed areas and waste places in the lowland zone; infrequent in SW BC, known from S Vancouver Island, the Gulf Islands and the adjacent mainland; S to OK, TX, NM, AZ and CA.

Description:

General: Annual herb from a slender taproot; stems erect, 15-60 cm tall, simple below, freely branched above, hairy to nearly glabrous.

Leaves: Basal leaves egg-shaped, pinnately cut or merely toothed, 2-15 cm long, 0.5-5 cm wide, usually lacking at anthesis; stem leaves gradually reduced upwards, entire or rarely toothed.

Flowers: Racemes numerous, elongate, many-flowered; flower stalks spreading, 2-6 mm long, round in cross section to somewhat flattened, glabrous to finely short-hairy; petals white, 1-3 mm long, conspicuous, as long as or slightly longer than the sepals; sepals about 1 mm long, glabrous or slightly soft-hairy.

Fruits: Silicles, orbicular to nearly round, 2.5-4 mm long, scarcely margined, shallowly notched (to 0.2 mm), usually glabrous; beaks short, well included within the notch.

LESQUERELLA

1. Silicles 5-8 mm long, slightly longer than wide; stem leaves few; plants of extreme N BC .... *L. arctica*
1. Silicles 3-4 mm long, as wide as long; stem leaves usually numerous; plants of S BC .... *L. douglasii*

Lesquerella arctica* (Wormsk. *ex* Hornem.) S. Wats. var. *arctica (*L. purshii* [S. Wats.] Fern. var. *arctica*)

arctic bladderpod

Habitat/Range:

Mesic to dry gravel bars, river terraces, scree slopes and rocky slopes in the montane to alpine zones; rare in extreme N BC; amphiberingian, N to AK, YT and NT; NW Asia.

Description:

General: Perennial from short taproot and a woody stem-base; stems decumbent to erect, 4-20 (25) cm tall.

Leaves: Basal leaves in a rosette, egg-shaped to oblanceolate, stalked, entire, (1) 2-6 (15) cm long; stem leaves much reduced, 0.5-1.5 (3) cm long, unstalked or short-stalked, hairy, the hairs starlike.

Flowers: Inflorescences loose; flower stalks ascending to erect, 5-20 (40) mm long; petals yellow, 5-6 (7) mm long; sepals 3-4.5 mm long, hairy, the hairs starlike.

Fruits: Silicles, subglobose to ellipsoid, 5-8 mm long, 4-5.5 (7) mm wide, glabrous or hairy, the hairs starlike; beaks 1-2.5 mm long.

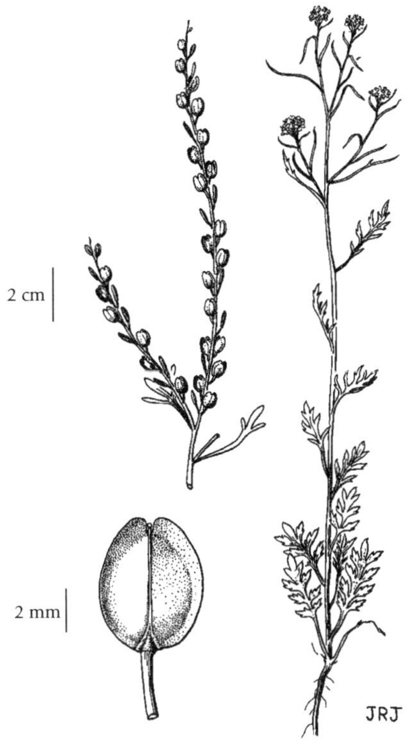

Lepidium sativum

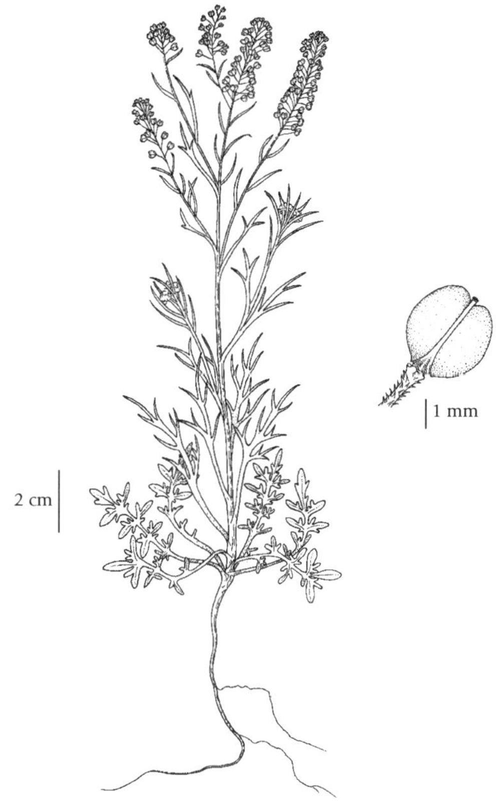

Lepidium virginicum

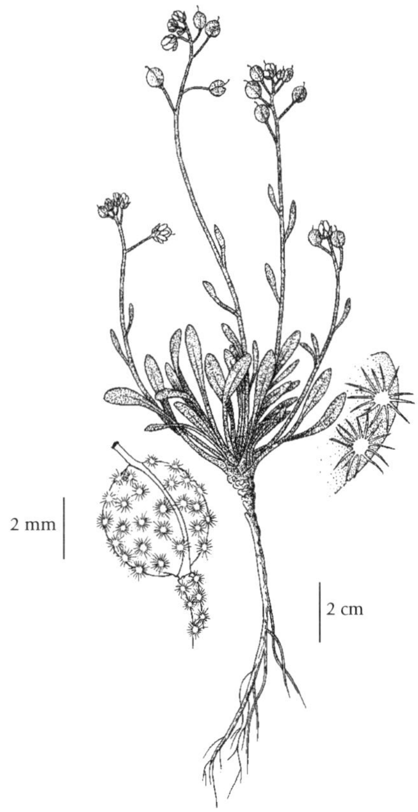

Lesquerella arctica var. *arctica*

Lesquerella douglasii **S. Wats.**

Columbia bladderpod

HABITAT/RANGE:

Dry grasslands, shrublands, meadows and rocky slopes in the steppe to alpine zones; infrequent in extreme SC and SE BC; S to ID and N OR.

DESCRIPTION:

General: Perennial from a taproot and a simple stem-base; stems several, erect, rarely branched, (10) 20-40 (50) cm tall, densely hairy.

Leaves: Basal leaves suborbicular to elliptic, stalked, entire to remotely toothed, 3-12 cm long, 2-15 mm wide, hairy; stem leaves many, usually linear to linear-oblanceolate, entire to sinuate or remotely toothed, 1-4.5 cm long, hairy.

Flowers: Flower stalks erect and nearly straight to S-shaped or slightly recurved, 6-17 mm long; petals yellow to yellowish-orange, 6-9 mm long.

Fruits: Silicles, very short-stalked, inflated, egg-shaped to nearly globose, 3-4 mm long and as wide, sparsely hairy or glabrous; beaks about equalling the silicle in length; seeds flattened, not winged.

LOBULARIA

Lobularia maritima **(L.) Desv.**

sweet alyssum

HABITAT/RANGE:

Dry waste places in the lowland zone; rare in SW BC; known only from SE Vancouver Island; introduced from Europe.

DESCRIPTION:

General: Perennial herb from a taproot; stems usually branched from the base, prostrate or spreading, 10-30 cm tall, greyish, hairy, the hairs pick-shaped or nearly glabrous.

Leaves: Basal leaves lacking; stem leaves linear to linear-lanceolate, sharp-pointed, 1-4 cm long, 1-4 mm wide.

Flowers: Very fragrant; flower stalks spreading, slender, 5-10 mm long; petals white or bluish-tinged, widely egg-shaped and short-clawed, 3-4 mm long; sepals 1.5-2 mm long, soon deciduous.

Fruits: Silicles, broadly egg-shaped or suborbicular, 2-3.5 mm long, sharp-pointed, often sparsely hairy; beaks about 0.5 mm long; seeds wingless.

LUNARIA

Lunaria annua **L.**

honesty

HABITAT/RANGE:

Mesic to dry roadsides and waste places in the lowland, steppe and montane zones; rare in S BC; introduced from Europe.

DESCRIPTION:

General: Annual or biennial from a taproot; stems 0.5-1.0 cm tall, freely branched, hairy.

Leaves: Stem leaves heart-shaped, coarsely toothed, 4-10 cm long, stalked, unstalked to short-stalked and clasping above, hairy.

Flowers: In loose racemes, scentless; petals bluish-purple, about 2 cm long.

Fruits: Silicles, oblong-oval, strongly compressed, 3.5-4.5 cm long, about 2/3 as wide; flower stalks slender, 7-12 mm long; beaks 6-8 mm long; seeds 3-5 per chamber, 7-9 mm wide, winged.

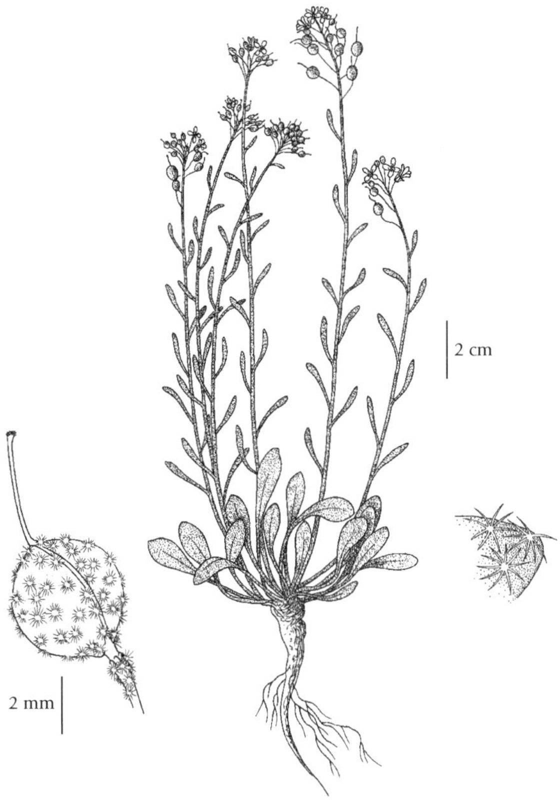

Lesquerella douglasii

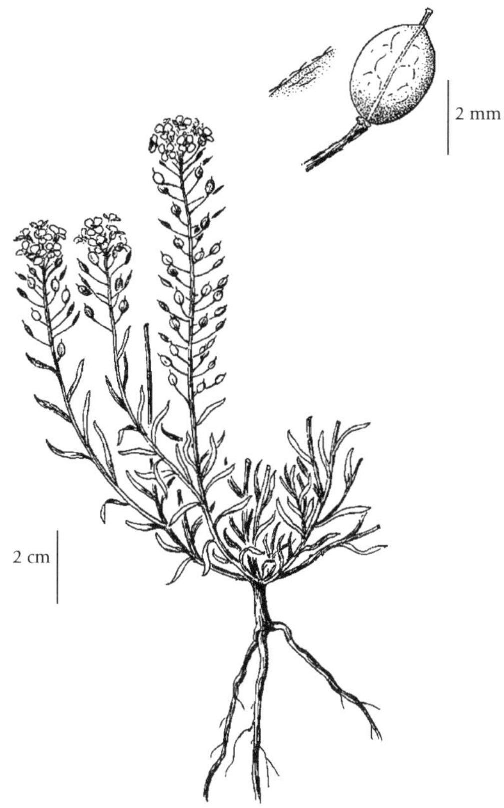

Lobularia maritima

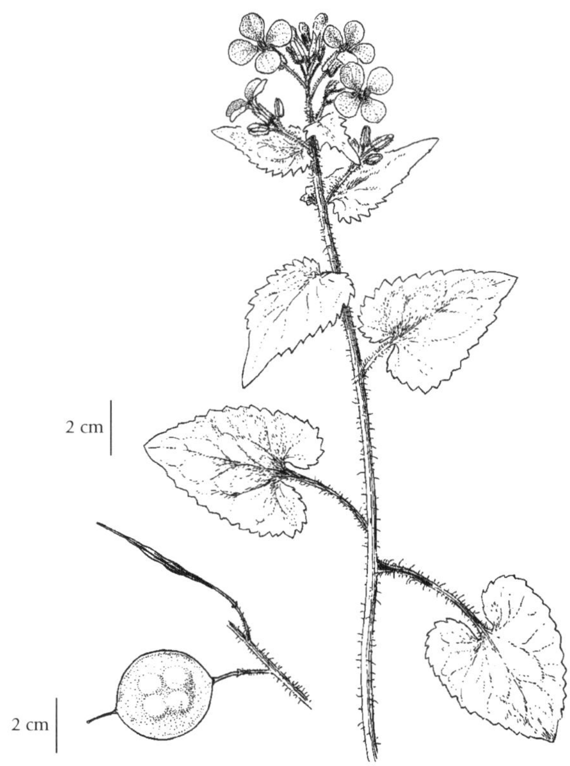

Lunaria annua

NESLIA

Neslia paniculata **(L.) Desv.**
ball mustard

Habitat/Range:

Mesic to dry fields and waste places in the lowland, steppe and montane zones; infrequent in S BC, rare northward; introduced from Eurasia.

Description:

General: Annual herb from a taproot; stems erect, single, usually branched above, 20-90 cm tall, more or less stiff-hairy with stalked, starlike, and branched hairs.

Leaves: Basal leaves soon deciduous; stem leaves lanceolate to elliptic, entire, 1.5-7 cm long, 2-20 mm wide, hairy with starlike or branched hairs, unstalked, with ear-like lobes at the base, becoming narrowly oblong-lanceolate and reduced above.

Flowers: Inflorescence much elongating in fruit; flower stalks slender, 6-10 mm long, ascending, straight; petals light yellow, 2-2.5 mm long; sepals greenish, 1.2-2 mm long, glabrous or short-hairy.

Fruits: Silicles, almost spherical, about 2 mm long, slightly broader, only slightly compressed, walls hardened, conspicuously wrinkled, glabrous; beaks 0.7-1 mm long, persistent or deciduous.

PARRYA

Parrya nudicaulis **(L.) Boiss.** (*P. nudicaulis* ssp. *interior* Hult.)
northern parrya

Habitat/Range:

Moist to mesic meadows in the alpine zone; rare in extreme NW BC; amphiberingian, N to AK, YT and NT; Asia.

Description:

General: Tufted perennial from a stout root, up to 50 cm long and 1.5 cm thick and a usually branched, enlarged stem-base, the branches clothed in old leaf bases; stems 5-35 cm tall, leafless, erect or ascending, simple, glabrous or hairy with stalked glandular hairs.

Leaves: Spoon-shaped or linear-oblong to broadly oblanceolate or nearly egg-shaped, 1.5-12 cm long, 0.4-3 cm wide, entire to more or less toothed, stalked, with stalked glandular hairs or glabrous.

Flowers: Inflorescence racemose; flower stalks 3-50 mm long, ascending to erect, with stalked glandular hairs to glabrous; petals rose-purple or rarely white, 10-20 mm long, 5-9 mm wide; sepals often tinged purplish, white-membranaceous along margins, 5.5-9 mm long, with stalked glandular hairs or glabrous.

Fruits: Siliques, 1.5-6.5 cm long, 3.5-7.5 mm wide, flat, tapering at both ends, constricted between the seeds, glabrous or glandular hairy; beaks 1-2.5 mm long; seeds light brown, 4-5.5 mm long, 3-4.5 mm wide, elliptic, flattened, covered with loose, netlike seedcoats that form wide wings.

Neslia paniculata

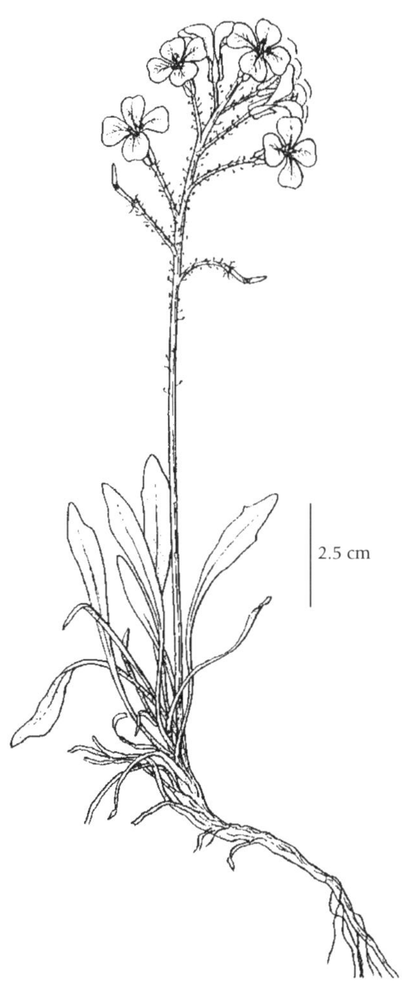

Parrya nudicaulis

PHYSARIA

Physaria didymocarpa **(Hook.) A. Gray var. *didymocarpa***

common twinpod

HABITAT/RANGE:

Dry grasslands, shrublands and rocky slopes in the steppe and montane zones; rare in SE BC; E to SW AB and S to WA, ID, and WY.

DESCRIPTION:

General: Tufted perennial from a heavy-rooted, often branched stem-base; stems numerous, decumbent, 2-17 cm long, with silvery starlike hairs, these not closely appressed.

Leaves: Basal leaves numerous, in a rosette, withering and persistent, 2-8 cm long, blades egg-shaped, mostly obtuse, with few inconspicuous teeth or toothed, rarely entire, with narrow to wide winged stalks, these about as long as the blades, silvery; stem leaves several, reduced, oblanceolate, entire or with occasional tooth.

Flowers: Racemes congested, elongating in fruit; flower stalks spreading, straight or slightly curved, 8-12 mm long; petals yellow, 9-12 mm long, broadly spoon-shaped or egg-shaped.

Fruits: Silicles, paired, much-inflated, 1-2 cm long, at least as wide, base very slightly heart-shaped, apical sinus narrow, nearly closed, 2-4 mm deep, loosely hairy with rays of hairs spreading; beaks 6-9 mm long; seeds 2-3 per chamber.

RAPHANUS

1. Flowers usually yellow; siliques usually 4- to 12-seeded, rarely 2- or 3-seeded, 3-6 mm wide .... *R. raphanistrum*
1. Flowers usually purple; siliques 1- to 3-seeded, rarely 4- or 5-seeded, 5-10 mm wide .... *R. sativus*

Raphanus raphanistrum **L.**

wild radish

HABITAT/RANGE:

Mesic to dry fields and waste places in the lowland, steppe and montane zones; frequent in SW BC, rare in SC BC; introduced from Eurasia.

DESCRIPTION:

General: Annual or biennial herb from a slender, whitish taproot; stems 30-80 cm tall, freely branched above, sparsely stiff-hairy

Leaves: Basal leaves pinnately cut with the terminal lobe largest, 6-20 cm long, 1-4 pairs of distant lateral lobes; stem leaves several, reduced, stalked.

Flowers: Inflorescence of ascending flower stalks, 1-2.5 cm long; petals yellow fading to white, often purplish-veined or -tinged, 15-20 mm long, twice the length of sepals.

Fruits: Siliques, 4-8 cm long (including the 1-2 cm beaks), 3-6 mm wide, moniliform, finally breaking between the 1-seeded segments, prominently several-grooved lengthwise; seeds 4-12.

Raphanus sativus **L.**

garden radish

HABITAT/RANGE:

Mesic to dry disturbed areas and waste places in the lowland zone; rare in SW BC; introduced from Eurasia.

DESCRIPTION:

General: Annual or biennial herb from a fleshy, elongated taproot; stems 0.3-1.2 m tall, simple or branched, sparsely stiff-hairy.

Leaves: Basal leaves pinnately lobed, 3-20 cm long, 1-8 cm wide, stalked, often round-toothed, terminal segment larger, orbicular; stem leaves reduced upwards.

Flowers: Inflorescence of spreading to ascending stalks, 8-27 mm long; petals pink, pink-purple to white or yellow with darker venation, 13-20 mm long; sepals 7-11 mm long, glabrous.

Fruits: Siliques, 2-6 cm long, 5-10 mm wide, continuous or sometimes slightly moniliform, lightly several-grooved, usually not breaking transversely, beaks conical, sharp, 1/4 to 1/2 as long as the body, glabrous; seeds 1-3 (5).

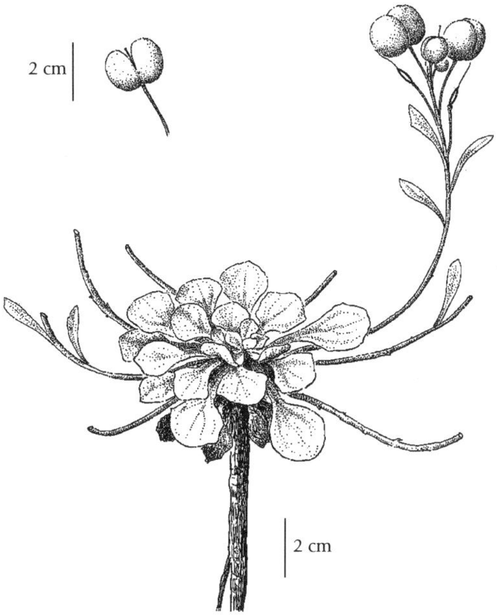

Physaria didymocarpa var. *didymocarpa*

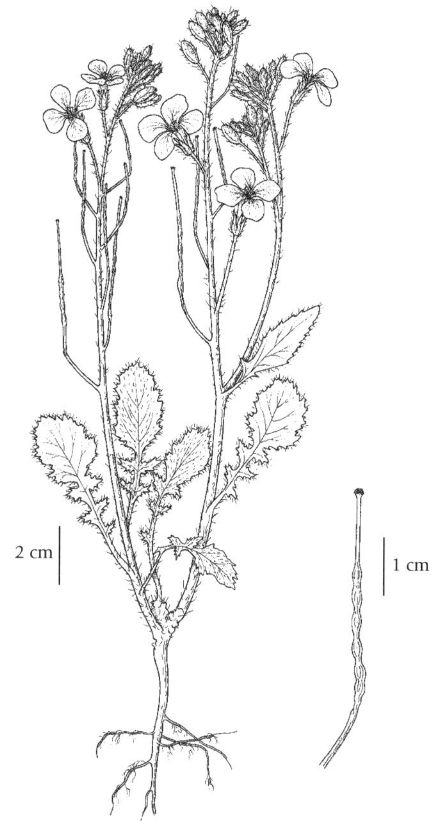

Raphanus raphanistrum

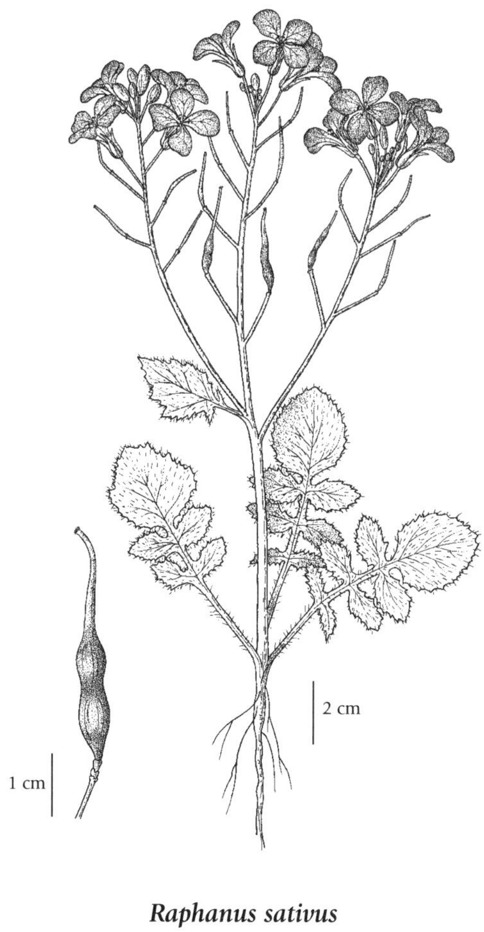

Raphanus sativus

RORIPPA

Notes: This is a difficult genus with many forms having been given specific status by various authors. There is a great deal of overlap in morphological characters among the taxa.

1. Petals white, sometimes purplish-tinged.
   2. Beaks usually over 1 mm long; seeds more or less 1-rowed with about 100 pits on each face .... *R. microphylla*
   2. Beaks lacking or less than 1 mm long; seeds distinctly 2-rowed with about 25 pits on each face .... *R.nasturtium-aquaticum*
1. Petals yellow.
   3. Plants rhizomatous perennials .... *R. sylvestris*
   3. Plants annuals or biennials, lacking rhizomes.
      4. Siliques 4- to 6-segmented, globose or pear-shaped; rare in N BC .... *R. barbareifolia*
      4. Siliques 2-segmented, egg-shaped to linear; mostly in S BC.
         5. Stalks of the silicles mostly 4-12 mm long, longer than siliques; stems usually erect .... *R. palustris*
         5. Stalks of the silicles mostly 2-4 mm long, shorter than siliques; stems often spreading or decumbent.
            6. Siliques linear-oblong; 6-15 mm long, 1-1.5 mm thick, usually curved .... *R. curvisiliqua*
            6. Siliques oval to oblong, 2-8.5 mm long, 2-2.5 mm thick; silicles pimpled or smooth, curved or not curved
               7. Seeds rough, minutely honeycombed and pimpled .... *R. tenerrima*
               7. Seeds smooth .... *R. curvipes*

***Rorippa barbareifolia* (DC.) Kitagawa**
hoary yellow cress

HABITAT/RANGE:
Mesic to dry roadsides in the montane zone; rare in N BC, known only from the Alaska Highway, where probably introduced; AK and NT; E Siberia.

DESCRIPTION:
General: Annual or biennial herb from a taproot; stems 30-70 (90) cm tall, erect, usually single, branched above, densely long-hairy below, moderately long-hairy to glabrous above.

Leaves: Basal leaves in rosettes, short-stalked, pinnately-cut with the terminal lobe the largest, obtuse; stem leaves often densely overlapping, oblong, oblanceolate or lanceolate, irregularly saw-toothed, raggedly divided or slightly wavy-margined, densely to sparingly long-hairy to stiff-hairy; lower stem leaves short wing-stalked to unstalked with ear-like lobes at the bases; upper stem leaves unstalked, with prominent clasping, ear-like lobes at the bases.

Flowers: Racemes terminal and axillary, 10-20 cm long; flower stalks ascending, 5-13.5 mm long, stiff-hairy or glabrous; petals yellowish, egg-shaped to broadly spoon-shaped, 1.8-3.5 mm long.

Fruits: Siliques, globose or pear-shaped, (3.2) 4-5.7 mm long; segments 4, rarely up to 6; beaks 0.6-1.4 mm long, abruptly attached to the rounded silique tips; seeds dark reddish-brown, oblong, finely wrinkled and pitted.

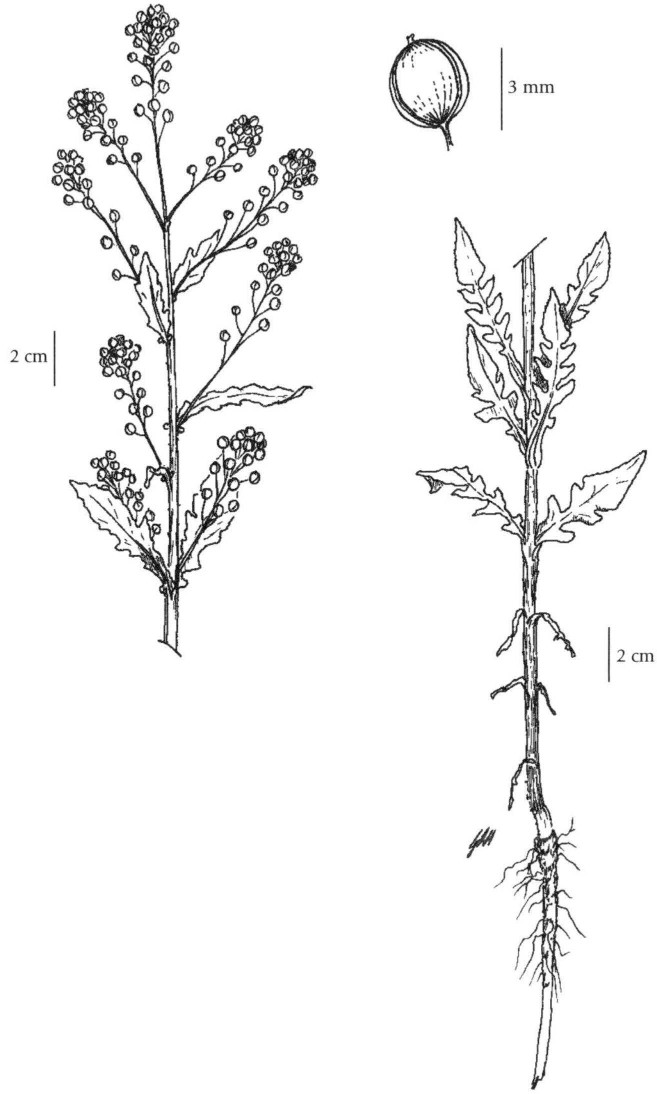

Rorippa barbareifolia

Rorippa curvipes **Greene var.** ***truncata*** **(Jeps.) Rollins** (*R. curvipes* var. *integra* [Rydb.] Stuckey)
blunt-leaved yellow cress

HABITAT/RANGE:

Moist sandy soils in the lowland to lower montane zones; infrequent in S BC east of the Coast-Cascade Mountains; E to MI, MO and TX, and S to CA and N MX.

DESCRIPTION:

General: Annual, or rarely, short-lived perennial; stems erect to somewhat decumbent, 10-50 cm long, single to several, usually branched at base, sparingly to moderately stiff-hairy below.

Leaves: Basal leaves egg-shaped to oblanceolate, 1.5-6 cm long, 5-20 mm wide, entire to shallowly wavy-margined and lobed to pinnately-cut with egg-shaped, wavy-margined to entire segments, generally stalked, lower surfaces glabrous, upper with few hairs; stem leaves oblong, egg-shaped, or oblanceolate, 4-8 cm long, 5-15 mm wide, generally with clasping, ear-like lobes, entire or round-toothed to saw-toothed.

Flowers: Inflorescence of 2-5 (7) mm long, divergent to ascending stalks; petals very pale yellow, oblong to spoon-shaped, 1-1.5 mm long, 0.2-1 mm wide; sepals 1-1.5 mm long, glabrous, not pouched at base, deciduous shortly after anthesis.

Fruits: Siliques, oval to oblong-lanceolate, slightly curved inward, 3-6 (8) mm long, 1.5-2.5 mm wide, blunt to obtuse at the tips; beaks 0.2-1 mm long; seeds brown, about 0.5 mm long, minutely honeycombed and pimpled, 10-80.

Rorippa curvisiliqua **(Hook.) Bess.** ***ex*** **Britt.**
western yellow cress

HABITAT/RANGE:

Moist waste places, meadows and open woods in the lowland and lower montane zones; common in S BC; E to MT and CO and S to Baja CA.

DESCRIPTION:

General: Annual or biennial herb from a taproot; stems single or diffusely branched above, 10-40 cm tall, decumbent to ascending, glabrous to sparsely hairy throughout, the hairs appressed.

Leaves: Variable, 2-7 cm long, 8-20 mm wide, oblong to oblanceolate, toothed, lobed or pinnately-cut, the segments linear to egg-shaped, entire or toothed, glabrous or short-hairy; stem leaves reduced upwards, clasping with earlike lobes basally.

Flowers: Inflorescence of 1-5 (7) mm long, spreading to ascending or slightly recurved stalks; petals yellow, 1-2 mm long; sepals 1-2 mm long, often spreading, promptly deciduous or somewhat persistent until the silicles are well formed, glabrous.

Fruits: Siliques, linear to oblong, 6-15 mm long, 1-1.5 mm wide, curved upward to sometimes straight, nearly round in cross section, 2-carpelled; beaks 0.3-1 mm long; seeds finely pimpled.

Notes: Several poorly defined varieties are sometimes recognized, but are not treated here.

Rorippa microphylla **(Boenn.** ***ex*** **Reichenb.) Hyl.** ***ex*** **A.& D. Love** (*Nasturtium microphyllum* Boenn. *ex* Reichenb., *N. officinale* Ait. f. var. *micropyllum* Boenn. *ex* Reichenb.)
one-rowed water cress

HABITAT/RANGE:

Shallow streams and ponds in the lowland and montane zones; rare in NW and SW BC, known only from S Vancouver Island and Atlin; probably introduced from E Canada.

DESCRIPTION:

General: Aquatic or semi-aquatic perennial herb from a fibrous root, usually forming colonies; main stems thick, greatly elongated, prostrate or decumbent, branched, rooting at nodes, 10-50 cm long.

Leaves: Basal leaves pinnate with 5-9 leaflets, stalks usually with small ear-like lobes at the base, terminal leaflets similar to lateral ones or sometimes larger, lateral ones mostly oblong, 1-2 cm long, unstalked, wedge-shaped at the base or with short stalks; stem leaves similar.

Flowers: Inflorescences loose, elongated; flower stalks widely spreading, 1-1.5 cm long; petals white, narrowly spoon-shaped, 4-5 mm long; sepals erect, outer pair pouched at base.

Fruits: Siliques, widely spreading to ascending, slightly curved upward or inward, 1.5-2.5 cm long, 1-1.5 mm wide, nearly round in cross section, faintly torulose; beaks about 1 mm long; seeds more or less in 1 row, reddish-brown, 1-1.2 mm long, about 75-100 pits per side.

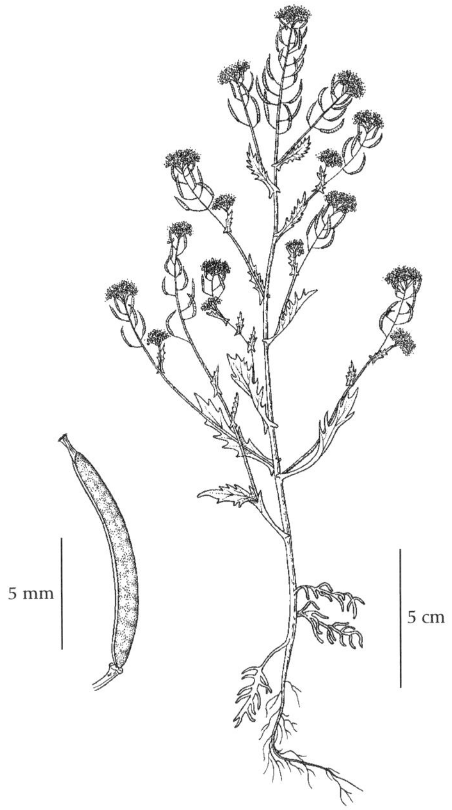

Rorippa curvipes var. *truncata*

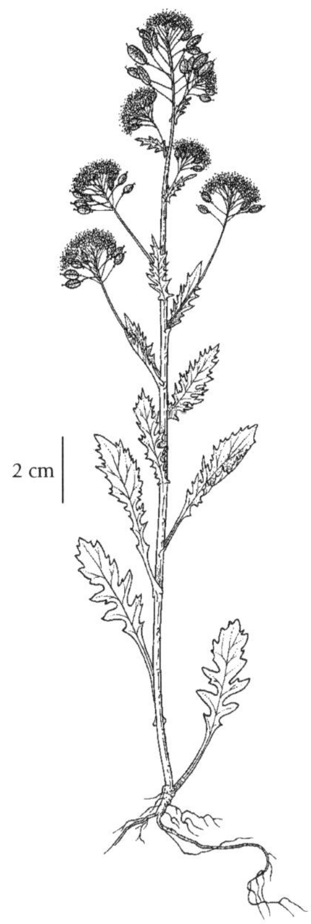

Rorippa curvisiliqua

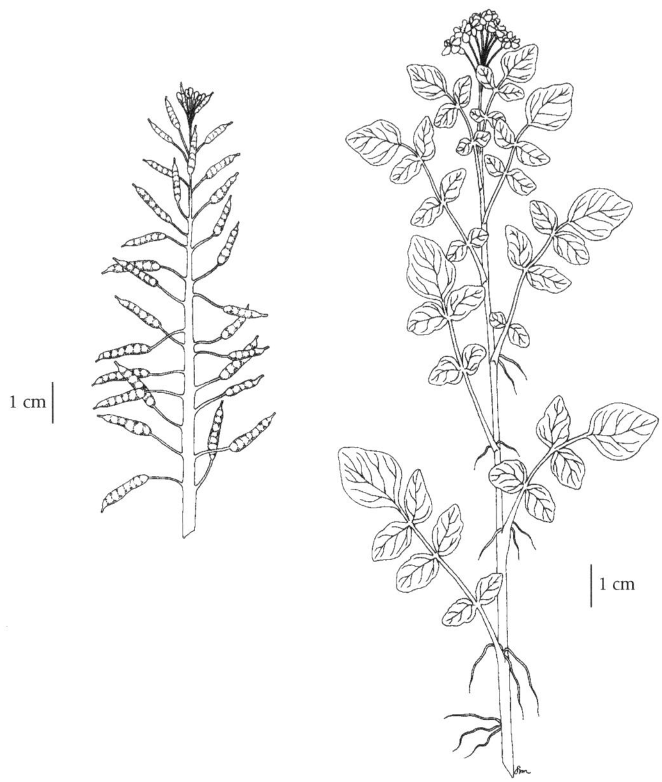

Rorippa microphylla

***Rorippa nasturtium-aquaticum* [L.] Hayek** (*Nasturtium officinale* Ait. f.)

common water cress

HABITAT/RANGE:

Streams and shallow ponds in the lowland and montane zones; frequent in S BC, rare northward; introduced from Europe.

DESCRIPTION:

General: Aquatic or semi-aquatic perennial herb from a fibrous root; stems leafy, succulent, submersed or partly floating, sometimes prostrate on mud, 10-60 cm long, only the tip usually erect, rooting at the nodes, glabrous or sparsely hairy.

Leaves: Basal leaves pinnate with 3-7 leaflets, egg-shaped to broadly oblong, unstalked, without ear-like lobes at the bases, entire or wavy-margined, terminal leaflets larger than lateral, stalked, base of the stalks usually with small ear-like lobes at the bases, glabrous; stem leaves similar.

Flowers: Racemes corymbose, elongate in fruit; flower stalks 6-20 mm long, spreading or even slightly recurved to ascending; petals white to purplish-tinged, 3-4 mm long, broadly spoon-shaped; sepals 2.5-3.5 mm long, glabrous, the outer pair pouched at base.

Fruits: Siliques, widely spreading to somewhat ascending, slightly curved upward to straight, narrowly oblong, 10-25 mm long, 2-3 mm wide; beaks obscure, rarely to 0.5 mm long; seeds distinctly 2-rowed, flattened and very slightly wing-margined, coarsely wrinkled, about 1 mm long, about 24-60 pits per side.

***Rorippa palustris* (L.) Bess.** (*R. islandica* [Oed.] Borbas)

marsh yellow cress

HABITAT/RANGE:

Wet meadows and boggy areas in the lowland and lower montane zones; common throughout BC; circumboreal, throughout North America.

DESCRIPTION:

General: Annual, biennial, or short-lived perennial herb from a slender taproot; stems erect, 15-100 cm tall, simple to much branched from base, glabrous to stiff-hairy.

Leaves: Basal and lower stem leaves oblong to oblanceolate, (4) 6-20 cm long, 0.5-4 cm wide, short-stalked to unstalked, irregularly saw-toothed, incised, deeply cleft, repand, or variously pinnately divided, with slightly clasping, ear-like lobes at the bases, sharp- or long-pointed at the tips; glabrous or stiff-hairy.

Flowers: Inflorescence of spreading-ascending stalks, 3-14 mm long, glabrous to sparsely hairy; petals light yellow, 1-3.5 mm long, oblong to broadly spoon-shaped; sepals often yellowish or purplish, 1-2 mm long, glabrous.

Fruits: Siliques, spreading to ascending, straight to somewhat upcurved, 3-14 mm long, 2-3 mm wide, egg-shaped to cylindric or clavate, 1- to 4-carpelled; beaks 0.2-1 mm long, straight, abruptly attached to apex; seeds brown to reddish-brown, mostly 0.4-0.6 mm long, minutely honeycombed and pimpled, 20-80.

Notes: At least six poorly defined varieties have been attributed to the BC flora, but are not treated here.

***Rorippa sylvestris* (L.) Bess.**

creeping yellow cress

HABITAT/RANGE:

Lawns, waste places and fields in the lowland and montane zones; frequent on S Vancouver Island, the Gulf Islands and adjacent mainland, rare eastward and northward; introduced from Europe.

DESCRIPTION:

General: Perennial herb from slender rhizomes, often mat-forming; stems nearly erect to decumbent, 15-60 cm tall, usually branching from base, nearly glabrous.

Leaves: Basal leaves in rosettes, these disappearing after the first year; stem leaves egg-shaped or oblong-oblanceolate to oblong egg-shaped, 4-10 cm long, deeply lobed to pinnately-cut, the divisions narrow, incised or sharply toothed to entire, slender-stalked.

Flowers: Racemes numerous, dense and short; flower stalks spreading to ascending, (3) 6-12 mm long; petals yellow, 3-5 mm long; sepals 2-3 mm long, deciduous shortly after anthesis, the outer pair not pouched at base.

Fruits: Siliques, linear, often curved upward, 1-2 cm long, 1-2 mm wide; beaks 0.5-1 mm long; seeds reddish-brown, 0.6-0.8 mm long, finely honeycombed, not margined.

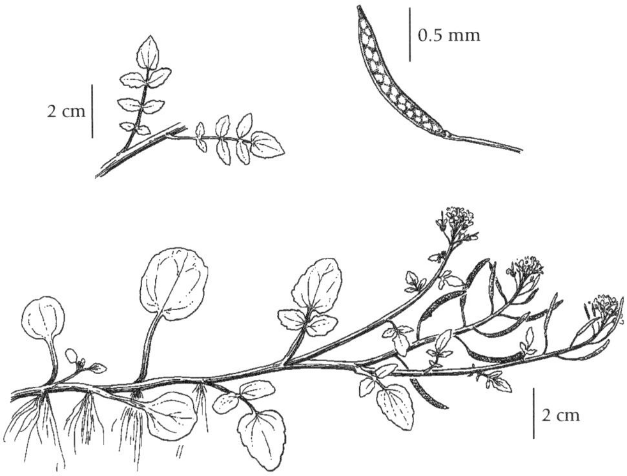

Rorippa nasturtium-aquaticum

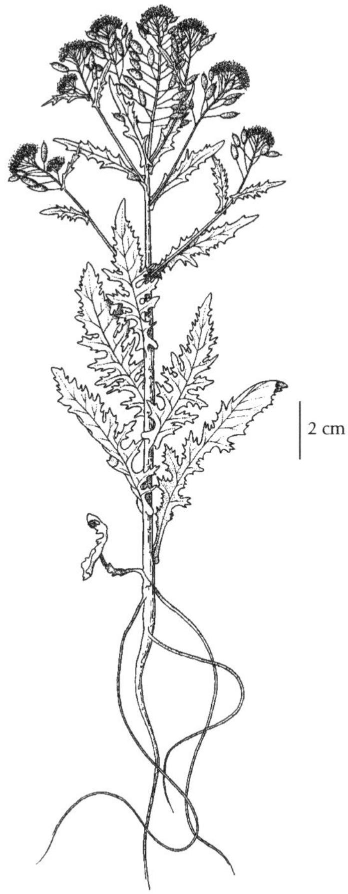

Rorippa palustris

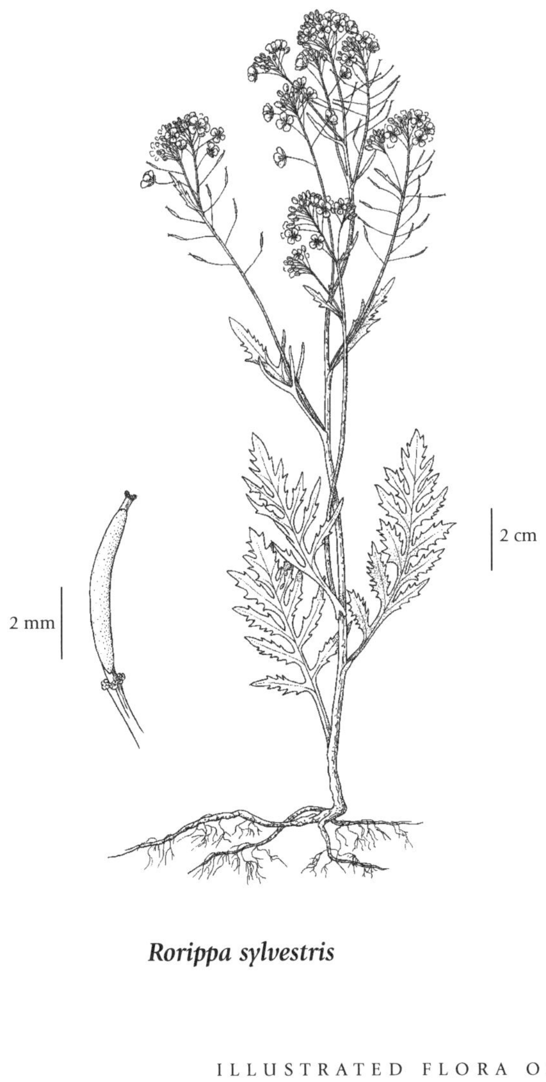

Rorippa sylvestris

Rorippa tenerrima **Greene** (*R. obtusa auct. non* [Nutt.] Britt.)
yellow cress
Habitat/Range:
Moist to wet streambanks, marshes, wet meadows, mud flats and lake margins in the lowland, steppe and montane zones; rare in S BC; E to SK and S to Nd, SD, OH, NM, AZ, CA and N MX.
Description:
General: Annual herb from a taproot; stems decumbent to prostrate, 10-20 cm long, much branched from base; nearly glabrous.

Leaves: Stem leaves oblong, oblanceolate to spoon-shaped, 2-5 (8) cm long, 8-15 mm wide, short-stalked, without earlike lobes at the bases, generally deeply pinnately lobed, middle lobes oblong to elliptic, 3-6 (10) mm long, 1-4 (9) mm wide, entire.

Flowers: Inflorescence of ascending, glabrous to rough with minute papillae, 2-4 mm long stalks; petals yellow, 0.6-0.8 mm long, 0.1-0.3 mm wide, oblong to narrowly spoon-shaped; sepals 0.7-1.2 mm long, 0.4-0.6 mm wide, egg-shaped to oblong, not persistent.

Fruits: Siliques, elongate-cylindric, curved slightly inward, occasionally outward, 3.5-8.5 mm long, obtuse below, tapering to apices, sparsely to densely minutely pimpled; beaks 0.5-1 mm long, broadened below, gradually merging into the pointed apex; seeds reddish-brown, 0.5-0.7 mm long, widely kidney-shaped, smooth, 20-40.

SCHOENOCRAMBE

Schoenocrambe linifolia **(Nutt.) Greene**
plains mustard; rush mustard
Habitat/Range:
Dry grasslands and shrublands in the steppe zone; infrequent in SC BC; S to MT, NM and NV.
Description:
General: Perennial herb from a rhizome and often with a large, branched stem-base; stems one to several or many, erect to somewhat decumbent, (10) 20-70 cm tall, simple or loosely branched above; glabrous to sparsely soft-hairy, glaucous.

Leaves: Basal leaves lacking; lower stem leaves mostly linear to linear-oblanceolate, 3-7 cm long, 1-4 (6) mm wide, from entire to remotely toothed to pinnately-cut with slender, linear, lateral lobes; middle and upper stem leaves reduced.

Flowers: Few- to many-flowered racemes, elongate; flower stalks ascending, 6-12 mm long, expanded at summit; petals yellow, 6-10 mm long, spoon-shaped, narrowed gradually; sepals yellowish, 4-5 mm long.

Fruits: Siliques, nearly erect, 2.5-6 cm long, about 1 mm wide, round in cross section, slightly alternately constricted and expanded, segments indistinctly nerved; beaks stout, nearly the diameter of silique, up to 1 mm long; seeds slightly over 1 mm long, about 0.5 mm in diameter, oblong, plump.

SISYMBRIUM

1. Siliques closely and tightly appressed to the axis of the inflorescence; petals 3-4 mm long .... *S. officinale*
1. Siliques spreading to erect, not appressed; petals 5-8 mm long.
    2. Fruiting stalks slender, much thinner than the silicles; siliques 2-3.5 cm long .... *S. loeselii*
    2. Fruiting stalks stout, nearly as thick as the silicles; siliques 3-10 cm long.
        3. Upper stem leaves with lobes or leaflets threadlike to linear; petals 6-8 mm long ..... *S. altissimum*
        3. Upper stem leaves more or less entire or with lobes or leaflets lanceolate to triangular; petals 8-10 mm long .... *S. orientale*

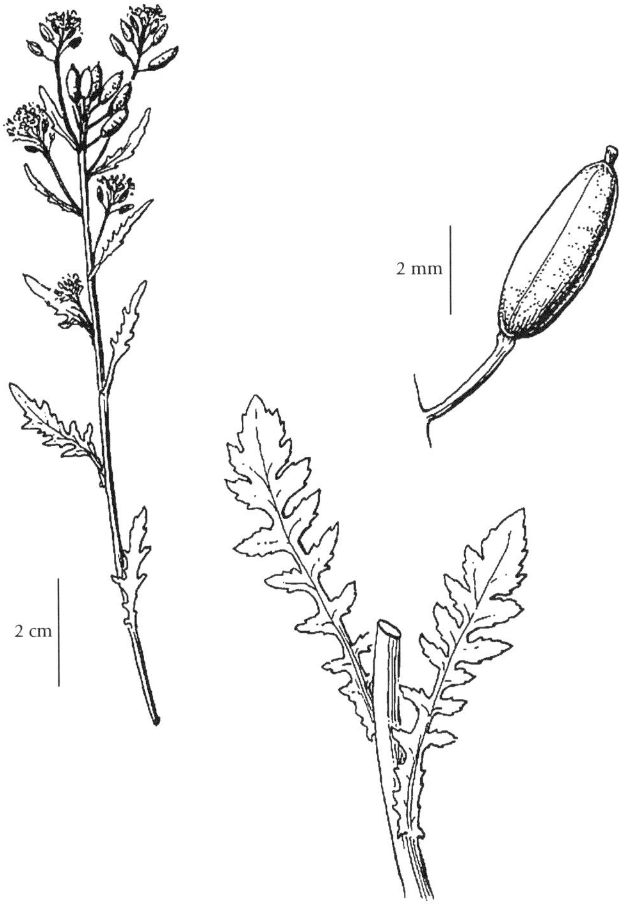

Rorippa tenerrima

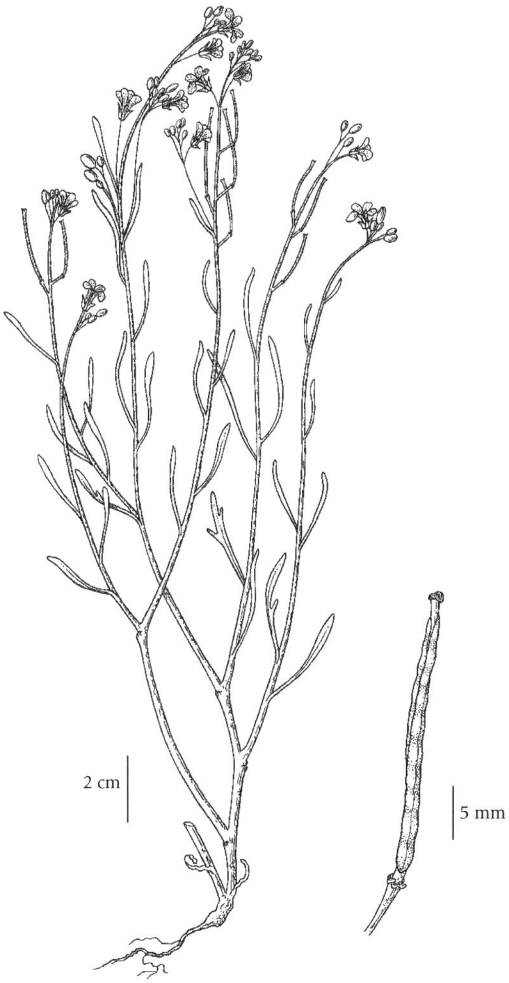

Schoenocrambe linifolia

***Sisymbrium altissimum* L.**
tall tumble-mustard
HABITAT/RANGE:
Mesic to dry fields and waste places in the lowland, steppe and montane zones; common in S BC; introduced from Eurasia.
DESCRIPTION:
General: Annual herb from a taproot; stems erect, 0.2-1.5 m tall, freely branched, sparsely to heavily stiff-hairy near the base but usually glabrous above.

Leaves: Basal leaves broadly lanceolate to oblanceolate, 3-20 cm long, 0.8-5 cm wide, stalked; lower stem leaves lobed or pinnately-cut, toothed; upper stem leaves with lobes or leaflets threadlike to linear.

Flowers: Inflorescence of spreading or ascending stalks, 4-10 mm long, stout; petals pale yellowish, 6-8 mm long; sepals 3-4.5 mm long, glabrous, outer 2 with erect horns at tip.

Fruits: Siliques, spreading, linear, 5-10 cm long, 1-2 mm wide, about same diameter as fruiting stalks, round in cross section, rigid and branchlike; segments with prominent midnerve, evident lateral nerves; stylar beaks none; seeds uniseriate, about 1 mm long, slightly gelatinous when wet.

***Sisymbrium loeselii* L.**
Loesel's tumble-mustard
HABITAT/RANGE:
Mesic to dry fields and waste places in the lowland, steppe and montane zones; frequent in S BC; introduced from Eurasia.
DESCRIPTION:
General: Annual herb from a taproot; stems erect, 0.4-1.2 m tall, freely branched, sparsely to densely hairy with spreading hairs or the hairs pointing backward, glabrous above.

Leaves: Broadly deltoid-lanceolate to lanceolate, to 15 cm long, short-stalked, pinnately-cut, the segments pointing backwards, usually with large, long-pointed, terminal lobes irregularly saw-toothed to toothed.

Flowers: Racemes to 40 cm long in fruit; petals lemon yellow, 6-8 mm long, broadly egg-shaped with long narrow claw; sepals about 3.5 mm long, lanceolate.

Fruits: Siliques, ascending to erect but not appressed, straight or slightly arcuate, 2-3.5 cm long, round in cross section, segments prominently 3-nerved; beaks about 0.5 mm long; fruiting stalks spreading to ascending, 7-20 mm long, slender; seeds uniseriate, about 0.7 mm long, not gelatinous when wet.

***Sisymbrium officinale* (L.) Scop.**
hedge mustard
HABITAT/RANGE:
Mesic to dry fields and waste places in the lowland zone; frequent in SW BC, infrequent in SC BC; introduced from Eurasia.
DESCRIPTION:
General: Annual herb from a taproot; stems 0.2-1.0 m tall, simple to (more commonly) loosely branched, stiff, usually strongly stiff-hairy with backward-pointing or spreading hairs.

Leaves: Basal leaves oblanceolate, 4-25 cm long, 1-6 cm wide, pinnately-cut, terminal lobe the largest and more or less egg-shaped and shallowly round-toothed, glabrate to sparsely hairy; stem leaves usually unstalked, much reduced, generally with 4-6 linear to narrowly lanceolate lobes, terminal lobes more deltoid-lanceolate.

Flowers: Racemes elongate in fruit, as much as 35 cm long; petals pale yellow, 3-4 mm long; sepals 1.5-2.5 mm long, oblong-oblanceolate.

Fruits: Siliques, erect, 8-15 long, 1-1.5 mm wide, linear and awn-like, tightly appressed, beaked, sparsely hairy to glabrous; segments 3-nerved, tardily dehiscent; fruiting stalks erect, 1-3 mm long, stout, enlarged at the tips and about as thick as siliques; beaks 1-2 mm long; seeds about 1.3 mm long, plump, not gelatinous when wet.

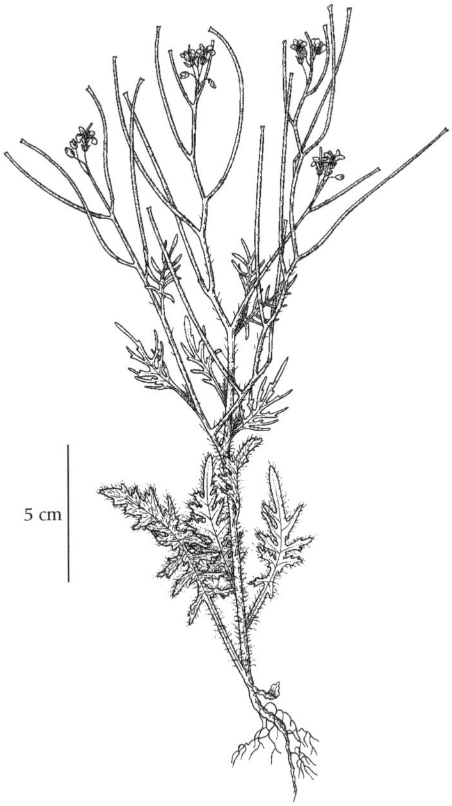

Sisymbrium altissimum

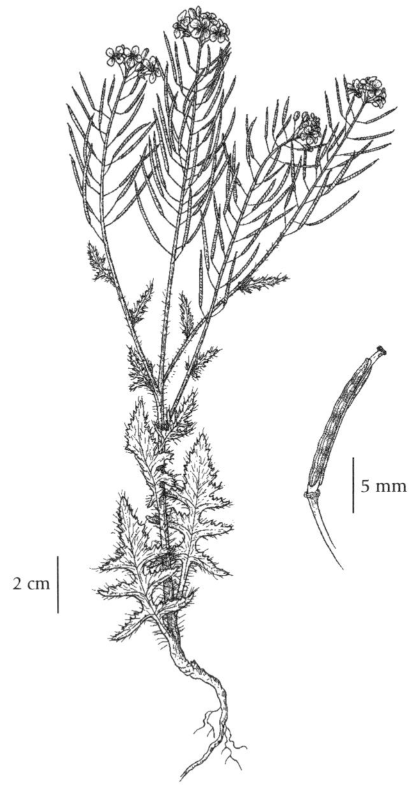

Sisymbrium loeselii

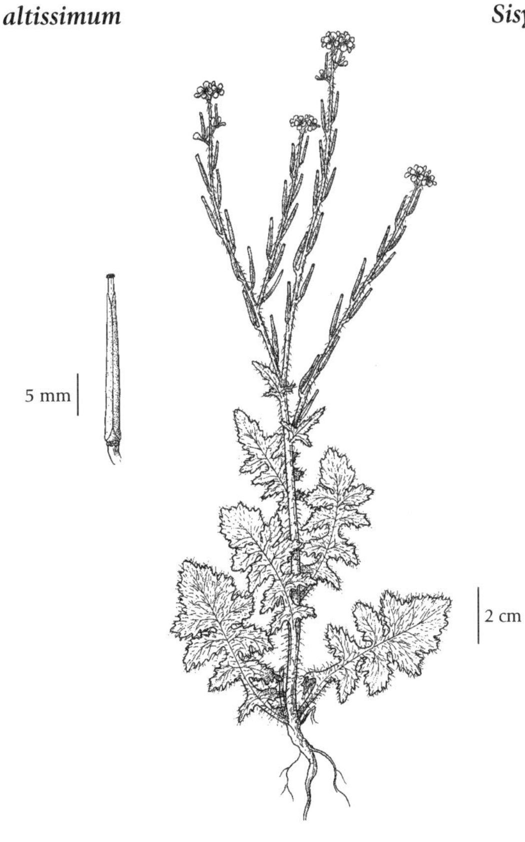

Sisymbrium officinale

***Sisymbrium orientale* L.**
eastern tumble-mustard
HABITAT/RANGE:
Mesic to dry fields, roadsides and waste places; rare in SW BC, known only from Vancouver; introduced from Eurasia.
DESCRIPTION:
General: Annual herb from a taproot; stems erect, 10-30 cm tall, branched, short soft-hairy.

Leaves: Basal leaves clustered, withering before flowering, deeply pinnately lobed or compound, long-stalked with large terminal lobe and about 4 pairs of broadly triangular lateral lobes; stem leaves with linear, arrowhead-shaped terminal lobe and fewer and smaller lateral lobes; upper stem leaves more or less entire or with lobes or leaflets lanceolate to triangular.

Flowers: Inflorescences much elongating in fruit; petals pale yellow, 8-10 mm long, twice length of sepals.

Fruits: Siliques, 3-10 cm long, 1-2 mm wide, straight, round in cross section, stiff-hairy becoming glabrous at maturity; segments 3-nerved; fruiting stalks ascending to erect, 3-6 mm long, stout, about as thick as the silique; beaks 1-3 mm long, club-shaped; seeds yellowish-brown, uniseriate, narrowly egg-shaped, about 0.9 mm long.

SMELOWSKIA

1. Siliques linear to narrowly oblong, tapering at both ends, 5-12 mm long; basal leaves strongly hairy along the margins of the stalks .... *S. calycina*
1. Siliques oblong to egg-shaped, tapering only at the tips, 2-6 mm long; basal leaves not hairy along the margins of the stalks .... *S. ovalis*

***Smelowskia calycina* (Steph.) C.A. Mey.**
alpine smelowskia
HABITAT/RANGE:
Dry scree and talus slopes in the alpine zone; rare in extreme SE BC; amphiberingian, N to AK, YT and NT, E to SW AB and S to CO, VT and NV; NE Asia.
DESCRIPTION:
General: Tufted perennial herb from an elongate rhizome and a branched stem-base, up to 10 cm or longer, mostly less than 5 mm in diameter, covered with withered and persistent leaf bases; stems 1.5-20 cm tall, few to numerous, simple, sparsely to densely hairy with branched and simple hairs.

Leaves: Basal leaves linear to wedge-shaped or egg-shaped, 0.4-10 cm long, 1-15 mm wide, entire to pinnately divided, segments oblong to wedge-shaped, sparsely to densely hairy with fine, freely-branched hairs, stalks slender, conspicuously ciliate near the base with long, stiff, simple hairs; stem leaves several, reduced, nearly unstalked, lobes more linear.

Flowers: Inflorescences head-like, elongating in fruit to 10 cm long; flower stalks ascending to spreading, 5-10 mm long, hairy with generally simple hairs; petals white to purplish-tinged, 5-7 mm long, 3-4 mm wide; sepals often pinkish- or purplish-tinged, 2-3 mm long, soft-hairy, deciduous just after anthesis.

Fruits: Siliques, linear to narrowly oblong, tapering at both ends, 5-12 mm long, 1-2.5 mm wide, glabrous (rarely hairy), slightly flattened or nearly round in cross section; beaks 0.3-1.5 mm long; seeds about 2 mm long, wingless, 4-10.

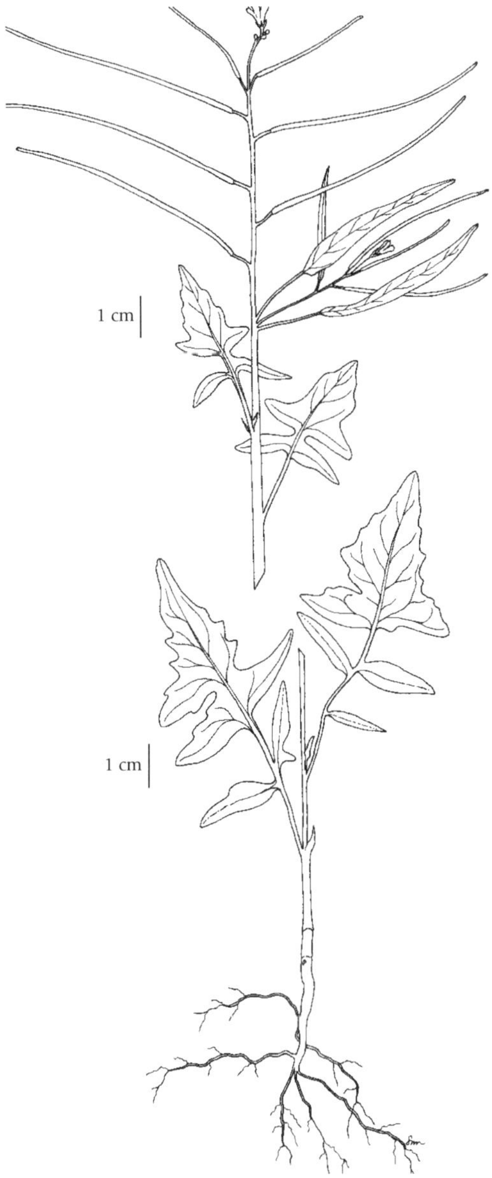

Sisymbrium orientale

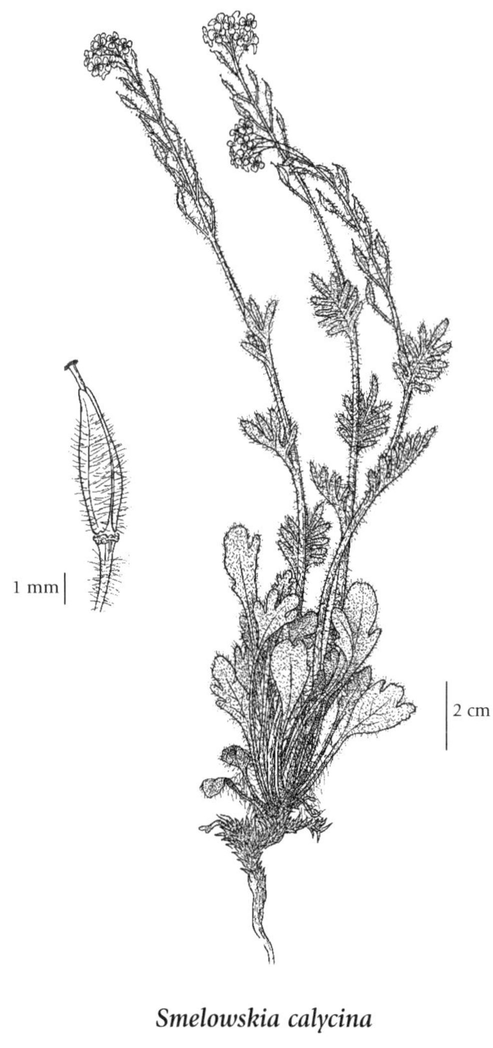

Smelowskia calycina

***Smelowskia ovalis* M.E. Jones**
short-fruited smelowskia
Habitat/Range:
Mesic to dry rocky slopes and fellfields in the alpine zone; locally frequent in extreme SC BC; S to N CA.
Description:
General: Perennial herb from a taproot and a several-branched stem-base; stems 5-15 cm long, simple or branched, densely hairy with long, simple and short, branched hairs.

Leaves: Basal leaves egg-shaped or oblong, 2-6 cm long, pinnately divided, segments egg-shaped, hairy with whitish, mostly branched hairs, stalks slender and usually at least as long as the blades; stem leaves similar, few, reduced, 1-3 cm long.

Flowers: Inflorescences corymbose, elongating but dense in fruit; flower stalks ascending to erect, 4-8 mm long, densely hairy with long, simple hairs; petals white or tinged with purple, 4-5 mm long, about 2 mm wide, rounded at the tips; sepals 2.5-3.5 mm long, usually persistent for some time after anthesis.

Fruits: Siliques, erect, egg-shaped to narrowly oblong, 2-6 mm long, 2-3 mm wide, usually distinctly obcompressed, glabrous, segments fairly prominently nerved most of their length; beaks less than 1 mm long; seeds about 2 mm long.

SUBULARIA

***Subularia aquatica* L. ssp. *americana* Mulligan & Calder**
awlwort
Habitat/Range:
Wet, usually shallow water along streams, shorelines and shallow ponds in the lowland, steppe and montane zones; infrequent in S BC; N to AK, YT and NT, E to NF and S to WY and CA.
Description:
General: Annual herb from a fibrous root; stems erect, 1-10 cm tall, glabrous.

Leaves: Basal leaves erect, linear and awl-like, 1-5 cm long, about 1 mm wide, entire; stem leaves similar, few, reduced.

Flowers: Inflorescences racemose, open; flower stalks 1-2 mm long, ascending to erect; petals white, 0.2-1 mm long; sepals green, 0.5-0.8 mm long, erect, glabrous.

Fruits: Silicles, narrowly elliptic to broadly egg-shaped, 2-3 mm long, 1.5-2 mm wide, inflated, glabrous; beaks obsolete; seeds oblong, about 1 mm long, wingless.

TEESDALIA

***Teesdalia nudicaulis* (L.) Ait. f.**
shepherd's cress
Habitat/Range:
Mesic to dry disturbed areas and waste places in the lowland zone; locally frequent in SW BC; introduced from Europe.
Description:
General: Annual herb from a slender taproot; stem erect, solitary, simple or freely branched from base, 5-25 cm tall, glabrous.

Leaves: Basal leaves in a rosette, 1.5-5 cm long, slender-stalked, blades oval to egg-shaped or oblanceolate, entire to pinnately-cut with a few short, obtuse, lateral lobes and larger, broader, often 3-lobed terminal segment; stem leaves, if present, less lobed or entire.

Flowers: Inflorescence of slender, divergent, 4-8 mm long stalks; petals about 1 mm long; sepals often purplish-tinged, about 0.5 mm long.

Fruits: Silicles, 3-3.5 mm long, slightly narrower, glabrous; beaks very short.

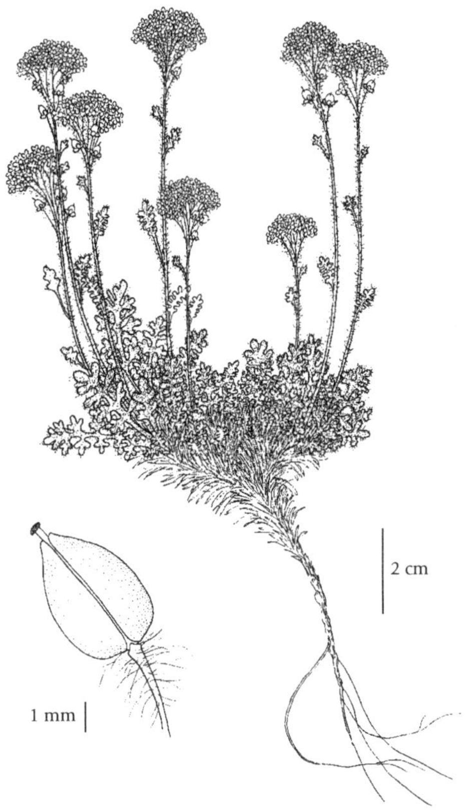

Smelowskia ovalis

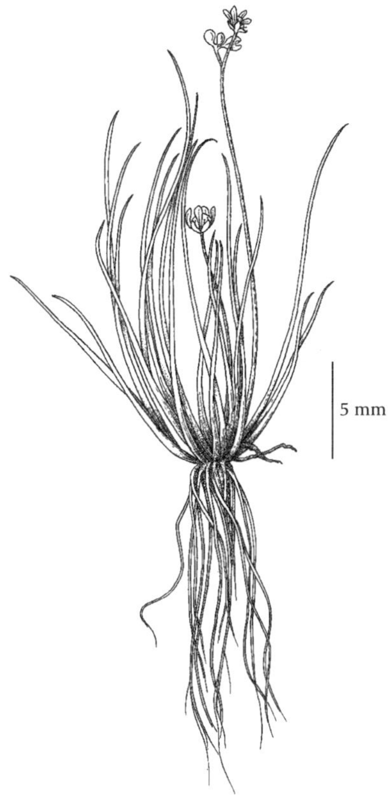

Subularia aquatica ssp. *americana*

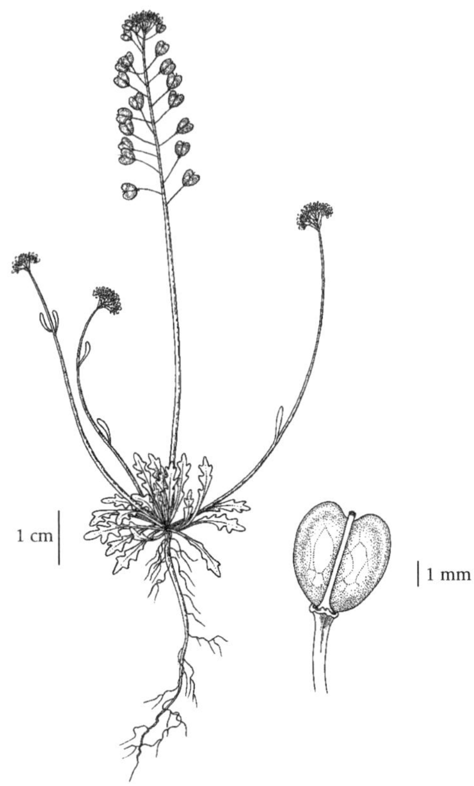

Teesdalia nudicaulis

THELYPODIUM

Thelypodium laciniatum **(Hook.) Endl. *ex* Valp.**

thick-leaved thelypody (var. *lacinatum*); **many-flowered thelypody** (var. *milleflorum*)

Habitat/Range:

Dry talus and rocky slopes in the steppe and lower montane zones; rare in SC BC; S to ID, NV and CA.

Description:

General: Biennial herb from a taproot; stems erect, 0.2-1.5 (2.1) m tall, usually freely branched, solid, glabrous and somewhat glaucous.

Leaves: Basal and lower stem leaves lanceolate to deltoid-lanceolate, less often oblong to egg-shaped, (4) 6-24 (45) cm long, laciniate to pinnately lobed, long-stalked; stem leaves numerous, reduced upwards, lanceolate to linear-lanceolate, often long-pointed, entire to toothed or laciniate, stalked, glabrous and somewhat glaucous.

Flowers: Inflorescences densely flowered racemes, usually branched, slightly elongate in fruit; flower stalks horizontal to spreading, (1.5) 2.5-6.5 (15) mm long, straight or sometimes curved, stout, often flattened at base; calyces tubular to campanulate; petals white to purplish, 6-20 mm long, linear to narrowly spoon-shaped; sepals greenish-white to purplish, 3.5-7 mm long, slightly or not at all pouched at the base.

Fruits: Siliques, widely spreading to ascending, 3-10 (12) cm long, about 1 mm thick, straight to strongly curved, very slightly flattened; beaks (0.5) 1-3 (4) mm long.

Notes: Two often difficult to separate varieties occur in BC.

1. Stalks of the silicles straight, horizontal; stems solid, not inflated; stalks of basal leaves glabrous; siliques spreading; petal blades linear .... var. *milleflorum* (A. Nels.) Pays.
1. Stalks of the silicles strongly curved upward, with tips usually erect; stems almost always hollow, at least to inflorescence; stalks of basal leaves ciliate at least near base; siliques erect to spreading, often appressed to axis of the inflorescence; petal blades oblanceolate to spoon-shaped .... var. *laciniatum*

THLASPI

Thlaspi arvense **L.**

field pennycress

Habitat/Range:

Mesic to dry fields and waste places in the lowland, steppe and montane zones; common throughout BC except rare on the Queen Charlotte Islands and adjacent mainland; introduced from Eurasia.

Description:

General: Annual herb from a taproot; stems 10-50 cm tall, simple to freely branched, glabrous.

Leaves: Basal leaves few, shed early, oblanceolate, 2-6 cm long, 1-2 cm wide, sinuate to almost lyrate, narrowed to short stalks, glabrous; upper stem leaves unstalked, with earlike lobes at the bases, lanceolate to oblanceolate, toothed to wavy-margined and lobed, glabrous.

Flowers: Flower stalks widely spreading to upcurved, 7-15 mm long, slender; petals white, 3-4 mm long; sepals 1.5-2.2 mm long, glabrous.

Fruits: Silicles, oval to nearly heart-shaped, 10-17 mm long, 9-15 mm wide, strongly obcompressed, slightly notched, sinus 1.5-2.5 mm deep, wing-margined all around; beaks 0.1-0.2 mm long; seeds about 2 mm long, concentrically corrugated.

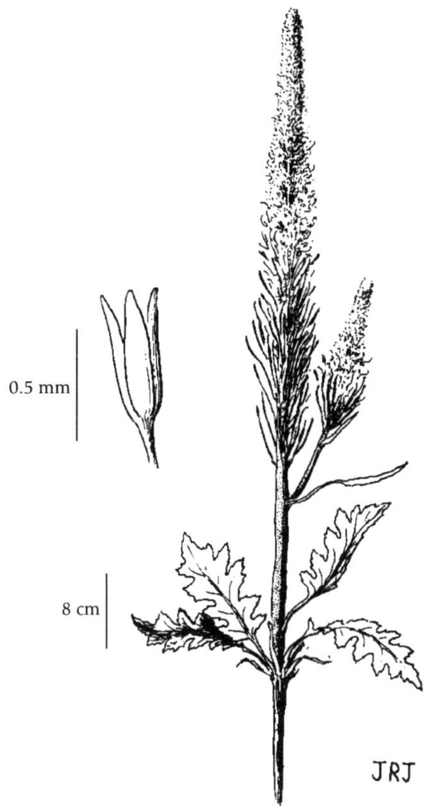

Thelypodium laciniatum var. *milleflorum*

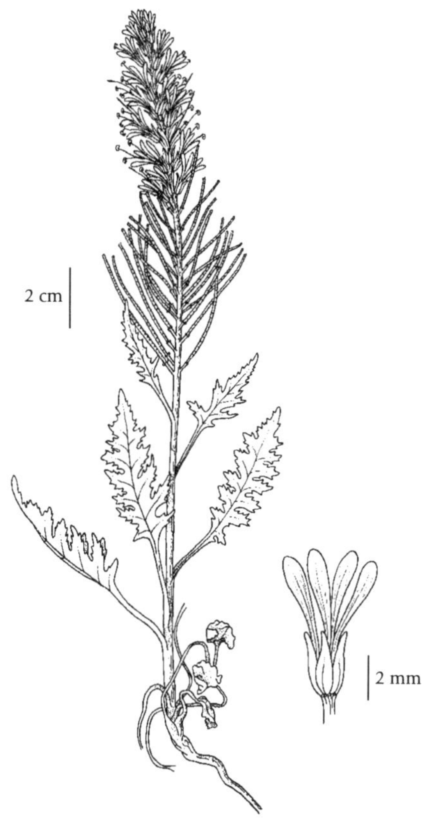

Thelypodium laciniatum var. *laciniatum*

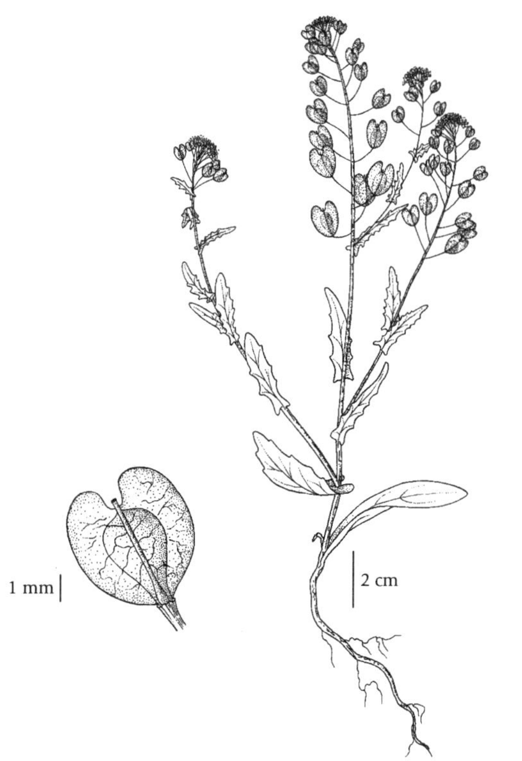

Thlaspi arvense

THYSANOCARPUS

Thysanocarpus curvipes **Hook.**

sand lacepod

Habitat/Range:

Mesic to dry disturbed areas and waste places in the lowland zone; locally infrequent in SW BC, known only from SE Vancouver Island; S to ID and CA.

Description:

General: Annual herb from a taproot; stems 15-80 cm tall, simple to freely branched, usually stiff-hairy below, glabrous above.

Leaves: Basal leaves several, shed early, oblanceolate, 1.5-5 (7) cm long, toothed to shallowly lobed, the lobes pointing backwards, short-stalked; lower stem leaves similar to basal; middle to upper stem leaves smaller, oblong to narrowly lanceolate, unstalked, with ear-like lobes at the bases, entire to toothed.

Flowers: Inflorescence of 4-7 mm long, slender (almost threadlike), recurved stalks; petals white or sometimes purplish-tinged, narrow, about 1 mm long; sepals often purplish, white-margined, about length of petals.

Fruits: Silicles, elliptic to round, 5-8 mm long, glabrous to stiff-hairy or somewhat woolly-hairy, wing entire, wavy-margined or round-toothed, radiately nerved; beaks 0.5-1.5 mm long, often exserted.

BUDDLEJACEAE

BUDDLEJA

Buddleja davidii **Franch.**

butterfly-bush

Habitat/Range:

Mesic to dry disturbed areas and rocky slopes in the lowland zone; frequent in SW BC, known from SE Vancouver Island and the lower Fraser Valley; introduced from China.

Description:

General: Deciduous shrub 1-5 m tall; branches hairy with hairs radiating in a starshape.

Leaves: Opposite, egg-shaped to lanceolate, 3-25 cm long, saw-toothed to more or less entire, grey woolly-hairy beneath, green above.

Flowers: Inflorescence a large, terminal, branched raceme 15-25 cm long; corollas narrowly tubular below then spreading to 4 lobes, 7-9 mm wide, purplish with an orange eye.

Fruits: Sharp-pointed, 2-segmented capsules, 5-8 mm long.

CABOMBACEAE

BRASENIA

Brasenia schreberi **J.F. Gmel.**

Water shield; water target

Habitat/Range:

Ponds and slow-moving streams in the lowland, steppe and montane zones; common in SW BC, less frequent in SC BC; N to SE AK, E to NB and NS and S to FL, TX, and CA; Cuba, S America.

Description:

General: Perennial aquatic herb from a slender rhizome; stems several, branched above, reddish-brown, covered with a gelatinous sheath, 5-40 cm long.

Leaves: Leaves floating, long-stalked, elliptic or oval, centrally attached to the stems, 5-12 cm long, 3-8 cm wide, gelatinous below.

Flowers: Solitary on 5-20 cm stalks; sepals and petals 3 (4) each, similar, purplish, linear-oblong, 10-15 mm long; stamens 12-18, purplish, slightly shorter than or equal to the petals.

Fruits: Oblong, leathery, 6-8 mm long, soon deciduous.

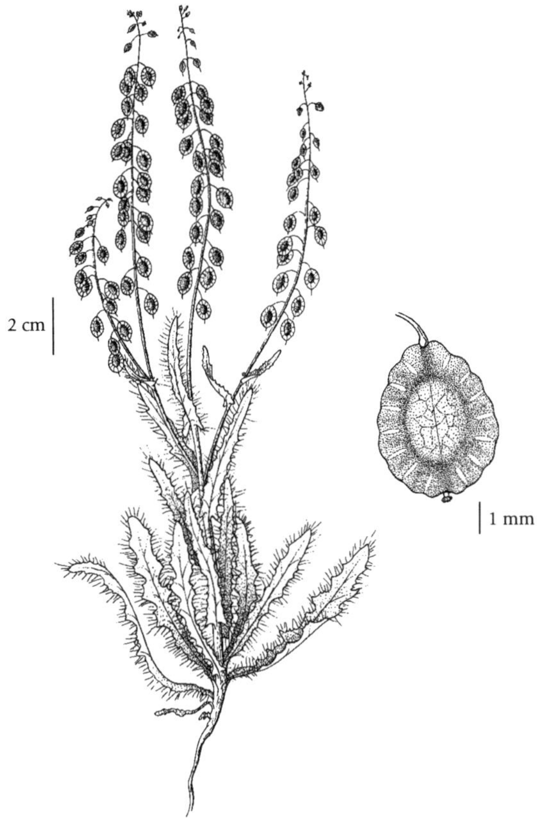

Thysanocarpus curvipes

Buddleja davidii

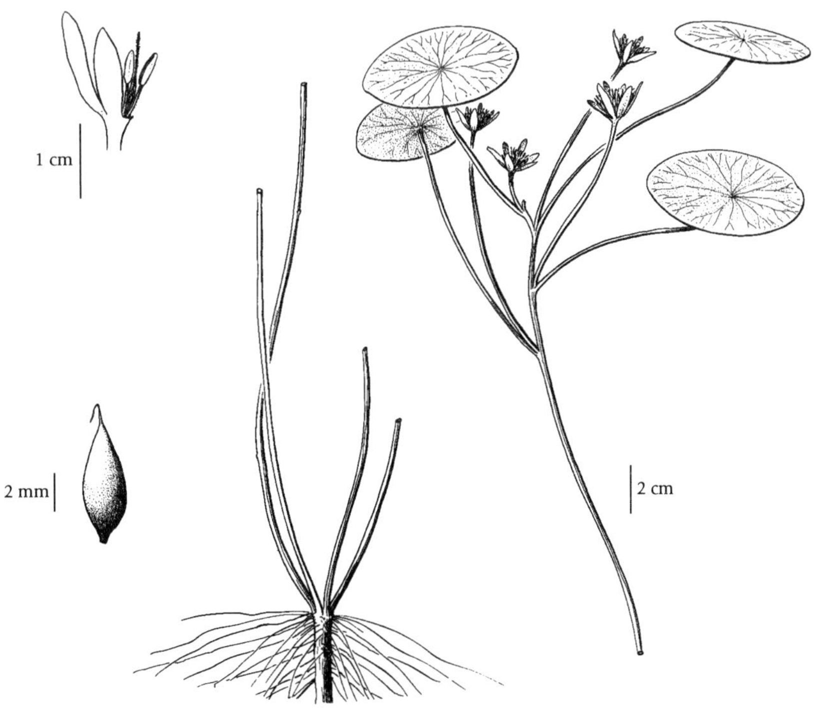

Brasenia schreberi

CACTACEAE

OPUNTIA

1. Stem segments not much flattened, readily detaching from the plant; spines strongly barbed; areoles white-woolly .... *O. fragilis*
1. Stem segments flattened, not readily detaching from the plant; spines only slightly barbed; areoles rusty-woolly or not woolly .... *O. polyacantha*

Opuntia fragilis **(Nutt.) Haw.**
brittle prickly-pear cactus
HABITAT/RANGE:
Dry grassy, sandy or gravelly slopes in the lowland, steppe and montane zones; common in S BC east of the Coast-Cascade Mountains, less frequent on SE Vancouver Island and the Gulf Islands, rare along the Peace River in NE BC; E to ON and S to KS, IL, TX and N CA.
DESCRIPTION:
General: Perennial herb from a fibrous root; mat-forming; stems prostrate, succulent, subglobose to rounded, fleshy, 5-20 cm tall; stem segments not much flattened, readily detaching from the plant, round to egg-shaped, 2-5 cm long; areoles coarsely white-woolly with a few yellowish, spiny bristles and 2-7 straight, yellowish to brownish, 1-3 cm long spines.

Leaves: Reduced to 2-7, large yellowish to brownish spines 1-3 cm long, strongly barbed.

Flowers: Inflorescence of single flowers with paper-thin petals, borne on previous year's pits, yellow, 3-5 cm across, stamens numerous with reddish stalks.

Fruits: Dry, spiny berry, 1.5-2 cm long, pear-shaped.

Opuntia polyacantha **Haw.**
plains prickly-pear cactus
HABITAT/RANGE:
Dry sandy or gravelly sites in the steppe zone; locally frequent in S BC east of the Coast-Cascade Mountains; E to S SK and S to MO, TX, AZ and E OR.
DESCRIPTION:
General: Perennial herb from a fibrous root; mat-formin; stems prostrate, succulent, egg-shaped to rounded, fleshy, 10-40 cm tall; stem segments flattened, not readily detaching from the plant, orbicular, 5-12 cm long; aereoles often with greyish-yellow woolly with tawny, spiny bristles and 5-11 straight, 1-5 cm long spines.

Leaves: Reduced to 5-11 large spines 1-5 cm long, straight or only slightly barbed.

Flowers: Inflorescence of single flowers with paper-thin petals, borne on previous year's aereole, yellow turning pinkish with maturity, rarely reddish, 5-7 cm wide, stamens numerous with reddish stalks.

Fruits: Dry, spiny berry, 2-2.5 cm long, pear-shaped.

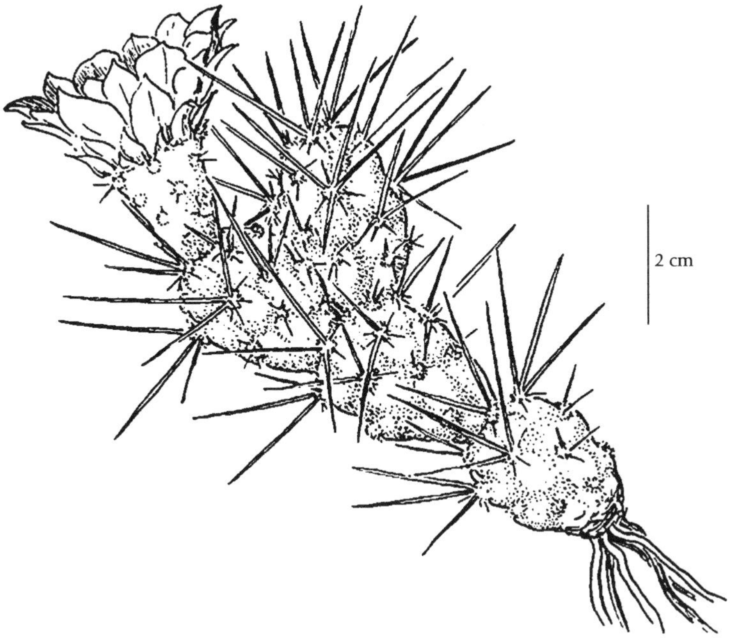

Opuntia fragilis

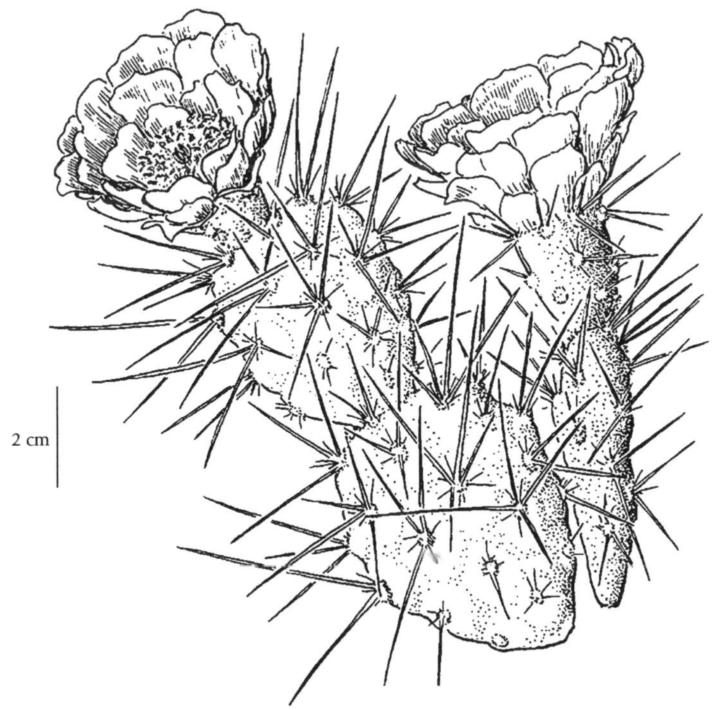

Opuntia polyacantha

CALLITRICHACEAE

CALLITRICHE

1. Female flowers borne on distinct stalks, much longer than the fruits; styles sharply reflexed .... *C. marginata*
1. Female flowers stalked or shortly-stalked; styles ascending (sometimes reflexed in *C. stagnalis*).
   2. Fruits encircled by a conspicuous wing margin .... *C. stagnalis*
   2. Fruits wingless or wing-margined toward the summit.
      3. Leaves all linear, single-nerved, their bases not joined by a ridge or wing .... *C. hermaphroditica*
      3. Leaves linear below, lanceolate, egg- or spoon-shaped above, their bases joined by a small, winglike ridge.
         4. Carpel faces with rather regular vertical lines of tiny, pitlike markings; fruits slightly wing-margined above .... *C. palustris*
         4. Carpel faces with irregularly distributed tiny pitlike markings; fruits scarcely if at all wing-margined .... *C. heterophylla*

***Callitriche hermaphroditica* L.**
northern water-starwort

HABITAT/RANGE:

Slow-moving streams, lakes and sloughs in the lowland, steppe and montane zones; common in SW BC, less frequent in SC and SE BC; circumboreal, N to AK, YT and NT, E to NF and S to MN, MI, NM and CA; Eurasia.

DESCRIPTION:

General: Aquatic or semi-aquatic perennial rooting in mud; stems slender, 5-15 cm long, with widely spaced internodes.

Leaves: All opposite; linear, 5-20 mm long, single-nerved, their bases not joined by a ridge or wing, light green.

Flowers: Inflorescence of tiny, inconspicuous flowers borne without bracts, stalkless or shortly-stalked; styles soon deciduous, ascending.

Fruits: Achene-like, 4, each 1-seeded, wingless, less than 1 mm long, egg-shaped, irregularly pitted.

***Callitriche heterophylla* Pursh ssp. *bolanderi* (Hegelm.) Calder & Tayl.** (*C. anceps* Fern. = ssp. *heterophylla*)
diverse-leaved water-starwort (ssp. *bolanderi*), **two-edged water-starwort** (ssp. *heterophylla*)

HABITAT/RANGE:

Shallow ponds, slow-moving streams and shorelines in the lowland and montane zones; ssp. *bolanderi*–frequent on S Vancouver Island, rare in SE BC, ssp. *heterophylla* - rare in coastal BC; N to AK, YT, NT, E to NF and S to GA, TX, OK and CA; South America (ssp. *bolanderi*).

DESCRIPTION:

General: Aquatic or semi-aquatic perennial rooting in mud; stems leafy, 2-40 cm long, with or without small membranous ridges between the leaf-bases.

Leaves: All opposite; submerged leaves linear, single-veined, notched at the tip, 5-25 mm long; floating and emergent leaves broadly egg-shaped or oblanceolate, 3-veined, up to 1 cm wide.

Flowers: Inflorescence of tiny, inconspicuous flowers borne with a pair of tiny bracts, stalkless or shortly-stalked; styles 1-5 mm long, ascending.

Fruits: Achene-like, 4, each 1-seeded, wingless or scarcely wing-margined, rounded, egg-shaped or heart-shaped, irregularly pitted.

Notes: Two subspecies occur in our region.

1. Fruits egg-shaped or slightly heart-shaped in outline, broadest above the middle; submersed leaves usually wider than the stem; plants long-stemmed .... ssp. *bolanderi* (Hegelm.) Calder & Tayl.
1. Fruits usually more nearly round or egg-shaped in outline; submersed leaves much narrower than the flattened stem; plants short-stemmed .... ssp. *heterophylla*

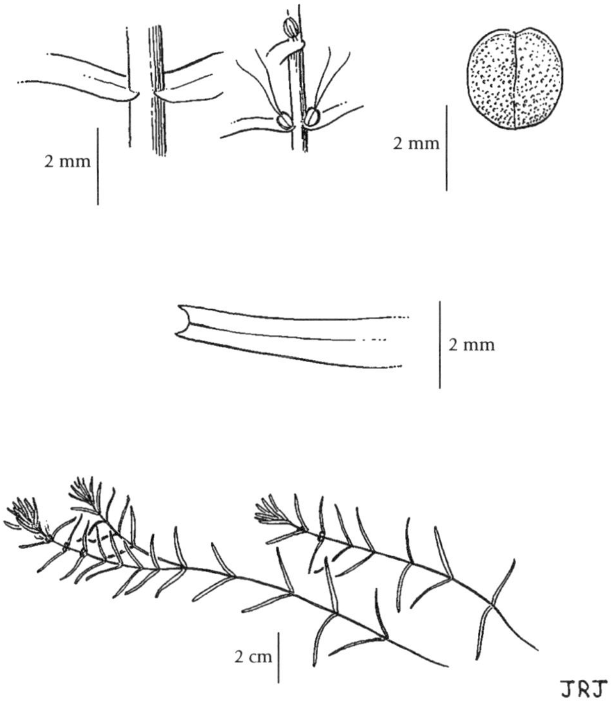

Callitriche hermaphroditica

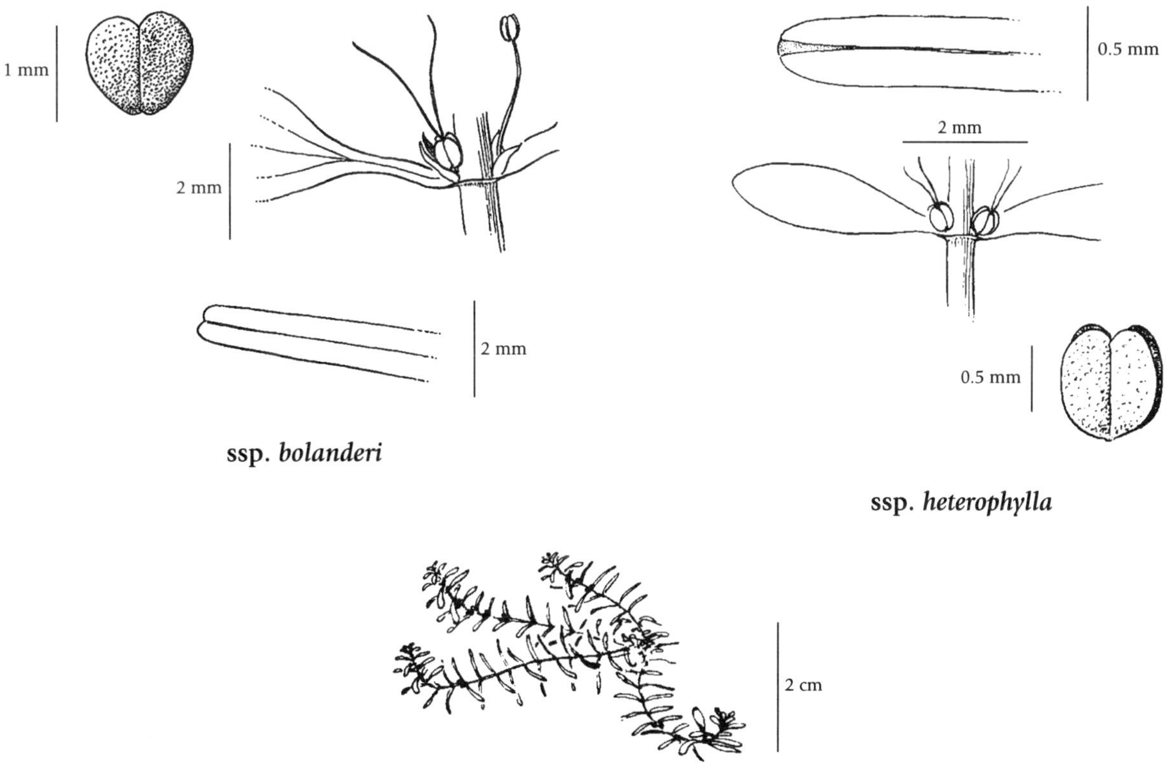

Callitriche heterophylla

Callitriche marginata **Torr.**
winged water-starwort
Habitat/Range:
Vernal pools in the lowland zone; rare, known only from S Vancouver Island and Mittlenatch Island; S to CA and MX.
Description:
General: Aquatic or semi-aquatic perennial rooting in mud; stems matted, very slender, 5-10 cm long, leaf-bases joined by a tiny, winged ridge.

Leaves: All opposite; submerged leaves linear, single-nerved, very slightly notched terminally; emergent leaves wider, incompletely 3-nerved.

Flowers: Inflorescence of tiny, inconspicuous flowers borne without bracts or with tiny, brown, linear bracts; female flowers with distinct stalks, much longer than the fruits; styles sharply reflexed.

Fruits: Achene-like, 4, each 1-seeded, conspicuously wing-margined, oval, irregularly pitted in a fine network.

Callitriche palustris **L.** (*C. verna* L.)
spring water-starwort
Habitat/Range:
Slow-moving streams and shorelines in the lowland, steppe and montane zones; frequent in S BC; circumboreal, N to AK, YT and NT, E to NF and S to NE, VA, NM and CA; Eurasia.
Description:
General: Aquatic or semi-aquatic perennial rooting in the mud; stems slender, 5-20 cm long, submerged, partially floating or stranded on mud, leaf bases joined by small, winged ridges.

Leaves: All opposite; submerged leaves linear, 5-20 mm long, thin, single-nerved, narrowly margined, terminally notched, veins thickened but not protruding; floating and emergent leaves broadly spoon-shaped, 3-nerved.

Flowers: Inflorescence of tiny, conspicuous flowers borne with bracts; female flowers stalkless or shortly-stalked; styles ascending.

Fruits: Achene-like, 4, each 1-seeded, slightly wing-margined above, with regular vertical lines of small pitlike markings.

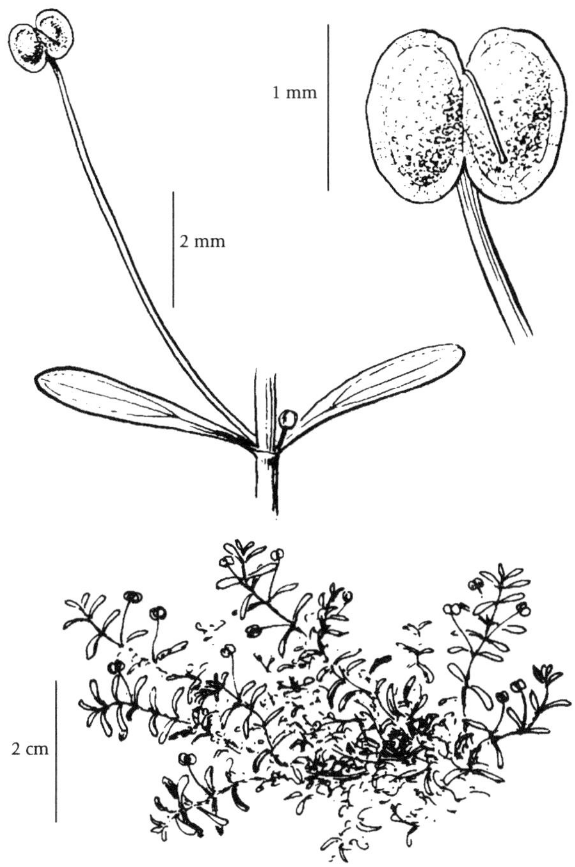

Callitriche marginata

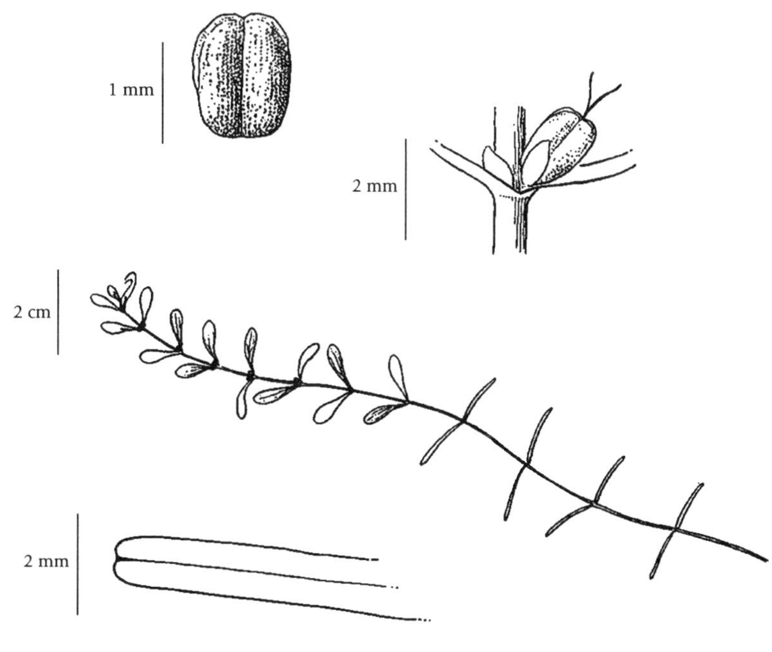

Callitriche palustris

***Callitriche stagnalis* Scop.**

pond water-starwort

HABITAT/RANGE:

Slow-moving streams and ponds in the lowland, steppe and montane zones; infrequent in S BC, locally frequent in WC BC (Skeena River estuary); introduced from Europe.

DESCRIPTION:

General: Aquatic or semi-aquatic perennial rooting in mud; stems 3-15 cm long, leaf bases joined by small winged ridges.

Leaves: All opposite, linear and single-nerved to broadly egg- or spoon-shaped and 3- to 5-nerved, 1 cm wide.

Flowers: Inflorescence of tiny inconspicuous flowers borne with bracts, white; female flowers stalkless or short-stalked; styles sometimes reflexed.

Fruits: Achene-like, 4, each 1-seeded, 1-1.5 mm long, conspicuously wing-margined all around, minutely pitted.

CAMPANULACEAE

1. Filaments and anthers united into a tube; corollas bilaterally symmetric.
   2. Flowers stalked, in leaf axils .... *Downingia*
   2. Flowers stalked, in terminal bracted racemes .... *Lobelia*
1. Filaments and anthers distinct; corollas radially symmetric.
   3. Perennials (except for one biennial species); flowers stalked .... *Campanula*
   3. Annuals; flowers stalked or nearly unstalked.
      4. Corollas shorter than the calyx lobes; leaves lanceolate .... *Githopsis*
      4. Corollas longer than the calyx lobes; leaves rounded-oblong or rounded egg-shaped.
         5. Upper flowers blue, 3-6 mm long, inserted opposite the bracts .... *Heterocodon*
         5. Upper flowers pale lavender to purple, 8-13 mm long, inserted in the axils of the middle and upper stem leaves .... *Triodanis*

CAMPANULA

1. Plants usually over 50 cm tall; flowers in a terminal cluster; introduced species.
   2. Plants biennials, stiff-hairy; calyces with a conspicuous appendage at each notch; corollas usually 3.5-5 cm long .... *C. medium*
   2. Plants perennials, (or sometimes biennial in *C. rapunculoides*) glabrous or sparsely hairy; calyces without appendages; corollas usually 2-3.5 cm long.
      3. Leaves linear to linear-lanceolate, finely toothed, the lower stem leaves nearly unstalked .... *C. persicifolia*
      3. Leaves lanceolate to egg-shaped, regularly toothed or saw-toothed, the lower stem leaves stalked.
         4. Stems sparsely rough-hairy; leaves regularly toothed; corollas lobed 1/2 to 2/3 their length, white or pale blue .... *C. lactiflora*
         4. Stems glabrous or if sparsely hairy then not rough-hairy; leaves irregularly saw-toothed; corollas not deeply lobed, blue .... *C. rapunculoides*

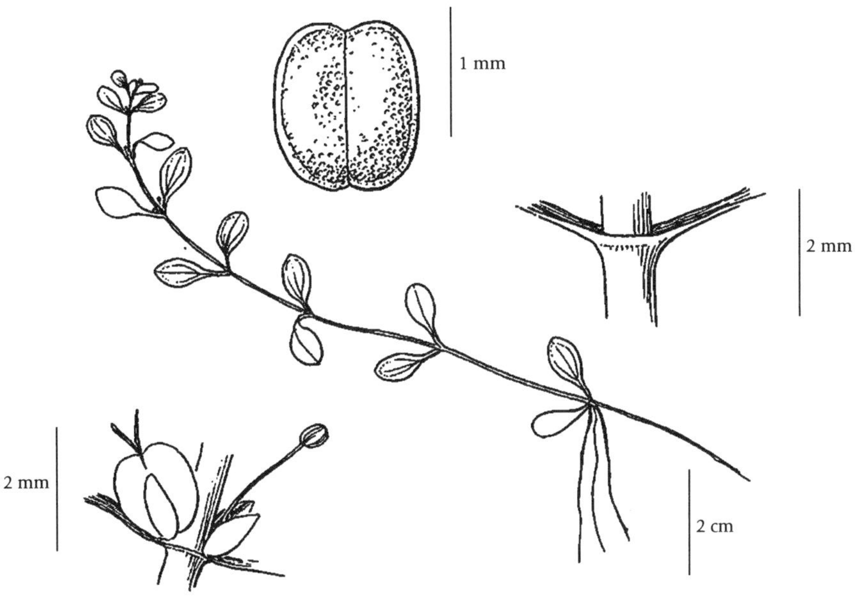

Callitriche stagnalis

1. Plants usually less than 50 cm tall (except some forms of *C. rotundifolia*); flowers solitary or in loose clusters; plants native.
    5. Corollas 15-35 mm long.
        6. Calyx tubes glabrous; sepals entire .... *C. rotundifolia*
        6. Calyx tubes hairy; sepals deeply cut .... *C. lasiocarpa*
    5. Corollas 2-18 mm long.
        7. Flowers usually several in loose clusters .... *C. scouleri*
        7. Flowers solitary.
            8. Styles exserted from the corolla; corolla lobes spreading or recurved, equalling or longer than the tube .... *C. aurita*
            8. Styles shorter than or equalling the corolla; corolla lobes erect to ascending, usually shorter than the tube .... *C. uniflora*

***Campanula aurita* Greene**
Alaskan harebell; Yukon bellflower
HABITAT/RANGE:
Mesic to dry meadows, forests and rock outcrops in the montane and subalpine zones; frequent in extreme NE BC; N to AK, YT and NT.
DESCRIPTION:
General: Perennial herb from an elongate rhizome; stems single to several, erect or decumbent to ascending, 10-35 cm tall, short rough-hairy or glabrous.

Leaves: Basal leaves absent; stem leaves alternate, linear to oblong-elliptic or oblanceolate, 0.8-3.5 cm long, 1-6 mm wide, smooth or irregularly toothed, glabrous or fringed with hairs near the base.

Flowers: Inflorescence a solitary bell-shaped flower on a rough-haired stalk; calyces glabrous; corolla 12-18 mm long, blue, the lobes spreading or recurved, equalling or longer than the tube; styles exserted from the corolla.

Fruits: Capsules, inverted cone-shaped to subcylindric.

***Campanula lactiflora* Bieb.** (*Gadellia lactiflora* [Bieb.] Schulkina)
European bellflower
HABITAT/RANGE:
Dry wooded river banks and forest openings in the lowland zone; rare in SW BC, known only from the lower Fraser River valley; introduced from Europe.
DESCRIPTION:
General: Perennial herb from a rhizome; stems erect, solitary, branched, sparsely rough-hairy, 0.6-1.5 m tall.

Leaves: Basal leaves soon deciduous; stem leaves alternate, lanceolate, regularly toothed, stalked, becoming reduced and unstalked upwards.

Flowers: Inflorescence a solitary, broadly bell-shaped flower; corollas 15-25 mm long, white to pale blue, deeply lobed 1/2 to 2/3 towards the base; sepals lanceolate to oblanceolate.

Fruits: Capsules, cylindric, hairy.

Campanula aurita

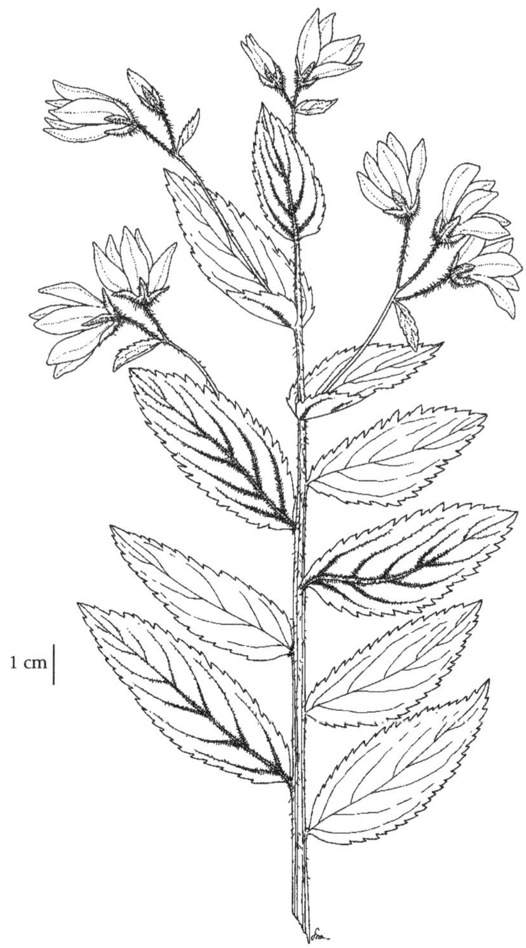

Campanula lactiflora

Campanula lasiocarpa **Cham.**
mountain harebell
HABITAT/RANGE:
Moist to mesic meadows, forests and rocky slopes in the montane to alpine zones; common N of 52° N; amphiberingian, N to AK, YT, and NT, E to W AB and S to N WA; E. Asia.
DESCRIPTION:
General: Perennial herb from a thin, branching rhizome; stems single to several, erect or ascending, 2-20 cm tall, smooth or slightly hairy.

Leaves: Mainly basal, oblanceolate to oblong, stalked, 1-6 cm long, 2-13 mm wide, sharply toothed; stem leaves fewer, alternate, narrower, reduced, without stalks.

Flowers: Inflorescence a solitary bell-shaped flower; corollas 1.5-3.5 cm wide, blue; sepals deeply cut, hairy.

Fruits: Capsules, 3-12 mm long, subcylindric, hairy.

Campanula medium **L.**
Canterbury-bells
HABITAT/RANGE:
Mesic disturbed areas in the lowland zone; rare garden escape in coastal SW BC; introduced from Europe.
DESCRIPTION:
General: Biennial herb from a fibrous root or rhizome; stems single to several, erect or ascending, 35-90 cm tall, bristly-hairy.

Leaves: Basal leaves stalked, egg-shaped, hairy; stem leaves more linear, alternate, toothed, reduced upwards.

Flowers: Inflorescence of terminal clusters of large, bell-shaped flowers; corollas 3.5-5 cm long, blue; calyces with a conspicuous appendage at each notch.

Fruits: Capsules, 8-10 mm long, cylindric.

Campanula persicifolia **L.**
peach-leaved bellflower
HABITAT/RANGE:
Mesic disturbed areas in the lowland zone; rare garden escape in coastal SW BC; introduced from Eurasia.
DESCRIPTION:
General: Perennial herb from a branching rhizome; stems single to several, erect, leafy, 35-90 cm tall, glabrous or sparsely hairy.

Leaves: Basal leaves lacking; stem leaves alternate, linear to linear-lanceolate, finely toothed, lower ones nearly unstalked.

Flowers: Inflorescence of terminal clusters of relatively few, bell-shaped flowers; corollas 2-3.5 cm long, blue; calyces without appendages.

Fruits: Capsules, opening above the middle.

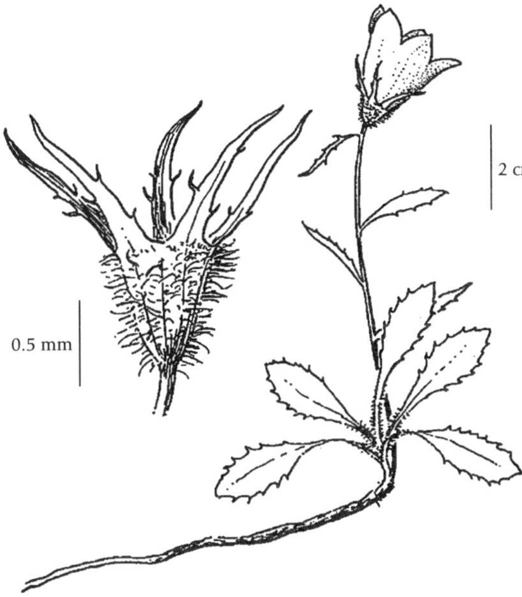

Campanula lasiocarpa

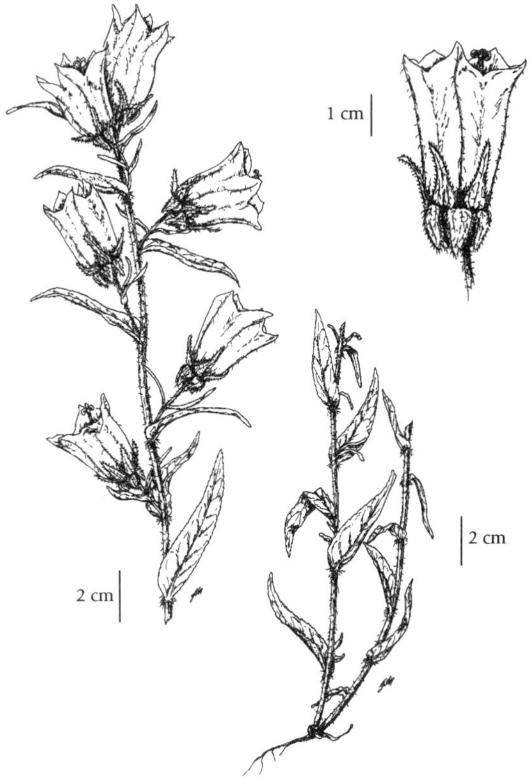

Campanula medium

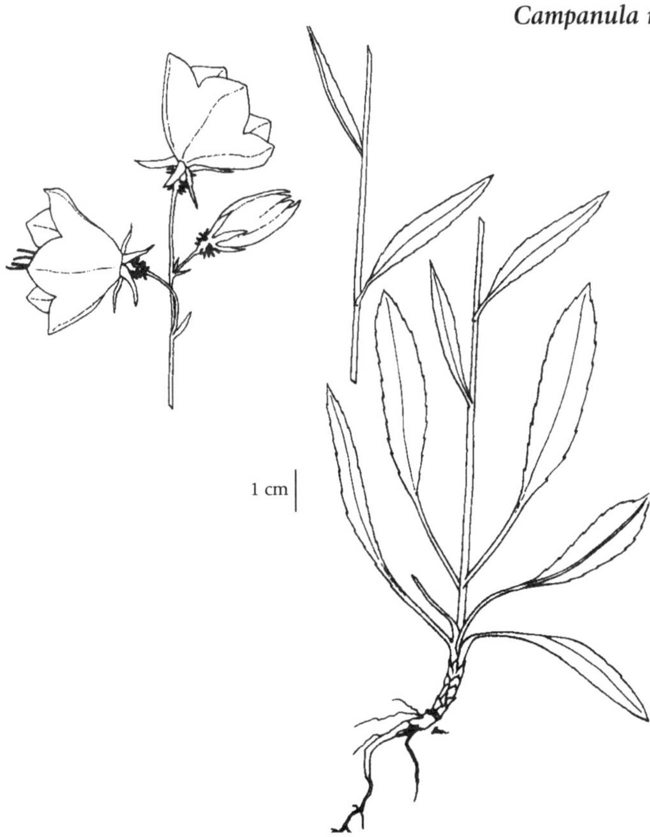

Campanula persicifolia

***Campanula rapunculoides* L.**
creeping bellflower
HABITAT/RANGE:
Disturbed areas in the lowland and lower montane zones; frequent, scattered across extreme S BC; introduced from Eurasia.
DESCRIPTION:
General: Perennial or sometimes biennial herb from a branching rhizome; stems single to several, erect, over 50 cm tall, glabrous or sparsely hairy.
Leaves: Basal leaves lacking; stem leaves alternate, lanceolate to egg-shaped, irregularly saw-toothed, lower ones stalked.
Flowers: Inflorescence of terminal clusters of numerous bell-shaped flowers; corollas 2-3.5 cm long, blue; calyces without appendages.
Fruits: Capsules, nodding, opening near the base.

***Campanula rotundifolia* L.** (*C. alaskana* [A. Gray] Wight, *C. rotundifolia* var. *alaskana* A. Gray)
common harebell; bluebells-of-Scotland
HABITAT/RANGE:
Moist to dry meadows and rock outcrops in all vegetation zones; common throughout BC; circumboreal, N to AK, YT and NT, E to NF and S to TX, NM and N CA.
DESCRIPTION:
General: Perennial herb from a taproot, slender rhizomes or branched stem-base; stems single to several, erect or ascending, 10-80 cm tall, more or less glabrous.
Leaves: Basal leaves stalked, oval to heart-shaped, often deciduous; stem leaves alternate, toothed, linear to linear-lanceolate or egg-shaped, 1.5-8 cm long.
Flowers: Inflorescence of several terminal clusters of numerous, bell-shaped flowers, erect or nodding; corollas 15-35 mm long, blue; calyx tubes glabrous; sepals entire.
Fruits: Capsules, 5-8 mm long, nodding, cylindric to inverted cone-shaped.
Notes: There is little evidence to suggest that *C. alaskana* is anything other than a coastal race of the highly variable *C. rotundifolia.* Many taxonomists (e.g., Shetler 1963, Welsh 1974, Scoggan 1979, Kartesz 1994) do not recognize any infraspecific taxa within this species.

***Campanula scouleri* Hook. *ex* A. DC.**
Scouler's harebell
HABITAT/RANGE:
Mesic to dry forests, rock outcrops and talus slopes in the lowland zone; locally common on Vancouver Island and the Gulf Islands; N to SE AK and S to N CA.
DESCRIPTION:
General: Perennial herb from a branching, slender rhizome; stems lax, often curved at the base, 5-40 cm tall, glabrous to inconspicuously short-hairy.
Leaves: Basal leaves egg-shaped, stalked; stem leaves alternate, sharply toothed, 1-6 cm long, progressively narrower and less stalked upwards.
Flowers: Inflorescence of several terminal, loose clusters of bell-shaped flowers; corollas 8-18 mm long, blue to whitish; corolla lobes recurved, spreading, longer than the basal tube.
Fruits: Capsules, roundish, 3-6 mm long, opening near the middle.

Campanula rapunculoides

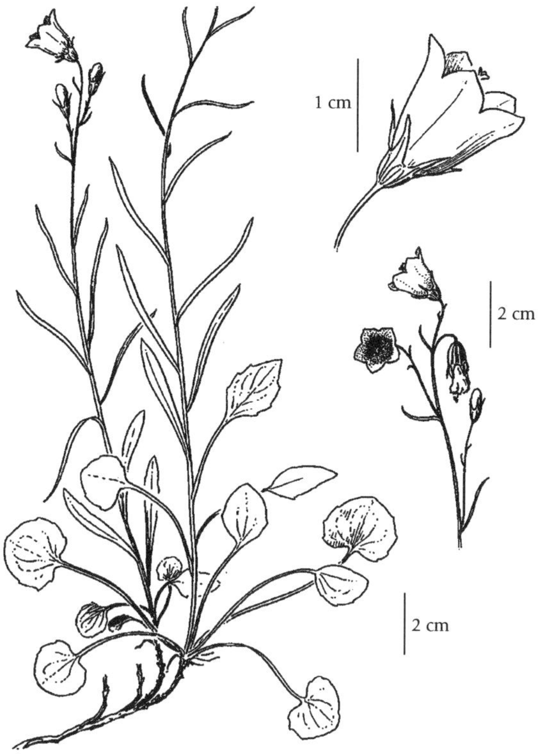

Campanula rotundifolia

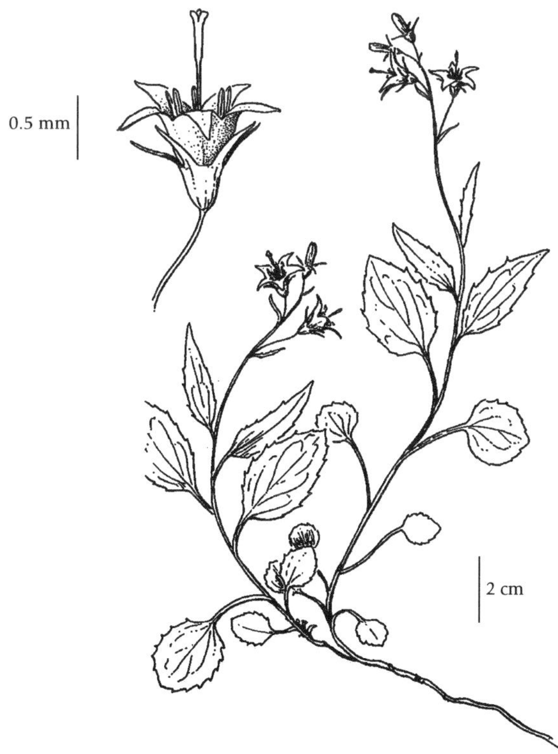

Campanula scouleri

***Campanula uniflora* L.**

arctic harebell

HABITAT/RANGE:

Mesic to dry meadows, rocky slopes and fellfields in the montane to alpine zones; infrequent and scattered throughout BC east of the Coast-Cascade Mountains; circumboreal, N to AK, YT and NT, E to PQ, and S to CO.

DESCRIPTION:

General: Perennial herb from a taproot and slenderly branched stem-base or rhizome; stems several, lax, 3-30 cm tall, glabrous except for a few long, loose hairs.

Leaves: Basal leaves oblanceolate or wideer, stalked; stem leaves alternate, sharply toothed, elliptic to oblanceolate or linear, 0.5-4.5 cm long, 0.1-0.7 cm wide, becoming stalked above.

Flowers: Inflorescence of solitary, erect, bell-shaped flowers; corollas 6-12 mm long, blue; corolla lobes erect to ascending, usually shorter than the tube; styles shorter than or equalling the corolla.

Fruits: Capsules, 12-20 mm long, elongate, tapering to the base, opening near the summit.

DOWNINGIA

***Downingia elegans* (Dougl.) Torr.**

common downingia

HABITAT/RANGE:

Wet meadows, lakeshores and ponds in the steppe and lower montane zones; rare in SE BC, known only from Leach Lake (Creston area); S to NV and N CA.

DESCRIPTION:

General: Annual herb from fibrous roots; stems erect or curved-ascending, simple or branched, 10-50 cm tall, glabrous.

Leaves: Basal leaves lacking; stem leaves alternate, stalked, 0.5-2.5 cm long, lower ones linear, middle and upper ones lanceolate.

Flowers: Inflorescence of solitary flowers in the axils of the middle and upper leaves; corollas blue to pink or white with a yellow-ridged white eye, mostly 8-18 mm long; corolla tubes short and widely flaring.

Fruits: Capsules, 2-5 cm long, 1-2 mm thick, broadest near the base, opening by long slits down the sides.

Notes: This species was probably extirpated when the water level of Leach Lake was raised during a Ducks Unlimited waterfowl enhancement project.

GITHOPSIS

***Githopsis specularioides* Nutt.**

common bluecup

HABITAT/RANGE:

Moist seepage areas on rock outcrops in the lowland zone; rare on S Vancouver Island; S to CA.

DESCRIPTION:

General: Annual herb from a fibrous root; stems branched, spreading-hairy or glabrous, 2-15 cm tall.

Leaves: Basal leaves lacking; stem leaves alternate, oblong to lanceolate, 4-20 mm long, 1-3.5 mm wide, remotely toothed.

Flowers: Inflorescence of solitary terminal flowers; calyx lobes elongate, linear; corollas tubular bell-shaped, blue with a whitish throat, shorter than the calyx lobes.

Fruits: Capsules, 6-15 mm long, prominently ribbed, opening at the summit.

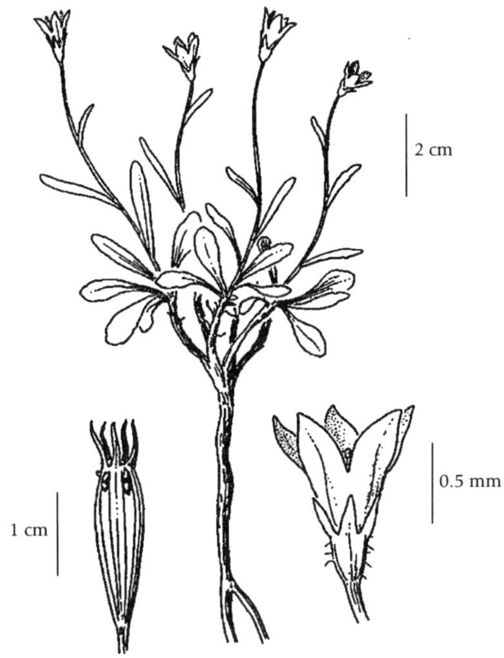

Campanula uniflora

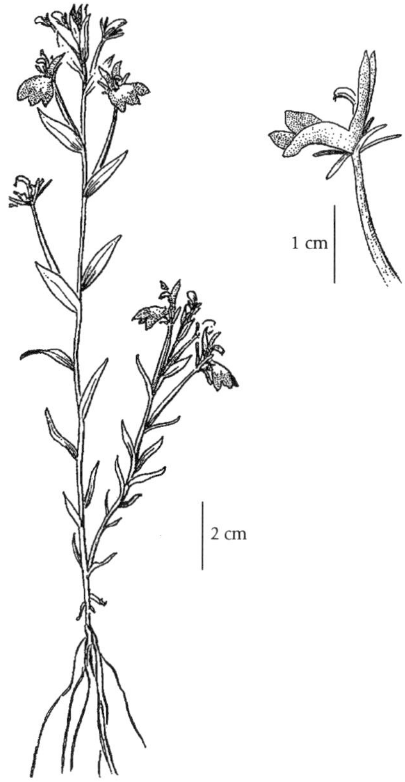

Downingia elegans

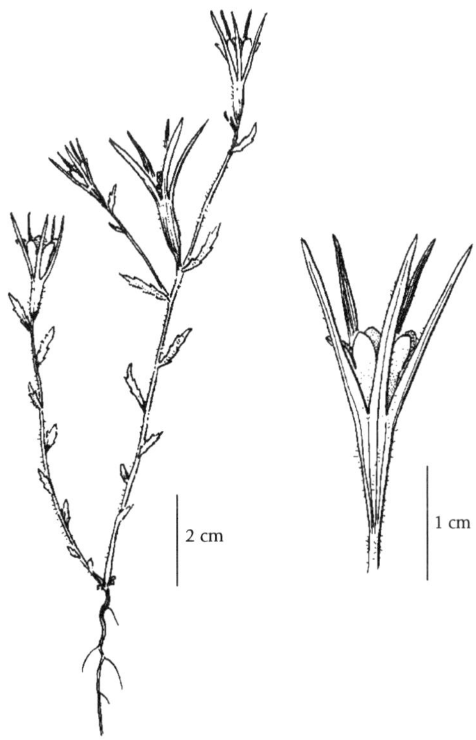

Githopsis specularioides

HETEROCODON

Heterocodon rariflorum **Nutt.**
heterocodon
HABITAT/RANGE:
Moist seepage areas in the lowland and lower montane zones; rare on S Vancouver Island and SE BC; S to WY, NV and CA.
General: Annual herb from a fibrous root; stems lax, simple or sparingly branched, 5-30 cm tall, glabrous or hairy on the leaf margins and angles of the stem.
Leaves: Basal leaves lacking; stem leaves alternate, stalked, toothed, round to egg-shaped, 2-10 mm long.
Flowers: Inflorescence of lax, solitary flowers borne opposite the leaflike bracts; calyx lobes leaflike, veiny, egg-shaped; corollas of upper flowers, tubular bell-shaped, blue, others abortive.
Fruits: Capsules, short, opening by inconspicuous irregular pores near the base.

LOBELIA

1. Stems leafless .... *L. dortmanna*
1. Stems leafy.
   2. Leaves entire to obscurely toothed; flowers blue with a large white eye .... *L. kalmii*
   2. Leaves conspicuously toothed; flowers whitish to pinkish or pale violet .... *L. inflata*

Lobelia dortmanna **L.**
water lobelia
HABITAT/RANGE:
Sandy or gravelly lake and pond margins in the lowland zone; locally common in coastal SW BC; E to NF and S to PA and N OR; NW Europe.
DESCRIPTION:
General: Aquatic perennial herb from a fibrous root; stems hollow, upright, mostly unbranched, 0.1-1 m tall, glabrous.
Leaves: Basal leaves in submerged rosette, linear, somewhat fleshy, hollow, curved, 2-8 cm long; stem leaves few, alternate, inconspicuous or reduced to threadlike bracts.
Flowers: Inflorescence of stalked, terminal racemes; corollas pale blue or white, 1-2 cm long, lower lip larger and hairy at the base, 3-lobed, upper lip 2-lobed.
Fruits: Capsules, 5-10 mm long, seeds roughened, squared at the base.

Lobelia inflata **L.**
Indian-tobacco
HABITAT/RANGE:
Mesic to dry roadsides and disturbed areas in the lowland zone; rare in SW BC, known only from the lower Fraser Valley area; introduced from E North America.
DESCRIPTION:
General: Perennial herb from a fibrous root; stems erect, simple or sometimes branched, slender, 10-40 cm tall, glabrous or inconspicuously hairy.
Leaves: Basal leaves often deciduous, spoon-shaped, 1-3 cm long, toothed; stem leaves similar, linear.
Flowers: Inflorescence of stalked, terminal racemes, loose, few-flowered; corollas whitish to pinkish or pale violet, 7-13 mm long, lower lip glabrous, longer than the tube.
Fruits: Capsules, 4-8 mm long, seeds roughened, abruptly sharp-pointed at both ends.

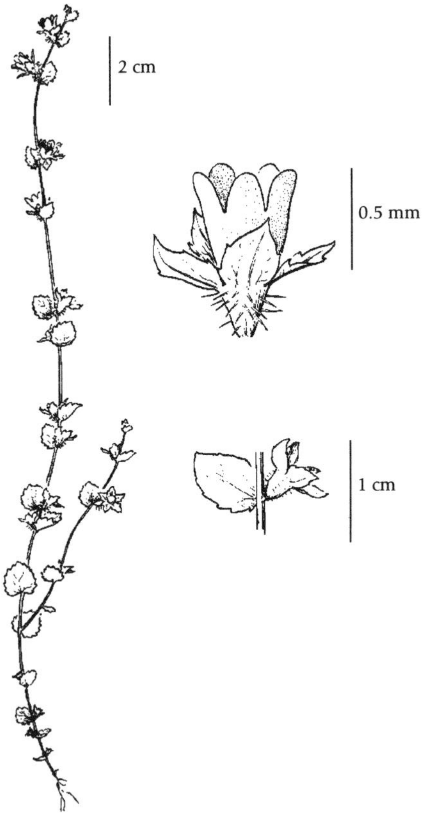

Heterocodon rariflorum

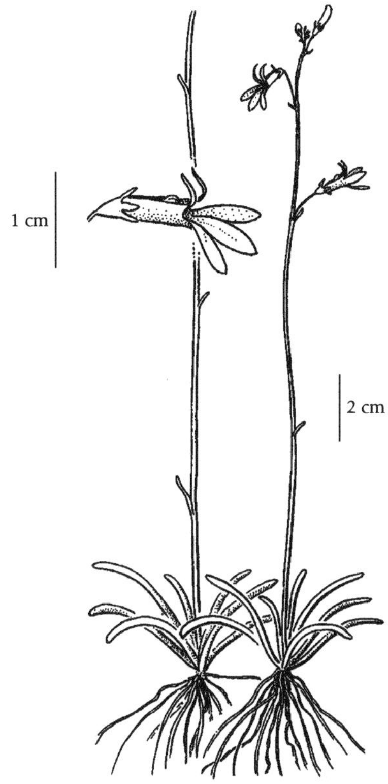

Lobelia dortmanna

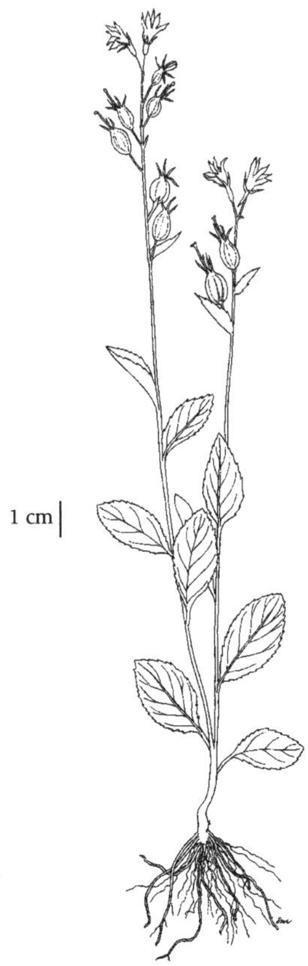

Lobelia inflata

***Lobelia kalmii* L.**
Kalm's lobelia
Habitat/Range:
Wet to moist calcareous fens, shorelines and meadows in the montane zone; frequent in SE and N BC; infrequent in SC BC; E to NF and S to PA, SD, MT and WA.
Description:
General: Perennial herb from a fibrous root; stems erect, simple or sometimes branched, 10-50 cm tall, glabrous or inconspicuously hairy.

Leaves: Basal leaves soon deciduous, spoon-shaped or rounded, stalked, 1-3 cm long, entire to obscurely toothed; stem leaves similar, linear, 1-5 cm long, not much reduced upwards.

Flowers: Inflorescence of stalked, terminal racemes, loose, few-flowered; corollas blue with a large white eye, 7-13 mm long, the lower lips glabrous, longer than the tube.

Fruits: Capsules, 4-8 mm long, seeds roughened, abruptly sharp-pointed at both ends.

TRIODANIS

Triodanis perfoliata* [L.] Nieuwl. var. *perfoliata (*Legousia perfoliata* (L.) Britt., *Specularia perfoliata* [L.] A. DC.)
Venus' looking-glass
Habitat/Range:
Dry grasslands, rocky slopes, open forests and disturbed areas in the lowland and lower montane zones; infrequent in S BC; E to PQ and S to FL, TX, NM and CA.
Description:
General: Annual herb from a taproot; stems simple or occasionally few-branched, erect, 10-60 cm tall, spreading short-hairy to bristly-hairy.

Leaves: Basal leaves lacking; stem leaves alternate, stalkless, heart-shaped or rounded, 0.5-3 cm long, clasping, palmately nerved, lowermost narrower with short stalks.

Flowers: Inflorescence of false spikes borne in the axils of the middle and upper leaves; corollas deep purple to pale lavender, 8-13 mm long, 5-lobed, lobes longer than the tube.

Fruits: Capsules, oblong or narrowly egg-shaped, opening above the middle.

CANNABACEAE

HUMULUS

***Humulus lupulus* L.**
common hops
Habitat/Range:
Ditches and disturbed areas in the lowland zone; infrequent in SW and SC BC, known only from the lower Fraser Valley and N Okanagan; introduced from Europe or elsewhere in North America.
Description:
General: Perennial herbaceous vine; stems twining, rough, up to 10 m long.

Leaves: Opposite, heart-shaped at the base, stipulate, 3-5 lobed, strongly toothed, rough-hairy, 4-10 cm long.

Flowers: Inflorescence of unisexual axillary flowers; female flowers in cylindric spikes, greenish-yellow, 3-5 cm long, yellow-glandular especially at the base of the bracts; male flowers in panicles.

Fruits: Achenes, hidden by enlarged bracts.

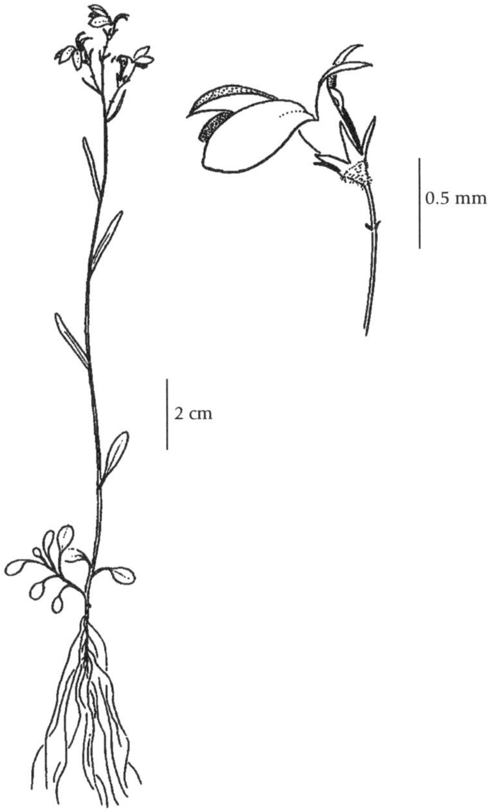

Lobelia kalmii

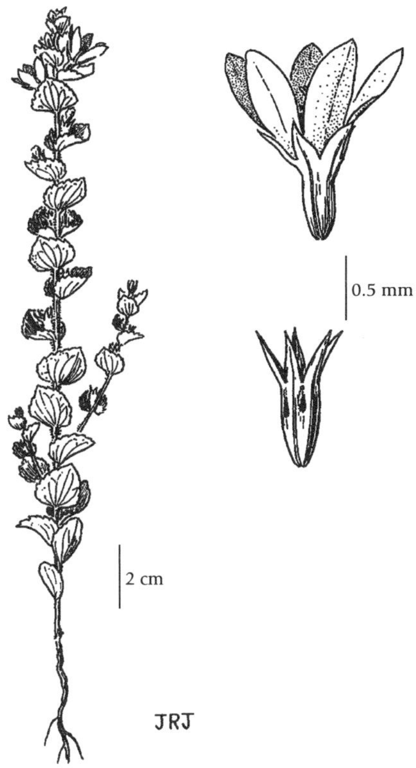

Triodanis perfoliata var. *perfoliata*

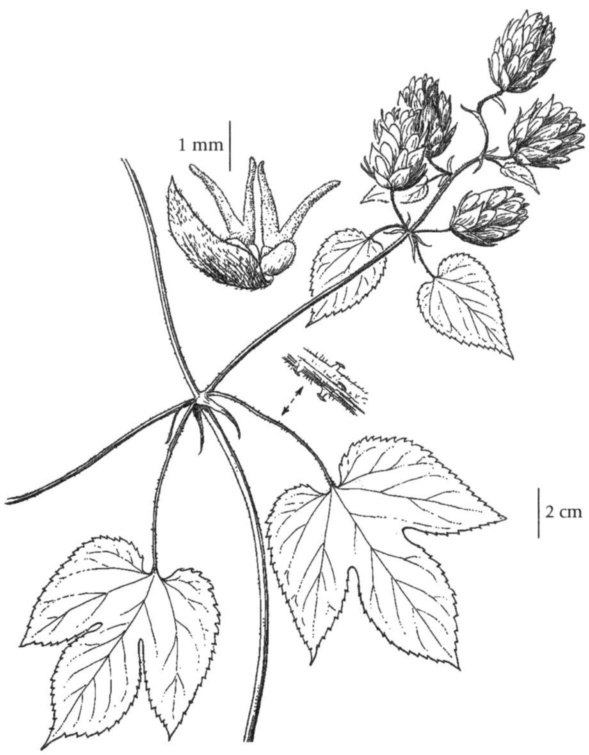

Humulus lupulus

CAPPARIDACEAE

CLEOME

***Cleome serrulata* Pursh**

stinking-clover; spider-flower; Rocky Mountain bee-plant

HABITAT/RANGE:

Moist to mesic sandy slopes and waste places in the steppe and lower montane zones; rare in SC and SE BC; E to MN and S to IL, NM and CA.

DESCRIPTION:

General: Annual herb from a fibrous root; stems erect, branched, 0.1-1.5 m tall, glabrous to sparsely soft-hairy.

Leaves: Basal leaves lacking; stem leaves with 3 leaflets, these linear-elliptic to narrowly lanceolate or oblanceolate, 1-7 cm long, stalked.

Flowers: Inflorescence of bracted racemes which elongate with maturity; corollas reddish-purple to pale pink or rarely white; petals 4, narrowly egg-shaped, 8-11 mm long; calyx lobes 4, abruptly pointed, not much longer than the tube; stamens 6, twice as long as the petals, with pink filaments.

Fruits: Podlike capsules, spreading to drooping, 3-6 cm long; seeds roughened with blisters.

CAPRIFOLIACEAE[23]

1. Leaves pinnately compound .... *Sambucus*
1. Leaves simple.
    2. Stems extensively creeping, slender .... *Linnaea*
    2. Stems erect or climbing, stout.
        3. Leaves lobed .... *Viburnum*
        3. Leaves entire or sometimes slightly lobed or wavy.
            4. Corollas regular or nearly so, bell-shaped; fruits 2-seeded white drupes .... *Symphoricarpos*
            4. Corollas irregular (bilaterally symmetric), funnel-shaped to tubular; fruits several-seeded berries .... *Lonicera*

[23] Key adapted from Mulligan (1961).

Cleome serrulata

LINNAEA

***Linnaea borealis* L.** (*L. borealis* ssp. *americana* [Forbes] Hult., *L. borealis* var. *americana* [Forbes] Rehd. = ssp. *longiflora*)

twinflower

HABITAT/RANGE:

Mesic forests in the lowland and montane zones; ssp. *longiflora* common throughout BC; *ssp. borealis* rare in extreme NW BC; ssp. borealis – circumboreal, N to AK and YT, *ssp. longiflora* – N to AK, YT and NT, E to NF and S to IN, NM and CA.

DESCRIPTION:

General: Evergreen, creeping, semiwoody herb or subshrub; stems long, slender, weakly rooting, thinly hairy and often also glandular, producing numerous short (less than 10 cm tall), more or less erect, leafy stems or branchlets with 2-4 pairs of leaves.

Leaves: Opposite, simple, short-stalked, egg-shaped to nearly circular, with a few shallow teeth along the upper half, 7-25 mm long, 5-15 mm wide, leathery, dark green above, paler below, glabrous or the stalks and margins of the leaf bases often fringed with long hairs.

Flowers: Inflorescence of nodding pairs of flowers on slender, erect, Y-shaped stalks that terminate the leafy branchlets; base of flowers cradled by two tiny bracts; corollas slenderly funnel-shaped, 9-16 mm long, pink to whitish, flaring above to 5 lobes, hairy in the throat; styles about as long as corollas, with a tiny knoblike stigma; stamens 4.

Fruits: Dry 1-seeded achenes (nutlets), flanked by a pair of tiny glandular bracts.

Notes: Two subspecies occur in BC.

1. Corollas 9-11 mm long, narrowly bell-shaped, flaring from within the calyces, tubes very short or sometimes absent .... ssp. *borealis*
1. Corollas 10-16 mm long, funnel-shaped, tubes about equalling or surpassing the calyces .... ssp. *longiflora* (Torr.) Hult.

LONICERA

1. Flowers in pairs (appearing united in *L. caerulea)* on axillary stalks; leaves distinct, plants more or less erect shrubs.
    2. Involucral bracts 4, relatively large, broad and leafy; ovaries and fruits wholly distinct .... *L. involucrata*
    2. Involucral bracts 2, small, linear to lance-oblong, ovaries and fruits wholly or partly united.
        3. Ovaries and fruits wholly united; corolla lobes about as long as the tube .... *L. caerulea*
        3. Ovaries and fruits united (sometimes scarcely) at the base; corolla lobes much shorter than the tube .... *L. utahensis*
1. Flowers several to many in terminal or terminal and axillary inflorescences; leaves united in pairs around the stem; plants vines or climbing shrubs.
    4. Leaves, or at least some of them with well-developed, egg-shaped to rounded stipules .... *L. hispidula*
    4. Leaves without stipules.
        5. Leaves long-fringed on the margins .... *L. ciliosa*
        5. Leaves not fringed on the margins.
            6. Corollas 1.5-2.5 cm long, the tubes densely hairy within .... *L. dioica*
            6. Corollas 3-5 cm long, the tubes glabrous within .... *L. etrusca*

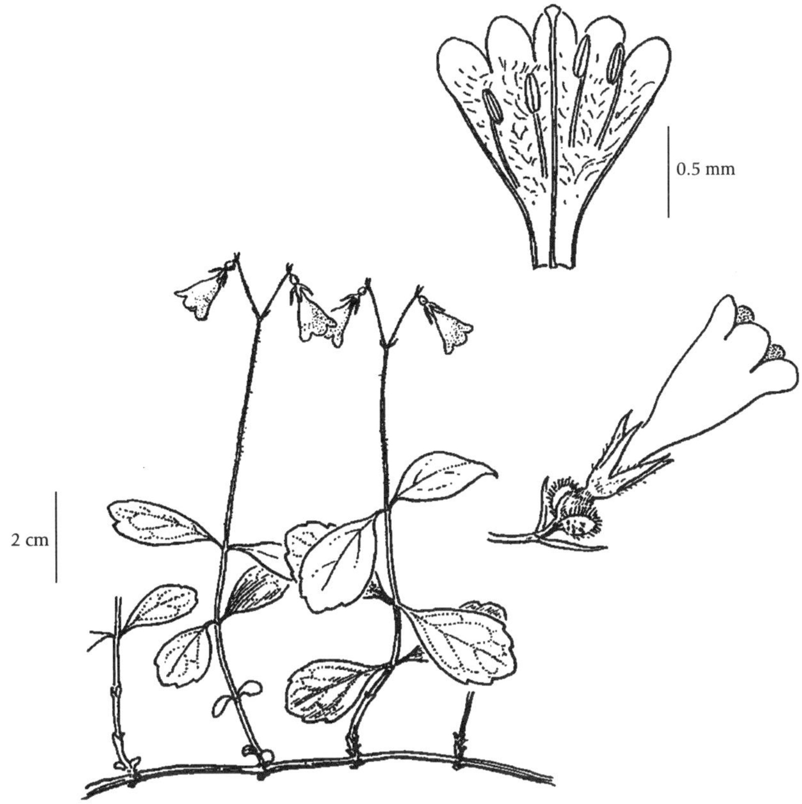

Linnaea borealis* ssp. *longiflora

Lonicera caerulea **L. var.** ***cauriana*** **(Fern.) Boivin**
bluefly honeysuckle
HABITAT/RANGE:
Moist to dry meadows, streambanks and forests in the montane zone; locally common in SE BC; circumboreal, E to AB and S to PA, MN, WY, NV and CA; Eurasia.
DESCRIPTION:
General: Deciduous shrub, erect, 0.2-2 m tall.

Leaves: Opposite, distinct, deciduous, short-stalked, elliptic to egg-shaped or oblong, entire, rounded at the tip, bristly-hairy on the undersurface.

Flowers: Inflorescence of paired trumpet-shaped flowers on axillary stalks; corollas yellow, the lobes about as long as the tubes, which are hairy inside and out; involucral bracts 2, small, linear to lance-oblong; ovaries and fruits appearing wholly united.

Fruits: Berries, 1 cm long, bluish-glaucous to red.

Lonicera ciliosa **(Pursh) Poir.** ***ex*** **DC.**
western trumpet; orange honeysuckle
HABITAT/RANGE:
Mesic to dry thickets and forests in the lowland, steppe and montane zones; common in S BC; S to MT and CA.
DESCRIPTION:
General: Deciduous woody vine, twining, trailing or climbing, sometimes to 6 m high, widely branching; twigs hollow, glabrous, glaucous when young.

Leaves: Opposite, short-stalked to nearly stalked, elliptic to oval, 4-10 cm long, 2.5-5 cm wide, entire, rounded to pointed at tip, glabrous except for a fringe of long hairs along the edges, glaucous beneath; end pair on each flowering twig joined to form a disk around the stem below the inflorescence.

Flowers: Inflorescence a tight, headlike cluster of flowers on very short stalks in what looks like a whorl, above the disk leaves at branch ends; corollas orange-yellow to red, 2.5-4 cm long, the tubes densely hairy within, bulging on one side at the base, 3-4 times as long as the weakly 2-lipped limbs; upper lips 4-lobed, lower lips narrow, 1-lobed.

Fruits: Berries, clustered, to 1 cm thick, orange-red, translucent, several-seeded.

Lonicera dioica **L. var.** ***glaucescens*** **(Rydb.) Butters**
glaucous-leaved honeysuckle; smooth-leaved honeysuckle; red honeysuckle
HABITAT/RANGE:
Moist to dry rocky slopes and forests in the lowland and montane zones; frequent in SC and E BC; N to NT, E to PQ and S to NC, OK, SC and GA.
DESCRIPTION:
General: Deciduous, semi-erect shrub or twining, trailing or climbing woody vine, sometimes to 5 m high, widely branching; twigs hollow, glabrous, green or purplish, glaucous when young; bark light coloured and shreddy.

Leaves: Opposite, nearly stalked, egg-shaped or oval, 5-8 cm long, entire, rounded or blunt at tip, dark green and glabrous above, pale and often short-hairy beneath, especially on veins; margins not fringed; upper leaf pairs joined at the base, forming a disk around the stem.

Flowers: Inflorescence a head-like cluster at branch tips, on very short stalks in what looks like a whorl above the disk leaves; corollas yellow to orange (often purplish or reddish with age), tubular to funnel-shaped, 2-3 cm long, the tubes often short-hairy on the outside, slightly bulging on one side at the base, 1.2-2 times as long as the 2 spreading lips; upper lips 4-lobed.

Fruits: Berries, 8-12 mm thick, red, clustered in disk leaves.

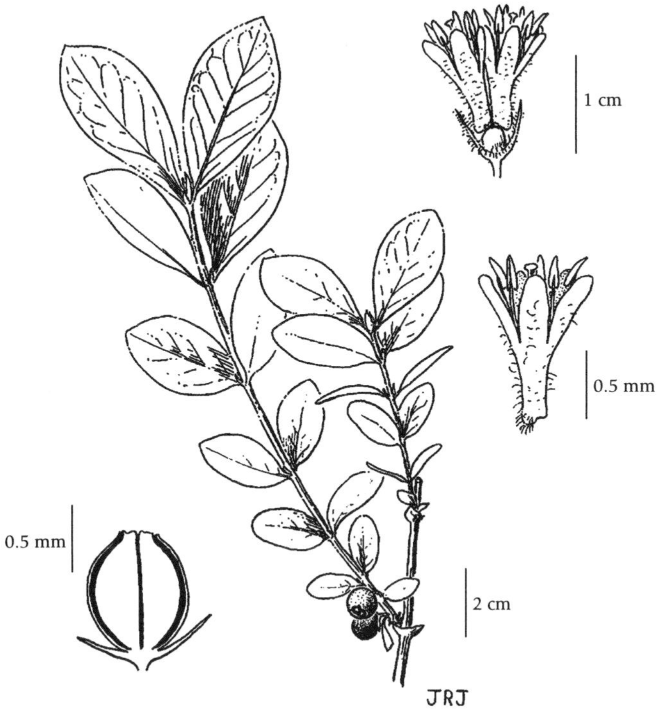

Lonicera caerulea var. *cauriana*

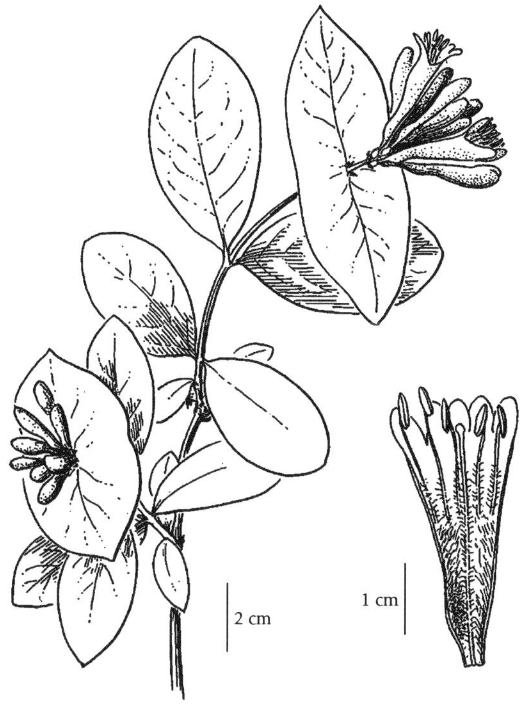

Lonicera ciliosa

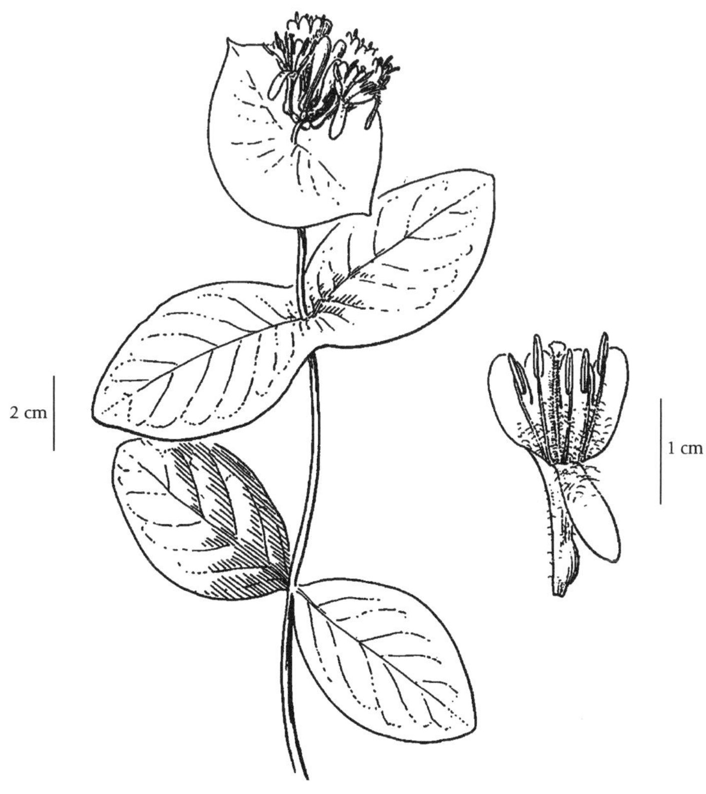

Lonicera dioica var. *glaucescens*

***Lonicera etrusca* Santi**

Etruscan honeysuckle

HABITAT/RANGE:

Moist to mesic thickets and disturbed areas in the lowland zone; frequent in SW BC, known from S Vancouver Island and the adjacent lower mainland, also rare on the Queen Charlotte Islands; introduced from Europe.

DESCRIPTION:

General: Partially evergreen, woody vine, twining to climbing.

Leaves: Opposite, broadly oval to egg-shaped, 1-3 cm long, entire, usually obtuse at tip, glabrous above, glaucous and often hairy beneath; upper leaf pairs fused into an oval disk around the stem, or all leaves distinct.

Flowers: Inflorescence of long-stalked, head-like whorls in terminal or axillary groups of 3; corollas trumpet-shaped, yellowish-white usually tinged with red, 3-5 cm long; tubes very slender, sometimes glandular outside but glabrous within, flaring to 2 lips that are sometimes as long as the tubes, the upper lips 4-lobed.

Fruits: Berries, red, several-seeded, flanked by bractlets nearly as long as the berries.

***Lonicera hispidula* (Lindl.) Dougl.**

hairy honeysuckle

HABITAT/RANGE:

Mesic to dry forests in the lowland and montane zones; common locally on S Vancouver Island, adjacent lower mainland and the Gulf Islands; S to CA.

DESCRIPTION:

General: Partially evergreen, woody vine or sprawling shrub, 1-6 m long/tall; twigs hollow, usually short-hairy, glaucous when young.

Leaves: Opposite, on stalks 3-5 mm long, some with paired, fused stipules between the leaf stalks, oblong to egg-shaped, 2-8 cm long, 1.5-5 cm wide, often somewhat heart-shaped at the base, obtuse at the tip, usually stiffly hairy, sometimes glabrous, glaucous beneath; upper pairs joined to form disks around the stem.

Flowers: Inflorescence of stalked and clustered flowers in compact whorls forming short, interrupted spikes terminal and axillary on long stalks that are glabrous to glandular hairy; corollas whitish, pink, or yellowish, sometimes purplish, 12-20 mm long, tubular below, the tubes bulging on one side above the base 2-lipped above, the upper lips 4-lobed, the lobes recoiled, lips as long as or longer than the tubes.

Fruits: Berries, red, 0.8-1 cm thick, several-seeded.

***Lonicera involucrata* (Richards.) Banks *ex* Spreng.**

black twinberry; bearberry honeysuckle

HABITAT/RANGE:

Moist forests and thickets in the lowland, steppe and montane zones; common in BC south of 58° N; N to YT, E to PQ and S to MI, WI, NM, CA and MX.

DESCRIPTION:

General: Deciduous, erect to straggly shrub, 0.5-5 m tall; young twigs 4-angled, greenish, glabrous; older stems with yellowish-grey, shredding bark.

Leaves: Opposite, short-stalked, elliptic to oblanceolate or broadly lanceolate, 3-16 cm long, 2-8 cm wide, entire, rounded or tapered at base, abruptly sharp-pointed at tip, pale and glabrous or somewhat hairy beneath.

Flowers: Inflorescence of short-stalked pairs of flowers cupped by large (2-4 cm across) involucres of 2 pairs of fused bracts, on 1-4 cm long stalks in leaf axils; corollas tubular bell-shaped, yellow, 1-2 cm long, glandular-hairy on outside, scarcely 2-lipped, with 5 roughly equal lobes above; corolla tubes with short thick spurs at bases.

Fruits: Berries, shiny black, 0.8-1.2 cm across, paired, loosely surrounded while young by paired bracts, which darken and redden (to purplish-maroon) as they recurve from the maturing fruits.

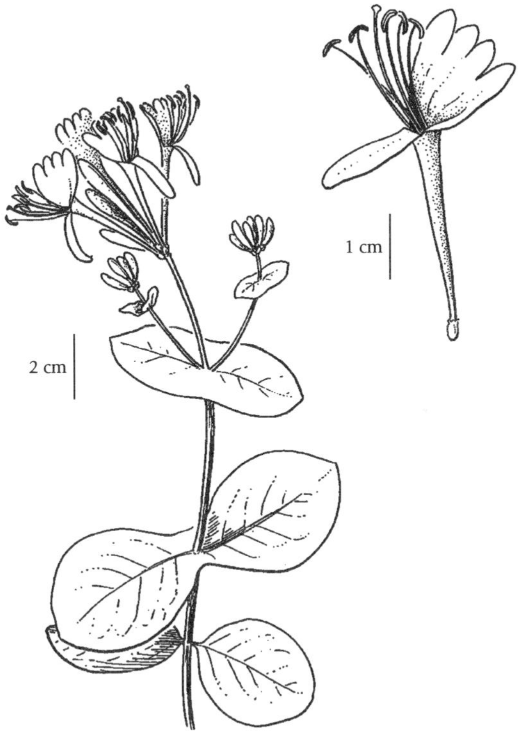

Lonicera etrusca

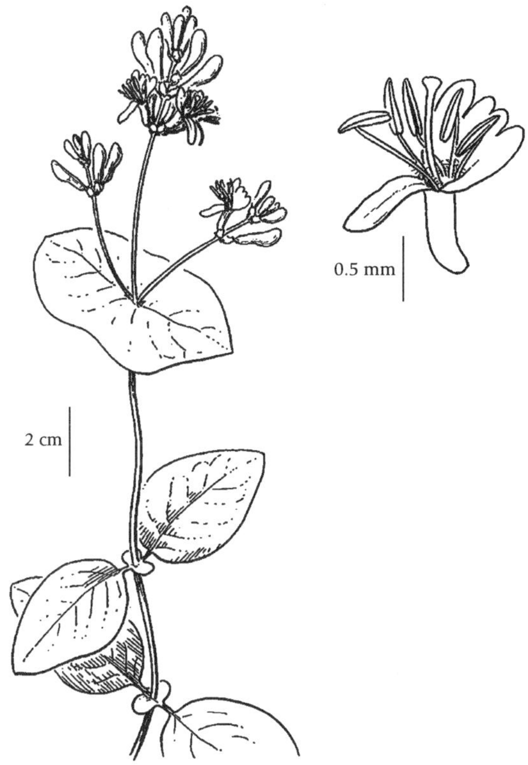

Lonicera hispidula

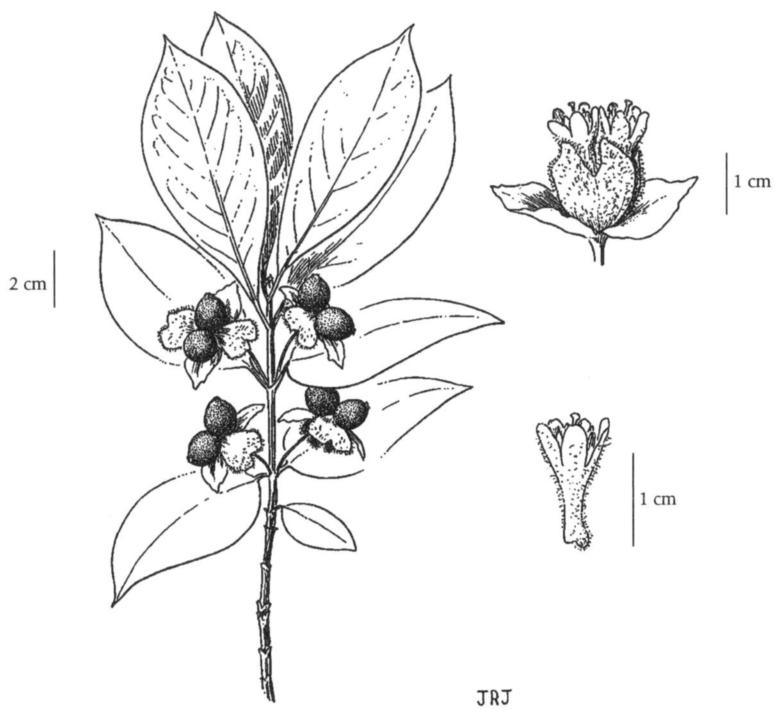

Lonicera involucrata

Lonicera utahensis **S. Wats.**

Utah honeysuckle; red twinberry

HABITAT/RANGE:

Moist to mesic meadows, grasslands, shrublands and open forests in the lowland, steppe and montane zones; common in SC and SE BC, rare in SW BC; E to SW AB and S to MT, WY, UT and N CA.

DESCRIPTION:

General: Deciduous, ascending or erect shrub, 0.6-2 m tall; twigs glabrous or nearly so.

Leaves: Opposite, short-stalked, elliptic to somewhat egg-shaped or oblong, rounded to somewhat heart-shaped at the base, obtuse to rounded at the tip, 2-8 cm long, 1-4 cm wide, glabrous above, often thinly hairy beneath and on stalks.

Flowers: Inflorescence of pairs of flowers in leaf axils, on the ends of 0.5-3-cm long stalks; involucres of 2 pairs of very small bracts; corollas cream or light yellow, funnel-shaped, 1-2 cm long, scarcely 2-lipped, the 5 slightly unequal lobes much shorter than the tubes, which are hairy within and have short thick spurs at the bases.

Fruits: Berries, red, 0.7-1 cm across paired, united at the base, 2- to 4-seeded.

SAMBUCUS

1. Inflorescence pyramidal or strongly convex; fruits not glaucous .... *S. racemosa*
1. Inflorescence flat-topped; fruits strongly glaucous .... *S. cerulea*

Sambucus cerulea **Raf. var. *cerulea*** (*S. nigra* L. ssp. *cerulea* [Raf.] R. Bolli)

blue elderberry; elder

HABITAT/RANGE:

Moist to mesic meadows, grasslands, shrublands, disturbed areas and open forests in the lowland, steppe and montane zones; common in S BC; S to MT, AZ, NM and CA.

DESCRIPTION:

General: Deciduous, erect shrub, occasionally a small tree, 1-5 (sometimes to 10) m tall, the trunk diameter to 25 cm, often several-stemmed from basal suckers; twigs soft, pithy, glaucous; old bark finely fissured.

Leaves: Opposite, stalked, large, compound, pinnately divided into 5-9 leaflets, which are lanceolate to narrowly egg-shaped, 5-15 cm long and 2-6 cm wide, often asymmetric at the base, abruptly sharp-pointed at the tip, saw-toothed on the margins, and usually glabrous.

Flowers: Inflorescence of numerous, small flowers in a flat-topped, 4-20 cm wide, umble-like cluster with 4-5 spokes; corollas wheel-shaped, white or cream; petals fused at base into short, flat tubes that spread (4-7 mm across) to 5 lobes that are longer than the tubes.

Fruits: Berrylike, globose, 4-6 mm across, powder blue (bluish-black with a heavy, waxy bloom), with 3-5 wrinkled stones, each enclosing a seed.

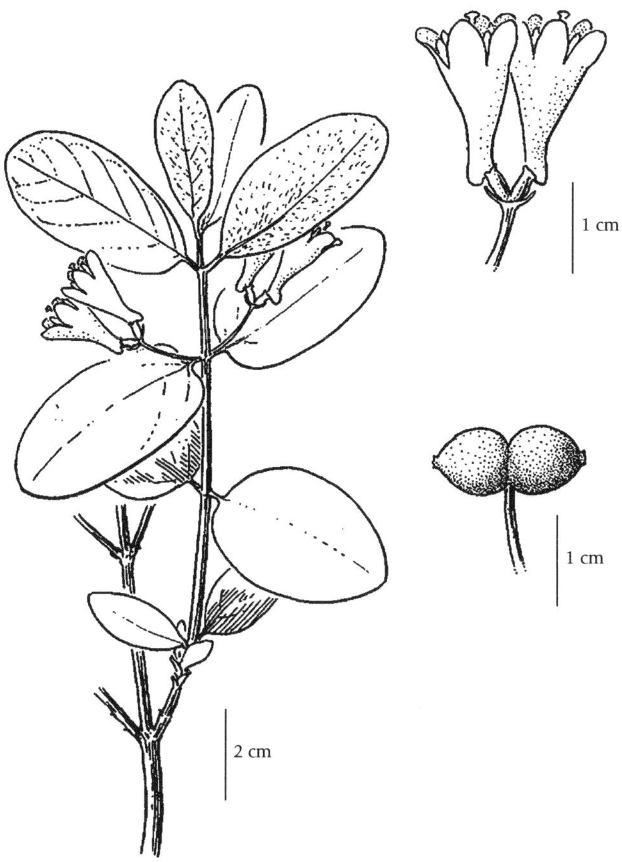

Lonicera utahensis

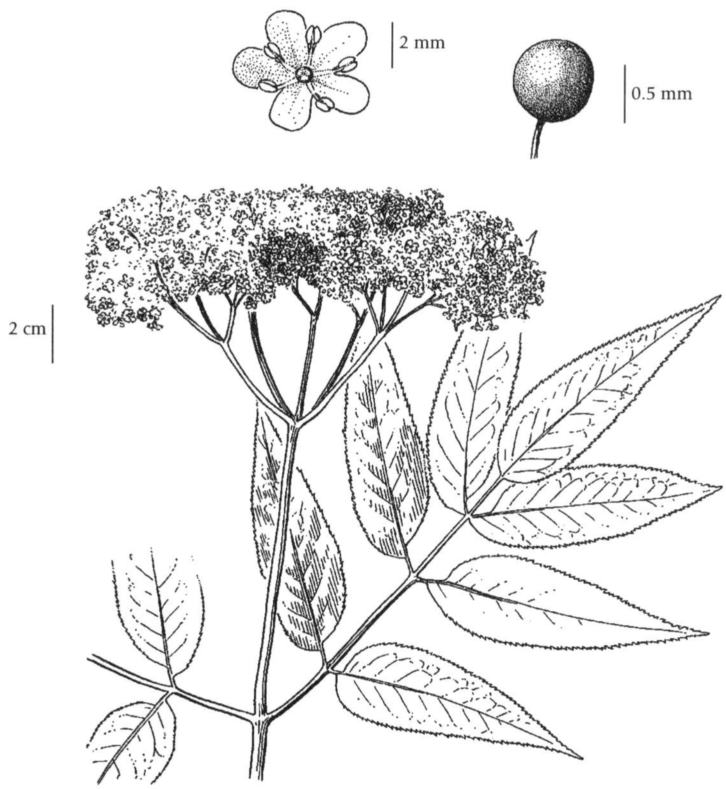

Sambucus cerulea var. *cerulea*

***Sambucus racemosa* L. ssp. *pubens* (Michx.) House**

black elderberry, black elder (var. *melanocarpa);* **coastal red elderberry**, coastal red elder (var. *arborescens*); **eastern red elderberry**, eastern red elder (var. *leucocarpa)*

HABITAT/RANGE:

Moist to mesic meadows, ditches, streambanks, grasslands, shrublands, disturbed areas and forests in the lowland, steppe and montane zones; var. *arborescens* is common in and W of the Coast-Cascade Mountains, var. *melanocarpa* is common E of the Coast-Cascade Mountains and less frequent to the W of them, var. *leucocarpa* is infrequent in SC and SE BC; var. *arborescens* – N to AK and YT and S to CA; var. *melanocarpa* – E to AB and S to NM, AZ, NV and N CA; var. *leucocarpa* – E to PQ and NB and S to TN and GA.

DESCRIPTION:

General: Deciduous, erect, large shrub, 1-6 m tall; twigs thick, soft, pithy, usually somewhat glaucous; bark dark reddish-brown, warty.

Leaves: Opposite, stalked, large, compound, pinnately divided into 5-7 elliptic to lanceolate leaflets, often asymmetric at the base, abruptly sharp-pointed at the tip, saw-toothed on the margins, and usually somewhat hairy beneath.

Flowers: Inflorescence of numerous, small flowers in a 4-10 cm long, egg-shaped to conical cluster with short lateral branches on a stronger central axis; corollas white or cream, wheel-shaped; petals fused at base into a short flat tube that spreads (3-6 mm across) at the top to 5 lobes, the lobes longer than the tubes and becoming reflexed.

Fruits: Berrylike, globose, 5-6 mm across, shiny red or purplish-black, occasionally brown, yellow, or whitish, but not glaucous, with 3 small, smooth to wrinkled or pebbled stones, each enclosing a seed.

Notes: Three varieties of this North American subspecies occur in BC.

1. Fruits black or purplish-black; nutlets slightly wrinkled or pebbly....... var. *melanocarpa* (A. Gray) McMinn
1. Fruits bright red (sometimes yellow or white); nutlets mostly smooth or slightly wrinkled or pebbly.
    2. Nutlets mostly smooth; plants 2-6 m tall .... var. *arborescens* (T.& G.) A. Gray
    2. Nutlets slightly wrinkled or pebbly; plants 0.5-3 m tall .... var. *leucocarpa* (T.& G.) Cronq. In Cronq. & Hitchcock

SYMPHORICARPOS

1. Stems trailing and rooting at the nodes, the branches rising less than 50 cm .... *S. hesperius*
1. Stems erect, more or less branching, 0.3-3 m tall.
    2. Corollas relatively long and narrow, the lobes not more than half the length of the tubes .... *S. oreophilus*
    2. Corollas relatively short and wide, the lobes nearly equalling the tubes.
        3. Styles elongate, exserted slightly from the corolla tube, usually long-hairy near the middle........... .... *S. occidentalis*
        3. Styles short, not exserted from the corolla tube, glabrous .... *S. albus*

Sambucus racemosa ssp. *pubens*

Symphoricarpos albus **(L.) Blake**
common snowberry
HABITAT/RANGE:
Mesic to dry meadows, disturbed areas, grasslands, shrublands, and forests in the lowland, steppe and montane zones; extremely common in S BC, common in C BC, infrequent northward; N to NT, E to PQ, and S to NE, VA, CO and CA.
DESCRIPTION:
General: Deciduous, erect, much-branched shrub, often densely colonial from rhizomes, 0.5-3 m tall; older stems dark greyish-brown with shredding bark; branchlets glabrous; twigs thin, wiry.

Leaves: Opposite, on stalks 2-4 mm long, elliptic to oval, 1.5-5 cm long and 1-3.5 cm wide, obtuse at base and tip, margins entire or occasionally with a few wavy teeth; leaves often larger and irregularly lobed on sterile shoots; glabrous above, glabrous or spreading-hairy beneath.

Flowers: Inflorescence of short, dense clusters (racemes) of few, short-stalked to nearly stalked flowers, at ends of twigs and often also in upper leaf axils; corollas widely bell-shaped, 5-7 mm long, hairy within; petals pink to whitish, fused at bases into tubes that flare at top to 5 lobes, which are as long to half as long as the tubes; styles 2-3 mm long, glabrous; stamens shorter than corolla lobes, the anthers 1-1.5 mm long, about as long as the filaments.

Fruits: Berrylike drupes, densely clustered, waxy white, ellipsoid or nearly globose, 5-15 mm long; nutlets 2, each enclosing a seed; some fruits persist through the winter.

Notes: Two varieties occur in BC.

1. Plants 0.5-1 m tall; fruits usually less than 1 cm long; infrequent in E BC .... var. *albus*
1. Plants 1-3 m tall; fruits 1-1.5 cm long; our common phase .... var. *laevigatus* (Fern.) Blake

Symphoricarpos hesperius **G.N. Jones** (*S. mollis* Nutt., *S. mollis* var. *hesperius* [G.N. Jones] Cronq.)
trailing snowberry; creeping snowberry
HABITAT/RANGE:
Mesic to dry forests in the lowland zone; locally common on SE Vancouver Island, the Gulf Islands and the lower Fraser Valley; S to N ID and CA.
DESCRIPTION:
General: Deciduous, trailing shrub; stems reclining, 1-3 m long, often rooting at the nodes, forming small tubers, and producing erect branches to 60 cm high; twigs slightly short-hairy.

Leaves: Opposite, on stalks 1-3 mm long, egg-shaped to elliptic, 1-3 cm long, 0.5-2 cm wide, entire or sometimes with a few coarse teeth or shallow lobes; leaves on vigorous shoots often larger and more frequently toothed or lobed, slightly hairy at least beneath.

Flowers: Inflorescence a short terminal cluster of a few flowers, sometimes also a few flowers in the upper leaf axils; corollas widely and often asymmetrically bell-shaped, pale pink, 3-5 mm long, tubular at the bases, flaring to 5 lobes almost as long as the tubes; tubes with collars of short hairs around the top inside; styles 2-3 mm long, glabrous, about as long as the tubes; stamens as long as corolla lobes, the anthers 1 mm long, a little shorter than the filaments.

Fruits: Berrylike drupes, clustered, white, nearly globose, 5-6 mm long; nutlets 2, each enclosing a single seed.

Symphoricarpos albus

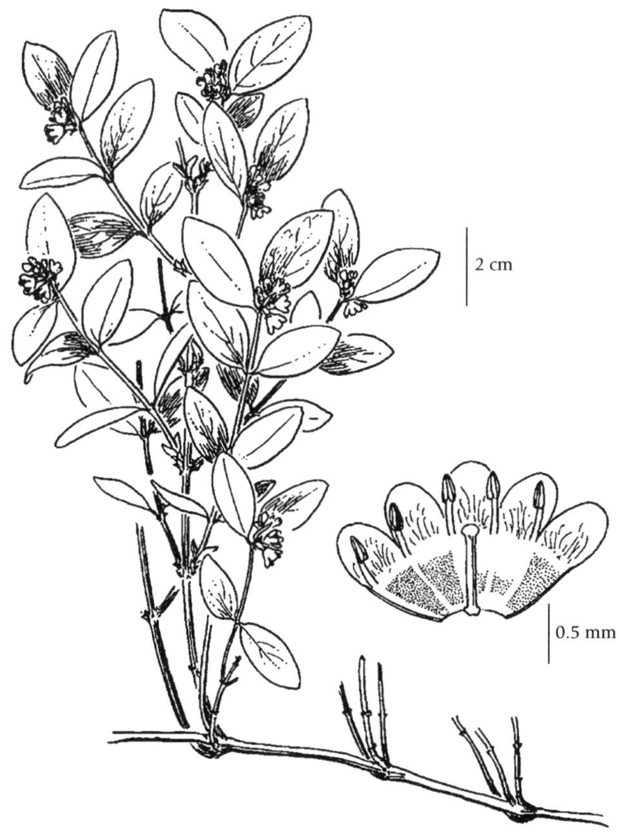

Symphoricarpos hesperius

***Symphoricarpos occidentalis* Hook.**

western snowberry; wolfberry

HABITAT/RANGE:

Moist streambanks and lakesides in the steppe and montane zones; frequent in BC east of Coast-Cascade Mountains; N to NT, E to PQ and S to KS, MO, VT, NM and WA.

DESCRIPTION:

General: Deciduous, low, erect shrub, 0.3-1 m tall, forming extensive colonies from rhizomes or by layering; stems dark grey, with shredding bark; twigs reddish-brown, short-hairy.

Leaves: Opposite, on stalks 3-10 mm long, elliptic or egg-shaped, mostly 2.5-8 cm long and 1.5-5 cm wide, entire or with a few blunt, irregular teeth; leaves often larger and more lobed on sterile shoots; glabrous above, hairy beneath at least along veins; similar to those of *S. albus*, but thicker and more leathery, and often pointed at the tip.

Flowers: Inflorescence of short, dense cluster of few to several flowers, stalked flowers at ends of twigs and in upper leaf axils; corollas bell-shaped, pale pink, tubular below, flaring above to 5 prominent, spreading lobes that equal or are a little longer than the tubes; styles 3-8 mm long, long-hairy near the middle or sometimes glabrous, equalling or slightly exceeding the corollas; anthers 1.5-2 mm long, shorter than the filaments.

Fruits: Berrylike drupes, clustered, white, nearly globose, 6-9 mm long; nutlets 2, each enclosing 1 seed.

***Symphoricarpos oreophilus* A. Gray var. *utahensis* (Rydb.) A. Nels.**

mountain snowberry

HABITAT/RANGE:

Dry meadows, grasslands, shrublands, rocky slopes and forest openings in the steppe to subalpine zones; locally common in extreme SC BC; S to NM, CA and N MX.

DESCRIPTION:

General: Deciduous, erect shrub, 0.5-1.5 m tall; twigs hairy to glabrous.

Leaves: Opposite, on stalks 1-4 mm long, egg-shaped to elliptic, 1-5 cm long, 0.5-2.5 cm wide, entire, glabrous or hairy.

Flowers: Inflorescence of short, terminal clusters of few flowers on short, drooping stalks, or in 1- to 3-flowered axillary clusters; corollas narrowly funnel-shaped, 7-11 mm long, pale pink to whitish, tubular below and usually hairy within, flaring above to 5 lobes 1/4 to 1/2 as long as the tube; styles glabrous, 2-4 mm long, shorter than the corolla tubes; stamens shorter than corolla lobes, the anthers 1-2 mm long, about as long as the filaments.

Fruits: Berrylike drupes, white, broadly ellipsoid, 7-10 mm long; nutlets 2, each with 1 seed.

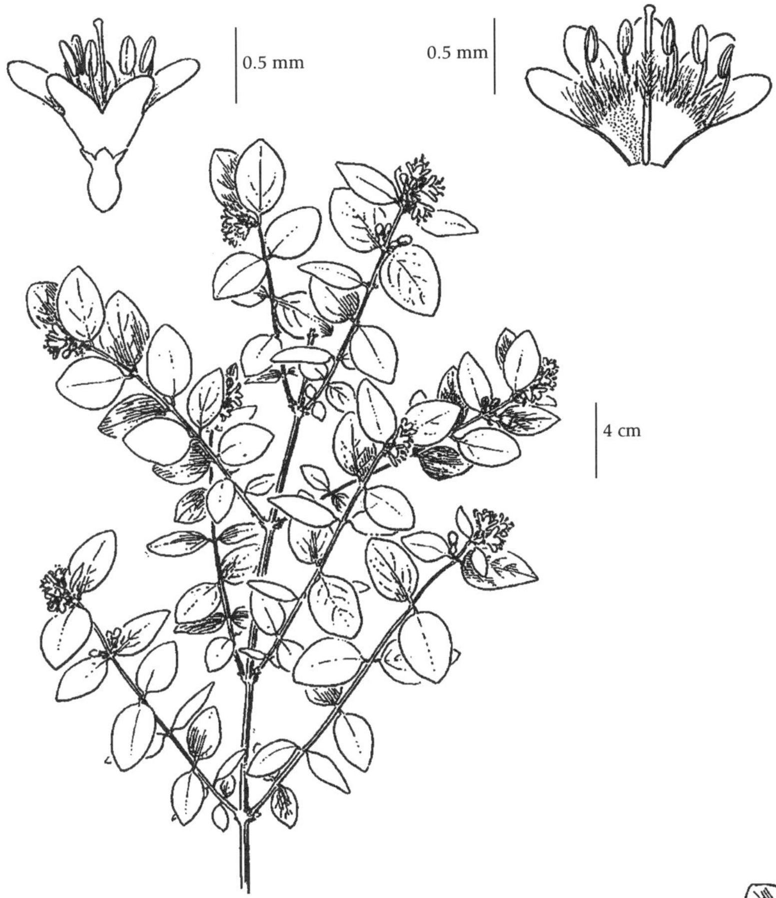

Symphoricarpos occidentalis

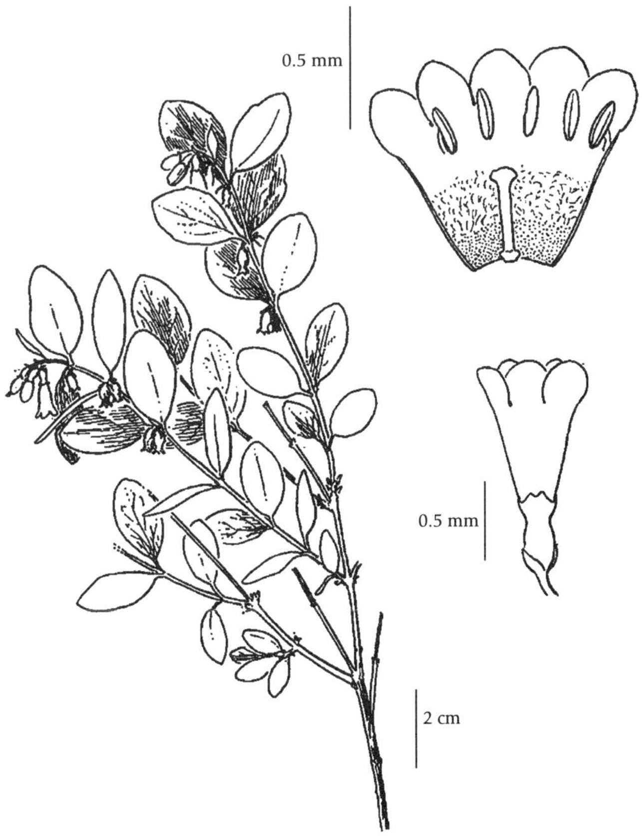

Symphoricarpos oreophilus var. *utahensis*

VIBURNUM

1. Leaves deeply 3-lobed; sterile corollas 1.5-2.5 cm across; stamens conspicuous, exserted from the corollas .... *V. opulus*
1. Leaves shallowly 3-lobed; corollas all 0.4-0.7 cm across; stamens inconspicuous .... *V. edule*

Viburnum edule **(Michx.) Raf.**

highbush-cranberry; squashberry; mooseberry

HABITAT/RANGE:

Wet to moist streambanks, swamps and forests in the lowland, steppe and montane zones; frequent throughout BC; N to AK, YT and NT, E to NF, and S to PA, ID, CO and OR.

DESCRIPTION:

General: Deciduous, sprawling to erect shrub, spreading from rhizomes and by layering, 0.5-3.5 m tall; twigs glabrous; bark smooth, reddish to grey.

Leaves: Opposite, stalked, elliptic in outline, most shallowly 3-lobed but some unlobed, jaggedly toothed, often with an isolated pair of glandular teeth near the junction of the blade and stalk, glabrous or hairy beneath, especially along the veins, turning crimson in the fall.

Flowers: Inflorescence a small, compact, flat or rounded cluster of few to several flowers, 1-3 cm across, borne on short axillary stalks with a pair of leaves; flowers all alike and fertile; petals whitish, fused at bases into tubes that flare (4-7 mm wide) at the tops to 5 lobes; stamens inconspicuous, the filaments 1 mm long or less.

Fruits: Berrylike drupes, clustered, red or orange, 0.8-1.5 cm long, each with a single, large, flattened stone.

Viburnum opulus **L. var. *americanum*** **Ait.** (*V. opulus* ssp. *trilobum* [Marsh.] Huft., *V. trilobum* Marsh.)

American bush-cranberry; wild guelder-rose

HABITAT/RANGE:

Moist streambanks and lakesides in the steppe and montane zones; infrequent in WC, SC and SE BC; E to NF and S to PA, WY, ID, and WA.

DESCRIPTION:

General: Deciduous, erect, large shrub, 1-4 m tall; twigs glabrous; bark smooth, grey.

Leaves: Opposite, stalked, broadly egg-shaped to nearly round in outline, 5-13 cm long, all deeply and sharply 3-lobed, coarsely blunt-toothed, nearly glabrous or hairy beneath on the veins; stalks with a shallow groove and small glands upward.

Flowers: Inflorescence a flattish cluster of several to many flowers, 5-15 cm across, with sterile marginal flowers much larger than the fertile inner ones; petals white, fused at bases into tubes that spread (1.5-2.5 cm wide) to 5 large lobes in the marginal flowers, to 3-4 mm wide in the small inner flowers; stamens conspicuous, exserted of corolla, the filaments 3-5 mm long.

Fruits: Berrylike drupes, clustered, scarlet, 0.8-1.5 cm long, each with a single, large, flattened stone.

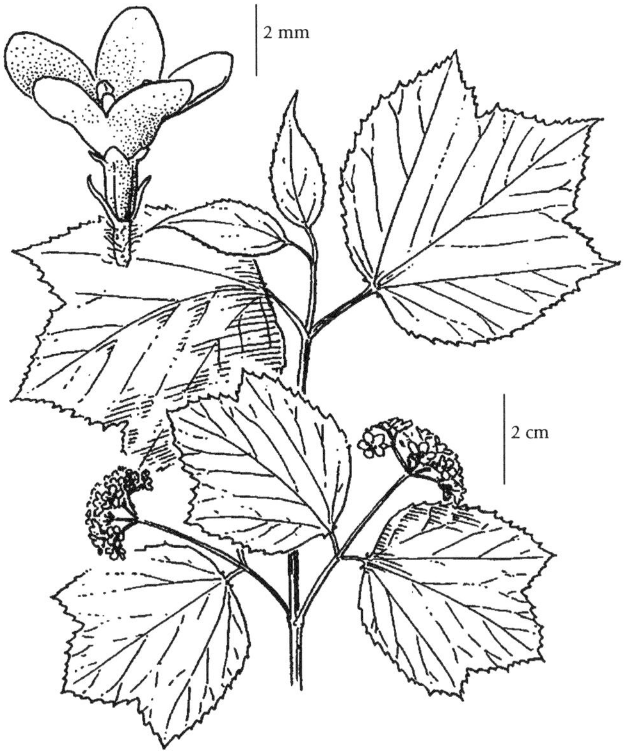

Viburnum edule

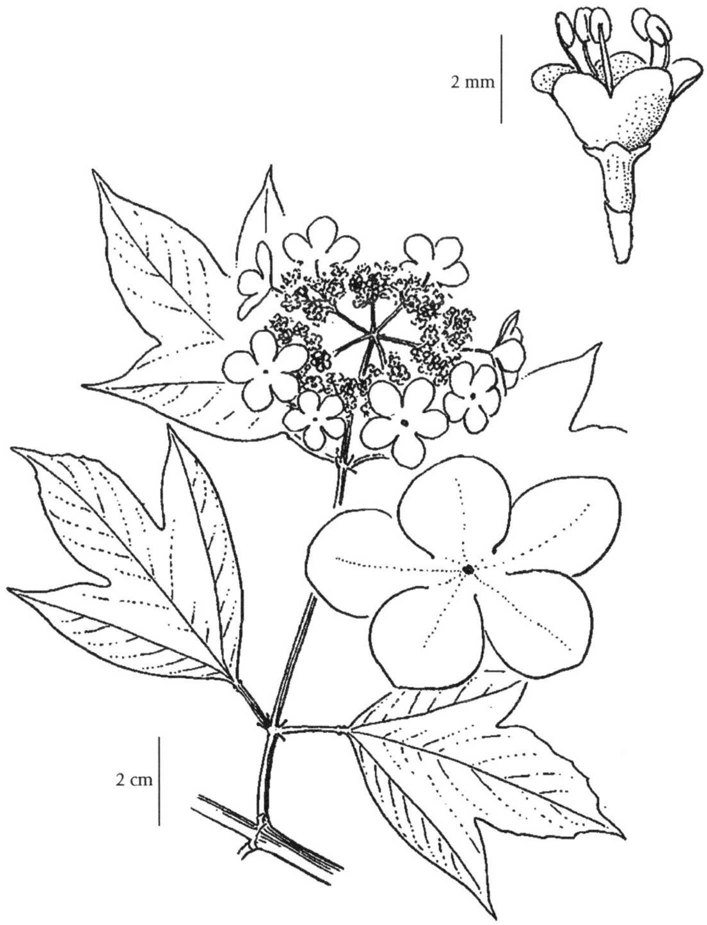

Viburnum opulus var. *americanum*

CARYOPHYLLACEAE [24]

1. Fruits 1-seeded, indehiscent.
    2. Leaves alternate; styles 3 .... *Corrigiola*
    2. Leaves opposite; styles 1-2 .... *Scleranthus*
1. Fruits several-seeded, dehiscent.
    3. Leaves whorled .... *Spergula*
    3. Leaves opposite.
        4. Calyces of separate sepals, or nearly so; petals lacking claws or narrow stalklike bases; ovaries not stalked.
            5. Stipules present, scarious.
                6. Styles 1, 3-cleft or -toothed .... *Spergularia*
                6. Styles 3, distinct .... *Polycarpon*
            5. Stipules absent.
                7. Capsules cylindric, often bent near summit.
                    8. Plants usually hairy; petals 5, sometimes 4, distinctly 2-parted .... *Cerastium*
                    8. Plants glabrous; petals 4, sometimes 5, entire or notched at the tip .... *Moenchia*
                7. Capsules egg-shaped or ellipsoid (or more or less cylindric in *Holosteum*).
                    9. Styles 4 or 5.
                        10. Leaves threadlike or awl-like; petals entire or none .... *Sagina*
                        10. Leaves lanceolate to heart-shaped or egg-shaped; petals deeply notched .... *Myosoton*
                    9. Styles 3.
                        11. Petals deeply notched or 2-cleft .... *Stellaria*
                        11. Petals entire to shallowly lobed or minutely toothed.
                            12. Plants markedly fleshy and succulent; seeds 3-4.5 mm long .... *Honkenya*
                            12. Plants of varied habit; seeds rarely over 1.5 mm long.
                                13. Inflorescence umbel-like; petals minutely toothed; plants introduced .... *Holosteum*
                                13. Inflorescence irregularly flat-topped; petals entire to shallowly lobed; plants native.
                                    14. Leaves elliptic to oblong-lanceolate, 1-4 cm long, usually more than 3 mm wide; seeds with a fleshy appendage (strophiole) .... *Moehringia*
                                    14. Leaves linear-lanceolate or awl-like, if egg-shaped to elliptic then less than 1 cm long and 3 mm wide; seeds without a strophiole.
                                        15. Capsules dehiscing by 6 valves or teeth .... *Arenaria*
                                        15. Capsules dehiscing by 3 entire valves .... *Minuartia*

[24] Contributed by G.W. Douglas and A. MacKinnon

4. Calyces of united sepals; petals clawed; ovaries stalked.
    16. Styles 2; capsules mostly 4-valved or -toothed.
        17. Calyces subtended by 1-3 pairs of bracts.
            18. Calyces 5-nerved, 4-6 mm long .... *Petrorhagia*
            18. Calyces 20- to 40-nerved, more than 10 mm long.
        17. Calyces without involucral bracts at the base.
            19. Calyces about 2 mm long .... *Gypsophila*
            19. Calyces more than 10 mm long.
                20. Calyces egg-shaped, 5-ribbed, wing-angled; petals not appendaged .... *Vaccaria*
                20. Calyces tubular, 20-nerved, not wing-angled; petals appendaged at base of blade .... *Saponaria*
    16. Styles 3-5; capsules 3-, 5-, 6-, or 10-valved or 10-toothed.
        21. Styles 3, sometimes 4 or 5; capsules opening by twice as many valves or teeth as styles .... *Silene*
        21. Styles 5, rarely 4 or 6; capsules opening by as many valves or teeth as styles.
            22. Calyces glandular, often inflated in fruit; inflorescence usually one to few-flowered, never contracted; plants never woolly-hairy .... *Silene*
            22. Calyces glabrous or hairy, not glandular, never inflated; inflorescence usually many-flowered and densely contracted, or, if few flowered then plants densely white woolly-hairy .... *Lychnis*

ARENARIA

1. Plants annuals, erect to ascending, not mat-forming or tufted .... *A. serpyllifolia*
1. Plants perennials, tufted or mat-forming.
    2. Stems 2-4 cm tall; flowers solitary .... *A. longipedunculata*
    2. Stems 5-20 cm tall; flowers several in a branching inflorescence .... *A. capillaris*

***Arenaria capillaris* Poir. ssp. *americana* Maguire** (*A. capillaris var. americana* [Maguire] R.J. Davis)
thread-leaved sandwort

HABITAT/RANGE:

Dry grasslands, shrublands, meadows and forest openings from the steppe to alpine zones; common in SC and SE BC, rare on Vancouver Island; E to AB and S to MT, NV and N OR.

DESCRIPTION:

General: Perennial herb from a taproot and slender, branched stem-base; stems numerous, erect to ascending, tufted, usually glaucous, glabrous below and glandular-hairy on the upper stems and in the inflorescence, 5-20 cm tall.

Leaves: Basal leaves many, grasslike, erect or ascending to somewhat lax, rather soft, sharply pointed, toothed along the margins, 2-6 cm long, 0.5-1 mm wide, 1-nerved, finely fringed to glabrous; stem leaves opposite, grasslike, reduced upwards, 2-5 pairs, to 4 cm long; stipules lacking.

Flowers: Inflorescence of 1-3 flowers in small open clusters; petals 5, white, egg-shaped, 5-9 mm long; sepals 5, 3-5 mm long, egg-shaped, the margins membranous and often violet-tinged, obtuse or abruptly sharp-pointed.

Fruits: Capsules egg-shaped, 5-8 mm long, sparsely glandular, 6-valved; seeds 1.5-2 mm long, kidney-shaped, flattened, slightly warty.

***Arenaria longipedunculata* Hult.** (*A. humifusa auct non* Wahl.)
low sandwort

HABITAT/RANGE:

Moist gravelly sites in the subalpine and alpine zones; rare in EC BC, known only from Mt. Robson Provincial Park; E to AB and N to AK, YT and NT.

DESCRIPTION:

General: Perennial herb from a taproot and branching stem-base, rhizomatous; forming mats to 10 cm across; flowering stems erect, 2-4 cm tall.

Leaves: Basal leaves oblanceolate to elliptic, 3-8 mm long, 1-2 mm wide, sharp-pointed; stem leaves opposite, 1-4 pairs, 3-5 mm long; stipules lacking.

Flowers: Inflorescence of erect, solitary flowers; petals 5, white, narrowly oblong, 3-4 mm long; sepals 5, egg-shaped, 3-4 mm long, more or less glandular-hairy.

Fruits: Capsules egg-shaped, 4-6 mm long, 3- or 6-valved; seeds brown, shiny, more or less kidney-shaped, 0.7-0.9 mm long, almost smooth.

***Arenaria serpyllifolia* L.**
thyme-leaved sandwort

HABITAT/RANGE:

Dry roadsides and disturbed areas in the lowland, steppe and montane zones; common in S BC; introduced from Eurasia.

DESCRIPTION:

General: Annual herb from a short, slender taproot; stems slender, ascending, usually much-branched, downy, often becoming glandular above, 8-30 cm tall.

Leaves: Leaves opposite, egg-shaped to lanceolate, 3-8 mm long, 1.5-4 mm wide, 3- to 5-nerved, needle-pointed, roughly hairy; stipules lacking.

Flowers: Inflorescence of usually several flowers, borne in typically open, diffuse, leafy-bracted clusters; petals 5, white, narrowly egg-shaped, to 2 mm long; sepals 5, free, narrowly egg-shaped, sharp-pointed, about 3 mm long, hairy.

Fruits: Capsules egg-shaped to pear-shaped, 3-4 mm long, 6-valved; seeds greyish-purple to black, 0.5 mm long, pimply in concentric rows.

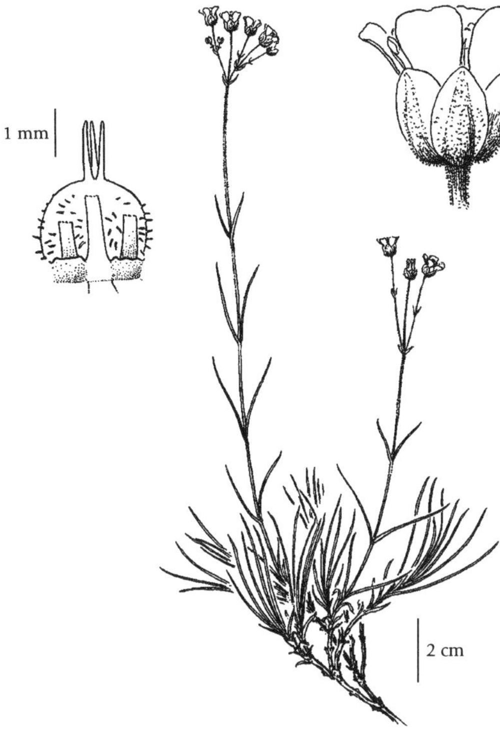

Arenaria capillaris ssp. *americana*

Arenaria longipedunculata

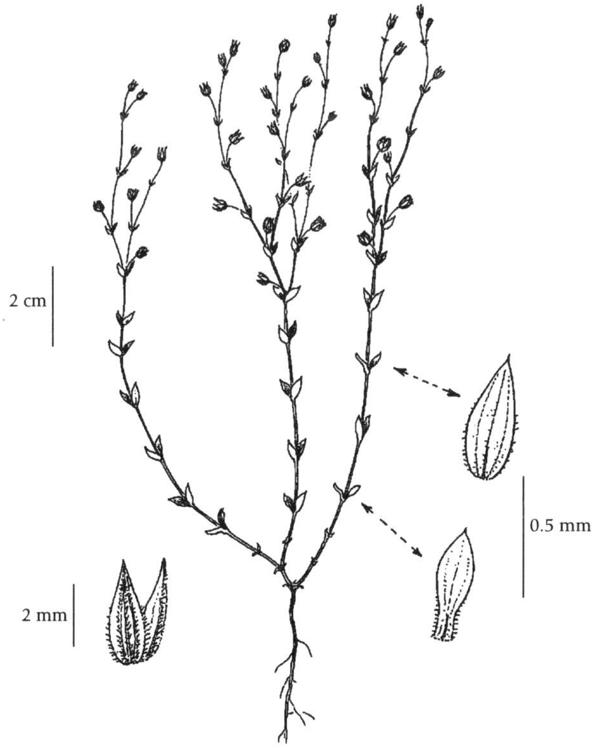

Arenaria serpyllifolia

CERASTIUM

1. Leaves and stems densely white woolly-hairy .... *C. tomentosum*
1. Leaves and stems never woolly-hairy.
   2. Plants native perennials; petals longer than the sepals.
      3. Leaves linear to lance-linear, oblong or narrowly elliptic (sometimes broader above); sterile branches arising in most leaf axils; bracts of the inflorescence papery-margined .... *C. arvense*
      3. Leaves broader, lanceolate to elliptic, oblong or oblanceolate; sterile branches lacking, or present only in the lowermost leaf axils; bracts of the inflorescence not or only slightly papery-margined.
         4. Stems coarse, 1-1.5 mm in diameter, yellowish stiff-hairy; stalks reflexed in fruit, yellowish stiff-hairy. .... *C. fischerianum*
         4. Stems slender, less than 1 mm in diameter, not yellowish stiff-hairy; stalks not reflexed in fruit, not yellowish stiff-hairy .... *C. beeringianum*
   2. Plants either annuals or introduced biennials or perennials; petals equal or nearly equal to the sepals.
      5. Flowers more or less tightly clustered; flower stalks shorter than the sepals .... *C. glomeratum*
      5. Flowers not tightly clustered; flower stalks conspicuously longer than the sepals.
         6. Plants not matted; sepals not broadly papery-margined .... *C. nutans*
         6. Plants matted; sepals broadly papery-margined.
            7. Plants annuals; flower stalks glandular-hairy; capsules 4.5-6.5 mm long .... *C. semidecandrum*
            7. Plants biennials or perennials; flower stalks generally stiff-hairy; capsules 7-10 mm long .... *C. fontanum*

***Cerastium arvense* L.**

field chickweed

Habitat/Range:

Moist to dry coastal bluffs, meadows, rockslides, rocky slopes and forest openings in all vegetation zones; frequent throughout BC; circumpolar, N to AK, YT and NT, E to NF and S to NE, GA, NM and CA; Eurasia.

Description:

General: Perennial herb from stolons or subrhizomes; often forming mats or clumps 10-40 cm wide; stems prostrate, branched, usually densely glandular-hairy above and in the inflorescence, glabrous to hairy below the nodes, the flowering stems decumbent to erect, 5-40 cm tall/long.

Leaves: Basal leaves lacking; stem leaves opposite, mostly linear to lance-linear, 1-3 cm long, 1-7 mm wide, 1-nerved, greyish-green; with sterile branches arising in most leaf axils; stipules lacking.

Flowers: Inflorescence of 3-6 flowers in an open cluster, rarely single; bracts of the inlorescence papery-margined; petals 5, white, egg-shaped, 8-12 mm long, deeply 2-cleft; sepals 5, oblong-lanceolate, 4-6 (7) mm long, glandular-hairy.

Fruits: Capsules cylindric, 6-10 mm long, 10-valved; seeds golden to reddish-brown, 0.8-1.2 mm long, pimply.

Notes: Highly variable, often confused with *C. beeringianum*.

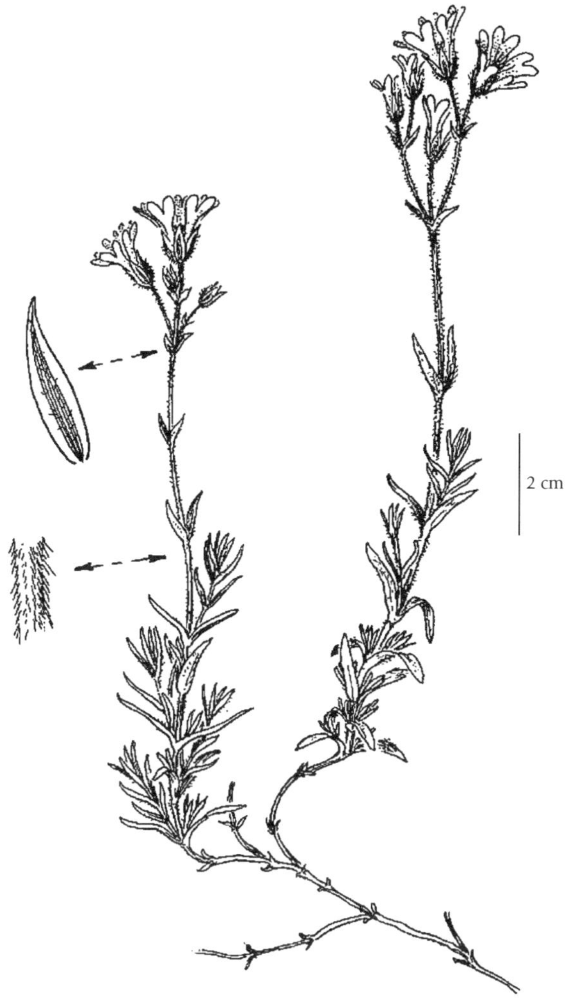

Cerastium arvense

***Cerastium beeringianum* Cham. & Schlecht.** (*C. beeringianum* var. *capillare* Fern & Wieg. = ssp. *earlei)*

Bering chickweed; alpine chickweed

HABITAT/RANGE:

Dry rocky slopes, talus slopes and cliffs in the upper subalpine and alpine zones; common E of the Coast-Cascade Ranges, rare on Vancouver Island; circumpolar, N to AK, YT and NT, E to NF and S to UT, CO, AZ and CA; Asia.

DESCRIPTION:

General: Perennial herb from a taproot and stoloniferous prostrate stems; forming mats 4-40 cm wide; stems prostrate, glandular-hairy, slender, less than 1 mm in diameter, flowering stems 3-25 cm long.

Leaves: Basal leaves lacking; stem leaves opposite, narrowly oblong or oblong-lanceolate to spoon-shaped, 5-35 mm long, 3-8 mm wide, sharp-pointed to rounded, hairy; stipules lacking.

Flowers: Inflorescence of usually 3-6 flowers in an open cluster; petals 5, white, 6-12 mm long, deeply 2-cleft; sepals 5, elliptic, 3-7 mm long, often purplish.

Fruits: Capsules cylindric, 8-12 mm long, 10-valved; seeds golden to reddish-brown, 0.8-1.2 mm long.

Notes: Sometimes confused with *C. fischerianum* where their ranges overlap. Two subspecies occur in BC.

1. Stalks with very fine, short, glandular hairs; plants of S BC .... ssp. *earlei* (Rydb.) Hult.
1. Stalks with long glandular hairs, plants of N BC.. ssp. *beeringianum*

***Cerastium fischerianum* Ser.**

Fischer's chickweed

HABITAT/RANGE:

Mesic gravelly shores and open slopes in the lowland and montane zones; rare in SW BC, known from N Vancouver Island and the Queen Charlotte Islands; amphiberingian, N to AK; E. Asia.

DESCRIPTION:

General: Perennial herb from a taproot and stolons; forming mats or clumps up to 50 cm across; stems mostly erect, branched, yellowish stiff-hairy; 8-50 cm tall.

Leaves: Basal leaves lacking; stem leaves opposite, lanceolate to oblong or elliptic, 7-50 mm long, 3-15 mm across, 1-nerved, sharp-pointed to rounded, somewhat hairy; stipules lacking.

Flowers: Inflorescence of 2-10 flowers in an open cluster; petals 5, white, 7-12 mm long, deeply 2-cleft; sepals 5, elliptic to lanceolate, 6-10 mm long, hairy.

Fruits: Capsules nearly cylindric, 10-14 mm long, 10-valved; seeds reddish-brown, 1-1.5 mm long.

***Cerastium fontanum* Baumg. ssp. *triviale* (Link) Jalas** (*C. vulgatum auct non* L.)

mouse-ear chickweed

HABITAT/RANGE:

Mesic to dry roadsides, waste places, gardens and fields in the lowland, steppe and lower montane zones; common in all but NC and NE BC; introduced from Eurasia.

DESCRIPTION:

General: Biennial or short-lived perennial herb from taproots and often from sprawling, nodally-rooting branches; forming clumps to 40 cm across; stems decumbent to erect, stiff-hairy and glandular to glandular-hairy, 5-40 cm tall/long.

Leaves: Basal leaves lacking; stem leaves opposite, oblanceolate or spoon-shaped, 7-25 mm long and 2-10 mm wide, 1-nerved, coarsely hairy; leaves of the flowering stems larger, up to 4 cm long and 15 mm wide; stipules lacking.

Flowers: Inflorescence several to many in an open cluster; flower stalks generally stiff-hairy; petals 5, white, 2-cleft, 5-7 mm long; sepals 5, 4-7 mm long, stiff-hairy, papery-margined.

Fruits: Capsules cylindric, 7-10 mm long, 10-valved; seeds 0.6-0.8 mm long, reddish-brown.

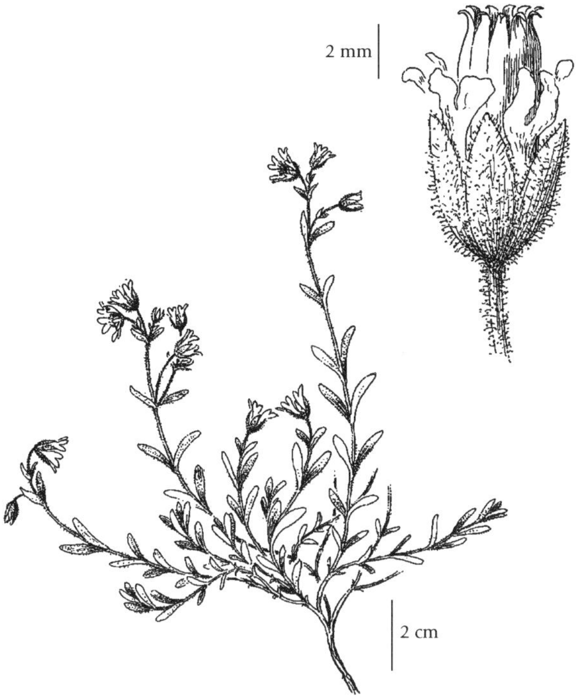

Cerastium beeringianum ssp. *earlei*

Cerastium fischerianum

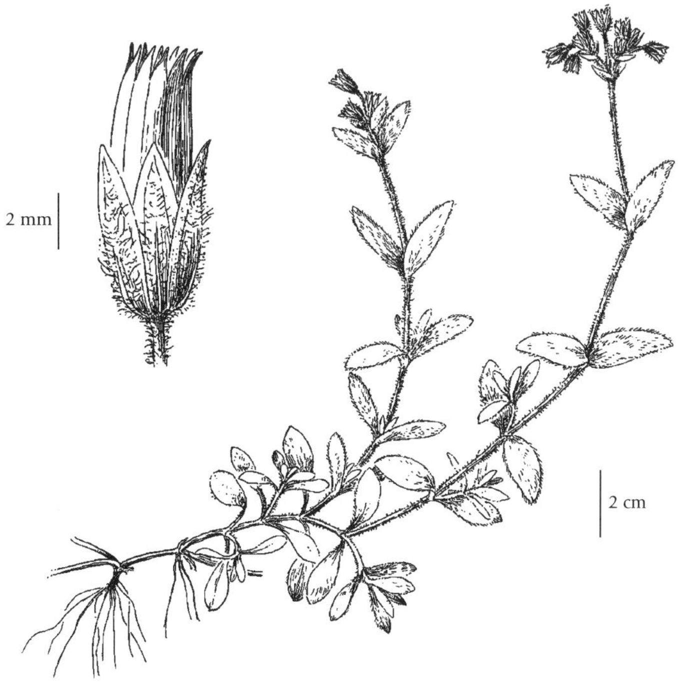

Cerastium fontanum ssp. *triviale*

Cerastium glomeratum **Thuill.** (*C. viscosum auct. non* L.)
sticky chickweed
HABITAT/RANGE:
Mesic to dry roadsides, waste places, lawns and pastures in the lowland, steppe and lower montane zones; common in SW BC, rare elsewhere in S BC and the Queen Charlotte Islands; introduced from Eurasia.
DESCRIPTION:
General: Annual herb from a slender taproot; stems decumbent to erect, usually dichotomously branched, very hairy, sticky, 5-50 cm tall/long.

Leaves: Basal leaves lacking; stem leaves opposite, pale yellowish-green, spoon-shaped to egg-shaped, 3-20 mm long, 2-10 mm wide, 1-nerved, slightly hairy above and below; upper stem leaves larger, to 25 mm long and 15 mm wide, egg-shaped; stipules lacking.

Flowers: Inflorescence numerous in a crowded cluster; flower stalks shorter than sepals; petals 5, white, 3-5 mm long, deeply bilobed; sepals 5, narrowly egg-shaped, 3-5 mm long, stiff-hairy and more or less glandular.

Fruits: Capsules cylindric, 6-9 mm long, 10-valved; seeds pale brown, 0.3-0.5 mm long, finely pimply.

Cerastium nutans **Raf.**
nodding chickweed
HABITAT/RANGE:
Moist to mesic streambanks, grasslands and open forests in the steppe and lower montane zones; infrequent in SC BC, rare in the Peace River area; N to S AK and S NT, E to PQ and NS, and S to SC, FL, UT, TX, AZ and OR.
DESCRIPTION:
General: Annual herb from a short, slender taproot; stems ascending to erect, glandular-hairy to somewhat soft-hairy, 1 to several, simple or basally branched, 5-25 cm tall.

Leaves: Basal and lower stem leaves spoon-shaped to oblanceolate, 6-30 mm long, broadly short-stalked; stem leaves opposite, several pairs, linear-lanceolate to egg-shaped, unstalked; stipules lacking.

Flowers: Inflorescence several in a loose, leafy, bracted cluster; stalks glandular soft-hairy; petals 5, white, usually sharply deflexed at the tips, 2-cleft, 6-10 mm long, occasionally lacking; sepals 5, elliptic, 4-5 mm long, generally glandular soft-hairy.

Fruits: Capsules cylindric, curved at maturity, 7-12 mm long, 10-valved; seeds light brown, 0.5-0.7 mm long.

Cerastium semidecandrum **L.**
little chickweed
HABITAT/RANGE:
Mesic to dry roadsides and fields in the lowland zone; rare in SC BC, frequent in SW BC, known from S Vancouver Island and lower Fraser Valley; introduced from Eurasia.
DESCRIPTION:
General: Annual herb from a slender taproot; stems erect to ascending, branched, hairy, sometimes glandular, 1-20 cm tall.

Leaves: Basal leaves oblanceolate with a stalk-like base; stem leaves opposite, egg-shaped to elliptic-oblong, 5-18 mm long, white-hairy; stipules lacking.

Flowers: Inflorescence in a spreading cluster; flowers stalks at first sharply deflexed at the base, later almost erect, glandular-hairy; petals 5, white, narrow, 2-4 mm long, slightly notched; sepals 5, narrowly lanceolate, sharp-pointed, 3-5 mm long, glandular-hairy, broadly papery-margined.

Fruits: Capsules cylindric, 4.5-6.5 mm long, 10-valved; stalks of the fruits curved downwards at first, later almost erect; seeds yellowish-brown, 0.4-0.6 mm long.

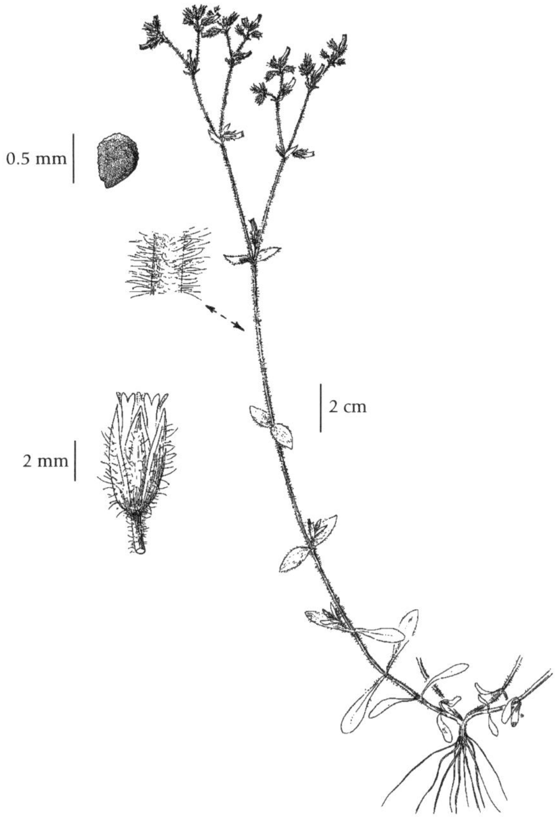

Cerastium glomeratum

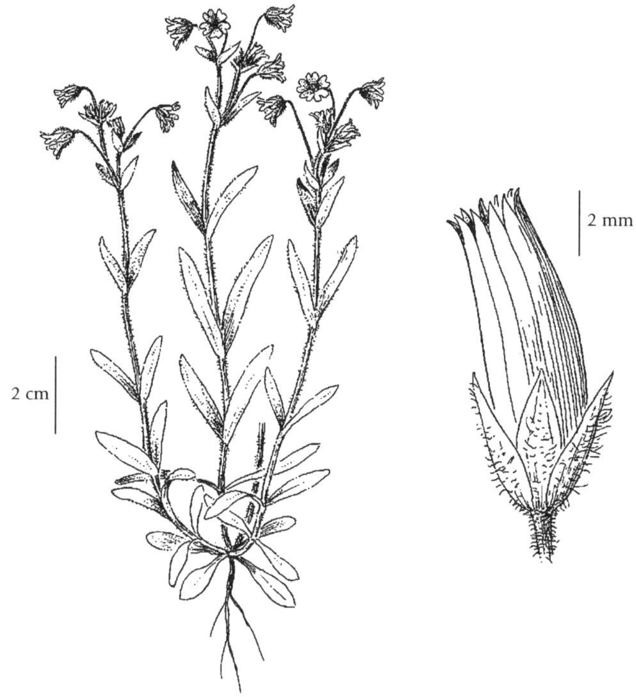

Cerastium nutans

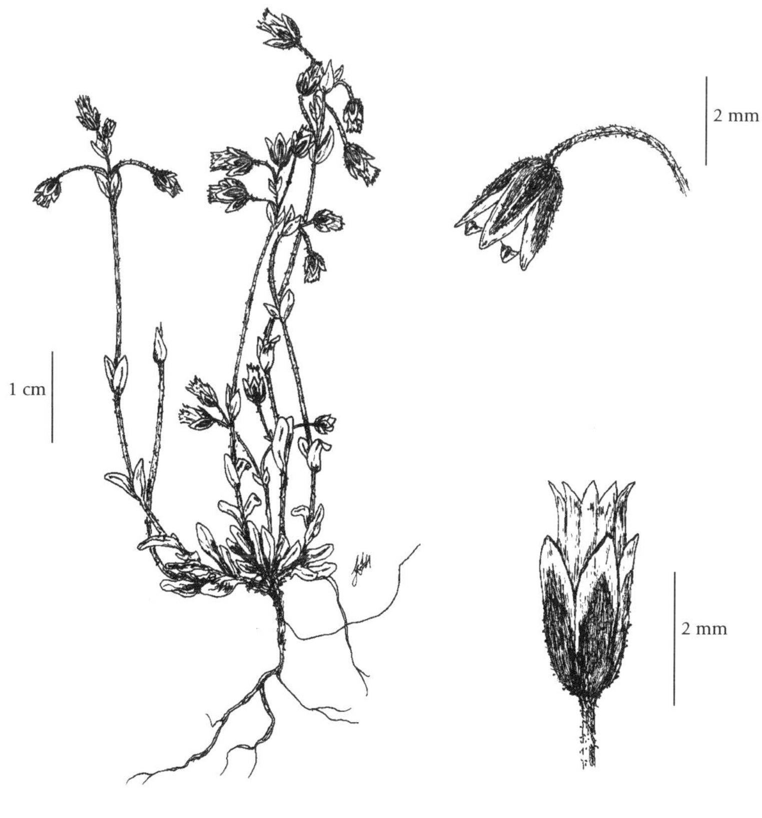

Cerastium semidecandrum

Cerastium tomentosum **L.**

snow-in-summer

HABITAT/RANGE:

Disturbed areas in the lowland zone; rare in SW BC, known only from SE Vancouver Island and Vancouver; introduced from Europe.

DESCRIPTION:

General: Biennial or more usually perennial herb from short, slender taproot; forming mats or clumps up to 80 cm across; stems usually numerous, decumbent, freely branched, densely white woolly-hairy, 10-35 cm long.

Leaves: Stem leaves opposite, oblanceolate, grey-hairy to white woolly-hairy, 3-20 mm long and 2-4 mm wide, 1-nerved, unstalked; stipules lacking.

Flowers: Inflorescence in an elongated cluster; petals 5, white, longer than the sepals, 5-7 mm long; sepals 5, 2-4 mm long, hairy.

Fruits: Capsules cylindric, 7-10 mm long, 10-valved; seeds 0.6-0.8 mm long.

CORRIGIOLA

Corrigiola litoralis **L.**

strapwort

HABITAT/RANGE:

Mesic to dry waste places in the lowland zone; rare in SW BC, known only from Vancouver and Langley; introduced from Europe.

DESCRIPTION:

General: Annual or biennial herb from a slender taproot; stems prostrate, branched, glabrous or glaucous, 10-25 cm long.

Leaves: Basal leaves lacking; stem leaves opposite, oblanceolate, slightly succulent, 5-30 mm long, 2-5 mm wide, gradually narrowed to short-stalked bases; stipules lacking.

Flowers: Inflorescence of several flowers in rounded clusters; petals 5, white or pinkish-tinged, less than 1 mm long; sepals 5, greenish or red, about 1 mm long, the 5 broadly white-margined lobes about equalling the tube.

Fruits: Capsules cylindric, about 1 mm long, purplish-black, 10-valved; seeds about 0.5 mm long.

DIANTHUS

1. Leaves 1.5-2 cm wide; plants glabrous .... *D. barbatus*
1. Leaves narrower, less than 0.8 cm wide; plants usually hairy.
    2. Flowers all or nearly all solitary; plants rough-hairy .... *D. deltoides*
    2. Flowers in a dense terminal cluster(s); plants usually more or less crisp-hairy .... *D. armeria*

Dianthus armeria **L.**

Deptford pink

HABITAT/RANGE:

Mesic to dry waste places in the lowland, steppe and montane zones; rare in extreme S BC; introduced from Eurasia.

DESCRIPTION:

General: Annual or biennial herb from a slender taproot; stems erect, 1-several, more or less crisp-hairy, 20-60 cm tall.

Leaves: Basal and stem leaves opposite, linear, appressed, 4-10 cm long, 2-5 mm wide, crisp short-hairy, reduced above; stipules lacking.

Flowers: Inflorescence in dense, 2- to many-flowered, terminal cluster(s); petals 5, pink to red with pale dots, the stalks 4-5 mm long, narrowly stalklike at the base, the blades more or less egg-shaped, toothed; sepals 5, united, forming a woolly-hairy tube about 15 mm long, 20- to 25-nerved, 5-toothed.

Fruits: Capsules cylindric, about 15 mm long, 4-valved; seeds compressed, concave on one side, 1-2 mm long.

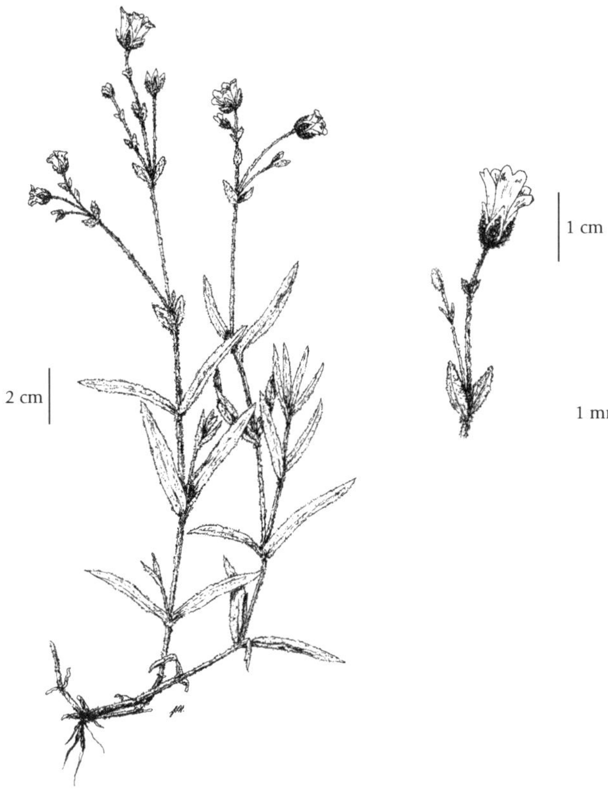

Cerastium tomentosum

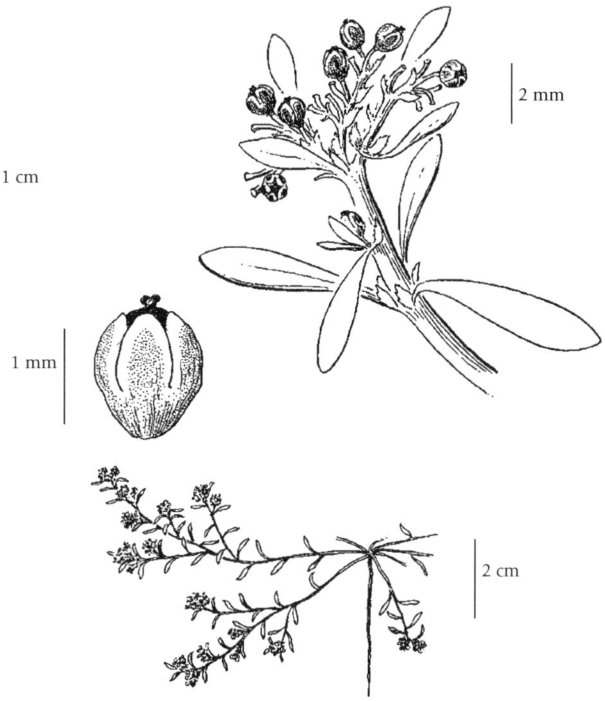

Corrigiola litoralis

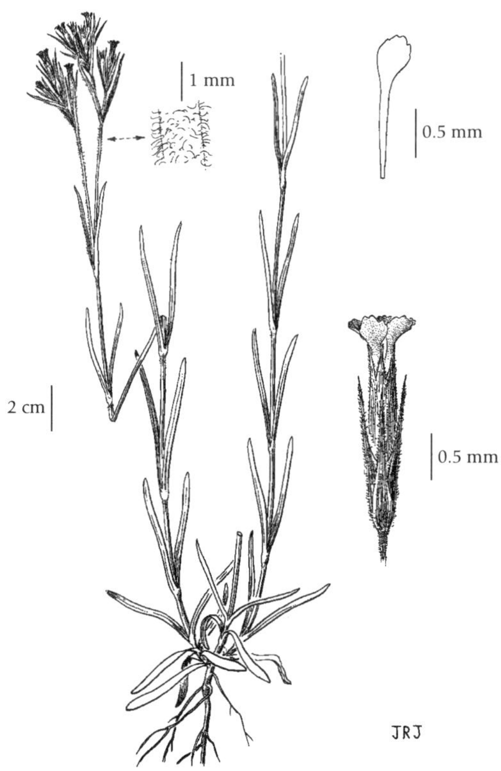

Dianthus armeria

***Dianthus barbatus* L.**

sweet William

HABITAT/RANGE:

Mesic to dry waste places in the lowland, steppe and montane zones; rare in S BC; introduced from Eurasia.

DESCRIPTION:

General: Perennial herb from a taproot; stems 1 or more, erect to decumbent, glabrous, 30-60 cm tall/ long.

Leaves: Basal leaves several, oblong-lanceolate, 7-9 cm long, 1-2 cm wide, glabrous; stem leaves opposite, 4-10 pairs, becoming lanceolate and only slightly reduced above, mostly 4-9 cm long, broadly short-stalked, the base sheathing; stipules lacking.

Flowers: Inflorescence of 3-30 flowers in densely crowded clusters; petals 5, white to more usually pink or red, with narrow claws about equalling the calyces, the blade more or less egg-shaped, 6-10 mm long, wavy-toothed; sepals 5, green below and often violet above, united, forming a glabrous tube about 15 mm long, about 40-nerved.

Fruits: Capsules oblong, 9-13 mm long, 4-valved; seeds compressed, concave on one side.

***Dianthus deltoides* L.**

maiden pink

HABITAT/RANGE:

Mesic to dry waste places in the lowland, steppe and montane zones; rare in S BC; introduced from Eurasia.

DESCRIPTION:

General: Perennial herb from a taproot and branched stem-base; stems trailing, much branched at the base with erect flowering shoots, copiously rough-hairy, 20-45 cm tall/long.

Leaves: Basal leaves few, soon deciduous; stem leaves opposite, numerous, linear to narrowly oblanceolate, often overlapping, 15-25 mm long, 2-8 mm wide, roughl-hairy on the margins; basal leaves somewhat shorter; stipules lacking.

Flowers: Inflorescence of solitary flowers, occasionally 2-3 in an open cluster; petals 5, pink to rose with pale spots or occasionally white, stalks of the petals about equalling the calyces, the blades egg-shaped, 6-10 mm long, shortly and broadly toothed, often sparsely long soft-hairy on the upper surface near the stalk; sepals 5, green or reddish, united, forming a hairy tube about 15 mm long, about 30-nerved.

Fruits: Capsules oblong, 15-18 mm long, 4-valved; seeds black, egg-shaped, 2-2.5 mm long, compressed, concave on one side.

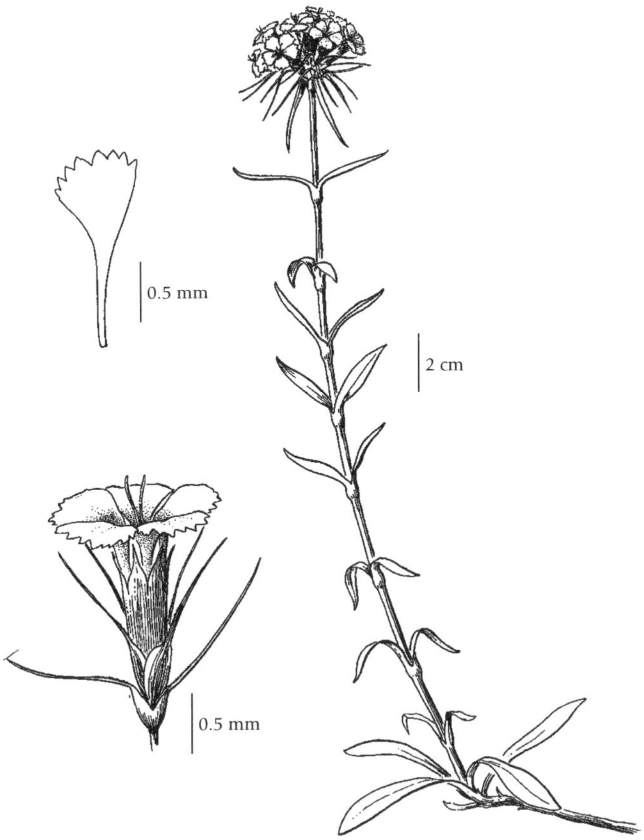

Dianthus barbatus

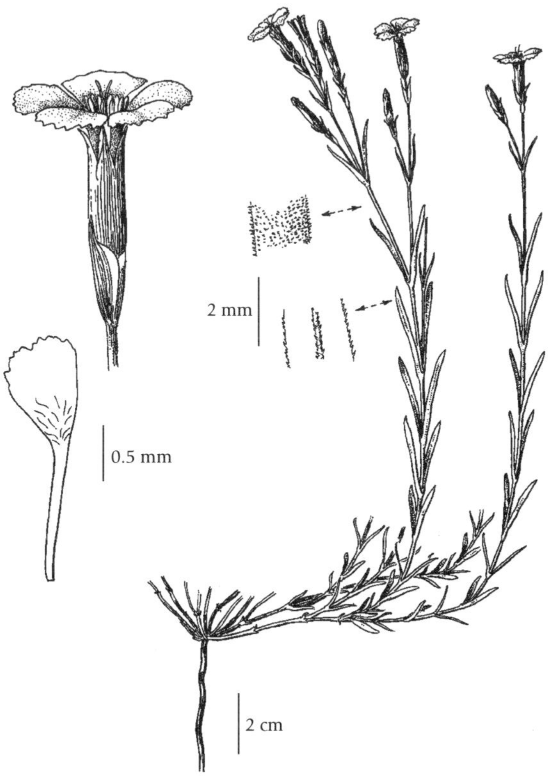

Dianthus deltoides

GYPSOPHILA

1. Inflorescence glandular-hairy; calyx teeth 3-4 mm long .... *G. scorzonerifolia*
1. Inflorescence glabrous; calyx teeth 1.5-2.0 mm long .... *G. paniculata*

Gypsophila paniculata **L.**

baby's breath

HABITAT/RANGE:

Dry roadsides, fields and waste places in the steppe and lower montane zones; infrequent SC and SE BC; introduced from Eurasia.

DESCRIPTION:

General: Perennial herb from a thick root; stems usually much-branched, glabrous and more or less glaucous, 0.4-1.2 m tall.

Leaves: Basal leaves lacking; stem leaves opposite, lanceolate to linear, 2-10 cm long, rarely over 2-9 mm wide, 1-nerved, unstalked; stipules lacking.

Flowers: Inflorescence of numerous flowers in a large, diffusely branched, leafy-bracted, panicle-like cluster; flowers stalks glabrous; petals 5, white, oblanceolate, rounded to somewhat notched, stalklike at the base, slightly exceeding the calyces, about equalling the 10 stamens; sepals 5, glabrous, the white membrane below the notches contrasting strongly with the usually purplish back of the 5 lobes, united, forming a top-shaped tube 1.5-2 mm long.

Fruits: Capsules egg-shaped, 2-3 mm long, 4-valved; seeds 2-5, blackish, 1.5-2 mm long, prominently cross-wrinkled and pimply in longitudinal, concentric rows.

Gypsophila scorzonerifolia **Ser.**

glandular baby's breath

HABITAT/RANGE:

Dry roadsides and waste places in the lowland zone; rare in SW BC, infrequent in SE BC; introduced from Europe.

DESCRIPTION:

General: Perennial herb from a stout rhizome; stems erect, several, usually much branched, glandular above, 0.3-1.0 m tall.

Leaves: Basal leaves lacking; stem leaves opposite, rounded to linear-oblong to lanceolate, 2-8 cm long, 1.0-3.5 cm wide, obscurely 1-3 nerved, unstalked; stipules lacking.

Flowers: Inflorescence of a few floers in a large, diffusely branched, leafy-bracted, panicle-like cluster; flowers stalks glandular-hairy; petals 5, white to pale pink, oblanceolate, rounded to somewhat notched, stalklike at the base, slightly exceeding the calyces, about equalling the 10 stamens; sepals 5, glandular-hairy, the white membrane below the notches contrasting strongly with the usually purplish back of the 5 lobes, united, forming a tube 1.5-2 mm long, fused nearly to the tip.

Fruits: Capsules egg-shaped, 2-3 mm long, 4-valved; seeds 2-5, nearly black, 1.5-2 mm long, prominently cross-wrinkled and pimply in longitudinal, concentric rows.

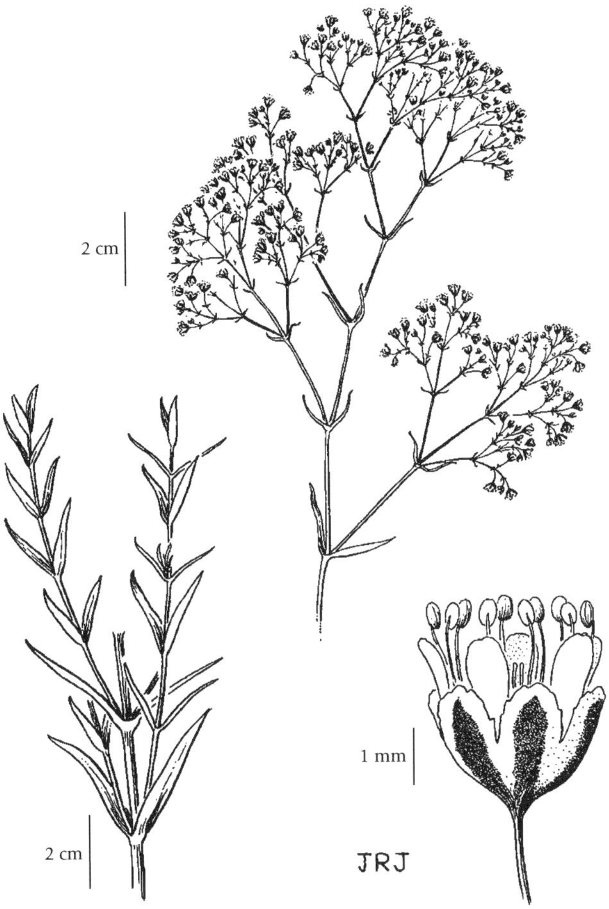

Gypsophila paniculata

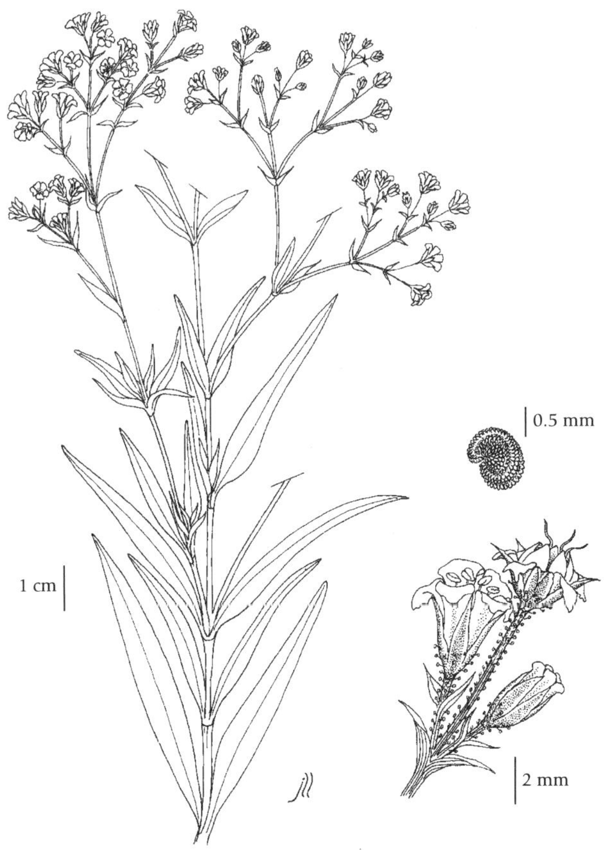

Gypsophila scorzonerifolia

HOLOSTEUM

***Holosteum umbellatum* L.**
umbellate chickweed
HABITAT/RANGE:
Dry disturbed areas in the lowland and steppe zones; infrequent in SC BC, rare in SW BC, known from the Saanich Peninsula and lower Fraser Valley; introduced from Eurasia.
DESCRIPTION:
General: Annual herb from a taproot; stems erect, 1-several, simple or basally branched, sticky above, glabrous below, stems 5-35 cm tall.

Leaves: Leaves mainly in a basal rosette, lanceolate to oblong, 1-2 cm long, short-stalked, soon deciduous; stem leaves opposite, 2-3 pairs remote from the inflorescence, narrowly egg-shaped to oblong, 1-3.5 cm long, unstalked; stipules lacking.

Flowers: Inflorescence of 2-16 flowers in a terminal umbel-like cluster with a tiny, bracted involucre; petals 5, oblanceolate, white or pale pink, raggedly-frilled, 5-7 mm long; sepals oblanceolate, distinct, 2.5-5 mm long.

Fruits: Capsules egg-shaped to cylindric, 4-6 mm long, 6-valved, the tips of the valves recurved to tightly curled down; seeds prominently pimply-warty, the warts with interlocking starlike bases.

HONKENYA

***Honkenya peploides* (L.) Ehrhart ssp. *major* (Hook.) Hult.** (*Arenaria peploides* L. var. *major* Hook., *Honkenya peploides* var. *major* [Hook.] Abrams)
seabeach sandwort
HABITAT/RANGE:
Moist sandy beaches and grassy shorelines in the lowland zone; common along the coast; N to AK, YT and NT and S to NW OR.
DESCRIPTION:
General: Succulent perennial herb from a deep taproot and slender rhizomes; forming mats to 80 cm across or more; stems numerous, trailing, usually freely branching, glabrous, flowering stems prostrate to erect, 5-40 cm tall/long.

Leaves: Basal leaves lacking; stem leaves opposite, 3-10 pairs or more, from lanceolate to egg-shaped, sharp-pointed to obtuse, 1-4 cm long, 3-20 mm wide, fleshy, yellowish-green, glabrous, often with sterile axillary shoots; stipules lacking.

Flowers: Inflorescence of solitary or a few flowers in the upper leaf axils, perfect or unisexual; petals 5, greenish-white, egg-shaped, 2-3 mm long; sepals 5, egg-shaped to lanceolate, 4-7 mm long, glabrous.

Fruits: Capsules globose or nearly so, 6-10 mm long, 3- to 5-valved; seeds chestnut brown, 3-4.5 mm long, smooth.

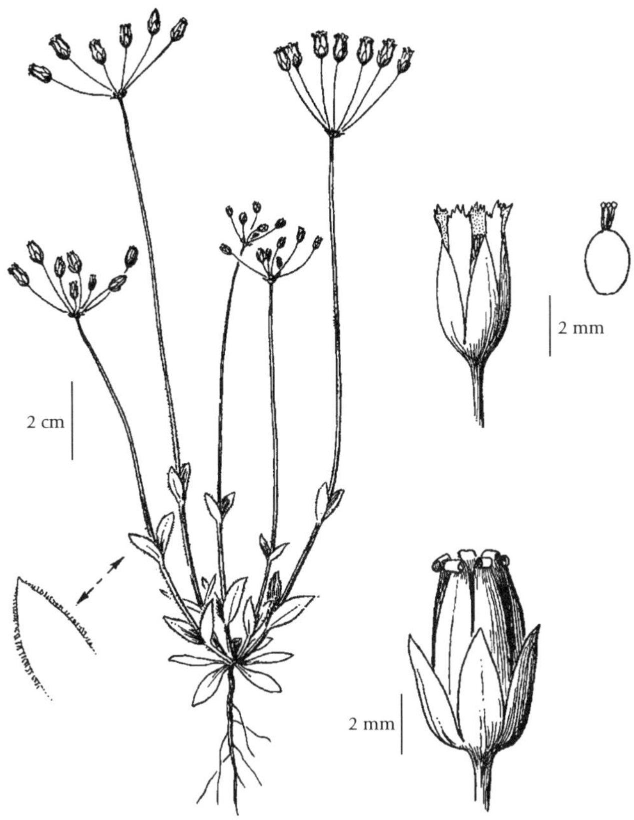

Holosteum umbellatum

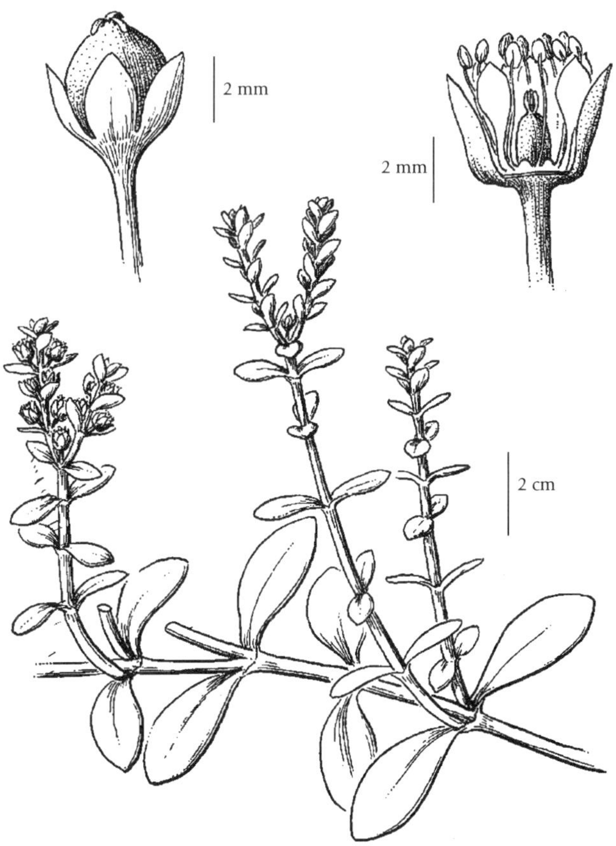

Honkenya peploides ssp. *major*

LYCHNIS

Lychnis coronaria **(L.) Desr.**

rose campion

HABITAT/RANGE:

Mesic to dry roadsides and waste places in the lowland zone; frequent in SW BC, known only from S Vancouver Island, the Gulf Islands and the lower Fraser Valley; introduced from Europe.

DESCRIPTION:

General: Perennial herb from a usually branched stem-base; stems erect, solitary to several, simple or branching, greyish woolly-hairy, 40-100 cm tall.

Leaves: Basal leaves lanceolate to oblanceolate, slenderly stalked, woolly-hairy, 5-10 cm long, 1-3 cm wide; stem leaves opposite, woolly-hairy; 5-10 pairs, often overlapping, gradually reduced upward and unstalked above; stipules lacking.

Flowers: Inflorescence of several flowers in a loosely branched, open, leafy-bracted cluster, perfect; petals 5, rose-magenta, the stalks of the petals about equalling the calyces, narrow, without earlike lobes at the base, the blade 10-15 mm long, broadly egg-shaped, with 2 linear, sharp-pointed, 2-4 mm long appendages near the base of the blades; sepals 5, lanceolate, united, forming a tube about 15 mm long at flowering time, enlarged but not inflated in fruit, woolly-hairy, 10-nerved, the lobes slender, contorted.

Fruits: Capsules elliptic, 12-16 mm long, 5-valved; seeds about 1 mm long, purplish-brown, evenly warty-pimply in concentric rows, the pimples bordered with minute teeth at the base.

MINUARTIA

1. Sepals elliptic to oblong, rounded to obtuse at the tips.
   2. Stems and leaves glabrous .... *M. stricta*
   2. Stems and leaves hairy, or the latter at least hairy-margined.
      3. Flowers on stalks less than 2 cm long; capsules 0.5-1.8 cm long, more than 2 times as long as the calyces .... *M. macrocarpa*
      3. Flowers on stalks more than 5 cm long; capsules 0.2-0.5 cm long, barely longer than the calyces.
         4. Leaves narrow, 7-8 times as long as wide, long-pointed .... *M. yukonensis*
         4. Leaves broader, 4-5 times as long as wide, obtuse.
            5. Flowering stems 3-10 cm tall; sepals 4-5 mm long; petals egg-shaped to spoon-shaped, usually much longer than the sepals .... *M. obtusiloba*
            5. Flowering stems 2-5 cm tall; sepals 2-4 mm long; petals narrowly oblong, usually not much longer than the sepals .... *M. biflora*
1. Sepals long-pointed or abruptly sharp-pointed.
   6. Stems glandular-hairy.
      7. Plants 10-30 cm tall, annuals (rarely persistent for more than a year) from a short, slight taproot, not matted .... *M. tenella*
      7. Plants 1-15 cm tall, perennials from a moderate to long, stout taproot, matted.
         8. Stem leaves numerous, with secondary leaves clustered in the axils of the primary leaves; sepals 4-6 mm long; seeds dark brown, about 1.0-1.3 mm long .... *M. nuttallii*
         8. Stem leaves few, without secondary leaves clustered in the axils of the primary leaves; sepals 2-4 mm long; seeds light reddish-brown, 0.4-0.6 mm long .... *M. rubella*
   6. Stems glabrous.
      9. Plants annuals; stems few; rare in SW BC .... *M. pusilla*

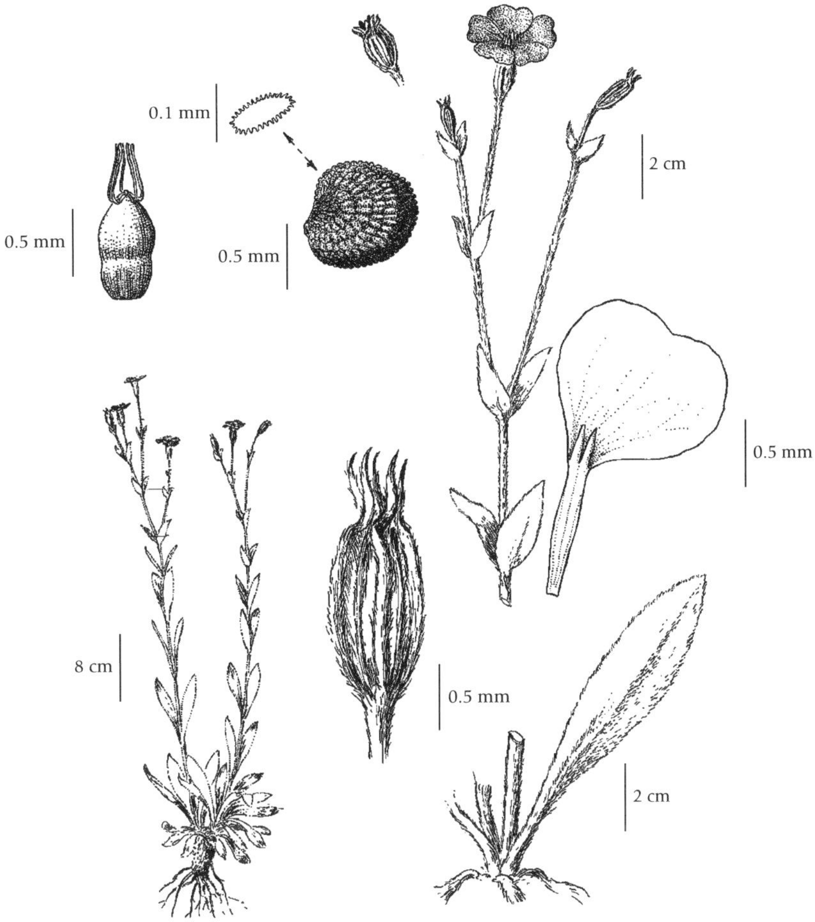

Lychnis coronaria

9. Plants perennials; stems several to many; rare to frequent in N or SE BC.

    10. Flowering stems 2-4 branched, sometimes with some simple stems; seeds black .... *M. dawsonensis*

    10. Flowering stems all simple, without branches; seeds light brown or reddish-brown.

        11. Sepals purple, weakly 3-nerved; petals as long as sepals, rarely lacking; seeds reddish-brown; rare in NE BC .... *M. elegans*

        11. Sepals green, strongly 3-nerved; petals shorter than sepals, more usually lacking; seeds light brown; frequent in SE BC .... *M. austromontana*

***Minuartia austromontana* Wolf & Packer**
Rocky Mountain sandwort

HABITAT/RANGE:

Dry, calcareous rocky slopes and scree in the alpine zone; rare in extreme SE BC; E to AB and S to WY, UT and NE OR.

DESCRIPTION:

General: Perennial herb from a taproot; forming cushions 5-15 cm wide; stems erect or ascending, simple, glabrous, 1-3 cm tall.

Leaves: Basal leaves linear to lance-linear, 3-6 mm long, less than 1 mm wide, 1-nerved, glabrous; stem leaves opposite, 2-3 pairs, 3-12 mm long, glabrous often with short, sterile axillary branches; stipules lacking.

Flowers: Inflorescence of solitary or occasionally a few flowers in a loose cluster; petals linear to oblong, 1-2 mm long or more usually lacking; sepals green, long-pointed or abruptly sharp-pointed, 2-3 mm long, strongly 3-nerved.

Fruits: Capsules 2-3 mm long, 3-valved; seeds light brown, 0.6-1.0 mm long.

Notes: Collections of this species were previously identified as *M. rossii* ssp. *columbiana* (*M. elegans*) (Wolf *et al.* 1979).

***Minuartia biflora* (L.) Schinz & Thell.** (*Arenaria sajanensis* Willd. *ex* Schlecht.)
mountain sandwort

HABITAT/RANGE:

Dry, calcareous gravelly to rocky slopes in the alpine zone; rare in N and SE BC; N to AK, YT and NT and E to PQ.

DESCRIPTION:

General: Perennial herb from a taproot and branching stem-base; forming mats up to 15 cm across; flowering stems erect, simple to few branched, glandular, 2-5 cm tall.

Leaves: Basal leaves linear, 2-10 mm long, less than 1 mm wide, 1-nerved, glabrous to slightly hairy; stem leaves opposite, 2-3 pairs, linear, 2-8 mm long; stipules lacking.

Flowers: Inflorescence of solitary or paired flowers; petals 5, white, narrowly oblong, 2-4.5 mm long; sepals 5, linear-oblong, rounded to obtuse and sometimes purplish at the tips, prominently 3-nerved, 2-4 mm long.

Fruits: Capsules cylindric, 2.5-5 mm long, 3-valved; seeds 0.6-0.8 mm long, reddish-brown, smooth to minutely roughened.

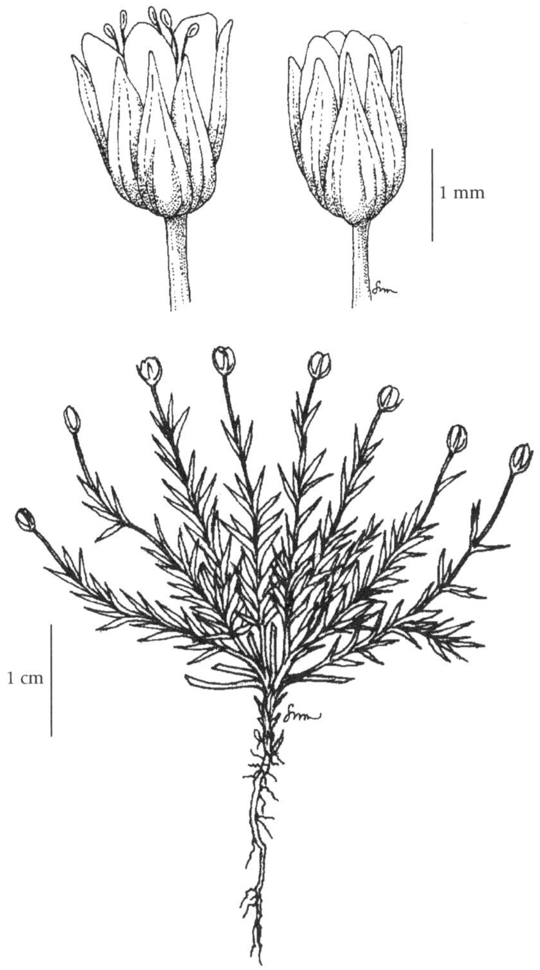

Minuartia austromontana

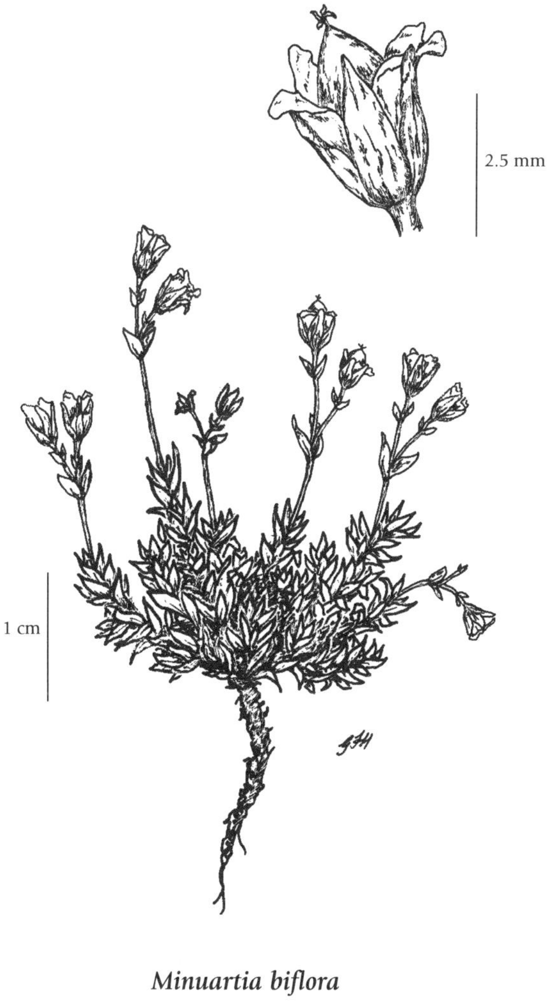

Minuartia biflora

***Minuartia dawsonensis* (Britt.) House** (*Arenaria dawsonensis* Britt., *A. stricta* Michx. var. *dawsonensis* [Britt.] Scog.)

rock sandwort

HABITAT/RANGE:

Moist to mesic forest openings and meadows in the montane and subalpine zones; frequent in NE BC; N to AK, YT and NT and E to PQ.

DESCRIPTION:

General: Short-lived perennial herb from a weak taproot and branching stem-base; tufted; stems erect, several, simple to 2- to 4-branched, glabrous, 5-20 cm tall.

Leaves: Basal leaves linear to oblong, 5-15 mm long, less than 1 mm wide, 1-nerved, glabrous; stem leaves opposite, 2-6 pairs, up to 14 mm long, often with sterile axillary branches; stipules lacking.

Flowers: Inflorescence of 2-7 flowers in a loose cluster, rarely solitary; petals 5, white, oblanceolate to oblong, 2-3.5 mm long; sepals 5, 3-nerved, margins translucent, egg-shaped, long-pointed or abruptly sharp-pointed, 3-4 mm long.

Fruits: Capsules oblong, 3-4.5 mm long, usually 3-valved; seeds 0.5-0.8 mm long, black, finely roughened.

***Minuartia elegans* (Cham. & Schlecht.) Schischk.** (*Arenaria elegans* Cham. & Schlecht., *A. rossii* R. Br. var. *columbiana* [Raup] Maguire, *A. rossii* ssp. *elegans* [Cham. & Schlecht.] Maguire, *Minuartia rossii* [R. Br.] Graebn. var. *orthotrichoides* [Schischk.] Hult.)

northern sandwort

HABITAT/RANGE:

Moist meadows and slopes in the alpine zone; rare in NE BC; amphiberingian, N to AK, YT and NT and E to WC AB; E. Asia.

DESCRIPTION:

General: Short-lived perennial herb from a taproot and branching stem-base; forming loose mats to 40 cm across; stems erect, several, branched, glabrous, 5-15 cm tall.

Leaves: Basal leaves linear to lance-linear, 3-10 mm long, less than 1 mm wide, 1-nerved; stem leaves opposite, 2-3 pairs, 3-12 mm long, often with short, sterile axillary branches; stipules lacking.

Flowers: Inflorescence of solitary flowers; flowers stalks glabrous; petals 5, oblong to egg-shaped, white, 2-4 mm long, or rarely lacking; sepals 5, purple, 2-4 mm long, weakly 3-nerved, long-pointed or abruptly sharp-pointed.

Fruits: Capsules 2-4 mm long, 3-valved; seeds 0.6-1 mm long, reddish-brown, slightly roughened.

***Minuartia macrocarpa* (Pursh) Ostenf.** (*Arenaria macrocarpa* Pursh)

large-fruited sandwort

HABITAT/RANGE:

Dry rocky slopes in the alpine zone; rare in NW BC, known only from Mount Mansfield; amphiberingian, N to AK, YT and NT; E. Asia.

DESCRIPTION:

General: Perennial herb from a taproot and branching stem-base; forming loose mats to 20 cm across; stems decumbent to ascending, several, branched, glabrous, 5-15 cm tall.

Leaves: Basal leaves linear to lance-linear, blunt, 0.4-1.5 cm long, less than 1 mm wide, 3-nerved, hairy-margined; stem leaves opposite, 1-6 pairs, 0.3-1.2 cm long, hairy-margined,often with secondary axillary leaves; stipules lacking.

Flowers: Inflorescence of solitary flowers on glandular stalks; petals 5, oblong to egg-shaped, white or sometimes purplish, 0.8-1.3 cm long; sepals 5, green or often purplish, glandular-hairy to nearly glabrous, 2.5-4.0 mm long, obtuse.

Fruits: Capsules 0.5-1.8 cm long, 3- to 4-valved; seeds 0.8-1.2 mm long, somewhat kidney-shaped, reddish-brown, roughened.

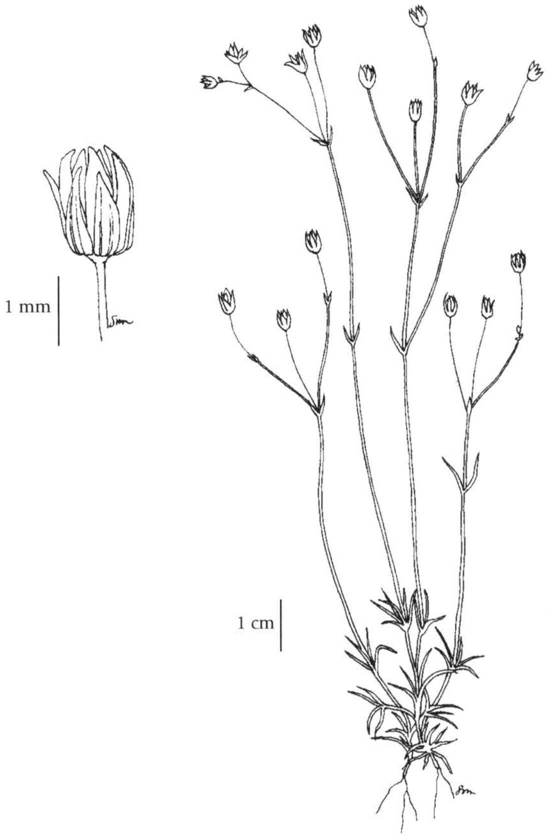

Minuartia dawsonensis

Minuartia elegans

Minuartia macrocarpa

Minuartia nuttallii **(Pax) Brig. ssp.** ***nuttallii*** (*Arenaria nuttallii* Pax)

Nuttall's sandwort

HABITAT/RANGE:

Dry grasslands, shrublands, forest openings and rocky slopes in the alpine zone; rare in SE BC; E to AB and S to WY, UT and OR.

DESCRIPTION:

General: Perennial herb from a deep, stout taproot and a much-branched stem-base; forming mats as much as 30 cm wide; stems trailing to ascending, several, branched, glandular-hairy, 4-12 cm tall/long; flowering stems brittle.

Leaves: Basal leaves lacking; stem leaves opposite, 6-10 mm long, about 1 mm wide, glandular-hairy, usually overlapping except on the upper part of the flowering stems, linear awl-like to narrowly lanceolate, 3-nerved, the middle nerve usually much more prominent, secondary leaves usually clustered in the axils of the numerous primary leaves along the stems; stipules lacking.

Flowers: Inflorescence few- to many-flowered in a compact, spreading cluster; petals oblong, obtuse, 3.5-4 mm long; sepals, 4-6 mm long, linear-lanceolate to lanceolate-egg-shaped, long-pointed or abruptly sharp-pointed, obscurely 3-nerved.

Fruits: Capsules egg-shaped, 3-5 mm long, 3-valved; seeds dark brown, plump, 1.0-1.3 mm long, prominently elongate-pimply in concentric rows.

Minuartia obtusiloba **(Rydb.) House** (*Arenaria obtusiloba* [Rydb.] Fern.)

alpine sandwort

HABITAT/RANGE:

Dry gravelly slopes, meadows and fellfields in the subalpine and alpine zones; common in S BC east of the Coast-Cascade Mountains; N to AK, YT and NT, E to W AB and S to ID, NM and NE OR.

DESCRIPTION:

General: Perennial herb from a taproot and branching stem-base; forming mats up to 40 cm across; stems prostrate to ascending or erect, short-hairy to hairy and more or less glandular above, 3-10 cm tall.

Leaves: Basal leaves oblong to linear, numerous, 1-nerved, 3-8 mm long, less than 1 mm wide, glabrous or glandular-hairy or hairy margined; stem leaves opposite, 1-2 pairs or occasionally lacking, similar to basal leaves but usually wider, glabrous or glandular-hairy or hairy margined; stipules lacking.

Flowers: Inflorescence usually of solitary flowers; flower stalks glandular-hairy, erect; petals 5, white, egg-shaped to spoon-shaped, 4-10 mm long; sepals 5, often purplish, especially at the tip, 4-5 mm long, mostly 3-nerved, oblong, rounded to obtuse, usually glandular-hairy.

Fruits: Capsules egg-shaped-cylindric, 4-8 mm long, 3- to 4-valved; seeds 0.8-1 mm long, smooth or nearly so.

Minuartia pusilla **(S. Wats.) Mattf.** (*Arenaria pusilla* S. Wats.)

dwarf sandwort

HABITAT/RANGE:

Dry rock cliffs in the lowland zone; rare in SW BC, known only from Rocky Point, S Vancouver Island; S to ID, AZ and NW CA.

DESCRIPTION:

General: Annual herb from a weak taproot; stems erect, solitary or or more usually few, simple to branched, glabrous and more or less glaucous, 2-5 cm tall.

Leaves: Basal and lower stem leaves opposite, linear, 2-4 mm long, less than 0.5 mm wide, glabrous, obtuse, 1-nerved; upper stem leaves few, similar, not much reduced; stipules lacking.

Flowers: Inflorescence of usually several flowers in an open, leafy-bracted cluster that is often 4/5 the total height of the plant; petals elliptic, 1-2 mm long, sometimes lacking; sepals lanceolate, long-pointed or abruptly sharp-pointed, 2-3 mm long, 3-nerved.

Fruits: Capsules egg-shaped, 1-2 mm long, 3-valved; seeds brown, about 0.3 mm long, minutely pimply.

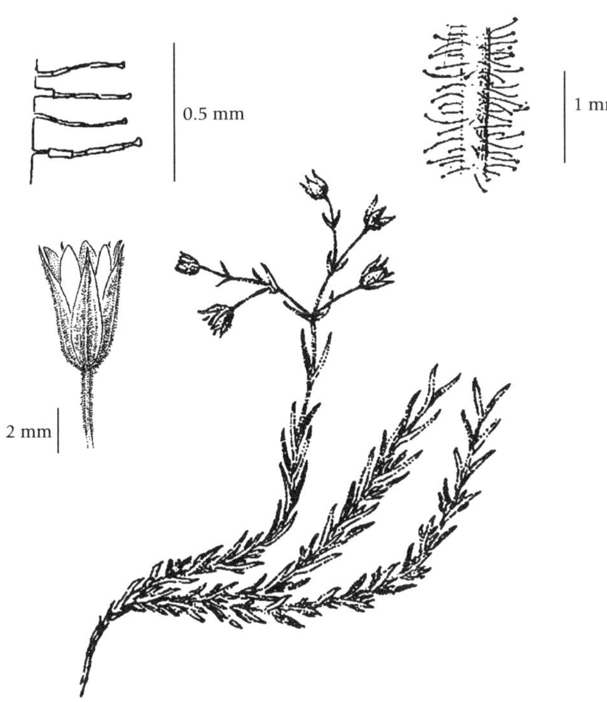

Minuartia nuttallii ssp. *nuttallii*

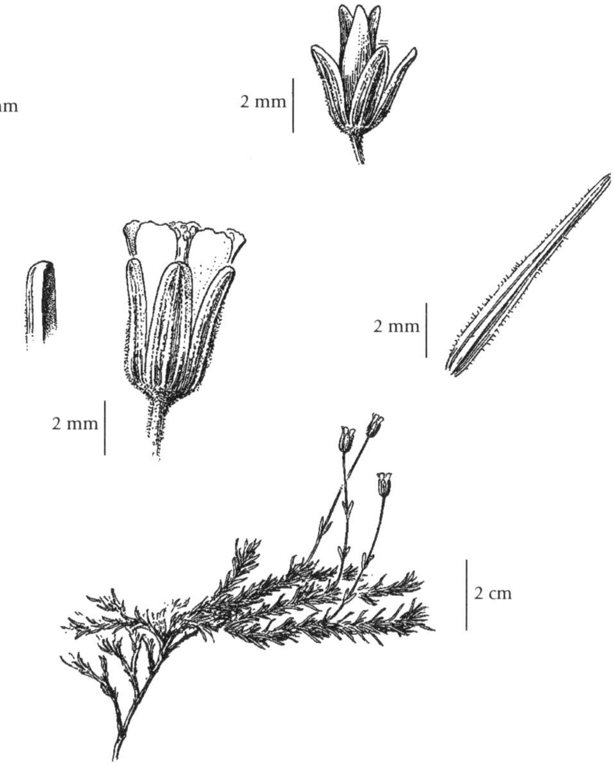

Minuartia obtusiloba

Minuartia pusilla

Minuartia rubella **(Wahl.) Hiern** (*Arenaria rubella* [Wahl.] Sm.)
boreal sandwort
HABITAT/RANGE:
Dry gravelly slopes, meadows and fellfields in the subalpine and alpine zones; common throughout BC, except rare on the Queen Charlotte Islands; circumpolar, N to AK, YT and NT, E to NF and S to NM, NV and CA; Eurasia.
DESCRIPTION:
General: Perennial herb from a small taproot and a branched crown; forming small cushions 4-20 cm across; stems slender, ascending, several, simple to few branched, glandular-hairy, 1-15 cm tall.

Leaves: Basal leaves linear-lanceolate, 3-12 mm long, less than 1 mm wide, densely glandular, densely crowded, prominently 3-nerved, bluish-green; stem leaves opposite, 2-6 pairs, 0.3-1.5 cm long, reduced upwards, densely glandular, bluish-green, without clustered secondary leaves in their axils; stipules lacking.

Flowers: Inflorescence of 2-4 flowers in an open cluster, occasionally flowers solitary; petals 5, white or sometimes reddish, oblanceolate, 2.5-5 mm long; sepals 5, often purplish, broadly lanceolate, long-pointed or abruptly sharp-pointed, 2-4 mm long, strongly 3-nerved, glandular.

Fruits: Capsules narrowly egg-shaped, 3-5 mm long, 3- to 4-valved; seeds kidney-shaped, 0.4-0.6 mm long, light reddish-brown, minutely roughened.

Minuartia stricta **(Sw.) Hiern.**
rock sandwort
HABITAT/RANGE:
Mesic gravel floodplains, rocky slopes and meadows in the subalpine and alpine zones; rare in N BC, known only from Mount Mansfield; circumpolar, N to AK, YT and NT; Eurasia.
DESCRIPTION:
General: Perennial herb from a much-branched taproot; forming cushions 10-50 cm across; stems slender, erect or ascending, many, simple or 2- to rarely 3-branched, glabrous, 0.6-2.5 cm tall.

Leaves: Basal leaves linear-lanceolate, 0.5-1.8 cm long, less than 1 mm wide, glabrous, densely crowded, 1- or obscurely 3-nerved; stem leaves opposite, 2-6 pairs, 0.2-1.4 cm long, glabrous, with clustered secondary leaves in their axils; stipules lacking.

Flowers: Inflorescence of solitary or 2, rarely 3 flowers in an open cluster; petals 5, white, oblanceolate, 2.0-3.5 mm long; sepals 5, greenish-purple, elliptic, obtuse, 2-4 mm long, faintly 1- to 3-nerved, glabrous.

Fruits: Capsules narrowly egg-shaped, 3-4.5 mm long, 3- to 5-valved; seeds 0.5-0.8 mm long, chestnut-brown, minutely roughened.

Minuartia tenella **(Nutt. in T & G.) Mattf.** (*Arenaria stricta* Michx. ssp. *macra* [A. Nels. & J.F. Macbr.] Maguire, *A. stricta* var. *puberulenta* [Peck] C.L. Hitchc., *A. tenella* Nutt.)
slender sandwort
HABITAT/RANGE:
Mesic to dry coastal bluffs and forest openings in the lowland and montane zones; common in SW BC, rare E of the Coast-Cascade Mountains; S to N OR.
DESCRIPTION:
General: Annual herb, rarely persistent for more than a year, from a short, slight taproot; not matted; stems decumbent to erect, solitary to several, much branched, more or less glandular-hairy, 10-30 cm tall.

Leaves: Basal leaves few, soon decciduous; stem leaves opposite, linear awl-like, 3-20 mm long, 3-nerved, glabrous, the primary ones mostly with clustered secondary leaves or short branches in their axils; stipules lacking.

Flowers: Inflorescence of several to numerous flowers in an open, often diffuse cluster frequently more than 1/2 the total height of the plant, with awl-like bracts; petals broadly egg-shaped, from only slightly to 1.5 times longer than the sepals; sepals lanceolate, long-pointed or abruptly sharp-pointed, 3-nerved, 2.5-3.5 mm long.

Fruits: Capsules egg-shaped, 3.5-4.5 mm long, 3-valved; seeds about 0.6 mm long, brown, finely pimply in a paving-block pattern.

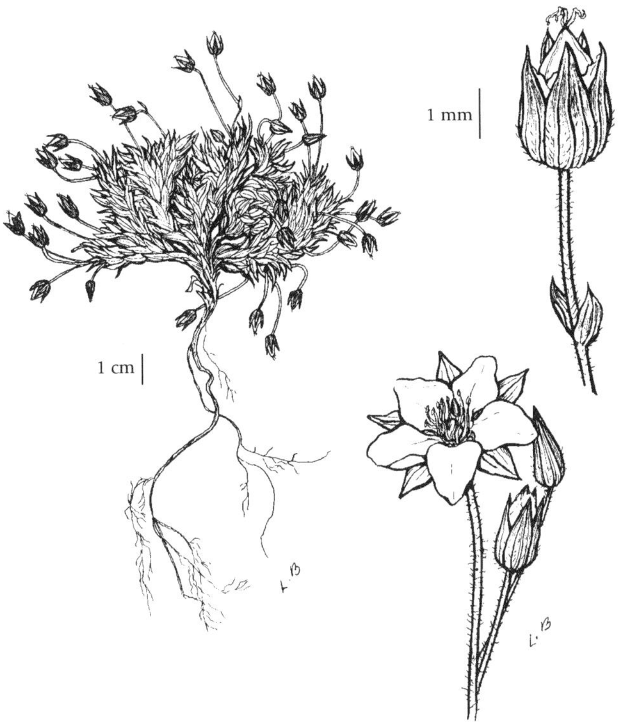

Minuartia rubella

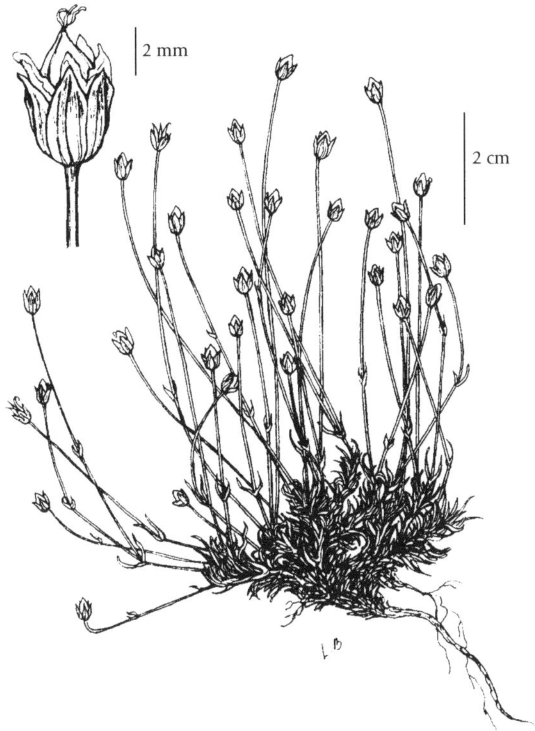

Minuartia stricta

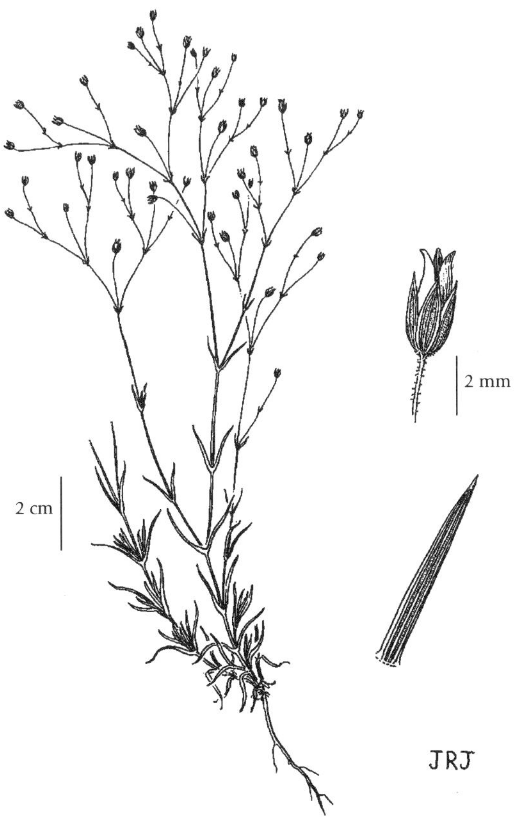

Minuartia tenella

***Minuartia yukonensis* Hult.** (*Arenaria laricifolia auct. non* B.P. Robbins)
Yukon sandwort
HABITAT/RANGE:
Dry rocky slopes and meadows in the alpine zone; rare in N BC, known only from Mount Mansfield and Toad River; amphiberingian, N to AK, YT and NT; E Asia.
DESCRIPTION:
General: Perennial herb from a much-branched taproot; forming small cushions 5-10 cm across; stems slender, erect or ascending, many, simple or more often 3- to 4-branched, glabrous, 0.2-1.7 cm tall.

Leaves: Basal leaves linear-lanceolate, 0.1-2.0 mm long, less than 1 mm wide, densely crowded, 1- or obscurely 3-nerved; stem leaves opposite, 2-6 pairs, 0.2-1.4 cm long, glandular-hairy or less often glabrous; stipules lacking.

Flowers: Inflorescence of solitary or more often 3-4 flowers in an open cluster; flower stalks glandular-hairy; petals 5, white, egg-shaped, 0.6-1.0 cm long; sepals 5, green, the papery margins sometimes purplish, spoon-shaped, obtuse to blunt, 3.5-6.0 mm long, 3-nerved, glandular-hairy.

Fruits: Capsules oblong, 6-8 mm long, 3- to rarely 5-valved; seeds somewhat kidney-shaped, 0.8-1.0 mm long, chestnut-brown, warty on the back.

MOEHRINGIA

1. Leaves abruptly sharp-pointed, essentially glabrous; sepals abruptly sharp-pointed or long-pointed .... *M. macrophylla*
1. Leaves mostly rounded to obtuse, margins and midrib short rough-hairy, usually short rough-hairy below as well; sepals oblong, rounded .... *M. lateriflora*

***Moehringia lateriflora* (L.) Fenzl** (*Arenaria lateriflora* L.)
blunt-leaved sandwort
HABITAT/RANGE:
Moist to dry forests and meadows in the lowland, steppe and montane zones; frequent E of the Coast-Cascade Mountains, rare on the Queen Charlotte Islands; N to AK, YT and NT, E to NF and S to PA, NM and CA.
DESCRIPTION:
General: Perennial herb from a branched stem-base and slender rhizome; stems ascending to erect, several, usually branched, more or less curly or downswept short-hairy or rarely glabrous, 4-20 cm tall.

Leaves: Basal leaves lacking; stem leaves opposite, 3-7 pairs, narrowly elliptic-oblong to oblong-lanceolate, 3-30 mm long, 2-12 mm wide, rounded to obtuse, lower ones smaller, margins and midrib short rough-hairy, usually short rough-hairy below as well, often with sterile shoots arising in the leaf axils; stipules lacking.

Flowers: Inflorescence of 5-6 flowers in terminal and lateral clusters, occasionally solitary; petals 5 or occasionally 4, white, egg-shaped, 3.5-8 mm long; sepals 5 or occasionally 4, green with white margins, oblong, rounded, 2-3 mm long, usually glabrous.

Fruits: Capsules nearly globose, 2-4 mm long, 6-valved; seeds smooth, black, shiny, 0.8-1.1 mm long, with a small, fleshy appendage.

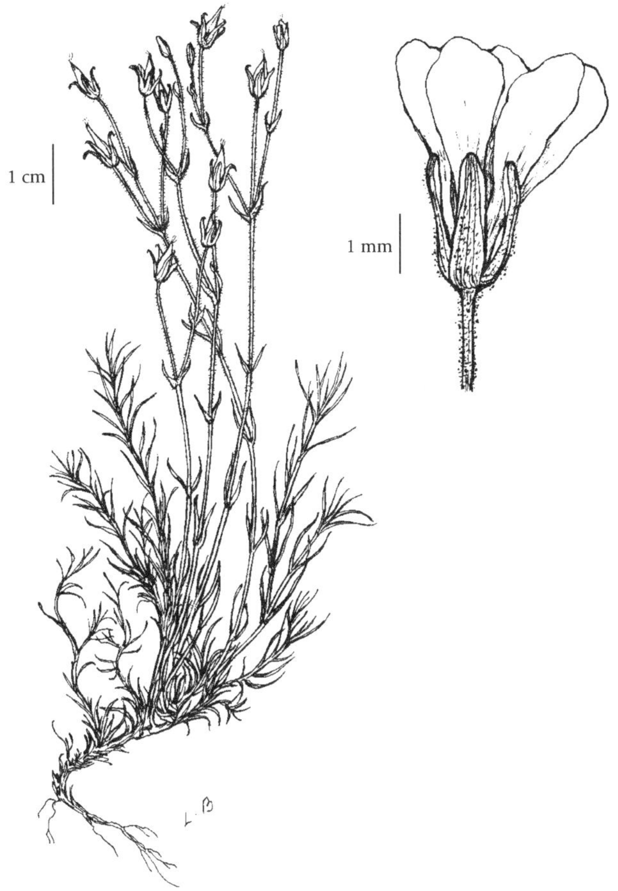

Minuartia yukonensis

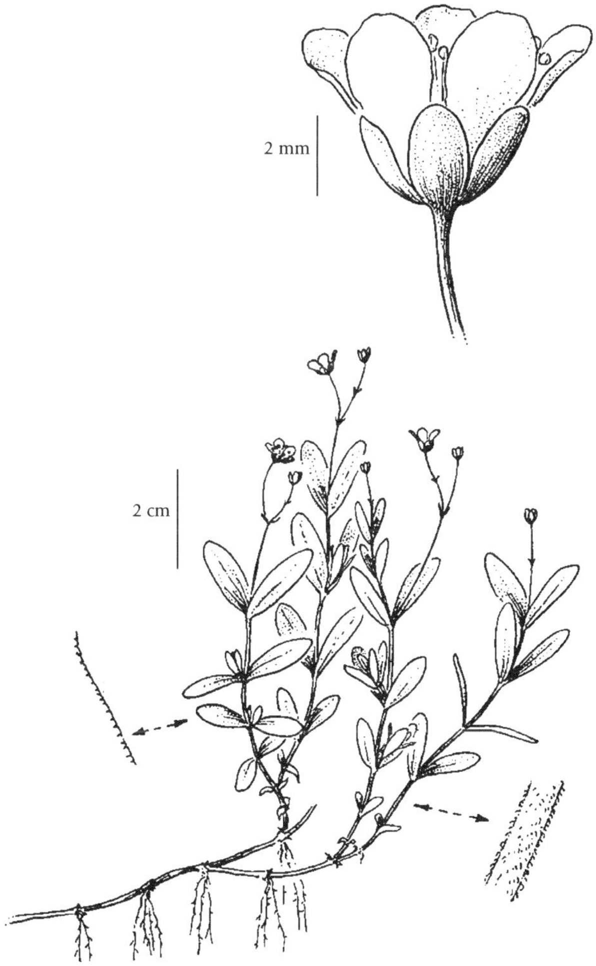

Moehringia lateriflora

***Moehringia macrophylla* (Hook.) Fenzl** (*Arenaria macrophylla* Hook.)

big-leaved sandwort

Habitat/Range:

Moist to dry forests and rocky slopes in the lowland, steppe and montane zones; common in extreme S BC, rare northward to 55° N; N to NT, E to PQ and S to NM and CA.

Description:

General: Perennial herb from an extensive slender rhizome; often forming loosely matted patches with many sterile as well as floriferous stems; stems decumbent to erect, several, usually branching, short rough-hairy, 3- or often 4-angled, 5-25 cm long.

Leaves: Basal leaves lacking; stem leaves opposite, numerous, linear-elliptic to lanceolate, abruptly sharp-pointed, 2-5 cm long, 2-12 mm wide, essentially glabrous; stipules lacking.

Flowers: Inflorescence of 2-5, rarely solitary, flowers in a slender, pedunculate, terminal and lateral, minutely bracted cluster; petals 5, oblong, blunt, white, 2-10 mm long; sepals 5, 2.5-6 mm long, narrowly egg-shaped, abruptly sharp-pointed or long-pointed, usually whiskery.

Fruits: Capsules globose-egg-shaped, 3-5 mm long, 6-valved; seeds few, smooth, with a prominent fleshy appendage.

MOENCHIA

Moenchia erecta* (L.) Gaertn., Mey. & Scherb. var. *erecta

upright chickweed

Habitat/Range:

Dry grassy meadows in the lowland zone; rare in SW BC, known only from the Victoria area and Saltspring Island; introduced from Europe.

Description:

General: Annual herb from a taproot; stems erect, solitary, simple or sometimes sparingly branched, glabrous, 2.5-15 cm tall.

Leaves: Basal leaves oblanceolate, 5-18 mm long, short-stalked, glabrous, glaucous; stem leaves 3-4 pairs, opposite, lanceolate, 3-12 mm long, unstalked; glabrous, glaucous; stipules lacking.

Flowers: Inflorescence of 1 to 3 flowers on long stalks; petals 4 or occasionally 5, white, 3-4 mm long; sepals lanceolate, 5-7 mm long, abruptly sharp-pointed, green with wide white margins.

Fruits: Capsules cylindric, 5-6 mm long, 8-valved; seeds reddish-brown, kidney-shaped, 0.5 mm wide, pimply.

MYOSOTON

***Myosoton aquaticum* (L.) Moench** (*Stellaria aquatica* [L.] Scop., *Malachium aquaticum* [L.] Fries)

water chickweed

Habitat/Range:

Moist to mesic shorelines and meadows in the lowland zone; infrequent on S Vancouver Island, the lower Fraser Valley and SC BC; introduced from Europe.

Description:

General: Perennial herb from overwintering stem-bases; stems decumbent or ascending, often trailing, 0.2-1.0 m long, glabrous below, glandular-hairy above.

Leaves: Leaves opposite, egg-shaped-long-pointed with a heart-shaped base and wavy margins, 2-8 cm long, lower leaves short-stalked, upper leaves unstalked; stipules lacking.

Flowers: Inflorescence in a 2-parted cluster with leaf-like bracts, the flowers so appearing solitary; petals white, to 12 mm long, deeply 2-cleft; sepals egg-shaped, 7-9 mm long, with membranous margins.

Fruits: Capsules egg-shaped, up to 1 cm long, drooping; seeds brown, 0.4 mm long, with dense, barbed pimples.

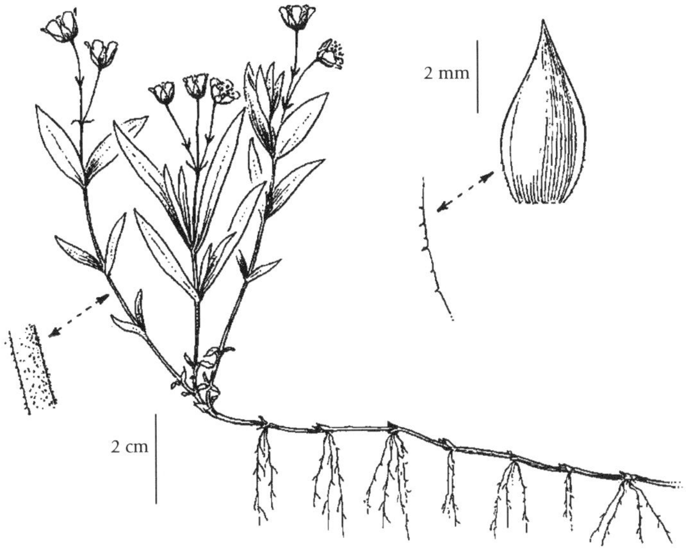

Moehringia macrophylla

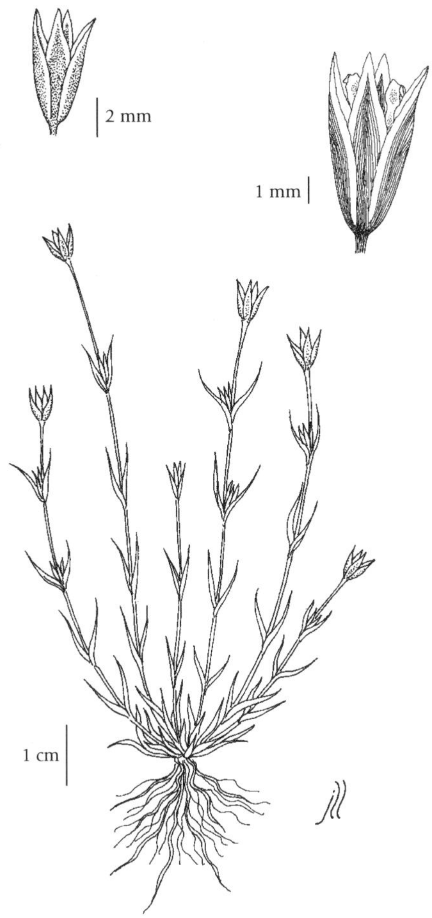

Moenchia erecta

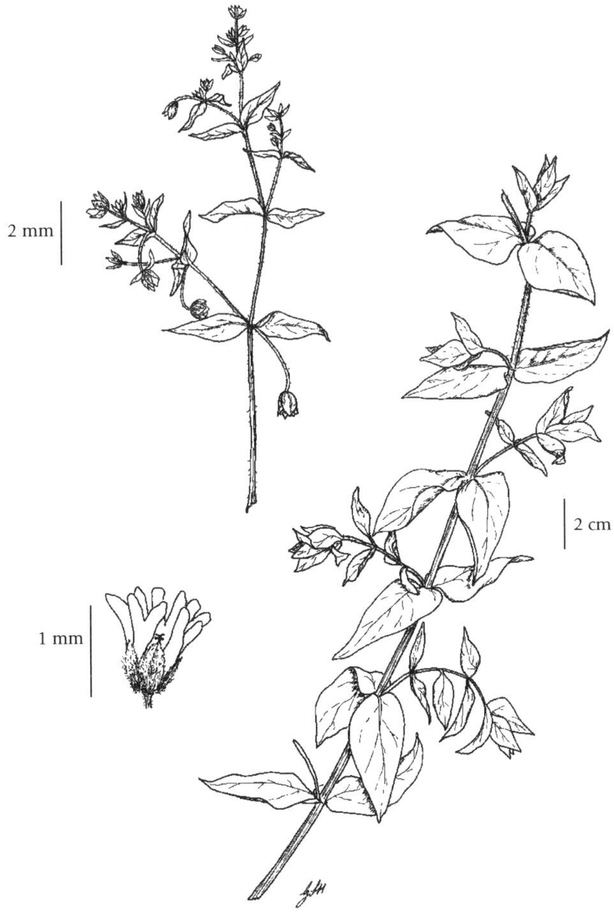

Myosoton aquaticum

PETRORHAGIA

1. Calyces 10-13 mm long; stem leaves linear to lanceolate, more than 1 cm long .... *P. prolifera*
1. Calyces 4-5 mm long; stem leaves linear and need-like, less than 1 cm long .... *P. saxifraga*

Petrorhagia prolifera **(L.) P.W. Ball & Heywood** (*Tunica prolifera* [L.] Scop.)
prolific petrorhagia

HABITAT/RANGE:

Dry, sandy disturbed areas in the lowland zone; rare in SW BC, known only from Comox and Surrey; introduced from Europe.

DESCRIPTION:

General: Annual herb from a slender taproot; stems ascending to erect, branched, glabrous, 30-50 cm tall.

Leaves: Basal leaves linear-oblanceolate to oblanceolate, 1.2-3 cm long, 1-3 mm wide; stem leaves opposite, 3 to many pairs, linear to linear lanceolate, fused at the base into a sheath.

Flowers: Inflorescence of solitary flowers or in a 3- to 7- flowered, headlike cluster, these subtended by 2-3 pairs of brown, papery, egg-shaped bracts; petals 5, 10-14 mm, pink, notched, exceeding the calyces; calyces 10-13 mm long; sepals 5.

Fruits: Capsules egg-shaped, 5-8 mm long; seeds brownish-black, 1.1-1.8 mm long, compressed, concave on 1 side.

Petrorhagia saxifraga **(L.) Link** (*Tunica saxifraga* [L.] Scop.)
tunic flower

HABITAT/RANGE:

Dry roadsides and waste places in the montane zone; rare garden escape in EC BC, known only from Prince George; introduced from Europe.

DESCRIPTION:

General: Perennial herb from a slender taproot; stems ascending to erect, much branched, glabrous or stiffly short-hairy, the lower internodes shortly stiff-hairy, 5-40 cm tall.

Leaves: Basal and stem leaves linear and needle-like, 0.5-1.0 cm long, 1-2 mm wide, 1-nerved, fringed with short hairs on the basal margins.

Flowers: Inflorescence of solitary or rarely 3 flowers at the stalked branch tips, these subtended by 2-4 pairs of papery, egg-shaped bracts; petals 5, 4.5-10 mm long, pink or white, notched, exceeding the calyces; calyces 4-5 mm long; sepals 5.

Fruits: Capsules egg-shaped, 5-8 mm long; seeds brownish-black, 0.9-1.6 mm long, compressed, concave on 1 side.

POLYCARPON

Polycarpon tetraphyllum **(L.) L.**
four-leaved all-seed

HABITAT/RANGE:

Dry to mesic roadsides and waste places in the lowland zone; rare in SW BC, known only from Victoria; introduced from Europe.

DESCRIPTION:

General: Annual herb from a slender taproot; stems prostrate to erect, much branched, glabrous, 5-15 cm tall.

Leaves: Basal leaves lacking; stem leaves opposite, egg-shaped, 8-13 mm long, narrowed at the base, blunt, the lower pairs so close as to appear like whorls of 4; stipules small, triangular and long-pointed, papery.

Flowers: Inflorescence of numerous flowers in a much-branched cluster; petals 5, white, narrowly oblong, 1-1.5 mm long, soon falling; sepals 5, 1-1.5 mm long, with a papery margin.

Fruits: Capsules egg-shaped, to 2 mm long; seeds brownish, 0.4-0.5 mm long, granular.

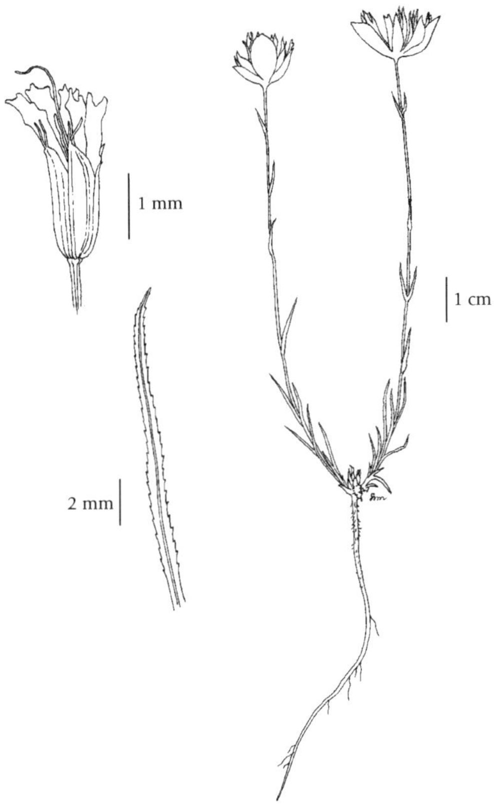

Petrorhagia prolifera

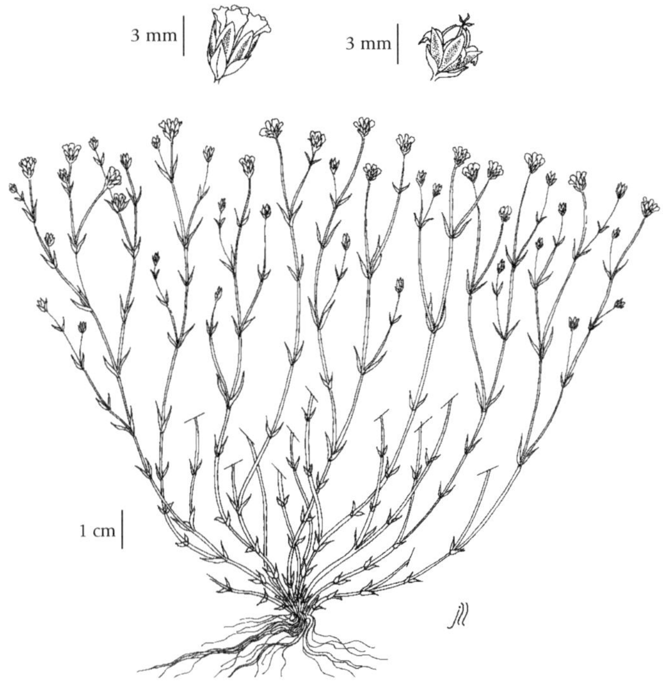

Petrorhagia saxifraga

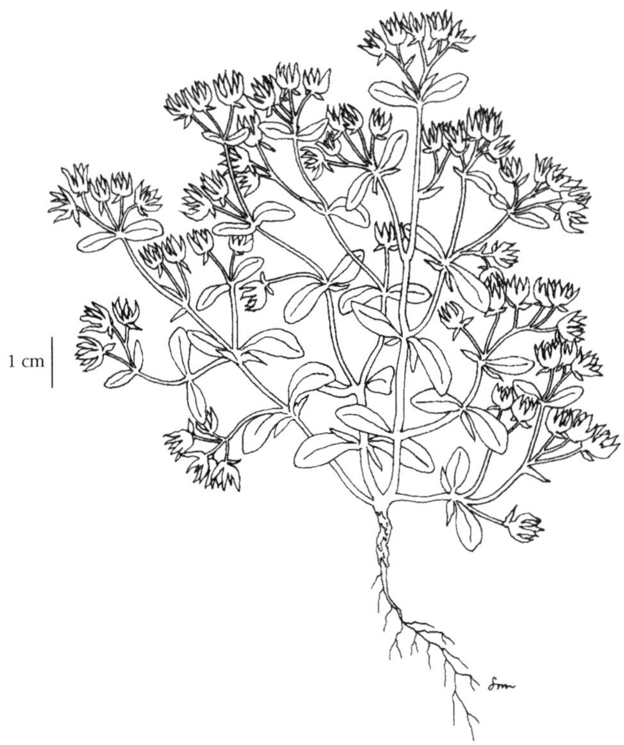

Polycarpon tetraphyllum

SAGINA [25]

1. Flowers 5-merous; leaves succulent; seeds kidney-shaped or nearly globose, lacking a dorsal groove.
    2. Seeds dark brown, with a small, distinct projection .... *S. japonica*
    2. Seeds reddish-brown, smooth to slightly pebbled .... *S. maxima*
1. Flowers 4- or 5-merous; leaves not succulent, or if so then flowers predominantly 4-merous; seeds obliquely triangular, dorsal groove present.
    3. Plants annual; stems threadlike.
        4. Upper leaves sparsely fringed near the base; sepals 4, rarely 5; petals lacking or rarely 4 .... *S. apetala*
        4. Upper leaves glabrous; sepals usually 5; petals usually 5 .... *S. decumbens*
    3. Plants perennial; stems not threadlike.
        5. Plants tufted, forming low cushions, stem leaves awl-like; sepal margins violet .... *S. nivalis*
        5. Plants ascending, spreading, decumbent or mat-forming, rarely tufted, except in some alpine plants; stem leaves linear to linear awl-like; sepal margins green or white.
            6. Petals less than 1 mm long, sometimes lacking; sepals spreading at time of capsules opening .... *S. procumbens*
            6. Petals more than 1 mm long; sepals appressed or at least loosely appressed at time of capsules opening .... *S. saginoides*

Sagina apetala **L. var.** ***barbata*** **Fenzl** ***ex*** **Ledeb.**
common pearlwort

HABITAT/RANGE:

Dry roadsides and waste places in the lowland zone; rare in SW BC, known from Vancouver and Victoria; introduced from Eurasia.

DESCRIPTION:

General: Annual herb from a slender taproot; stems ascending, numerous, branched, threadlike, finely glandular short-hairy above, especially in the inflorescence, 3-10 cm tall.

Leaves: Basal and stem leaves opposite, linear awl-like, 2-5 mm long, sparsely on their slightly papery bases, minutely needle-pointed; stipules lacking.

Flowers: Inflorescence of numerous flowers on 2-12 mm stalks; petals 4, minute or more often lacking; sepals 4, rarely 5, egg-shaped, 1.5-1.75 mm long, finely glandular short-hairy above, erect in fruit.

Fruits: Capsules oblong, 2-3 mm long, 4-valved; seeds 0.2-0.3 mm long, kidney-shaped, with a small projection.

Sagina decumbens **(Ell.) T.& G. ssp.** ***occidentalis*** **(S. Wats.) Crow** (*S. occidentalis* S. Wats.)
western pearlwort

HABITAT/RANGE:

Moist to mesic vernal pool margins, forest openings and open slopes in the lowland zone; rare in SW BC; S to AZ and CA.

DESCRIPTION:

General: Annual herb from a taproot; stems decumbent to erect or ascending, several, branched, glabrous to weakly glandular short-hairy above, 5-15 cm long/tall.

Leaves: Basal leaves linear, 1-3 cm long, less than 1 mm wide, 1-nerved, glabrous; stem leaves opposite, 2-6 pairs, 3-10 mm long, glabrous, becoming shorter towards the stem apex; stipules lacking.

Flowers: Inflorescence of usually several flowers per stem; petals 5, oblong, white, 1.2-2.6 mm long; sepals 5, greenish, rounded, glabrous to weakly glandular short-hairy, 1.5-2.5 mm long.

Fruits: Capsules obliquely triangular, compressed, 2.5-5 mm long, 5-valved; seeds about 0.2-0.3 mm long, obliquely triangular with a dorsal groove, reddish-brown, minutely roughened.

---

[25] Key adapted from Crow (1978).

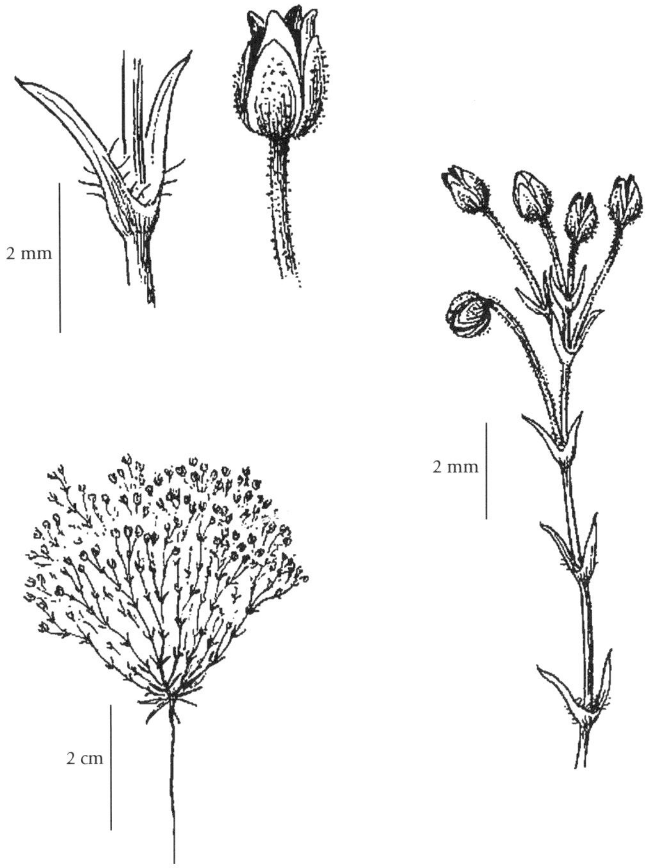

Sagina apetala var. *barbata*

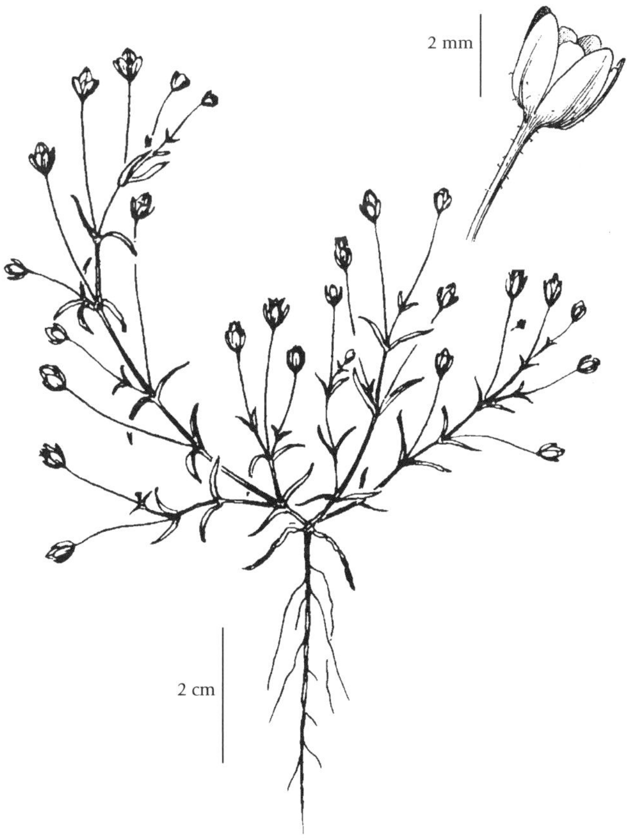

Sagina decumbens ssp. *occidentalis*

***Sagina japonica* (Sw.) Ohwi**
Japanese pearlwort
HABITAT/RANGE:
Dry waste places in the lowland zone; rare in coastal BC, known only from Prince Rupert and Nanaimo; introduced from E. Asia.
DESCRIPTION:
General: Annual or biennial herb from a taproot and branching stem-base; stems decumbent to erect, glabrous, 2-20 cm tall/long.

Leaves: Basal leaves succulent, linear, 0.7-1.8 cm long, up to 1.5 mm wide, 1-nerved, glabrous, in rosettes; stem leaves opposite, 2-6 pairs, linear, 5-15 mm long, often with clusters of secondary leaves in their axils; stipules lacking.

Flowers: Inflorescence of axillary, solitary flowers; petals 5, narrowly egg-shaped, white, 1.8-2.0 mm long, sometimes lacking; sepals 5, greenish, elliptic, rounded, about 2 mm long.

Fruits: Capsules egg-shaped, 2-2.2 mm long, 5-valved; seeds about 0.4 mm long, slightly pebbled, dark brown, with a small, distinct projection.

***Sagina maxima* A. Gray** (*S. crassicaulis* S. Wats. = ssp. *crassicaulis*)
coastal pearlwort
HABITAT/RANGE:
Mesic to dry rocky or sandy bluffs and gravelly beaches in the lowland zone; common in coastal BC; amphiberingian, ssp. *maxima* ranges S to N WA and ssp. *crassicaulis* N to AK and S to CA.
DESCRIPTION:
General: Biennial or perennial herb from a taproot and branching stem-base; forming clumps to 20 cm across; stems decumbent to erect, several, branched, glabrous or sometimes very sparsely glandular-hairy above, 2-12 cm tall/long.

Leaves: Basal leaves succulent, linear, 1-3 cm long, up to 2 mm wide, 1-nerved, glabrous, in rosettes; stem leaves succulent, opposite, 2-6 pairs, linear, 5-15 mm long, often with fascicles of secondary leaves in their axils; stipules lacking.

Flowers: Inflorescence of erect, solitary, terminal or lateral flowers; petals 5, oblong, white, 1.5-2.5 mm long, sometimes lacking; sepals 5, greenish, egg-shaped, 2.5-3.5 mm long.

Fruits: Capsules broadly egg-shaped, 3-5 mm long, 4- to 5-valved; seeds about 0.5 mm long, smooth to slightly pebbled, reddish-brown, shiny.

Notes: Two subspecies occur in BC.

1. Plants glandular above .... ssp. *maxima*
1. Plants glabrous .... ssp. *crassicaulis* (S. Wats.) Crow

***Sagina nivalis* (Lindbl.) Fries** (*S. intermedia* Fenzl)
snow pearlwort
HABITAT/RANGE:
Moist to mesic gravelly slopes and snowbeds in the subalpine and alpine zones; rare in WC and NE BC, known only from Morfee Mountain and Serb Creek; N to AK, YT and NT and E to PQ.
DESCRIPTION:
General: Perennial herb from a taproot and much-branched woody stem-base; forming low cushions to 10 cm across; stems decumbent to erect, numerous, branched, glabrous, 1-10 cm tall/long.

Leaves: Basal leaf rosette in first season only, the leaves linear, 4-8 mm long, less than 1 mm wide, glabrous; stem leaves opposite, 1-8 pairs, awl-like, usually with clustered secondary leaves in the axils; stipules lacking

Flowers: Inflorescence of solitary flowers on short stalks; petals 4 or 5, white, narrowly elliptic, to 1.5 mm long; sepals 4 or 5, with violet margins, egg-shaped, 1.5-2 mm long.

Fruits: Capsules 2.5-3 mm long, 4- or 5-valved, dull greenish- or whitish-yellow; seeds 0.3-0.4 mm long, reddish-brown.

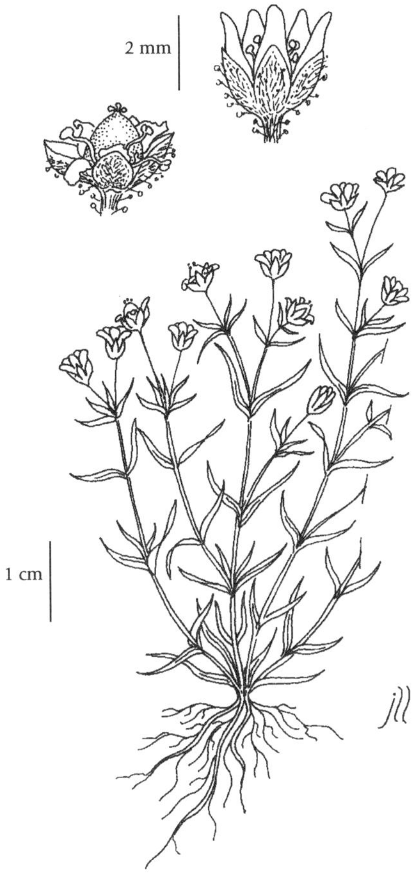

Sagina japonica

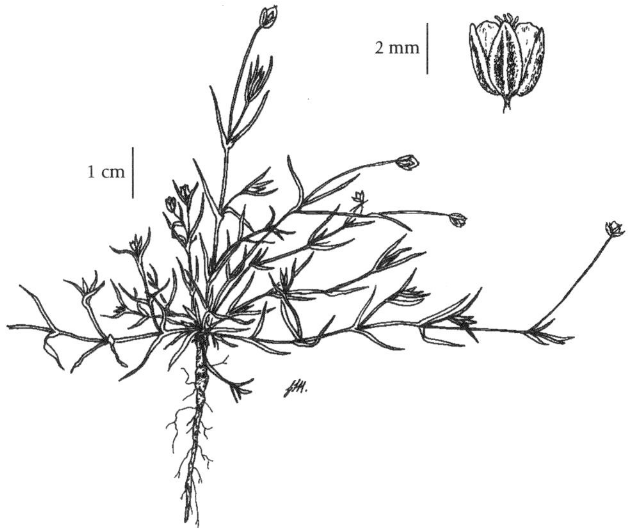

Sagina maxima ssp. *crassicaulis*

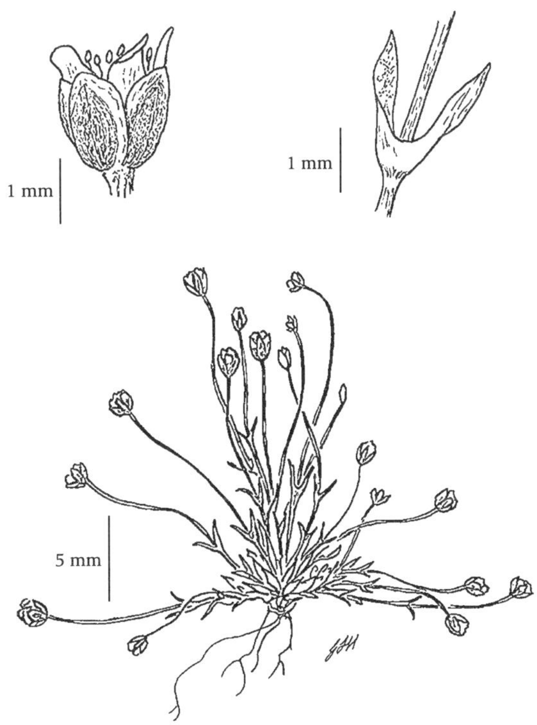

Sagina nivalis

Sagina procumbens **L.**
bird's-eye pearlwort
HABITAT/RANGE:
Wet to moist roadsides, lawns, coastal cliffs and pond margins in the lowland and steppe zones; common in SW BC, less frequent northward along the coast and in SC BC; N to AK, E to NF and S to MI, MN, UT, CA and MX.
DESCRIPTION:
General: Biennial or perennial (sometimes flowering the first year) herb from a taproot and freely rooting stems; stems prostrate, several, branched, glabrous, 2-20 cm long.

Leaves: Basal leaves 1-2.5 cm long, numerous in a dense persistent rosette; stem leaves opposite, usually linear, needle-pointed, 5-12 mm long, the upper ones sometimes minutely fringed, 2 to several of the pairs usually with secondary leaves clustered in their axils; stipules lacking.

Flowers: Inflorescence of usually 2 to several flowers per stem; petals 4 or sometimes 5, less than 1 mm long or lacking; sepals 5 or more usually 4, egg-shaped, 1.5-2.5 mm long, spreading at time of capsules opening.

Fruits: Capsules egg-shaped, 2-5 mm long, 4- or 5-valved; seeds about 0.3 mm long, not shiny.

Sagina saginoides **(L.) Karst.**
arctic pearlwort
HABITAT/RANGE:
Wet to moist pond margins, mud flats, streamsides and meadows from the subalpine to alpine zones; common in S BC, infrequent northward; N to AK, YT and NT, E to PQ and S to NM, AZ, CA and MX.
DESCRIPTION:
General: Biennial or more usually perennial (sometimes flowering the first year) herb from a taproot and slender branching stem-base; forming mats to 20 cm wide; stems decumbent to ascending, several, simple to few-branched, glabrous, 1-12 cm tall/long.

Leaves: Basal leaves in rosettes, linear, abruptly sharp-pointed, 4-20 mm long, less than 1 mm wide, 1-nerved, glabrous; stem leaves opposite, 2-6 pairs, linear, mostly 3-8 mm long, sometimes with secondary clusters of axillary leaves; stipules lacking.

Flowers: Inflorescence of erect, solitary, terminal or lateral flowers; petals 5, white, rounded, stalklike at the base, 1-1.5 mm long; sepals usually 5 or sometimes 4, greenish with a white margin, egg-shaped, 1.5-2 mm long, appressed or at least loosely appressed at time of capsules opening.

Fruits: Capsules egg-shaped, 2.5-3 mm long, 4- to 5-valved; seeds 0.2-0.4 mm long, reddish-brown, not shiny.

SAPONARIA

Saponaria officinalis **L.**
bouncing-bett; soapwort
HABITAT/RANGE:
Dry roadsides and waste places in the lowland, steppe and montane zones; infrequent in SW and SC BC; introduced from Eurasia.
DESCRIPTION:
General: Perennial herb from a stout, branched rhizome and long stolons; stems erect or ascending, several, simple or branched above, glabrous or nearly glabrous, forming large clumps, 30-90 cm tall.

Leaves: Basal leaves lacking; stem leaves opposite, as many as 20 pairs, lanceolate to oblanceolate, 4-10 cm long, 1.5-4 cm wide, short-stalked or the upper ones unstalked; stipules lacking.

Flowers: Inflorescence of numerous flowers in many tight clusters, fragrant, often double; petals 5, white to pink, stalklike at the base, stalk of the petals usually slightly exceeding the calyces, the blade mostly 10-15 mm long, egg-shaped to wedge-shaped with 2 linear appendages near the base, shallowly notched at the tip; sepals 5, green, united, forming a tube 15-20 mm long, with 5 teeth at the tip.

Fruits: Oblong to narrowly egg-shaped capsules, to 25 mm long, 4- or 5-valved; seeds black, about 1.7 mm long, concentrically pimply.

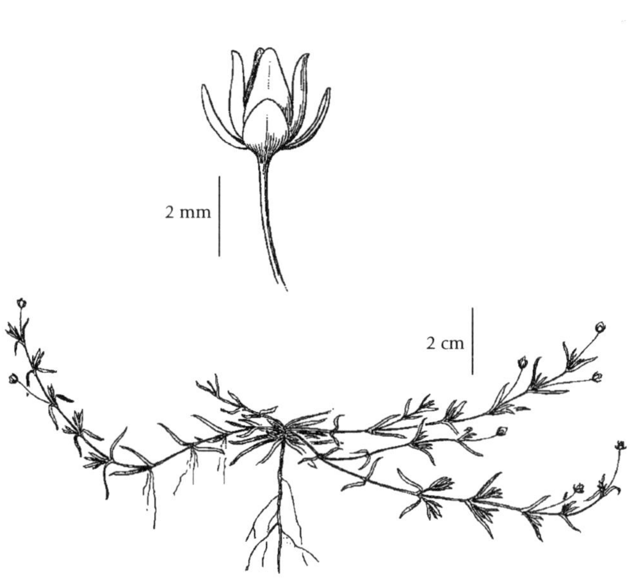

Sagina procumbens

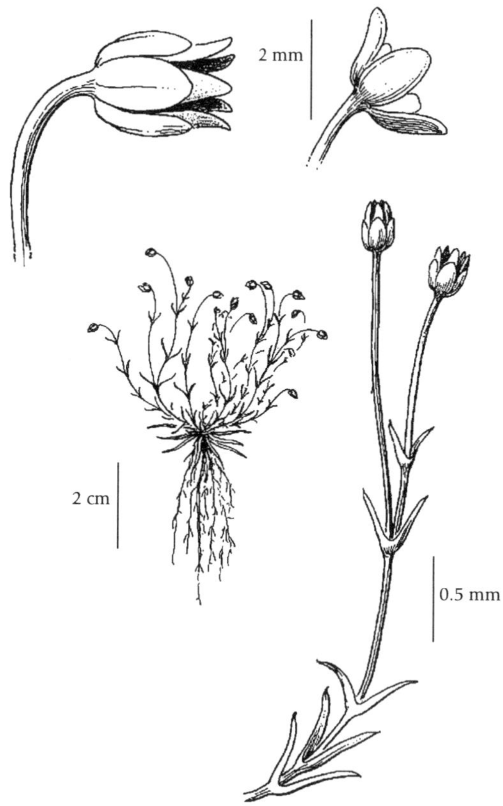

Sagina saginoides

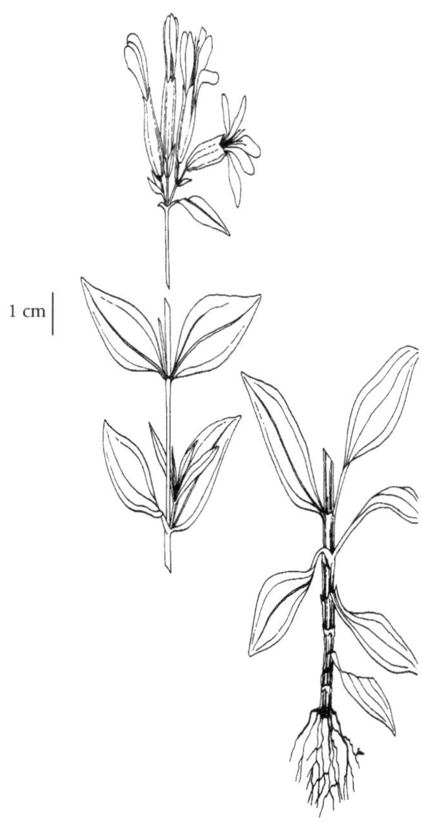

Saponaria officinalis

SCLERANTHUS

Scleranthus annuus **L.**

annual knawel

HABITAT/RANGE:

Dry roadsides, fields and waste places in the lowland, steppe and montane zones; infrequent in S BC; introduced from Eurasia.

DESCRIPTION:

General: Annual or biennial herb from a slender taproot; stems prostrate to ascending, several, few branched above, glabrous to short-hairy, 2.5-15 cm tall/long.

Leaves: Basal and stem leaves opposite, awl-like, 5-20 mm long, slightly papery-margined near their shortly fused bases; stipules lacking.

Flowers: Inflorescence of solitary flowers in stem forks, or in a terminal or axillary cluster; petals lacking; sepals usually 5, fused for 1/2 their length into a bell-shaped, 10-nerved tube, green or whitish, the lobes lanceolate, abruptly sharp-pointed, 1.5-2 mm long, glabrous.

Fruits: Inflated bladders, egg-shaped, 1-seeded, enclosed by the hardened calyx tube.

SILENE

1. Calyces glabrous.
   2. Plants cushion-forming, less than 10 cm tall .... *S. acaulis*
   2. Plants not cushion-forming, more than 10 cm tall.
      3. Petals white, 2-cleft.
         4. Calyces 15-20 mm long, inflated, prominently net-veined above the middle .... *S. vulgaris*
         4. Calyces 8-12 mm long, somewhat inflated, obscurely net-veined above the middle .... *S. cserei*
      3. Petals white, pink, rose or lavender, not 2-cleft.
         5. Petals white to pink; stems slender, the upper part usually with dark sticky bands below the nodes, otherwise glabrous above; calyces 4-10 mm long .... *S. antirrhina*
         5. Petals pink to rose or lavender; stems relatively stout, without sticky bands, glabrous to sparsely short-hairy above; calyces 12-17 mm long .... *S. armeria*
1. Calyces hairy.
   6. Plants introduced annuals; styles 3.
      7. Inflorescence irregularly flat-topped, the flowers 2 or more at each node; petal blades 2-cleft less than half their length .... *S. noctiflora*
      7. Inflorescence falsely racemose, the flowers solitary at each node; petal blades mostly entire or if 2-cleft then cleft more than half their length.
         8. Petal blades entire or shallowly notched at the tips, the basal appendages about 1 mm long and linear .... *S. gallica*
         8. Petal blades 2-cleft more than half their length, the basal appendages about 0.2 mm long and rounded .... *S. dichotoma*
   6. Plants native or introduced biennials or perennials; styles 4 or 5 or if 3 then plants native perennials.
      9. Flowers usually solitary, purplish or rarely pinkish.
         10. Petals included or slightly exserted; flowers nodding in bud .... *S. uralensis*
         10. Petals well exserted; flowers erect or spreading in bud .... *S. involucrata*
      9. Flowers 3 or more (rarely solitary in *S. taimyrensis*), white, pinkish or greenish or sometimes purplish-tinged.

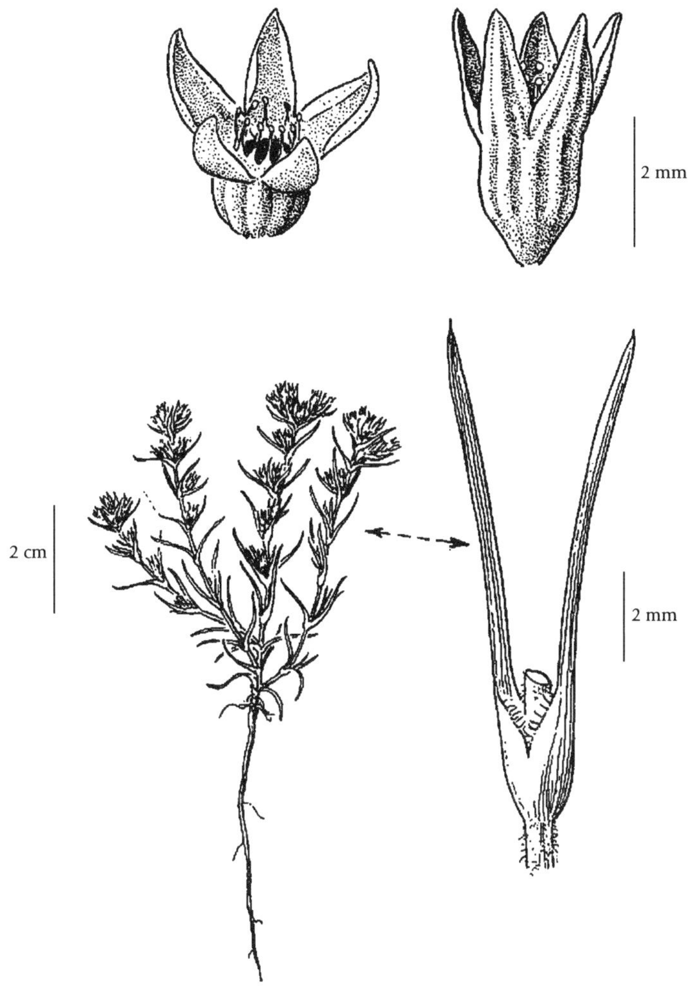

Scleranthus annuus

11. Petals with prominent lateral teeth on each blade lobe.

12. Leaves 6-20 cm long, those of the stem consisting of 3-12 pairs (rarely 1 or 2), progressively reduced upwards .... *S. scouleri*

12. Leaves 3-8 cm long, those of the stem consisting of 2-3 pairs, not greatly reduced upwards ..... *S. parryi*

11. Petals without prominent lateral teeth on each blade lobe.

13. Petals included or slightly exserted, not 2-cleft

14. Leaves unstalked and slightly fused above and below; petal blade appendages 4 (5-6); seeds 2.0 mm long .... *S. spaldingii*

14. Leaves unstalked above, stalked below, not fused; petal blade appendages 2; seeds 0.7 mm long .... *S. drummondii*

13. Petals well exserted, 2-cleft.

15. Calyces usually less than 1 cm long.

16. Styles 3 or sometimes 4; plants of S BC .... *S. menziesii*

16. Styles 5; plants of NW BC .... *S. taimyrensis*

15. Calyces more than 1 cm long.

17. Plants stout, erect; calyx teeth gradually tapering to abruptly sharp-pointed; petal blade appendages triangular, irregularly toothed; flowers unisexual (the plants dioecious).

18. Petals white, bilobed to midlength; leaves lanceolate to oblanceolate .... *S. latifolia*

18. Petals red or pink, irregulary and shallowly cleft .... *S. dioica*

17. Plants tufted, often matted; calyx teeth rounded; petal blade appendages oblong or egg-shaped; flowers perfect.

19. Capsules 3- (4)-celled nearly to the tip; calyces white to pink or purple, strongly stiff-hairy and more or less glandular .... *S. repens*

19. Capsules 1-celled; calyces creamy white, greenish, pink, or purplish tinged, not glandular or sparsely glandular on short-hairs .... *S. douglasii*

***Silene acaulis* (L.) Jacq.** (*S. acaulis* var. *exscapa* [All.] DC. = var. *acaulis*)

moss campion

HABITAT/RANGE:

Mesic to dry meadows, rock outcrops and fellfields in the alpine zone; common throughout BC, except infrequent on the Queen Charlotte Islands; var. *acaulis* – circumpolar, N to AK, YT, and NT, E to NF and S to OR, ID, MT and NH, var. subacaulescens – E to AB and S to MT, AZ, NM, NV and OR; Eurasia.

DESCRIPTION:

General: Perennial herb from a woody root and a branched stem-base; densely tufted, forming thick cushions to 40 cm wide; stems erect, solitary, simple, glabrous, 3-6 (rarely 15) cm tall.

Leaves: Basal and stem leaves opposite, linear to lance-linear, unstalked, 5-15 mm long, to 1.5 mm wide, glabrous to fringed at the base or stiffly short-hairy, channelled above, so dense as to obscure the stem; stipules lacking.

Flowers: Inflorescence of solitary flowers, usually unisexual (the plants either male or female); petals 5, pink to lavender (occasionally white), egg-shaped, 2-cleft, 8-12 mm long, with narrow stalklike bases about twice as long as the blades; sepals 5, green or frequently pinkish to purplish-tinged, united, forming a tube 4-9 mm long, 10-nerved, glabrous.

Fruits: Capsules subcylindric, 4-10 mm long, 6-valved; seeds light brown to yellowish, kidney-shaped, 0.8-1.2 mm long, weakly pimply.

Notes: Two varieties occur in BC.

1. Calyces 4-7 mm tall .... var. *acaulis*

1. Calyces 6-11 mm tall .... var. *subcaulescens* (F.N. Williams) Fern. & St. John

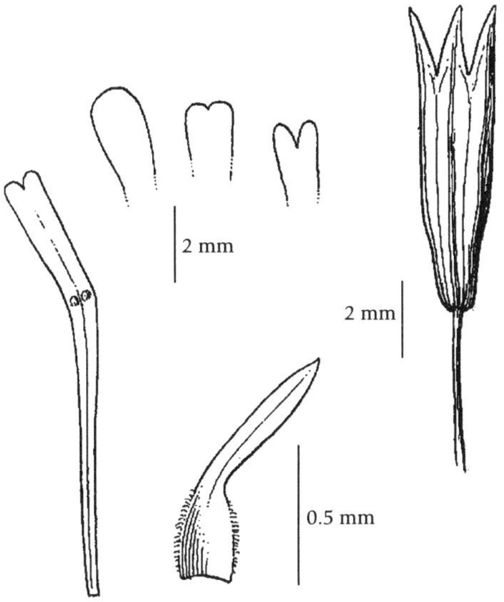

var. *subcaulescens*

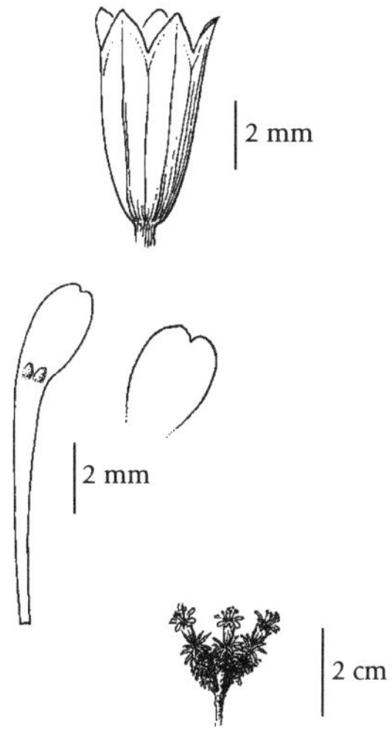

var. *acaulis*

Silene acaulis

***Silene antirrhina* L.**

sleepy catchfly

HABITAT/RANGE:

Mesic to dry roadsides, fields and open forests in the lowland, steppe and montane zones, frequent in S BC; E to PQ and NB and S to TX, AZ, NM and MX.

DESCRIPTION:

General: Annual herb from a taproot; stems erect, slender, simple to branched, more or less downsweptly short-hairy below, becoming glabrous above, usually purplish sticky-glandular in bands below the nodes or occasionally nonglandular, 20-80 cm tall.

Leaves: Basal leaves oblanceolate to spoon-shaped, 3-6 cm long and 2-20 mm wide; stem leaves opposite, several pairs, oblanceolate to linear, glabrous to rough-hairy or short-hairy; stipules lacking.

Flowers: Inflorescence of few to many flowers in a compact to open cluster; petals 5, white to pink, usually withered by day, stalklike at the bases, the narrow stalks about 2-3 times as long as the heart-shaped blades; sepals 5, united, forming a tube 4-10 mm long, 10-nerved, glabrous.

Fruits: Capsules egg-shaped, 3-chambered; seeds blackish, 0.6-1.0 mm long, with 3-4 concentric rows of low pimples.

***Silene armeria* L.**

sweet William catchfly

HABITAT/RANGE:

Mesic to dry roadsides and fields in the lowland zone; infrequent in SW BC, known only from S Vancouver Island and the lower Fraser Valley; introduced from Europe.

DESCRIPTION:

General: Annual herb from a taproot; glabrous to sparsely short-hairy, slightly sticky above; stems erect, solitary, branched, hollow, glabrous, 10-70 cm tall.

Leaves: Basal leaves linear to spoon-shaped, mostly 2-5 cm long and 0.5-1.5 cm wide, unstalked and more or less clasping, glabrous; stem leaves oblanceolate, unstalked and more or less clasping, glabrous, glaucous; stipules lacking.

Flowers: Inflorescence of usually several to many flowers in an open to contracted and often flat-topped cluster; petals 5, pink to rose or lavender, stalklike at the bases, the stalks 6-8 mm long, the blades egg-shaped to wedge-shaped, rounded to shallowly lobed, 4-7 mm long; sepals 5, reddish, united, forming a tube 12-17 mm long, 10-nerved, glabrous.

Fruits: Capsules cylindric egg-shaped, to 16 mm long, 6-valved; seeds dark brown to black, about 0.6 mm long, evenly wrinkled.

***Silene cserei* Baumg.**

biennial campion

HABITAT/RANGE:

Dry roadsides and waste places in the steppe and montane zones; rare in S BC, also known from Prince George; introduced from Europe.

DESCRIPTION:

General: Biennial or short-lived perennial herb from a taproot; stems erect, several, branched, usually glabrous and more or less glaucous, 20-80 cm tall.

Leaves: Basal leaves lacking; stem leaves opposite, oblanceolate to broadly lanceolate, stalkless, 5-10 cm long, glaucous; stipules lacking.

Flowers: Inflorescence of numerous flowers borne in whorls around the stem; petals 5, white, deeply 2-cleft at the tip, stalklike at the bases; sepals 5, united, forming a tube 8-12 mm long, pinkish, somewhat inflated and egg-shaped, obscurely net-veined above the middle.

Fruits: Capsules 10-14 mm long; seeds 1-1.5 mm long, conspicuously pimply.

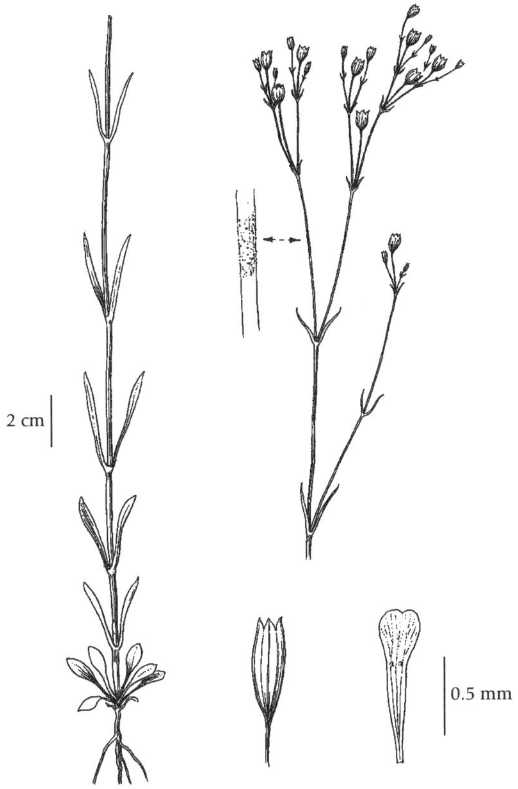

Silene antirrhina

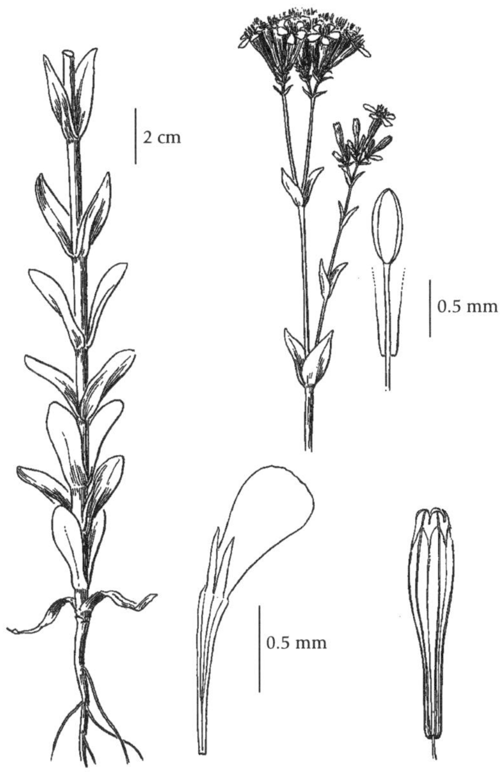

Silene armeria

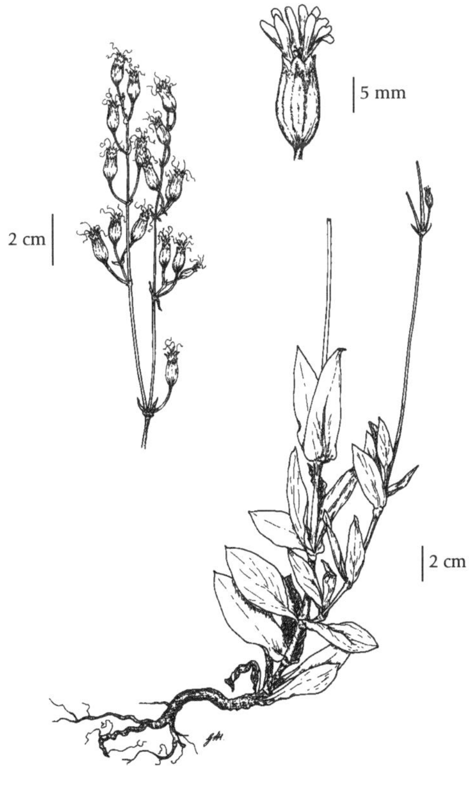

Silene cserei

Silene dichotoma **Ehrh.**
forked catchfly
HABITAT/RANGE:
Dry roadsides, fields and waste places in the steppe and montane zones; rare in SW BC (known from Vancouver) and SE BC; introduced from Eurasia.
DESCRIPTION:
General: Annual herb from a taproot; stems erect, solitary, simple to branched, strongly stiff-hairy, 20-80 cm tall.

Leaves: Basal leaves stalked, lanceolate to oblanceolate, short-hairy, 3-8 cm long, 0.5-3.5 cm wide; stem leaves opposite, short-hairy, slightly if at all reduced but becoming unstalked above; stipules lacking.

Flowers: Inflorescence of usually numerous flowers, falsely racemose, the flowers actually solitary at each node, usually perfect, but sometimes more or less with either male or female flowers; petals 5, white to occasionally pink, stalklike at the bases, the narrow stalks 10-15 mm long, thes blade 5-9 mm long, wedge-shaped to egg-shaped, 2-cleft more than half its length, the basal appendages about 0.2 mm long and rounded; sepals 5, united, forming a tube 10-15 mm long, 10-nerved, hairy.

Fruits: Capsules egg-shaped, 3-valved; seeds dark brown, kidney-shaped, 1-1.5 mm long, finely and regularly wrinkled.

Silene dioica **(L.) Clairville** (*Lychnis dioica* L., *Melandrium dioicum* [L.] Cosson & Germ.)
red campion
HABITAT/RANGE:
Mesic open forests in the lowland zone; rare in SW BC, known only from the lower Fraser Valley; introduced from Eurasia.
DESCRIPTION:
General: Biennial or perennial herb from a fibrous root; stems erect, several, branched, glandular above, 30-90 cm tall.

Leaves: Basal leaves egg-shaped, narrowed to winged stalks, hairy; stem leaves opposite, broadly elliptic to egg-shaped-elliptic, 4-10 cm long, 2-4 cm wide, hairy; stipules lacking.

Flowers: Inflorescence many-flowered, in clusters, flowers unisexual, plants both male and female; petals 5, red to purplish-red or pink, stalklike at the bases, deeply 2-cleft; sepals 5, usually purplish, hairy, the hairs with purple cross walls, united, forming a tube 10-15 mm long, slightly glandular.

Fruits: Capsules egg-shaped, with 5, 2-toothed valves; seeds black, densely roughened.

Notes: Hybrids between *S. dioica* and *S. latifolia* ssp. *alba* have been collected in the lower Fraser Valley and N to Prince George (McNeill 1978).

Silene douglasii **Hook. var.** ***douglasii***
Douglas' campion; Douglas' silene; Douglas' catchfly
HABITAT/RANGE:
Mesic to dry grasslands, shrublands, meadows and open forests in the lowland, steppe and montane zones; frequent in S BC, rare in N BC; S to MT, VT, NV and CA.
DESCRIPTION:
General: Perennial from a stout taproot and branched stem-base; tufted, often matted; stems decumbent to ascending, several, branched, finely and densely hairy, 10-70 cm tall/long.

Leaves: Basal leaves mostly matted at the base of the stems, oblanceolate to lanceolate, 2-5 cm long, hairy, stalked, 2-10 mm wide; stem leaves opposite, 1-8 pairs, oblanceolate, 2-8 cm long, becoming smaller and unstalked above, hairy; stipules lacking.

Flowers: Inflorescence of solitary or more usually 3 flowers; petals 5, creamy white to yellowish or greenish, often tinged with pink or purple, bilobed, 12-16 mm long, stalklike at the bases, the stalks 8-12 mm long; sepals 5, creamy white or pink, greenish or purplish, united, forming a tube 10-15 mm long, inflated and papery in fruit, 10-nerved.

Fruits: Capsules 1-celled, 8-12 mm long; seeds 1.1-1.5 mm long, reddish-brown, roughened.

Silene dichotoma

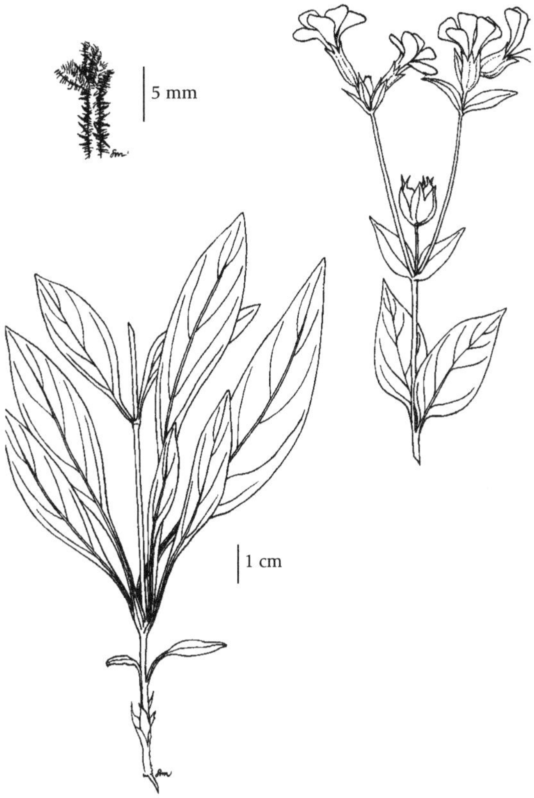

Silene dioica

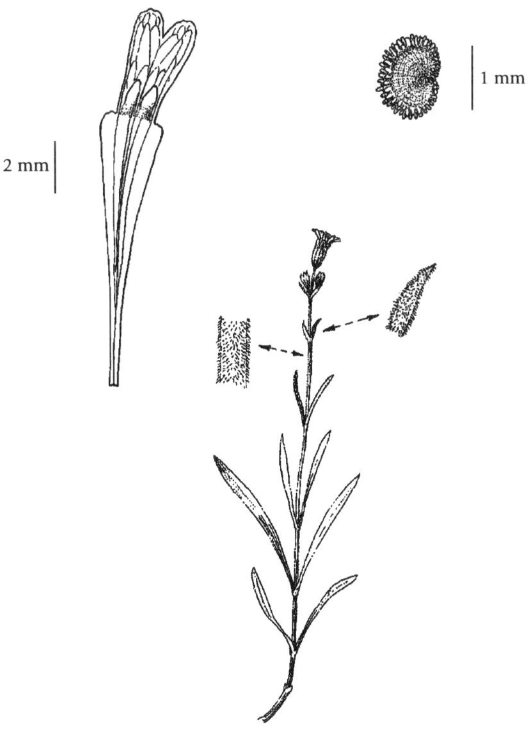

Silene douglasii* var. *douglasii

Silene drummondii* Hook. var. *drummondii (*Lychnis drummondii* [Hook.] S. Wats., *Melandrium drummondii* [Hook.] Hult.)

Drummond's campion

HABITAT/RANGE:

Dry shrublands, meadows and forest openings in the steppe to alpine zones; rare at scattered locations throughout BC east of the Coast-Cascade Mountains; E to S MB and S to NE, CO, AZ and NV.

DESCRIPTION:

General: Perennial herb from a taproot and simple or branched crown; stems erect, solitary or several, simple or branched, strongly hairy, the hairs downswept, becoming glandular above, often with purple cross walls above, 20-50 cm tall.

Leaves: Basal leaves in a rosette, the blades lanceolate or elliptic to oblanceolate, hairy, 2-10 cm long, 3-15 mm wide, narrowed to slender stalks; stem leaves opposite, usually linear, hairy, remote, 2-4 pairs, much reduced; stipules lacking.

Flowers: Inflorescence of perfect, solitary to several flowers inconspicuously bracted in a loose narrow cluster; petals 5, white or pinkish, often included in the calyces, or the blades exserted, stalklike at the bases, the stalks narrow and gradually flared above and usually broader than the blades, the blades 1-3 mm long, oblong, notched to shallowly 2- or sometimes 4-lobed, with 2 appendages; sepals 5, united, forming a tube 10-15 mm long, glandular, enlarged (but not inflated) in fruit, 10-nerved.

Fruits: Capsules narrowly oblong, 4- or (usually) 5-valved; seeds dark brown, about 0.4-0.7 mm long, prominently wrinkled and pimply concentrically.

***Silene gallica* L.**

small-flowered catchfly

HABITAT/RANGE:

Mesic to dry roadsides and waste places; common in SW BC, known from SE Vancouver Island, the Gulf Islands and the adjacent mainland; introduced from Europe.

DESCRIPTION:

General: Annual from a slender taproot; stems ascending, several, branched, conspicuously white-hairy and sticky, 10-50 cm tall.

Leaves: Basal leaves few, oblanceolate to spoon-shaped, hairy, broadly stalked; stem leaves opposite, lanceolate, hairy, unstalked, 1.5-4 cm long, 2-15 mm wide; stipules lacking.

Flowers: Inflorescence falsely racemose, the flowers solitary at each node; petals 5, whitish to pink or purple, stalklike at the bases, the stalks narrow, 4-6 mm long, the blades elliptic to egg-shaped, entire or shallowly notched at the apex, the basal appendages about 1 mm long and linear, slightly twisted, and the corolla therefore somewhat pinwheel-like in appearance; sepals 5, united, forming a tube 6-9 mm long at flowering time, inflating in fruit, 10-nerved, hairy and glandular.

Fruits: Capsules egg-shaped, to 10 mm long, 6-valved; seeds blackish, kidney-shaped, about 1 mm long, hairy, finely corrugate-wrinkled.

Silene involucrata* (Cham. & Schlecht.) Bocquet ssp. *involucrata (*Lychnis furcata* [Raf.] Fern., *Silene furcata* Raf.)

arctic campion

HABITAT/RANGE:

Mesic meadows and rocky cliffs in the alpine zone; rare in NE BC, known only from Nevis Creek; N to AK, YT and NT and E to SW AB and PQ.

DESCRIPTION:

General: Biennial or more usually perennial from a taproot and branching stem-base; stems more or less erect, several, simple or sometimes branched, white-hairy below, glandular-hairy above, 3-35 cm tall.

Leaves: Basal leaves oblanceolate to narrowly elliptic, 1.5-5.5 cm long, 2-8 mm wide, margins fringed; stem leaves opposite, 2-3 pairs, reduced upwards; stipules lacking.

Flowers: Inflorescence of usually solitary or sometimes 2 or 3, erect to spreading flowers; petals 5, white to pink or purple, well exserted, 2-cleft; sepals 5, united, forming a bell- to urn-shaped tube 10-17 mm long, hairy, purple-nerved.

Fruits: Capsules 12-18 mm long, 5- or 6-valved; seeds 0.9-1.5 mm long, with winged margins.

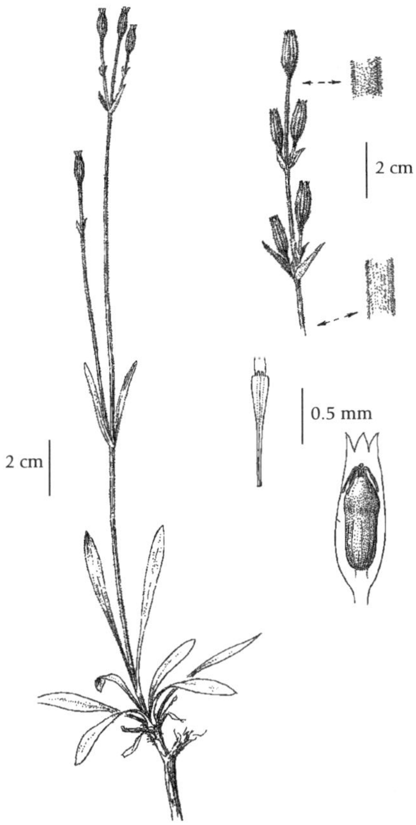

Silene drummondii* var. *drummondii

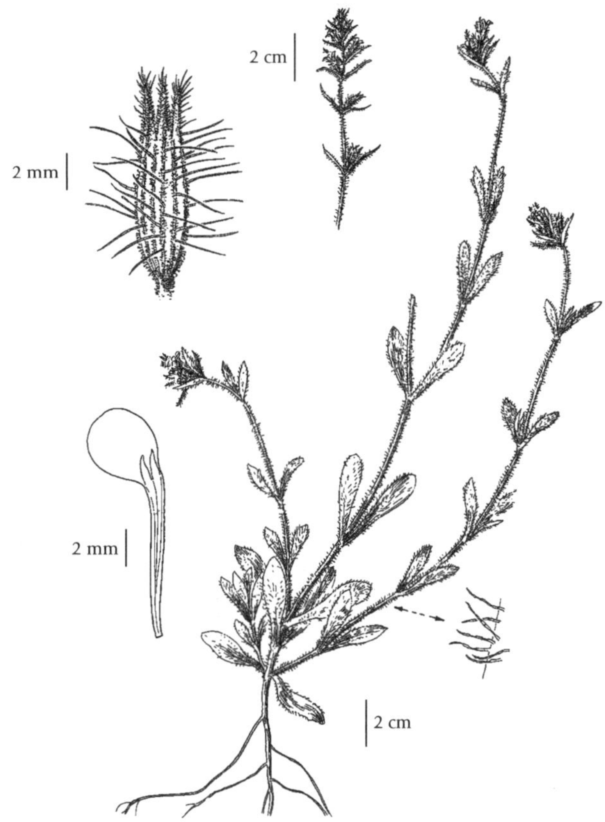

Silene gallica

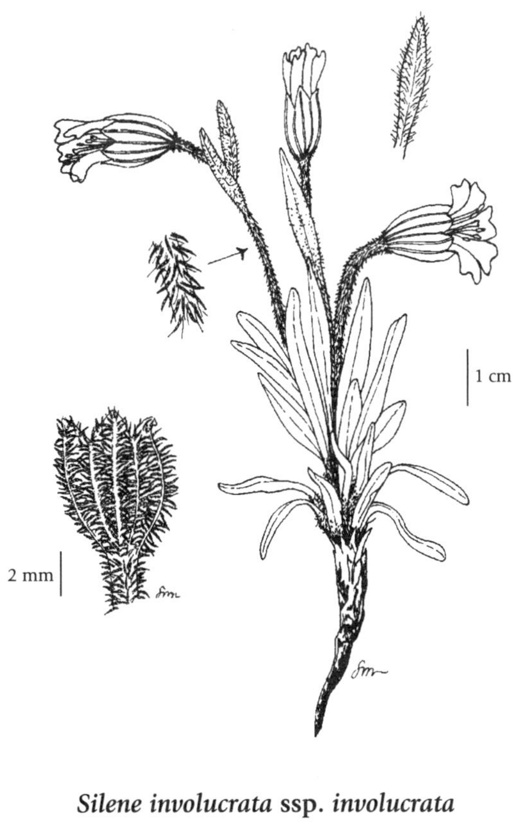

Silene involucrata* ssp. *involucrata

***Silene latifolia* Poir. ssp. *alba* (P. Mill.) Greuter & Burdet** (*Lychnis alba* P. Mill., *L. loveae* Boivin, *Melandrium album* [P. Mill.] Garcke, *Silene alba* [P. Mill.] Krause)

white cockle; white campion

HABITAT/RANGE:

Mesic to dry roadsides, fields and waste places in the lowland, steppe and montane zones; infrequent in SW BC; introduced from Europe.

DESCRIPTION:

General: Biennial or short-lived perennial herb from a branched stem-base; stems erect, several, branched, stiff-hairy below, becoming glandular in the inflorescence, 0.3-1.0 m tall.

Leaves: Basal leaves lanceolate to oblanceolate, 2-10 cm long, 2 cm wide, slenderly to broadly stalked; stem leaves opposite, 5-10 pairs, lanceolate, the lower ones often larger than the basal, becoming slightly reduced and unstalked above; stipules lacking.

Flowers: Inflorescence of unisexual (the plants either male or female) flowers, rather showy, evening blooming and scented, several to many in open, spreading, leafy-bracted clusters; petals 5, white, stalklike at the bases, the blades 7-10 mm long, narrowly egg-shaped, 2-cleft to about midlength, with triangular appendages at the base; sepals 5, united, forming a downy, sticky tube 15-20 mm long.

Fruits: Capsules egg-shaped-cylindric, to 20 mm long, with 5, 2-toothed valves; seeds about 1.5 mm long, dark bluish-brown, prominently warty-pimply in concentric rows.

Notes: See *S. dioicia* for discussion of hybridization.

***Silene menziesii* Hook.**

Menzies' campion; Menzies' silene; Menzies' catchfly

HABITAT/RANGE:

Mesic meadows and open forests in the lowland, steppe and montane zones; common in S BC; N to AK, YT and NT, E to MB and S to NM and CA.

DESCRIPTION:

General: Perennial herb from a long, slender, branched rhizome; stems decumbent to ascending, several, branched, hairy and usually glandular above, 5-70 cm tall/long.

Leaves: Basal leaves usually deciduous by flowering time; stem leaves opposite, lanceolate to narrowly egg-shaped, usually long-pointed, 2-10 cm long, 3-30 mm wide, glandular-hairy, unstalked or nearly so; stipules lacking.

Flowers: Inflorescence of several flowers in a leafy, open cluster or, less commonly, solitary; petals 5, white, stalklike at the bases, the blades 2-3 mm long, lobed halfway or more, the stalks about 5 mm long; sepals 5, purplish or greenish, united, forming a glandular-hairy tube 5-8 mm long, 10-nerved; styles 3 or sometimes 4.

Fruits: Capsules oblong, 5-8 mm long, 1-celled; seeds 0.7-0.9 mm long, black, shiny, smooth or lightly net-veined.

Notes: Two varieties occur in BC.

1. Stems densely hairy, without glandular hairs .... var. *menziesii*
1. Stems densely long soft-hairy with spreading glandular hairs .... var. *viscosa* (Greene) C.L. Hitchc. & Maguire

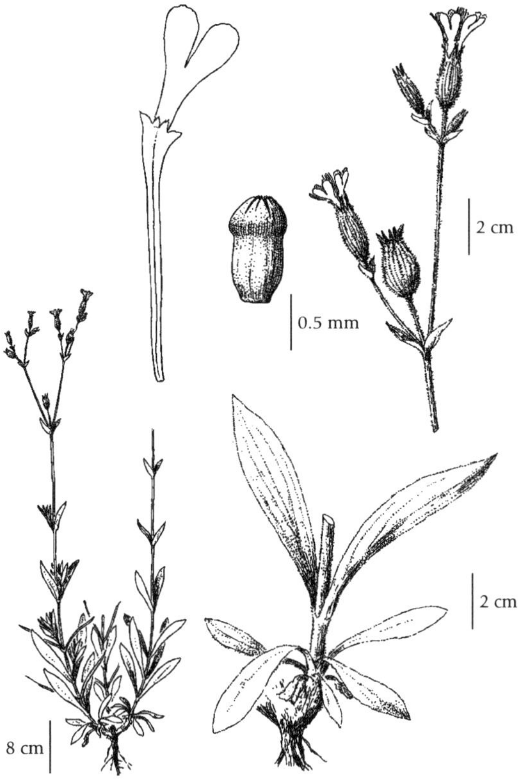

Silene latifolia ssp. *alba*

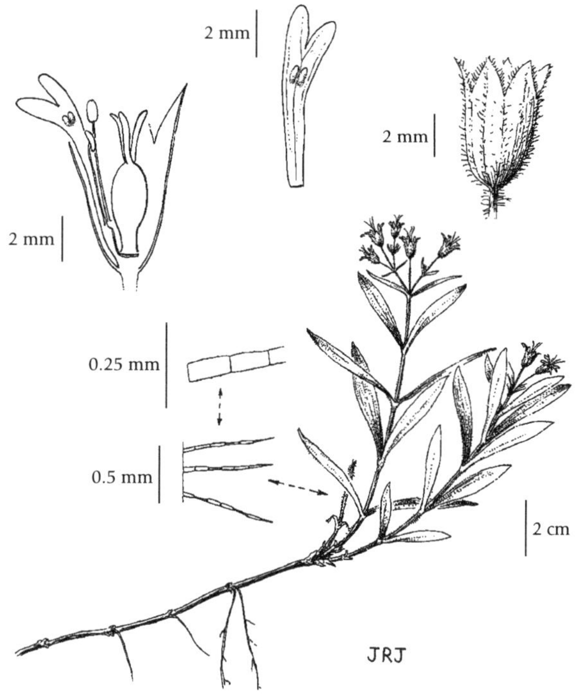

Silene menziesii var. *menziesii*

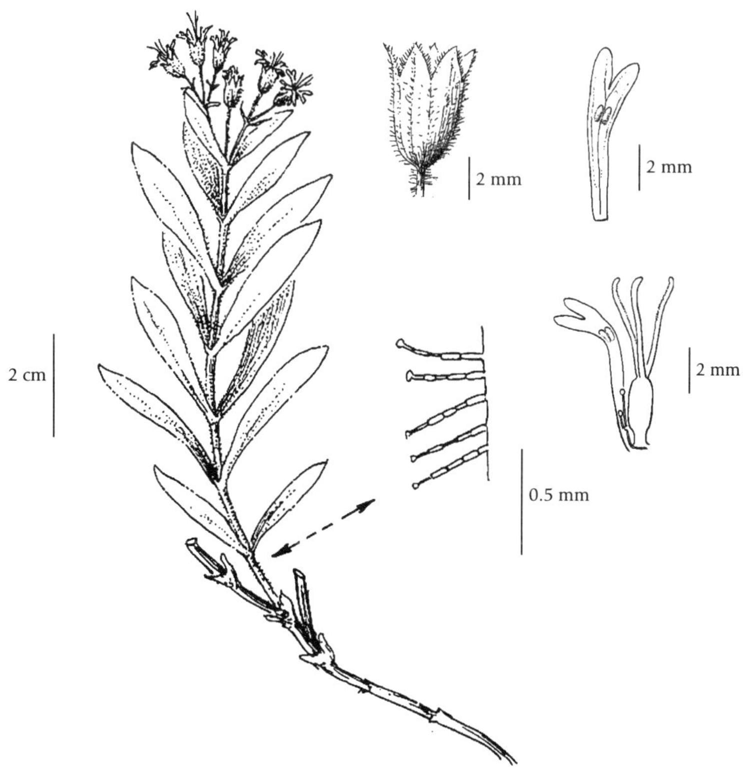

Silene menziesii var. *viscosa*

***Silene noctiflora* L.** (*Melandrium noctiflorum* [L.] Fries)
night-flowering catchfly; sticky cockle
HABITAT/RANGE:
Mesic to dry waste places and cultivated fields in the lowland, steppe and montane zones; common in S BC, rare on the Queen Charlotte Islands; introduced from Europe.
DESCRIPTION:
General: Annual herb from a taproot; stems erect, several, branched, hairy, becoming sticky-hairy above, 20-90 cm tall.

Leaves: Basal leaves spoon-shaped, 5-12 cm long, short-stalked; stem leaves opposite, lanceolate to elliptic, 2-10 cm long, up to 4 cm wide, unstalked, hairy above and below; stipules lacking.

Flowers: Inflorescence of 2-many flowers in a cluster, opening at night, fragrant; petals 5, yellowish beneath and rosy above, glabrous, stalklike at the bases, the blades 7-10 mm long, the stalks 12-25 mm long, 2-cleft less than 1/2 their length; sepals 5, greenish to pinkish, united, forming a woolly, glandular-hairy tube 12-25 mm long, 10-nerved.

Fruits: Capsules egg-shaped, 1.5-2.5 cm long, 6-valved; seeds 0.8-1.2 mm long, brown, roughened.

***Silene parryi* (S. Wats.) C.L. Hitchc. & Maguire**
Parry's campion; Parry's catchfly
HABITAT/RANGE:
Mesic to dry meadows and open forests from the montane to alpine zones; common in SC and SE BC; E to AB and S to WY.
DESCRIPTION:
General: Perennial herb from a simple to branched stem-base; stems erect, several, simple, hairy, more or less glandular above, 10-60 cm tall.

Leaves: Leaves mostly basal, linear-lanceolate to spoon-shaped, 3-8 cm long, 4-10 mm wide, long-stalked; stem leaves opposite, usually 2-3 pairs, not greatly reduced though narrower than the basal leaves; stipules lacking.

Flowers: Inflorescence of usually 3-7 flowers in a rather short, narrow cluster, the lateral clusters usually reduced to a single flower; petals 5, white or greenish- or purplish-tinged, stalklike at the bases, the stalks about 11 mm long, the blades 3-5 mm long, 2-cleft nearly to midlength, with prominent lateral teeth on each blade lobe, sometimes the blade nearly equally 4-lobed; sepals 5, united, forming a glandular-hairy tube 10-20 mm long, with 10 prominent greenish or purplish nerves.

Fruits: Capsules oblong, 1-celled; seeds brown, 1-1.5 mm long, somewhat wrinkled.

***Silene repens* Patrin *ex* Pers.**
pink campion
HABITAT/RANGE:
Dry meadows and open forests from the montane to alpine zones; rare in NE BC, known only from Nevis Creek; circumpolar, N to AK, YT and NT and S to MT and WY; Eurasia.
DESCRIPTION:
General: Perennial from a branched stem-base and a short rhizome; tufted, often matted; stems erect, several, simple or branched, densely crinkled-hairy, 7-35 cm tall.

Leaves: Basal leaves linear to narrowly lanceolate, 2-6 cm long, 2-6 mm wide, mostly deciduous by flowering time; stem leaves opposite, 2-6 pairs, about the same size as the basal leaves, slightly hairy; stipules lacking.

Flowers: Inflorescence of usually 2-15 flowers in a contracted cluster; petals 5, white to pink or occasionally purple, stalklike at the bases, the blades 3-7 mm long, 2-lobed, the stalks 5-8 mm long; sepals white to usually pink or purple, united, forming a tube 10-15 mm long, 10-nerved, hairy and more or less glandular.

Fruits: Capsules 10-14 mm long, 3- (4)-valved; seeds brown, 0.8-1.1 mm long, roughened.

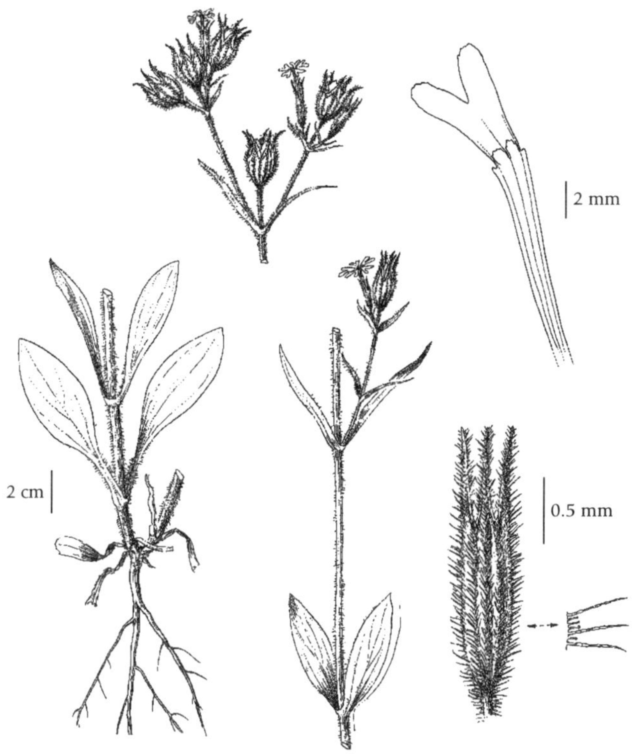

Silene noctiflora

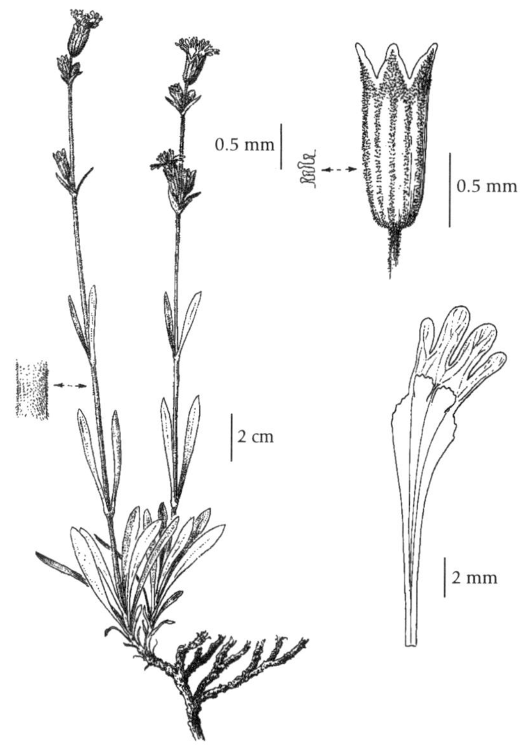

Silene parryi

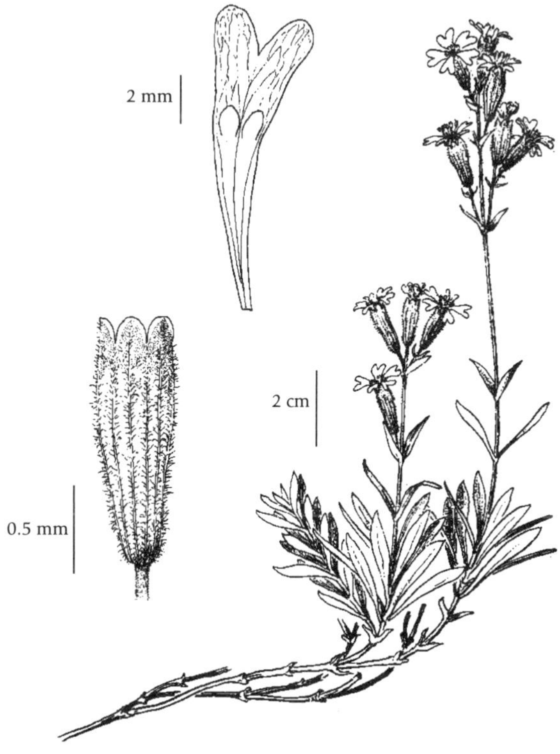

Silene repens

Silene scouleri (Eastw.) C.L. Hitchc. & Maguire (*S. grandis* Eastw., *S. pacifica* Eastw., *S. scouleri* var. *pacifica* [Eastw.] C.L. Hitchc. = ssp. *grandis*)

Scouler's campion; Scouler's catchfly

HABITAT/RANGE:

Dry coastal bluffs and gravelly or rocky slopes in the lowland and steppe zones; rare in SW BC, known only from Vancouver Island and the Gulf Islands (ssp. *grandis)* and infrequent in SC BC (ssp. *scouleri);* S to MT, NM, AZ and CA.

DESCRIPTION:

General: Perennial herb from a simple or branched stem-base; stems erect, 1 to several, simple, densely hairy throughout and glandular above, 15-80 cm tall.

Leaves: Basal leaves slender-stalked, the blades oblanceolate, hairy, 6-20 cm long and up to 3 cm wide; stem leaves opposite, hairy, 1-12 pairs, reduced gradually and becoming unstalked above; stipules lacking.

Flowers: Inflorescence of numerous flowers, the inflorescence elongate, narrow, the lateral clusters mostly congested; petals 5, greenish-white to purplish, stalklike at the bases, the stalks of the petals 7-16 mm long, the blades 4-8 mm long, from 2-cleft to nearly equally 4-lobed, with prominent lateral teeth on each blade lobe; sepals 5, united, forming a tube 10-18 mm long, not inflating in fruit, 10-nerved.

Fruits: Capsules elliptic to egg-shaped, hairy; seeds 0.9-1.5 mm long, greyish-brown, conspicuously pimply.

Notes: Two subspecies occur in BC.

1. Leaves less than l5 mm wide, those of the stem usually 4-6 pairs; plants of SC BC .... ssp. *scouleri*
1. Leaves, or at least some of them, more than 15 mm wide, those of the stem usually 3-11 pairs; plants of SW BC .... ssp. *grandis* (Eastw.) C.L. Hitchc. & Maquire

Silene spaldingii S. Wats.

Spalding's campion

HABITAT/RANGE:

Dry grasslands in the steppe zone; rare in SE BC, known only from Tobacco Plains; S to MT, ID and OR.

DESCRIPTION:

General: Perennial herb from a simple or branched stem-base; stems erect, several, branched above, long woolly-hairy and more or less glandular-hairy, 20-60 cm tall.

Leaves: Basal leaves few, soon deciduous; stem leaves opposite, 4-7 pairs, oblanceolate below to lanceolate above, long woolly-hairy and more or less glandular-hairy, 6-7 cm long, 0.5-1.5 cm wide, unstalked and slightly fused; stipules lacking.

Flowers: Inflorescence of several to many flowers in a leafy and usually compact cluster; petals 5, white, stalklike at the bases, the stalks about 15 mm long, the blades very short, egg-shaped, about 2 mm long, entire or shallowly notched at the tip, the appendages 4 (5-6); sepals 5, united, forming a tube about 15 mm long at flowering time, becoming more nearly club- or bell-shaped in fruit, 10-nerved.

Fruits: Capsules oblong, 1-celled; seeds light brown, 2.0 mm long, corrugate-wrinkled and inflated.

Silene taimyrensis (Tolm.) Bocquet (*Lychnis triflora* R. Br. *ex* Sommerf., *L. dawsonii* [B.L. Robins.] J.P. Anders., *Melandrium ostenfeldii* Pors., *M. taimyrensis* Tolm., *Silene sorensensis* [Boivin] Boquet)

Taimyr campion

HABITAT/RANGE:

Dry open calcareous slopes from the montane to alpine zones; rare in NW BC, known only from the Dease Lake area; amphiberingian, N to AK, YT and NT and E to Greenland; E. Asia.

DESCRIPTION:

General: Perennial from a taproot and branching stem-base; stems erect or ascending, many branched, white-hairy and glandular, 5-50 cm tall.

Leaves: Basal leaves narrowly oblanceolate, white-hairy and glandular, 1.5-6 cm long, 2-8 mm wide, 1-nerved; stem leaves opposite, 2-4 pairs, reduced upwards; stipules lacking.

Flowers: Inflorescence of usually 3-6 in a cluster, occasionally solitary; petals 5, white or pinkish, 2-cleft, stalklike at the bases; sepals 5, purplish-nerved, united, forming a tube 6-11 mm long; styles 5.

Fruits: Capsules 8-11 mm long; seeds dark brown, 0.6-1.0 mm long, sometimes slightly winged, granular.

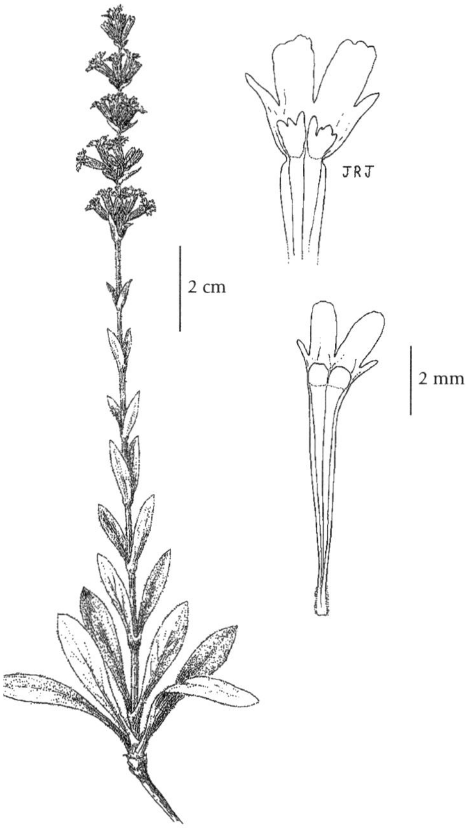

Silene scouleri ssp. *grandis*

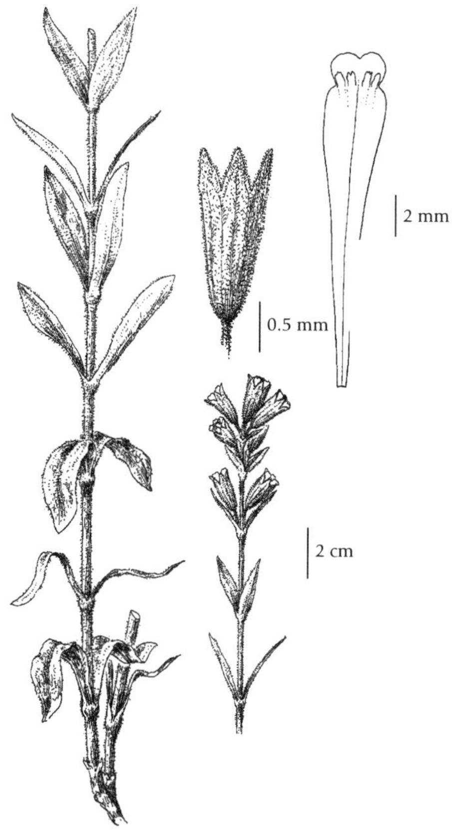

Silene spaldingii

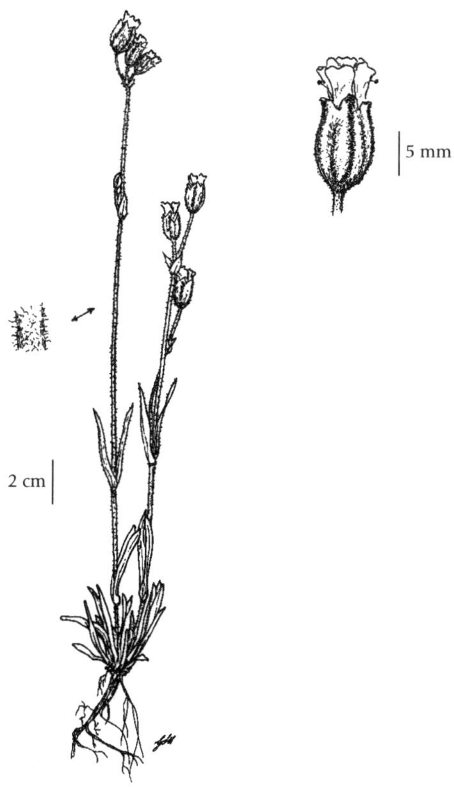

Silene taimyrensis

Silene uralensis **(Rupr.) Bocquet ssp. *attenuata*** **(Farr) McNeill** (*Lychnis apetala* L. ssp. *attenuata* [Farr] Maguire, *L. apetala* var. *attenuata* [Farr] C.L. Hitchc)

nodding campion

HABITAT/RANGE:

Mesic to dry river banks, talus slopes and meadows from the montane to alpine zones; infrequent, scattered throughout N and E BC; circumpolar, N to AK, YT and NT and E to AB; N Eurasia.

DESCRIPTION:

General: Perennial herb from a branched stem-base; stems ascending to erect, usually several, usually simple, finely white-hairy below, becoming glandular above, 20-60 cm tall.

Leaves: Basal leaves in tufts, narrowly oblanceolate, glabrous to nearly glabrous, 2-4 cm long, 1.5-4 mm wide, narrowed to slender stalks; stem leaves opposite, usually 2-3 (4) pairs, somewhat narrower than the basal leaves, hairy with multicellular hairs, unstalked; stipules lacking.

Flowers: Inflorescence of usually solitary, erect flowers, nodding in the bud, usually perfect; petals 5, purplish or rarely pinkish, stalked, the stalks about equalling or exceeded by the calyes, flared above and usually flanged at the bases, the blades generally narrower than the stalks, 3-4 mm long, 2-cleft; sepals 5, united, forming a tube about 1 cm long at flowering time, considerably inflated and up to 17 mm long in fruit, prominently 10-nerved, glandular-hairy.

Fruits: Capsules 1.2-2.2 cm long, 5-valved; seeds 1.5-2.5 mm long, prominently wrinkled-pimply in concentric rows, usually more or less flattened and wing-margined.

Silene vulgaris **(Moench) Garcke** (*S. cucubalus* Wibel, *S. cucubalus* var. *latifolia* [P. Mill.] G. Beck, *S. latifolia* [P. Mill.] Brill. & Rendle)

bladder campion

HABITAT/RANGE:

Dry roadsides, fields and waste places in the lowland, steppe and lower montane zones; common in S BC; introduced from Eurasia.

DESCRIPTION:

General: Perennial from a strong taproot; stems erect to ascending, several, branched, usually glabrous and glaucous, 20-80 cm tall, the lower branches sometimes decumbent and rhizome-like.

Leaves: Basal leaves lacking; stem leaves opposite, narrowly egg-shaped to oblanceolate, 3-8 cm long, 1-3 cm wide, glabrous, lower ones short-stalked, upper ones unstalked; stipules lacking.

Flowers: Inflorescence of several to many flowes in a usually open and loose cluster; petals 5, white, clawed, the claws expanded above, about 2 cm long, the blades 4-6 mm long, deeply 2-cleft, otherwise entire; sepals 5, pale green to purplish, united, forming a tube 15-20 mm long and becoming inflated, prominently net-veined, glabrous except for the fringed lobes, 15- or 20-nerved.

Fruits: Capsules egg-shaped to ellipsoid, 1.8-2 cm long, 6-valved; seeds 1-1.5 mm long, white becoming brown, warty-pimply.

SPERGULA

Spergula arvensis **L.**

corn-spurry; stickwort

HABITAT/RANGE:

Mesic to dry fields and waste places in the lowland and lower montane zones; common in SW BC, rare elsewhere in S BC; introduced from Eurasia.

DESCRIPTION:

General: Annual herb from a taproot; stems prostrate to erect, several, branched, more or less glandular-hairy, yellowish-green, 10-60 cm tall/long.

Leaves: Basal leaves few, soon deciduous; stem leaves in whorls of 6, narrowly linear, channelled beneath, 1.5-5 cm long, less than 1 mm wide, 1-nerved; stipules small, deciduous.

Flowers: Inflorescence of numerous and usually terminal flowers on branched clusters; petals 5, white, egg-shaped, 3-6 mm long; sepals 5, greenish, egg-shaped, 2-5 mm long, glandular-hairy.

Fruits: Capsules egg-shaped, 2.5-6 mm long, 5-valved; seeds black, plump, 1.2-1.5 mm long, minutely roughened.

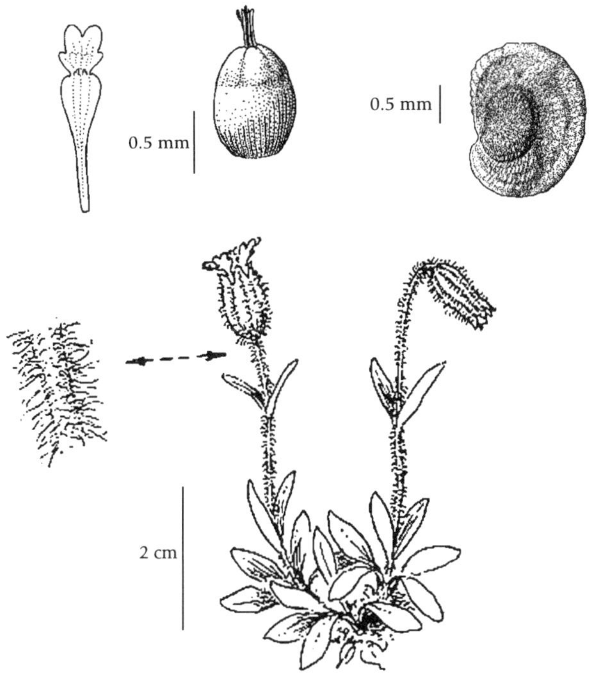

Silene uralensis ssp. *attenuata*

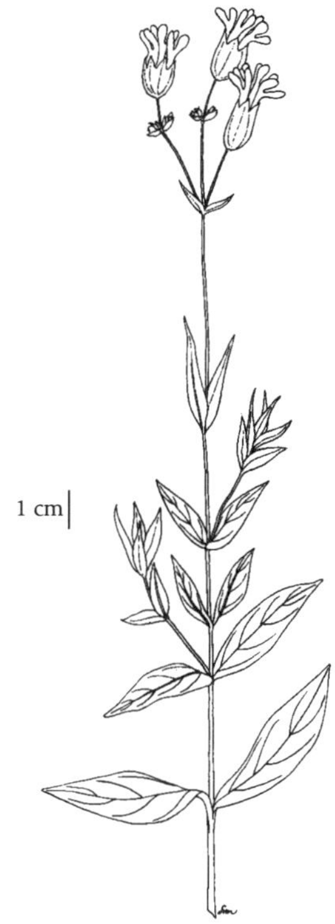

Silene vulgaris

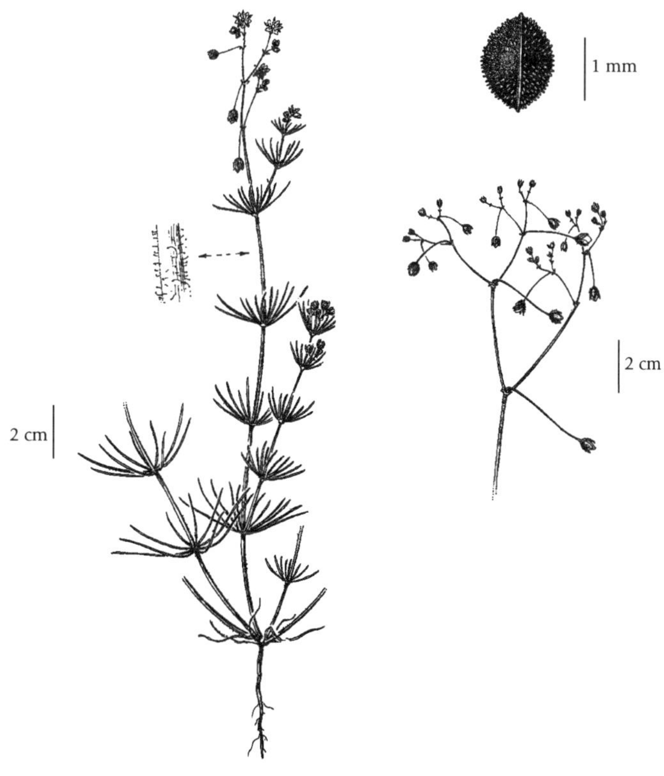

Spergula arvensis

SPERGULARIA

1. Stamens 2-5; sepals glabrous to moderately glandular-hairy.
   2. Leaves blunt; seeds usually wing-margined, 0.8-1.1 mm long .... *S. canadensis*
   2. Leaves abruptly slender-pointed; seeds wingless, 0.6-0.9 mm long .... *S. salina*
1. Stamens 6-10; sepals densely glandular-hairy.
   3. Seeds dark reddish-brown, nearly smooth, usually surrounded with a thin, white or brownish wing; sepals 5-9 mm long; plants native perennials .... *S. macrotheca*
   3. Seeds light or dark brown or greyish-brown, finely reticulate and pimply, wingless; sepals 3.5-5 mm long, or less; plants introduced annuals or short-lived perennials .... *S. rubra*

Spergularia canadensis **(Pers.) G. Don**
Canadian sand-spurry

HABITAT/RANGE:

Wet to moist coastal mud flats and sandy coastal beaches in the lowland zone; common in coastal BC; var. *canadensis* ranges N to AK and along the Atlantic Coast, var. *occidentalis* ranges S to N CA.

DESCRIPTION:

General: Annual herb from a taproot; stems erect or ascending to decumbent, several, freely branched, essentially glabrous, 4-30 cm tall/long.

Leaves: Basal leaves lacking; stem leaves opposite, linear, 1-4.5 cm long, 1-2 mm wide, more or less glabrous, fleshy; stipules squared-off, wider than long, membranous, sheathing at the base.

Flowers: Inflorescence of few to many flowers in a leafy cluster; petals 5, white or pink, 1-3 mm long; sepals 5, greenish, glabrous to moderately glandular-hairy, 2-3.5 mm long.

Fruits: Capsules 3-5 mm long, 3-valved; seeds 0.8-1.1 mm long, light brown, usually wing-margined.

Notes: Two sometimes poorly defined varieties occur in BC.

1. Plants prostrate or decumbent; plants occurring N of Vancouver Island .... *var. canadensis*
1. Plants erect or strongly ascending; plants of Vancouver Island and southward .... var. *occidentalis* Rossbach

Spergularia macrotheca **(Hornem.) Heynh. var.** ***macrotheca***
beach sand-spurry

HABITAT/RANGE:

Wet to moist salt marshes and sandy coastal beaches in the lowland zone; rare in SW BC, known only on SE Vancouver Island and adjacent Gulf Islands; S to CA and MX.

DESCRIPTION:

General: Perennial herb with a thick, rather woody root and branched stem-base; stems decumbent to ascending, several, few branched, usually glandular-hairy, 3-45 cm tall/long, 0.8-3.0 mm in diameter.

Leaves: Basal leaves lacking; stem leaves opposite, 2-several per node, mostly 7-50 mm long, 0.6-3 mm wide, fleshy, glabrous to densely glandular-hairy, bristle-tipped; stipules 4-11 mm long, triangular and tapering to a long-point, membranous, sheathing at the base.

Flowers: Inflorescence of solitary to few flowers in a terminal or axillary cluster, glandular-hairy; petals 5, dark to very pale pink, 3.5-8 mm long; sepals 5, 5-9 mm long, densely glandular-hairy, broadly lanceolate, usually more or less abruptly sharp-pointed.

Fruits: Capsules 5.5-10 mm long, 3-valved; seeds dark reddish-brown, nearly smooth, mostly 0.6-1 mm long, usually surrounded by a white or brownish wing about 0.1 mm wide.

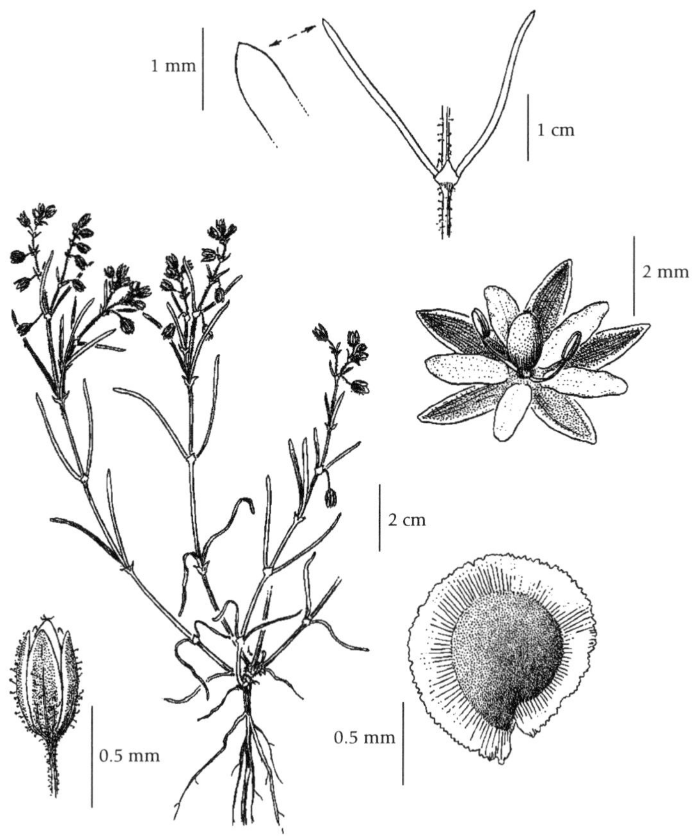

Spergularia canadensis var. *occidentalis*

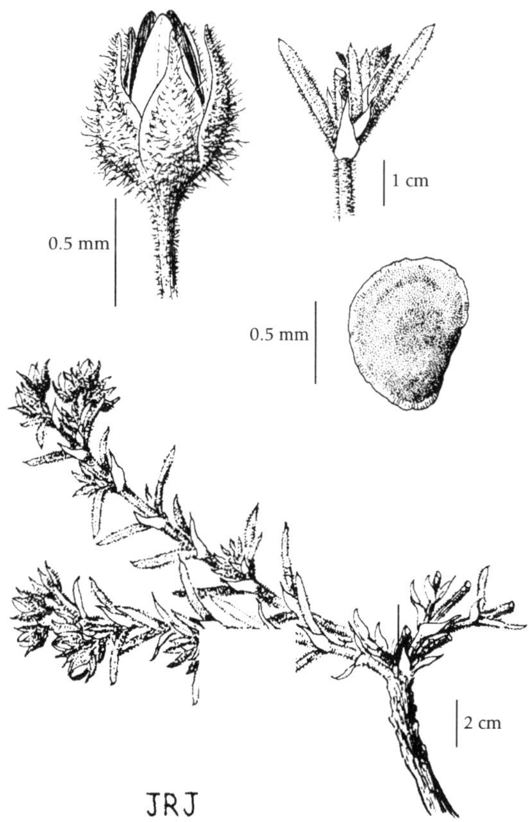

Spergularia macrotheca var. *macrotheca*

Spergularia rubra **(L.) J.& K. Presl**
red sand-spurry
Habitat/Range:
Dry to mesic disturbed areas and waste places in the lowland, steppe and lower montane zones; common in S BC, rare northwards; introduced from Europe.
Description:
General: Annual or short-lived perennial herb from a slender taproot; stems prostrate, several, branched, glandular-hairy, 5-30 cm long.

Leaves: Basal leaves lacking; stem leaves opposite, narrowly linear, tapering to an awned tip, 5-20 mm long, less than 1 mm wide, usually in bunches, 1-nerved; stipules silvery, lanceolate, long-pointed, 3.5-5 mm long.

Flowers: Inflorescence of a few flowers in a leafy cluster; petals 5, pink to pale red, egg-shaped, 2-3.5 mm long; sepals 5, greenish, narrowly egg-shaped, 3.5-5 mm long, glandular-hairy, with a wide, papery margin.

Fruits: Capsules 3-5 mm long, 3-valved; seeds light or dark brown or greyish-brown, finely reticulate and pimply, wingless, 0.4-0.6 mm long.

Spergularia salina **J.& K. Presl var. *salina*** (*S. marina* [L.] Griseb.)
salt marsh sand-spurry
Habitat/Range:
Wet to moist coastal mudflats or inland alkaline or saline ponds in the lowland and montane zones; common in SW BC, rare in SE and N BC; introduced from Eurasia.
Description:
General: Annual herb from a slender taproot; stems ascending or more often mostly prostrate to decumbent, several, much-branched, more or less glandular-hairy, 10-30 cm tall/long, 0.6-2.0 mm in diameter.

Leaves: Leaves opposite, linear, 1-3 cm long, about 1 mm wide, circular in cross section, abruptly slender-pointed; stipules egg-shaped, 2-4 mm long.

Flowers: Inflorescence several flowers in an open cluster; petals 5, pink with a white base, egg-shaped, blunt, 2.5-3 mm long; sepals 5, often pink-tinged, lanceolate, 3-4 mm long, glabrous to moderately glandular-hairy.

Fruits: Capsules 3-6.4 mm long, 3-valved; seeds brown, nearly smooth but often somewhat pimply, mostly 0.6-0.9 mm long, wingless.

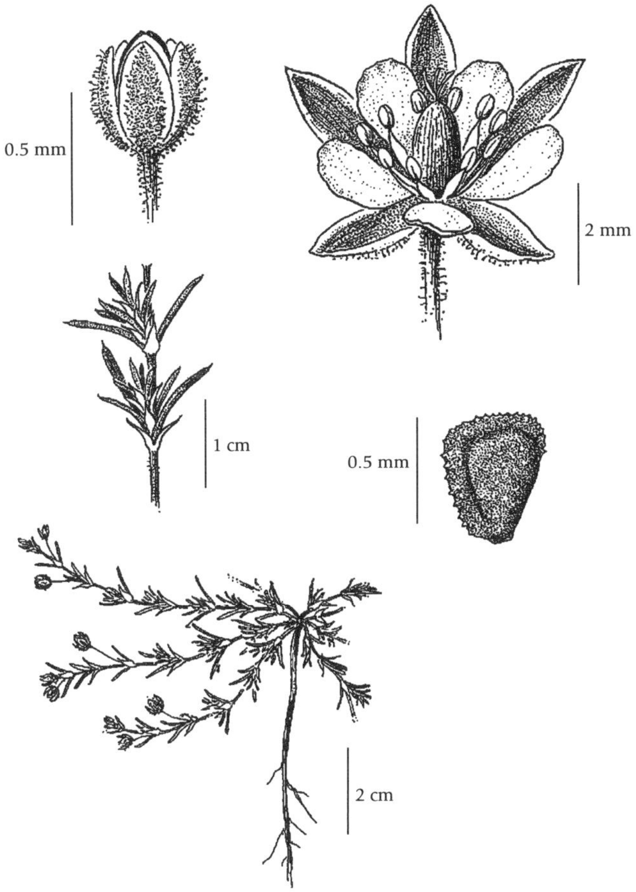

Spergularia rubra

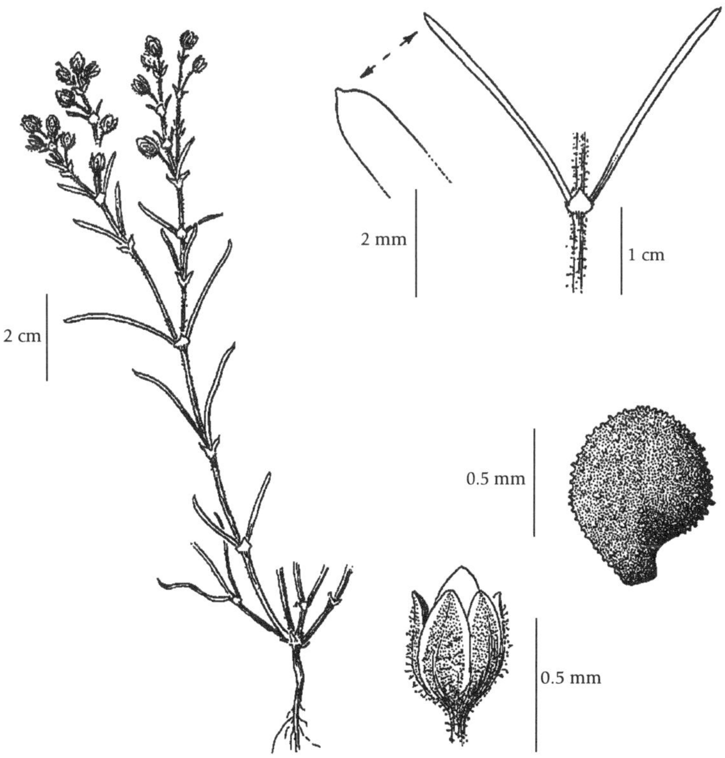

Spergularia salina var. *salina*

STELLARIA [26]

1. Leaves linear to lanceolate, usually unstalked.
   2. Petals rudimentary or lacking, less than half the length of the 2-3 mm sepals; stems and flower stalks slender .... *S. umbellata*
   2. Petals usually more than half the length of the more than 3 mm long sepals; stems and flower stalks stouter.
      3. Flowers axillary, not in terminal clusters; petals at least as long as the sepals.
         4. Leaves linear or linear-lanceolate, abruptly sharp-pointed, stiff; sepals strongly 3-nerved .... *S. longipes* (var. *longipes*)
         4. Leaves linear-lanceolate to oblong, abruptly sharp-pointed or obtuse, not stiff; sepals lightly 1- or 3-nerved.
            5. Sepals 3.5-5 mm long at flowering time, inconspicuously 3-nerved, longer than the capsules; leaves fleshy; plants of coastal BC .... *S. humifusa*
            5. Sepals 2-3 mm long at flowering time, inconspicuously 1-nerved, shorter than the capsules; leaves thin; plants not coastal .... *S. crassifolia*
      3. Flowers in terminal clusters, petals usually shorter than sepals.
         6. Flower clusters leafy-bracted; petals shorter than the sepals, sometimes lacking .... *S. calycantha*
         6. Flower clusters either membranous-bracted or the petals longer than the sepals.
            7. Leaf margins very finely roughly-pimpled (at 30X) .... *S. longifolia*
            7. Leaf margins smooth.
               8. Sepals usually prominently fringed; capsules green, about equalling the calyces .... *S. graminea*
               8. Sepals usually glabrous; capsules usually purplish, mostly longer than the calyces .... *S. longipes*
1. Leaves, or at least the basal ones, broadly lanceolate, elliptic, oblong or egg-shaped, stalked or unstalked.
   9. Leaves, or at least the basal ones, long-stalked.
      10. Stems decumbent, rooting nodally, hairy in longitudinal lines .... *S. media*
      10. Stems erect, more or less threadlike, not rooting nodally, glabrous or uniformly curly-hairy .... *S. nitens*
   9. Leaves unstalked or short-stalked.
      11. Flowers single and terminal .... *S. longipes* (var. *monantha*)
      11. Flowers either single and axillary or in terminal clusters, sometimes few flowered with leaf-like bracts.
         12. Flowers single and axillary.
            13. Sepals long-pointed or abruptly sharp-pointed, rarely obtuse, nerves prominent or inconspicuous; capsules egg-shaped; seeds distinctly wrinkled or very lightly net-veined.

---

[26] Key adapted from Hitchcock & Cronquist (1964) and Morton & Rabeler (1989).

14. Leaf margins crisped and translucent; seeds distinctly wrinked-warty .... *S. crispa*

14. Leaf margins entire, not crisped or translucent; seeds very lightly net-veined, appearing nearly smooth .... *S. humifusa*

13. Sepals obtuse, nerves mostly obscure; capsules globose or nearly so; seeds finely net-furrowed .... *S. obtusa*

12. Flowers in terminal clusters, sometimes few flowered with leaf-like bracts.

15. Flower clusters papery-bracted; seeds strongly pimply-wrinkled .... *S. alsine*

15. Flower clusters usually leafy-bracted; seeds smooth or very slightly to evidently wrinkled.

16. Leaves narrowly egg-shaped to elliptic, usually 2.5 cm long or less, net-veined; sepals in open flower usually shorter than 2.5 mm, nerves obscure, not prominently ridged; stems glabrous or hairy above, never pimply .... *S. calycantha*

16. Leaves usually lanceolate to linear-lanceolate, occasionally narrowly egg-shaped to elliptic, up to 6 cm long, veins not visible, except for the midrib; sepals in open flower to 4 mm long with 1-3 prominent nerves; stems glabrous or pimply above, rarely hairy .... *S. borealis*

Stellaria alsine **Grimm**

bog starwort

HABITAT/RANGE:

Wet to mesic ditches, lawns and disturbed sites in the lowland and steppe zones; rare in SW and SC BC; introduced from Eurasia.

DESCRIPTION:

General: Annual or sometimes perennial from slender rhizomes; stems weak and trailing, freely rooting, strongly 4-angled, glabrous, up to 50 cm long.

Leaves: Leaves opposite, elliptic to oblanceolate, 5-15 mm long, 1.5-5 mm wide, unstalked or some of the lower ones narrowed to distinct stalks, glabrous except for some hairs on the margins, the margins sometimes strongly wrinkled-curly; stipules lacking.

Flowers: Inflorescence of a few flowers in axillary or terminal, papery-bracted clusters; petals 5, white, deeply 2-cleft, shorter than the sepals; sepals 5, narrowly lanceolate, abruptly sharp-pointed, 2.5-3.5 mm long, glabrous.

Fruits: Capsules egg-shaped, 3-4 mm long, 6-valved; seeds about 0.6 mm long, strongly pimply-wrinkled.

Stellaria borealis **Bigelow** (*S. calycantha auct. non* [Ledeb.] Bong.= ssp. *borealis*; *S. calycantha* vars. *bongardiana* [Fern.] Fern., *isophylla* [Fern.] Fern. and *sitchana* [Steud.] Fern., *S. sitchana* Steud. = ssp. *sitchana*)

boreal starwort

HABITAT/RANGE:

Wet to moist meadows, streambanks, shrub thickets and open forests in the lowland and montane zones; common throughout BC except lacking along the coast (ssp. *borealis*), common along the coast and in S BC south of 52° N, infrequent elsewhere except lacking in NE BC (ssp. *sitchana*); N to AK, YT and NT, E to NF and NS and S to NY, MI, MN, NM, AZ and CA (ssp. *borealis*), N to AK and YT and S to SD, WY and CA (ssp. *sitchana*).

DESCRIPTION:

General: Perennial herb from a long rhizome; stems prostrate to ascending, glabrous or pimply above, 15-50 cm tall/long.

Leaves: Basal leaves lacking or soon deciduous; stem leaves linear-lanceolate to lanceolate, sometimes narrowly egg-shaped to elliptic, 2-6 cm long, opposite, unstalked, glabrous, not net-veined, margins minutely toothed.

Flowers: Inflorescence of solitary flowers in leaf axils or few to many flowers on branch tips; petals white, sometimes lacking, nearly as long as the sepals; sepals 5, sometimes 4, green, 1- to 3-nerved, lanceolate, abruptly sharp-pointed.

Fruits: Capsules egg-shaped, usually dark brown to straw-coloured, translucent to opaque, 4.5-7.5 mm high, more than 1.5 times as high as wide; seeds smooth to wrinkled, reddish-brown, 1.0-1.2 mm long.

Notes: This species belongs to the highly variable *S. calycantha* complex which was first monographed for North America by Fernald (1940). More recently, Morton and Rabeler (1989) have presented a revision of the complex that separates *S. borealis* from *S. calycantha*. It would appear most of our material in northwestern North America, according to Morton and Rabeler (1989), now belongs to *S. borealis*. In BC two subspecies of *S. borealis* are recognized:

1. Sepals egg-shaped to triangular, midrib extending to near the tip, lateral nerves only visible near the base; capsules dark brown and opaque as they open; seeds smooth or indistinctly wrinkled; leaves widest just below the middle, usually 2-3 cm long .... ssp. *borealis*

1. Sepals narrowly triangular, with three prominent nerves; capsules straw-coloured and translucent; seeds usually wrinkled; leaves widest near the base, usually 3-6 cm long .... ssp. *sitchana* (Steud.) Piper

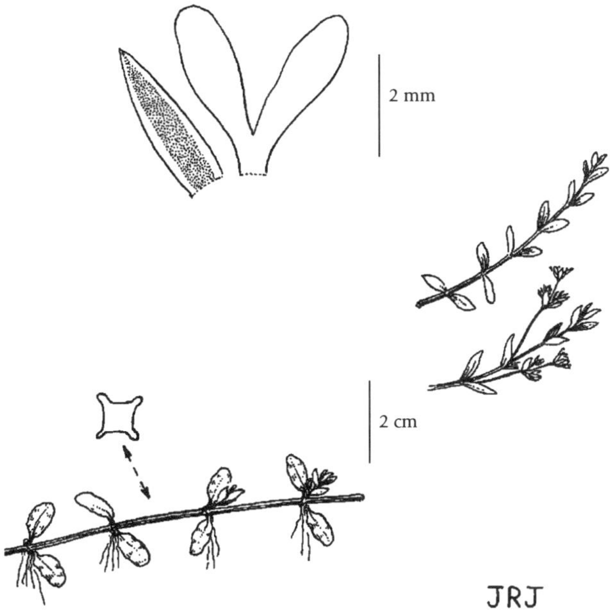

Stellaria alsine

Stellaria borealis ssp. *sitchana*

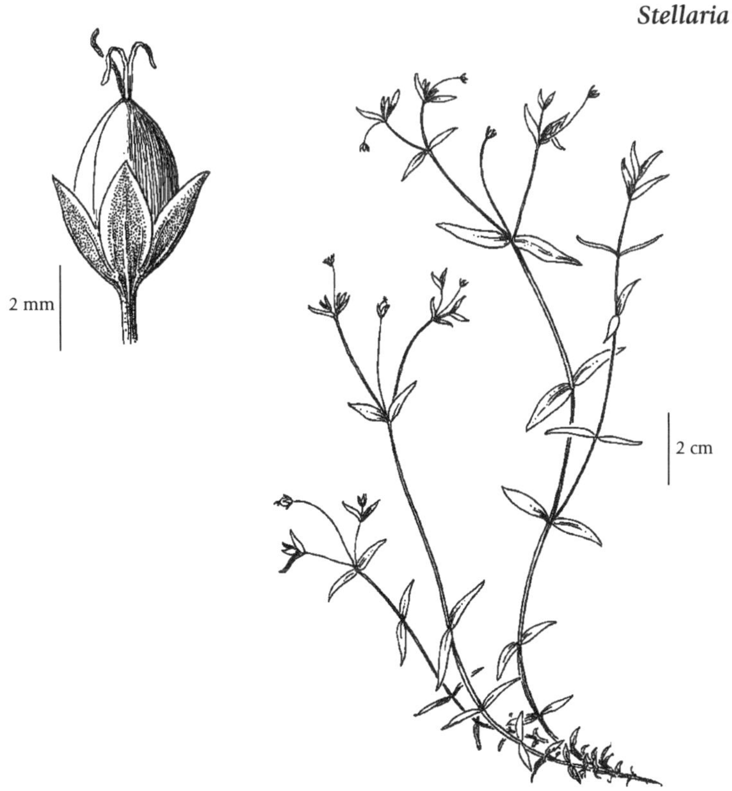

Stellaria borealis ssp. *borealis*

Stellaria calycantha **(Ledeb.) Bong.** (*S. calycantha* ssp. *interior* Hult., *S. calycantha* var. *simcoei* [Howell] Fern., *S.simcoei* [Howell] C. L. Hitchc.)

northern starwort

HABITAT/RANGE:

Wet to moist meadows, streambanks, shrub thickets and open forests in the lowland and montane zones; infrequent throughout BC except absent on Vancouver Island; amphiberingian, N to AK, YT and NT, and S to NM, UT, NV and CA, E Asia.

DESCRIPTION:

General: Perennial herb from a long rhizome; stems prostrate to ascending, glabrous to sparsely hairy, 5-25 cm tall/long.

Leaves: Basal leaves few, reduced, soon deciduous; stem leaves narrowly egg-shaped to elliptic, 1-2.5 cm long, opposite, unstalked, glabrous, net-veined, margins minutely toothed.

Flowers: Inflorescence of solitary flowers in leaf axils or few to many flowers on the branch tips; petals white, sometimes lacking, nearly as long as the sepals; sepals 5, sometimes 4, green, nerves obscure, lanceolate, abruptly sharp-pointed.

Fruits: Capsules rounded to broadly egg-shaped, usually straw-colored, 4.5-7.5 mm high, more than 1.5 times as high as wide; seeds smooth to wrinkled, 0.7-0.9 mm long.

Stellaria crassifolia **Ehrh.**

thick-leaved starwort

HABITAT/RANGE:

Wet to moist streambanks, meadows and lakeshores in the montane zone; infrequent throughout BC east of the Coast-Cascade Mountains; circumpolar, N to AK, YT and NT, E to NF and S to MN, ND and CO; Eurasia.

DESCRIPTION:

General: Perennial herb from a slender rhizome; stems decumbent to erect, many, branched, often matted, very slender, 3-30 cm tall/long.

Leaves: Leaves opposite, unstalked, lanceolate to lance-linear, 5-20 mm long, 0.5-5.0 mm wide, thin; stipules lacking.

Flowers: Inflorescence of solitary flowers in the leaf axils, occasionally several flowers in an open cluster; petals 5, white, deeply 2-cleft, 2.5-5 mm long; sepals 5, oblong-lanceolate, 2-3 mm long at flowering time, inconspicuously 1-nerved, glabrous or nearly so.

Fruits: Capsules straw-coloured, narrowly egg-shaped, 4-5 mm long, 6-valved; seeds about 0.7-1 mm long, reddish-brown, minutely roughened.

Stellaria crispa **Cham. & Schlecht.**

crisp starwort; crisp sandwort

HABITAT/RANGE:

Moist seepage sites, meadows, streambanks, fields and forests in the lowland and montane zones; common in S and coastal BC, infrequent northward; circumpolar, N to AK and YT, E to SW AB and S to WY and CA; Eurasia.

DESCRIPTION:

General: Perennial herb from a slender rhizome; stems prostrate to ascending, several, simple, matted, glabrous, 5-60 cm tall/long.

Leaves: Basal leaves lacking; stem leaves opposite, lanceolate to egg-shaped, unstalked or nearly so, 5-20 mm long, 2-15 mm wide, the margins curly; stipules lacking.

Flowers: Inflorescence of solitary flowers in the leaf axils; petals usually lacking; sepals usually 5, lanceolate, abruptly sharp-pointed, rarely obtuse, 3-4 mm long, prominently 3-nerved.

Fruits: Capsules straw-coloured or brownish, egg-shaped, 3.5-6 mm long, 6-valved; seeds brown, 0.7-1.1 mm long, minutely wrinkled-warty.

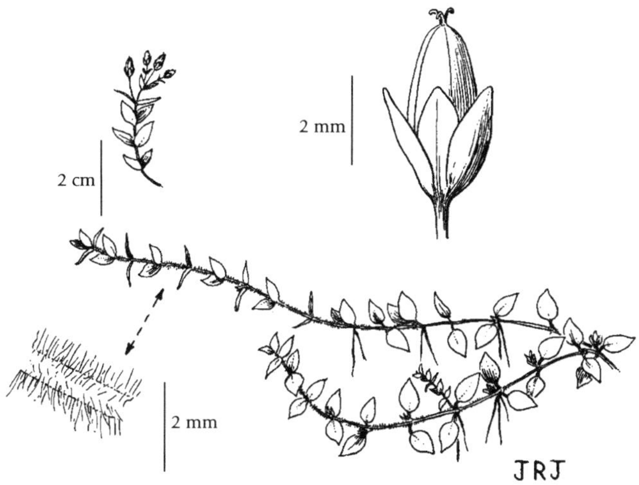

Stellaria calycantha

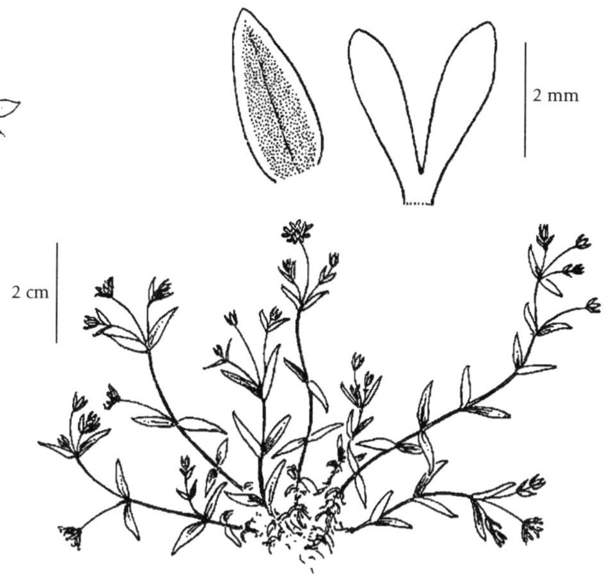

Stellaria crassifolia

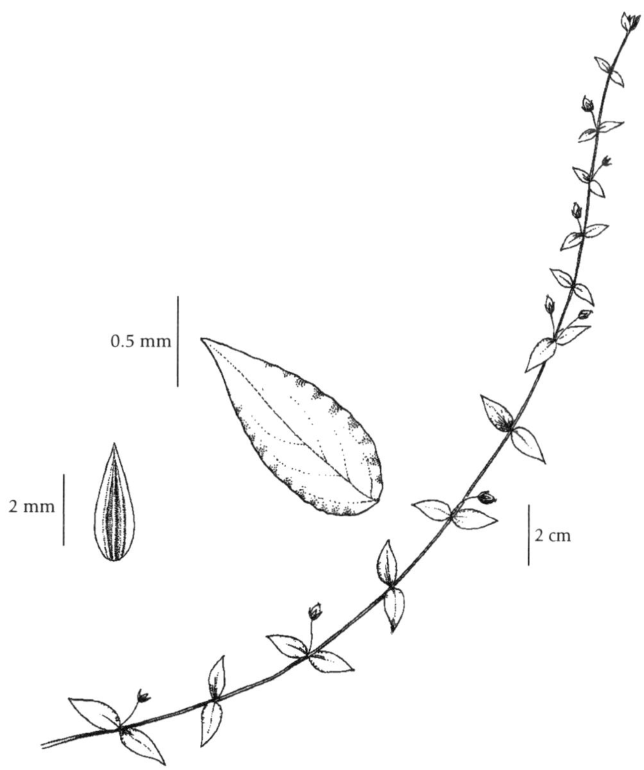

Stellaria crispa

Stellaria graminea **L.**
grass-leaved starwort
HABITAT/RANGE:
Mesic to dry roadsides, gardens, lawns and disturbed areas in the lowland, steppe and montane zones; infrequent throughout BC; introduced from Eurasia.
DESCRIPTION:
General: Perennial herb from a slender rhizome; stems decumbent to ascending, several, branched, 4-angled, glabrous or slightly hairy above, 20-90 cm tall/long.

Leaves: Basal leaves few, reduced, soon deciduous; stem leaves opposite, lanceolate to linear-lanceolate, unstalked, 1.5-4 cm long, usually long-fringed at the base, the margins entire; stipules lacking.

Flowers: Inflorescence of many flowers in a terminal, membranous-bracted, diffuse cluster; petals 5, white, 2-cleft, 4-8 mm long; sepals 5, lanceolate, abruptly sharp-pointed, 3.5-7 mm long, membranous-margined and usually strongly hairy-fringed, 3-nerved.

Fruits: Capsules green, oblong to egg-shaped, 4-8 mm long, 6-valved; seeds reddish-brown, 0.9-1.1 mm wide, wrinkled in concentric rows.

Stellaria humifusa **Rottb.**
salt marsh starwort
HABITAT/RANGE:
Wet salt marshes, mudflats and saline shores in the lowland zone; infrequent in caostal and SW BC, known from the Queen Charlotte Islands, Vancouver Island and Burrard Inlet; circumpolar, N to AK, YT and NT, E to NF and S to OR; Eurasia.
DESCRIPTION:
General: Perennial herb from slender rhizomes; stems prostrate, several, branched, matted, glabrous, 2-40 cm long.

Leaves: Basal leaves lacking; stem leaves opposite, unstalked, narrowly lanceolate to elliptic, 0.6-1.8 cm long, 1-5 mm wide, thick and fleshy; stipules lacking.

Flowers: Inflorescence of solitary flowers in the leaf axils, or in a few-flowered, leafy cluster; petals 5, white, 2-cleft, 4-6 mm long; sepals 5, lanceolate, 3.5-5 mm long at flowering time, inconspicuously 3-nerved.

Fruits: Capsules straw-coloured, egg-shaped, 4-5 mm long, 6-valved; seeds brown, 0.8-1 mm long, lightly roughened, nearly smooth.

Stellaria longifolia **Muhl. *ex* Willd.**
long-leaved starwort
HABITAT/RANGE:
Wet to moist meadows, marshes, streamsides and road banks in the lowland and montane zones; frequent throughout BC; circumpolar, N to AK, YT, NT, E to NF and S to SC, NM, AZ, CA and MX; Eurasia.
DESCRIPTION:
General: Perennial herb from slender rhizomes; stems decumbent to ascending, several, branched, matted, 4-angled, glabrous except minutely rough-hairy above, 3-40 cm tall/long.

Leaves: Basal leaves few, reduced, soon deciduous; stem leaves opposite, unstalked, linear to linear-lanceolate, 1.5-5.0 mm long, 1-4 mm wide, glabrous or with a few basal hairs on the margins, the margins very finely roughly-pimpled (at 30X).

Flowers: Inflorescence of solitary flowers, or a few to several flowers in an open cluster; petals 5, white, 2-cleft, 1.0-1.2 mm long; sepals 5, narrowly-elliptic to lanceolate, 3-4 mm long, usually glabrous, 3-nerved.

Fruits: Capsules greenish-yellow, oblong, 4-6 mm long, 6-valved; seeds 0.8-0.9 mm long, reddish-brown, lightly roughened.

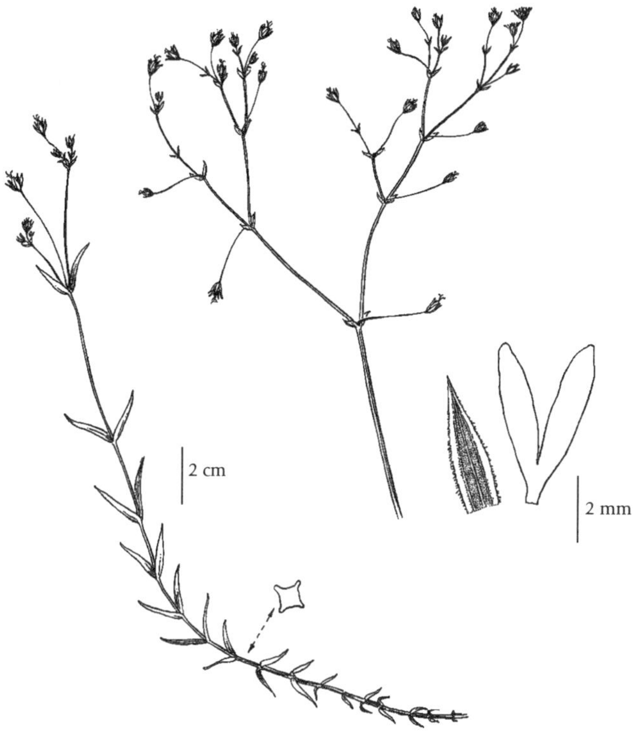

Stellaria graminea

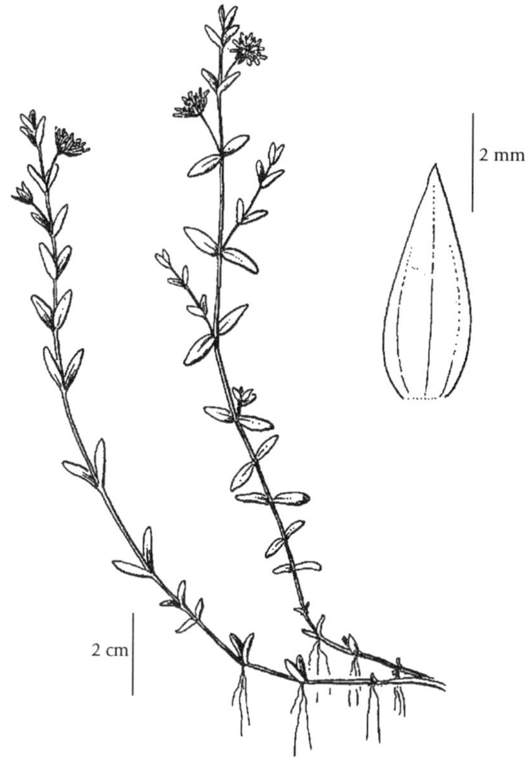

Stellaria humifusa

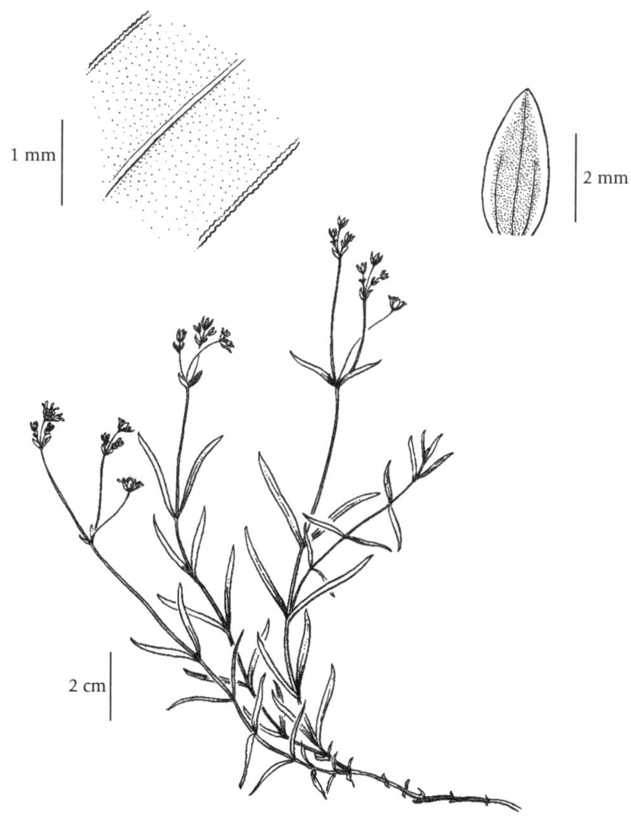

Stellaria longifolia

***Stellaria longipes* Goldie** (*S. alaskana* Hult., *S. edwardsii* R. Br., *S. laeta* Richards., *S. longipes* var. *edwardsii* [R. Br.] A. Gray, *S. longipes* var. *laeta* [Richards.] S. Wats., *S. longipes* var. *subvestita* [Greene] Pol., *S. stricta* Richards., *S. subvestita* Greene = var. *longipes*; *S. longipes* var. *altocaulis* [Hult.] C.L. Hitchc., *S. monantha* Hult. = var. *monantha*)

long-stalked starwort

HABITAT/RANGE:

Mesic to dry rocky slopes, lakeshores, meadows, scree slopes, fellfields and open forests in all vegetation zones; common throughout BC; circumpolar, N to AK, YT and NT, E to NF and S to SD, NM, AZ and CA; Eurasia.

DESCRIPTION:

General: Perennial herb from a slender rhizome; stems decumbent to more usually erect, several, branched, matted, 4-angled, glabrous to slightly hairy near the base, 3-30 cm tall/long.

Leaves: Basal leaves few, reduced, soon deciduous; stem leaves opposite, unstalked, egg-shaped or linear to linear-lanceolate, 5-30 mm long, 1-4 mm wide, abruptly sharp-pointed, rigid and stiff, often glaucous; stipules lacking.

Flowers: Inflorescence of solitary, or few to several flowers in an open cluster; petals 5, white, 2-cleft, 3-8 mm long; sepals 5, lanceolate, 3-5 mm long, usually glabrous, strongly 3-nerved.

Fruits: Capsules usually purplish, 4-6 mm long, 6-valved; seeds 0.6-0.9 mm long, reddish-brown, lightly roughened.

Notes: In an earlier treatment (Douglas *et al.* 1989) *Stellaria longipes* was treated as a single, highly variable taxon based on the research of Chinnappa and Morton (1976, 1984). Recently, Emery and Chinnappa (1994) have shown that it now best to recognize three varieties within the *S. longipes* complex. Two of these occur in BC.

1. Flowers solitary; leaves egg-shaped; plants mostly of the alpine zone .... var. *monantha* (Hult.) Emery & Chinnapappa

1. Flowers on most plants more than 1; leaves linear to linear-lanceolate; plants occurring in all vegetation zones .... var. *longipes*

***Stellaria media* (L.) Vill.**

common chickweed; common starwort

HABITAT/RANGE:

Moist to mesic waste places, fields and gardens in the lowland and montane zones; frequent in S BC; introduced from Eurasia.

DESCRIPTION:

General: Perennial herb from a slender rhizome; stems decumbent, rooting nodally, hairy in longitudinal lines, several, branched, often matted, 7-50 cm long.

Leaves: basal leaves lacking; stem leaves opposite, stalked below, unstalked above, egg-shaped to elliptic, 0.5-4.5 cm long, 0.5-1.5 mm wide, fleshy, hairy; stipules lacking.

Flowers: Inflorescence of single flowers in the leaf axils or terminal flowers in an open cluster; petals 5, white, deeply 2-cleft, 2-3 mm long; sepals 5, oblong-lanceolate, 3-4.5 mm long, hairy and more or less glandular.

Fruits: Capsules straw-coloured, egg-shaped, 4-5 mm long, 6-valved; seeds about 0.9-1.3 mm long, reddish-brown, minutely roughened.

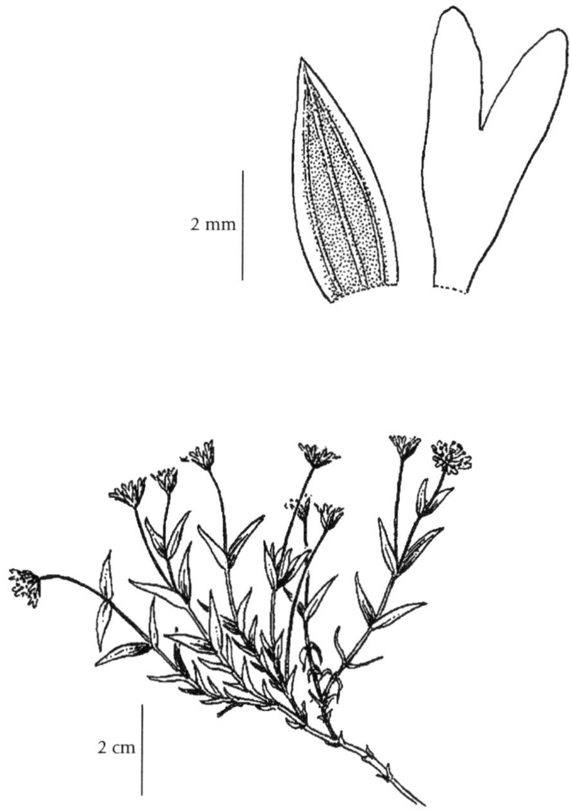

Stellaria longipes var. *monantha*

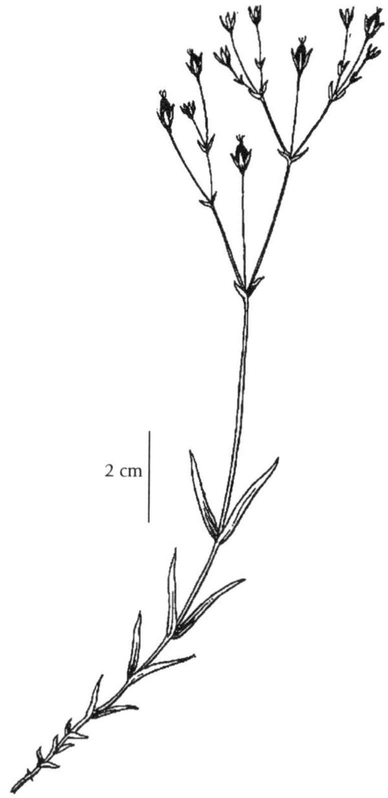

Stellaria longipes var. *longipes*

Stellaria media

Stellaria nitens **Nutt.**

shining starwort

HABITAT/RANGE:

Mesic to dry grassy bluffs, streambanks and open slopes in the lowland, steppe and montane zones; frequent on S Vancouver Island and the Gulf Islands, infrequent E of the Coast-Cascade Mountains in S BC; E to SW AB and S to MT, VT, CA and MX.

DESCRIPTION:

General: Annual herb from slender taproot; stems ascending to erect, simple, solitary to few branched, glabrous or uniformly curly-hairy, threadlike, 3-25 cm tall.

Leaves: Leaves opposite, confined to the lower half of the stem, the basal with a small egg-shaped blade 3-8 mm long and a slender stalk at least as long; stem leaves linear or linear-lanceolate, 5-10 mm long; stipules lacking.

Flowers: Inflorescence of several flowers in large membranous-bracted clusters; petals 5 or lacking, shorter than the sepals, 2-cleft nearly to the base; sepals 5, 3-4 mm long, lanceolate, abruptly sharp-pointed, margins translucent, 3-nerved.

Fruits: Capsules narrowly cylindric egg-shaped, 3-4 mm long, 6-valved; seeds 0.5-0.7 mm long, finely wrinkled-warty.

Stellaria obtusa **Engelm.**

blunt-sepaled starwort

HABITAT/RANGE:

Wet to moist meadows and streambanks in the steppe and montane zones; rare E of the Coast-Cascade Mountains in S BC; N to S AK, E to AB and S to CO and CA.

DESCRIPTION:

General: Perennial herb from a slender rhizome; stems prostrate to ascending, several, branched, matted, glabrous, 3-20 cm tall/long.

Leaves: Basal leaves lacking; stem leaves opposite, egg-shaped, abruptly sharp-pointed, 5-12 mm long, 2.5-8.0 mm wide, glabrous except for a few long basal, marginal hairs, unstalked or the lower ones short-stalked; stipules lacking.

Flowers: Inflorescence of solitary flowers in the leaf axils; petals lacking; sepals 4 or 5, egg-shaped to elliptic, 2-3 mm long, very slightly or not at all papery on the margins, more or less obtuse.

Fruits: Capsules globose-egg-shaped, 2-4 mm long, 6-valved; seeds 0.6-0.7 mm long, dark brown, finely though distinctly cross-wrinkled with undulate ridges.

Stellaria umbellata **Turcz. *ex* Kar. & Kinr.**

umbellate starwort

HABITAT/RANGE:

Moist meadows, streambanks and open forests in the montane to alpine zones; rare in SW and SC BC, known only from the Monashee Mountains and Nicomen Lake; circumpolar, N to AK, E to AB and S to MT, CO and CA; Eurasia.

DESCRIPTION:

General: Perennial herb from a slender rhizome; forming small clumps or mats; stems erect or ascending, several, branched, glabrous, 5-30 cm tall.

Leaves: Basal leaves lacking; stem leaves opposite, oblong to linear-lanceolate, 5-25 mm long, 2-5 mm wide, unstalked, glabrous or sometimes fringed at the base, margins curly; stipules lacking.

Flowers: Inflorescence of 2 to numerous flowers in terminal and axillary umbel-like clusters; petals reduced or more usually lacking; sepals 5, lanceolate, 2-3 mm long, 3-nerved.

Fruits: Capsules straw-coloured, egg-shaped, 3-4.5 mm long, 6- to 8-valved; seeds brown, 0.5-0.7 mm long, lightly roughened.

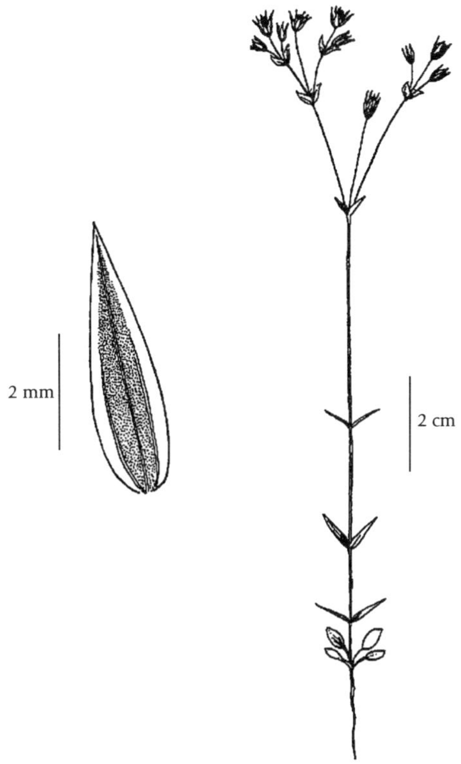

Stellaria nitens

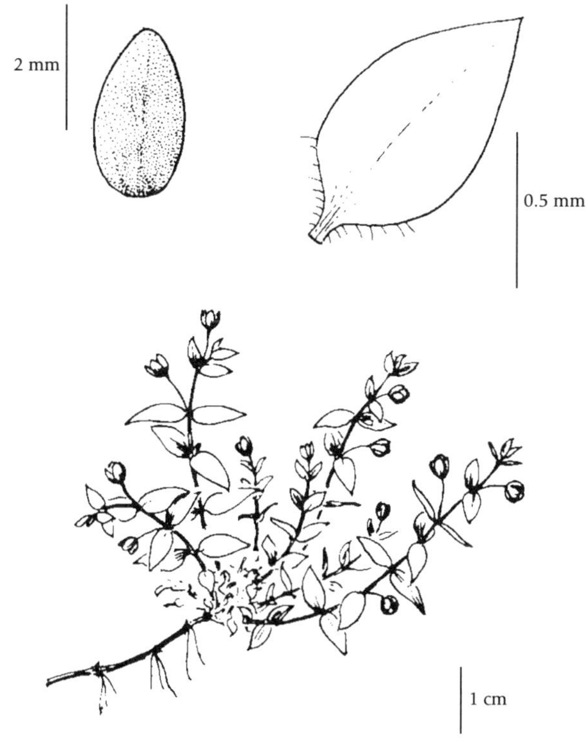

Stellaria obtusa

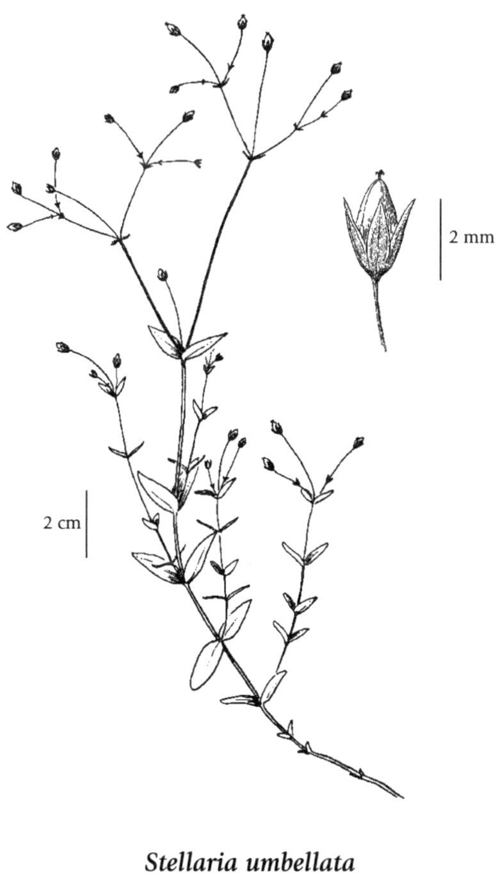

Stellaria umbellata

VACCARIA

***Vaccaria pyramidata* Medikus** (*Saponaria vaccaria* L., *Vaccaria segetalis* [Neck.] Garcke *ex* Aschers.)
cow-basil; cowherb; cow-cockle
Habitat/Range:
Dry fields and waste places in the steppe and lower montane zones; rare in SW and SC BC; introduced from Eurasia.
Description:
General: Annual herb from a slender taproot; stems erect, solitary, branched, glabrous and more or less glaucous, 15-80 cm tall.

Leaves: Basal leaves oblong-lanceolate, unstalked, glabrous, glaucous; stem leaves opposite, lanceolate to oblong, heart-shaped, unstalked, 3-9 cm long, 5-25 mm wide, unstalked, glabrous, glaucous; stipules lacking.

Flowers: Inflorescence of few to many flowers in open clusters; petals 5, pink, the blade wedge-shaped or egg-shaped, 5-8 mm long; sepals 5, often purplish, united, forming a tube 11-15 mm long, with 5 sharp angles or wings, glabrous.

Fruits: Capsules globular, 1-celled; seeds reddish-black, globose, 1.7-2.1 mm wide.

CELASTRACEAE

1. Leaves evergreen, 1-3 cm long; plants less than 1 m tall .... *Pachistima*
1. Leaves deciduous, usually well over 3 cm long; plants 2-5 m tall .... *Euonymus*

EUONYMUS

Euonymus occidentalis* Nutt. *ex* Torr. var. *occidentalis
western wahoo
Habitat/Range:
Mesic forests and thickets in the lowland and montane zones; rare in SW BC, known only from Courtenay, Vancouver Island; S to CA.
Description:
General: Deciduous shrub; stems glabrous, grooved, 2-5 m tall.

Leaves: Opposite, deciduous, leathery, oval to elliptic; leaf margins toothed, 5-10 cm long; stipules minute or lacking.

Flowers: Inflorescence of numerous flowers in axillary clusters; petals 4, greenish- and purplish-mottled to purplish-red, 4-5 mm long; stamens 4.

Fruits: Capsules; seeds covered by a reddish-orange, fleshy outgrowth.

Notes: The Tokyo (1994) edition of the International Botanical Code listed *Euonymus* as a nomenclaturally conserved name over the spelling '*Evonymus*', which was originally used by Linnaeus.

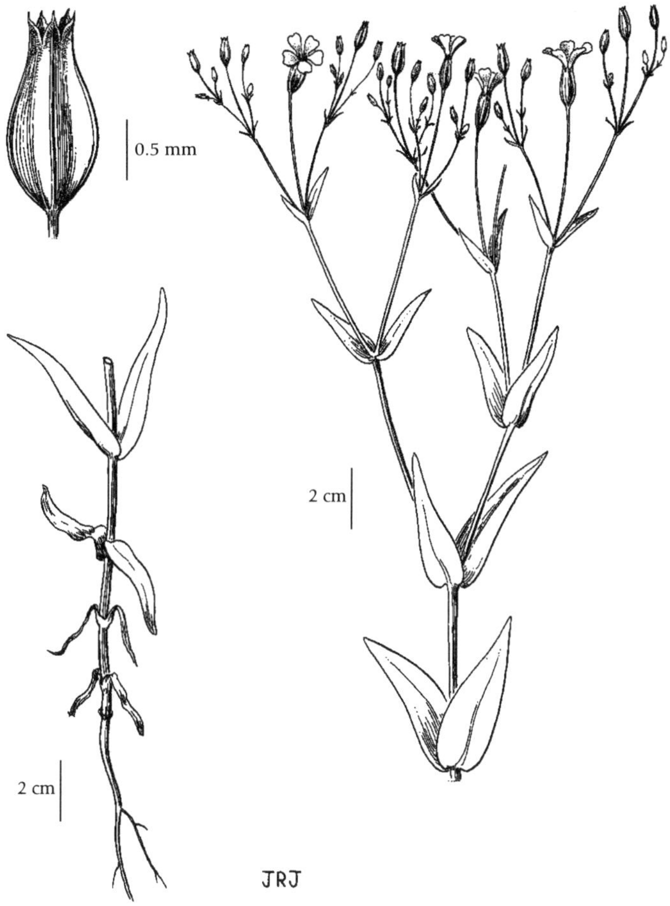

Vaccaria pyramidata

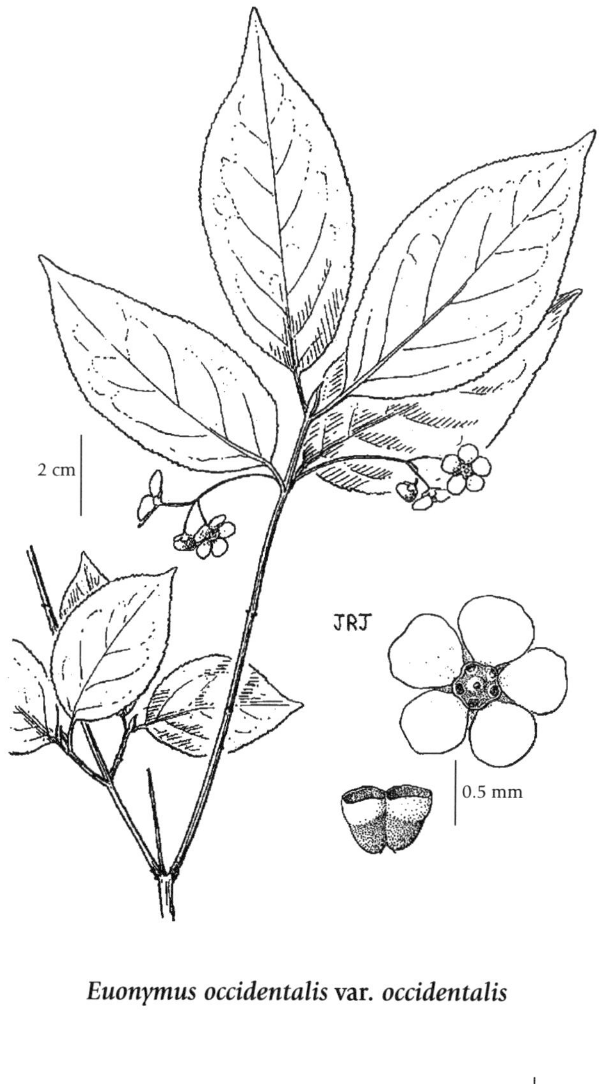

Euonymus occidentalis var. *occidentalis*

PAXISTIMA

***Paxistima myrsinites* (Pursh) Raf.**
falsebox; mountain boxwood; Oregon boxwood; mountain-box; mountain-lover
HABITAT/RANGE:
Mesic to dry slopes and forests in the lowland, steppe and montane zones; common in BC south of 56° N; E to SW AB and S to CA.
Description
General: Evergreen shrub; stems reddish-brown, 4-ridged or smooth, branches erect or prostrate, 0.2-1.0 m tall.

Leaves: Opposite, oblong-lanceolate, lanceolate or oblanceolate, evergreen, 1-3 cm long, leathery, glossy, toothed, margins slightly rolled under.

Flowers: Inflorescence of solitary or numerous, fragrant flowers in axillary leaf clusters all along the branches; petals 4, maroon, 3-4 mm wide; sepals 4.

Fruits: Capsules egg-shaped, 3-4 mm long 1-2 seeded; seeds dark brown, covered by a thin, lacerate, white membrane.

CERATOPHYLLACEAE

CERATOPHYLLUM

1. Fruits wing-margined, with several lateral spines; leaf divisions fine, threadlike, sometimes with fine teeth .... *C. echinatum*
1. Fruits wingless, with only two basal spines; leaf divisions stout, stiff, with distinct teeth .... *C. demersum*

***Ceratophyllum demersum* L.**
common hornwort; coontail
HABITAT/RANGE:
Ponds, lakes and slow-moving streams in the lowland, steppe and montane zones; common in S BC, less frequent northwards; N to AK, YT and NT, E to PQ, NS and NB and S to FL, TX, CA and MX.
DESCRIPTION:
General: Aquatic herb, usually without roots; stems freely branching, 0.3-4.0 m long.

Leaves: Stem leaves submerged, in whorls of 5-12, 2-3 times forked into linear or threadlike segments, these often stout, stiff and spiny-toothed.

Flowers: Inflorescence of solitary, minute, axillary flowers; petals lacking; sepals 3-15, sometimes petal-like; stamens 12-16.

Fruits: Achenes usually lacking, if present then much flattened, 4-6 mm long, ellipsoid, wingless with a persistent style and 2 basal spines.

***Ceratophyllum echinatum* A. Gray**
spring hornwort
HABITAT/RANGE:
Lakes and sloughs in the lowland and montane zones; locally frequent on S Vancouver Island and the Gulf Islands, rare in the lower Fraser Valley and along the Alaska Highway; E to MI and ME and S to FL and MX.
DESCRIPTION:
General: Aquatic herb, usually without roots; stems freely branching, 0.3-4.0 m long.

Leaves: Stem leaves submerged, in whorls of 5-12, 2-4 times forked into linear or threadlike segments, these usually entire or sometimes finely toothed.

Flowers: Inflorescence of solitary, minute, axillary flowers; petals lacking; sepals 3-15, sometimes petal-like; stamens 12-16.

Fruits: Achenes usually lacking, if present then much flattened, 5-7 mm long, ellipsoid, wing-margined with several lateral spines.

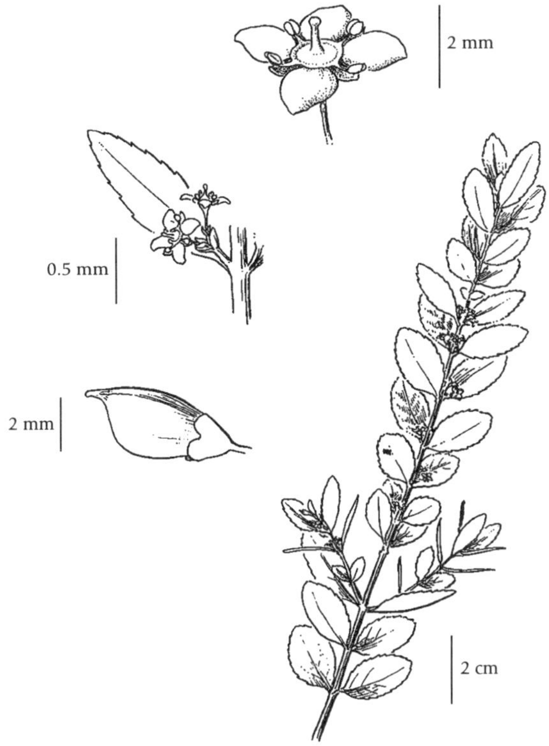

Paxistima myrsinites

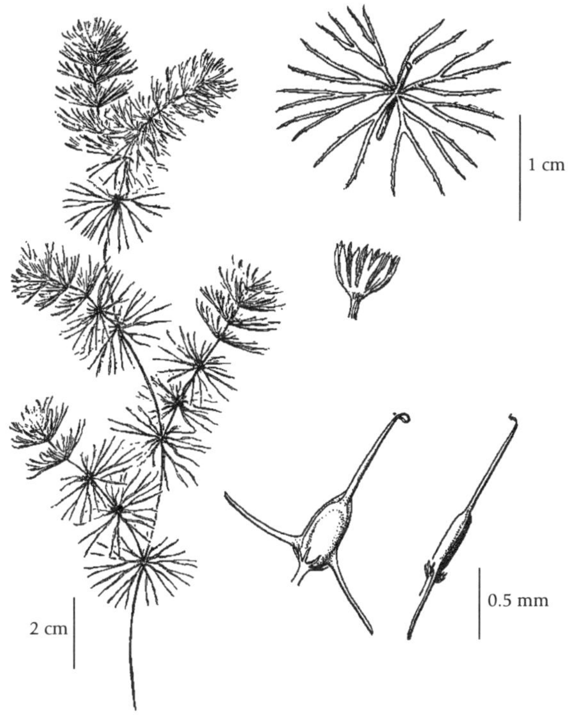

Ceratophyllum demersum

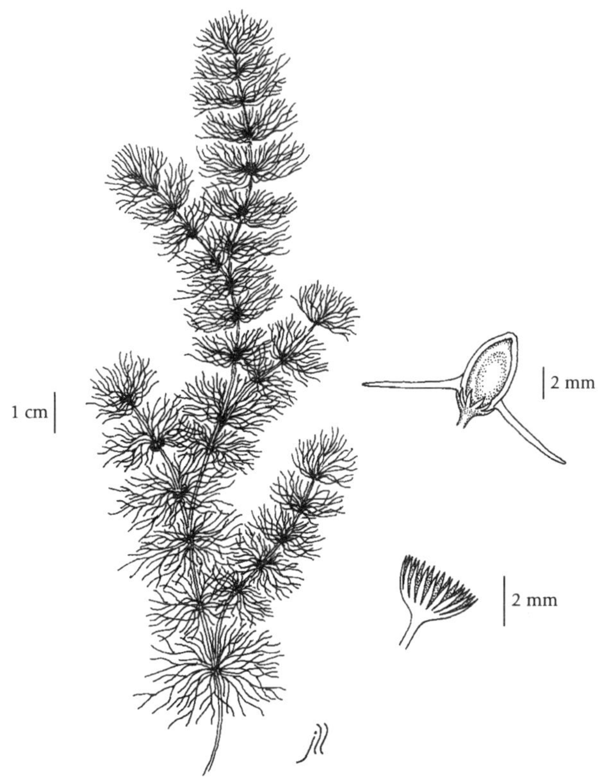

Ceratophyllum echinatum

CHENOPODIACEAE

1. Stems jointed, fleshy; leaves reduced to small, appressed scales; flowers sunken in hollows or pits .... *Salicornia*
1. Stems not jointed; leaves not scalelike; flowers not sunken.
    2. Leaves and bracts strongly spine-tipped; fruiting calyces broadly transversely-winged .... *Salsola*
    2. Leaves and bracts not spine-tipped; fruiting calyces usually not transversely winged.
        3. Leaves threadlike to linear, linear-lanceolate or linear-oblong, entire, unstalked.
            4. Calyx lobes each terminating in a slender hooked dorsal spine .... *Bassia*
            4. Calyx lobes without hooked dorsal spines.
                5. Leaves thick and fleshy, more or less round in cross section .... *Suaeda*
                5. Leaves relatively flat and non-fleshy.
                    6. Flowers forming short, dense, leafy spikes in the leaf axils; leaves usually finely fringed towards the base .... *Kochia*
                    6. Flowers solitary in the upper leaf axils or bractlet axils; leaves not fringed.
                        7. Flowers in the upper leaf axils, subtended by pairs of spinelike, papery, whitish bractlets; calyces deeply 5-lobed, the papery sepals erect, or nearly so .... *Polycnemum*
                        7. Flowers in the axils of the upper, papery-margined bractlets; calyces of a single delicate sepal opposite the bractlet .... *Corispermum*
        3. Leaves broader, toothed or lobed (except *Axyris*), stalked, at least below.
            8. Leaves with starlike hairs, entire .... *Axyris*
            8. Leaves without starlike hairs, toothed or lobed.
                9. Perianths consisting of a single bractlike segment, smaller than the fruit and not enclosing it; seeds vertical .... *Monolepis*
                9. Perianths 3- to 5-lobed; fruits at least partially enclosed by the perianth or by large subtending bracts; seeds often horizontal.
                    10. Flowers mostly either perfect or pistillate, or both, not enclosed between bractlets .... *Chenopodium*
                    10. Flowers imperfect, the pistillate ones enclosed by 2 bractlets .... *Atriplex*

ATRIPLEX

1. Plants perennial semi-shrubs .... *A. nuttallii*
1. Plants annual herbs.
    2. Plants copiously and permanently greyish-mealy, especially on the lower leaf surfaces and in the inflorescence.
        3. Leaves usually toothed, whitish-grey .... *A. rosea*
        3. Leaves usually entire, greyish-green.
            4. Bracteoles fused, 4-10 mm long, deeply toothed .... *A. argentea*
            4. Bracteoles wedge-shaped to oblong, 2.0-3.5 mm long, entire or undulating, never toothed .... *A. truncata*

2. Plants soon green and nearly glabrous, sometimes sparsely greyish-mealy on the lower leaf surfaces and in the inflorescence, but these still appearing green.
   5. Bracteoles more or less thickened toward the base by the presence of spongy inner tissue .... *A. subspicata*
   5. Bracteoles relatively thin, green or membranous throughout.
      6. Bractioles rounded, entire, without lateral teeth.
         7. Seeds both horizontal and vertical; bracteoles more or less heart-shaped or broadly rounded at the base, the veins merging above the base .... *A. hortensis*
         7. Seeds vertical; bracteoles shortly long-pointed at the base, the veins merging only at the base .... *A. micrantha*
      6. Bracteoles never rounded, frequently toothed, usually with lateral teeth.
         8. Largest bracteoles egg-shaped, deltoid, triangular or diamond-shaped.
            9. Lower leaves lanceolate to diamond-shaped, alternate; upper leaves green below .... *A. nudicaulis*
            9. Lower leaves oblong-lanceolate, opposite; upper leaves green or whitish-mealy below.
               10. Largest bracteoles egg-shaped to triangular, margins free to the base, lateral teeth lacking, loosely spaced in the inflorescence, of 2 sizes; upper leaves whitish below .... *A. oblongifolia*
               10. Largest bracteoles diamond-shaped, margins united almost to the middle, lateral teeth present, densely compressed in the inflorescence, more or less uniform in size; upper leaves green below .... *A. patula*
         8. Largest bracteoles strap-shaped or narrowly egg-shaped.
            11. Seeds 1.5-2.7 mm long .... *A. gmelinii*
            11. Seeds 2.8-3.7 mm long .... *A. alaskensis*

Atriplex alaskensis **S. Wats.** (*A. patula* ssp. *alaskensis* [S. Wats.] Hall & Clem., *A. patula* var. *alaskensis* [S. Wats.] Welsh)

Alaskan orache

HABITAT/RANGE:

Moist to mesic shorelines in the lowland zone; rare, known only from the Queen Charlotte Islands, N to AK.

DESCRIPTION:

General: Annual herb from a taproot; stems erect or curved, simple or branched, 0.5-1.0 m tall.

Leaves: Lower leaves opposite, upper leaves alternate, stalked or stalkless, 1-13 cm long, lanceolate to broadly lanceolate, 2-5 cm wide, rounded to arrowhead-shaped at the base, smooth or toothed, covered with a whitish mealy substance when young, but becoming glabrous and greenish with maturity.

Flowers: Inflorescence of terminal or axillary spikes or panicles; pistillate bracteoles somewhat basally thickened at maturity, the largest strap-shaped or narrowly egg-shaped, usually toothed, 3-6 mm long.

Fruits: Membranous pericarps; seed erect, 2.8-3.7 mm long.

Notes: This species is likely extirpated in BC since it is known only from an 1878 collection, even though the Queen Charlotte Islands have been fairly well-collected. It is also quite possible that the major difference between *A. gmelinii* and this species, the larger seeds, does not merit specific recognition for *A. alaskensis*.

Atriplex argentea **Nutt. ssp.** ***argentea***

silvery orache; silver saltbush

HABITAT/RANGE:

Saline disturbed areas and fields in the steppe and montane zones; rare in SC and SE BC; E to W MB and S to NM and N CA.

DESCRIPTION:

General: Annual herb from a taproot; stems erect, freely branching from the base, 15-80 cm tall.

Leaves: Lower leaves opposite, upper leaves alternate, stalked or stalkless, lanceolate to egg-shaped or diamond-shaped, 2-6 cm long, squared-off to wedge-shaped at the base, margins smooth to slightly toothed, covered with a white mealy substance and generally greyish-green overall, sometimes nearly glabrous on upper surfaces.

Flowers: Inflorescence of axillary spikes, greenish, numerous; pistillate bracteoles fused, broadest above the middle, 4-10 mm long, firmly enclosing the fruit, the margins green and deeply toothed.

Fruits: Membranous pericarps; seeds erect, about 1.5 mm long.

Atriplex gmelinii **C.A. Mey.** ***ex*** **Bong.** (*A. patula* L. ssp. *obtusa* [Cham.] Hall & Clem., *A. patula* var. *obtusa* [Cham.] Peck, *A. patula* ssp. *zosterifolia* [Hook.] Hall & Clem., *A. patula* var. *zosterifolia* [Hook.] C.L. Hitchc.)

Gmelin's orache

HABITAT/RANGE:

Moist to mesic shorelines in the lowland zone; infrequent along the coast; N to AK.

DESCRIPTION:

General: Annual herb from a taproot; stems decumbent to erect, freely branched, 0.1-1.0 m tall/long.

Leaves: Lower leaves opposite, upper leaves alternate, stalked or stalkless, lanceolate to oblong or rounded, 2-10 cm long, the bases smooth or toothed, covered with a whitish mealy substance when young but becoming glabrous and greenish with maturity.

Flowers: Inflorescence of terminal or axillary spikes or panicles; pistillate bracteoles only slightly thickened basally or not at all, largest ones strap-shaped or narrowly egg-shaped, 4-12 mm long.

Fruits: Membranous pericarps; seeds erect, 1.5-2.7 mm long.

Atriplex alaskensis

Atriplex argentea ssp. *argentea*

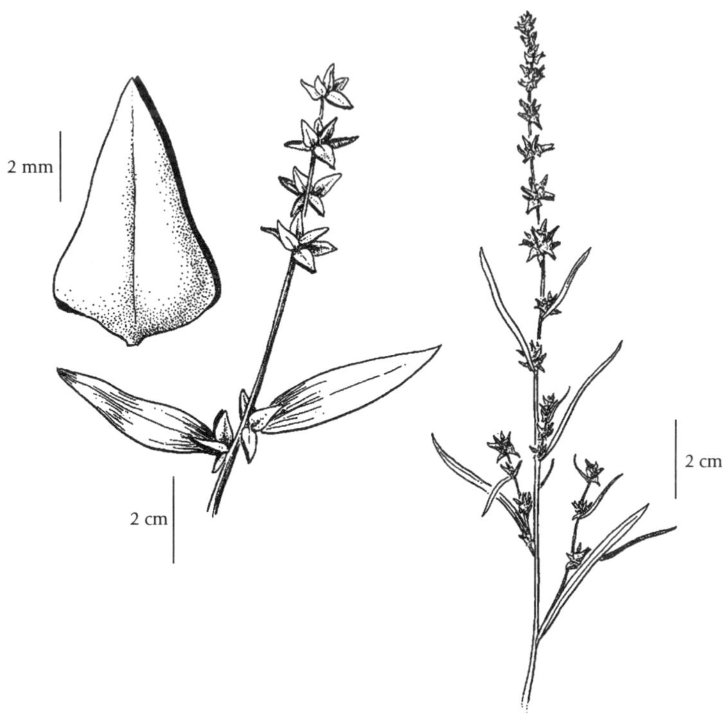

Atriplex gmelinii

***Atriplex hortensis* L.**

garden orache; French spinach

HABITAT/RANGE:

Mesic to dry roadsides and waste places in the lowland, steppe and montane zones; infrequent garden escape in S BC; introduced from Asia.

DESCRIPTION:

General: Annual herb from a taproot; stems decumbent to erect, freely branched, mostly 0.6-2.5 m tall/long.

Leaves: Lower leaves opposite, upper leaves alternate, stalked or stalkless, broadly lanceolate to egg-shaped, 5-20 cm long, 2-10 cm wide, blunt to arrowhead-shaped at the base, smooth or slightly toothed or undulate, covered with a whitish mealy substance when young but becoming glabrous and greenish with maturity.

Flowers: Inflorescence of terminal or axillary spikes or panicles; pistillate flowers of 3 kinds, one with a small calyx and no bracteoles, one with small rounded bracteoles and the third with no calyx but 2 large, 6-12 mm wide, rounded bracteoles, these without teeth, veins merging above the base.

Fruits: Membranous pericarps; seeds of the pistillate flowers without bracteoles, horizontal, small, black, those of the pistillate flowers with more or less heart-shaped bracteoles or the bracteoles rounded at the base, vertical, black or brown and about 2 mm wide.

***Atriplex micrantha* C. A. Mey.** (*A. heterosperma* Bunge)

Russian orache

HABITAT/RANGE:

Dry roadsides and waste places in the steppe zone; frequent in SC BC; introduced from Eurasia.

DESCRIPTION:

General: Annual herb from a taproot; stems erect, branched, 1.0-1.5 m tall.

Leaves: Lower leaves opposite, upper leaves alternate, stalked or stalkless, triangular, 1-5 cm long, the base arrowhead shaped, smooth or toothed, covered with a whitish mealy substance below but becoming glabrous and greenish on both surfaces above.

Flowers: Inflorescence of terminal or axillary spikes or panicles; pistillate flowers of 2 kinds, one with small bracteoles and one with large bracteoles, the bracteoles rounded and shortly long-pointed at the base, lacking teeth, 5-6 mm long, veins merging only at the base.

Fruits: Membranous pericarps; seeds of the small bracteoles black, 1.5 mm long, erect, those of the large bracteoles brown, 2-3 mm long, erect.

***Atriplex nudicaulis* Boguslaw** (*A. longipes* Drej.)

long-stalked orache

HABITAT/RANGE:

Moist to mesic marine shorelines in the lowland zone; rare in SW BC, known only from Surrey and White Rock; introduced from Eurasia.

DESCRIPTION:

General: Annual herb from a taproot; stems procumbent to erect, several, branched, 30-90 cm tall/long.

Leaves: Lower leaves alternate, lanceolate to diamond-shaped, with basal lobes, entire to shallowly toothed, green; upper leaves green above and below.

Flowers: Inflorescence of terminal or axillary spikes or panicles; smaller bracteoles triangular, fused at the base, green, 5-10 mm long; larger bracteoles triangular, leafy at the tip, 2.5-3.0 cm long, some of these with stalks 5-20 mm long.

Fruits: Membranous pericarps; seeds erect.

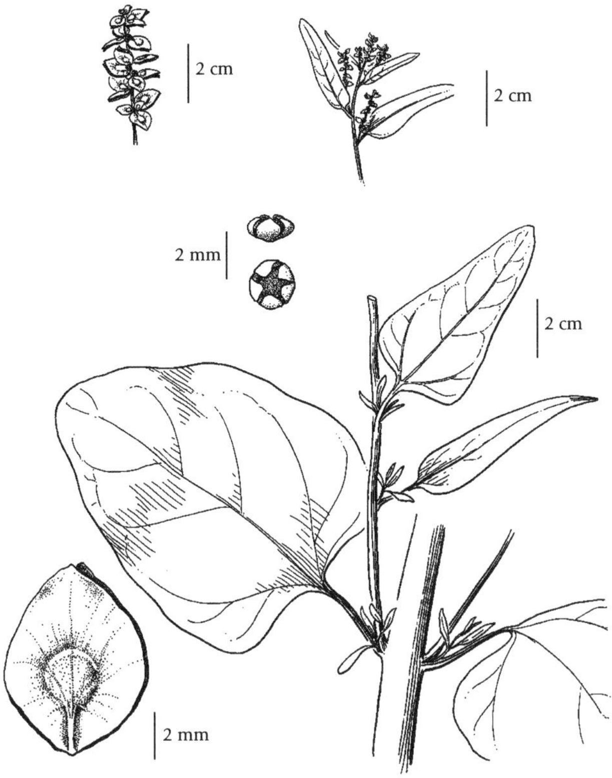

Atriplex hortensis

Atriplex micrantha

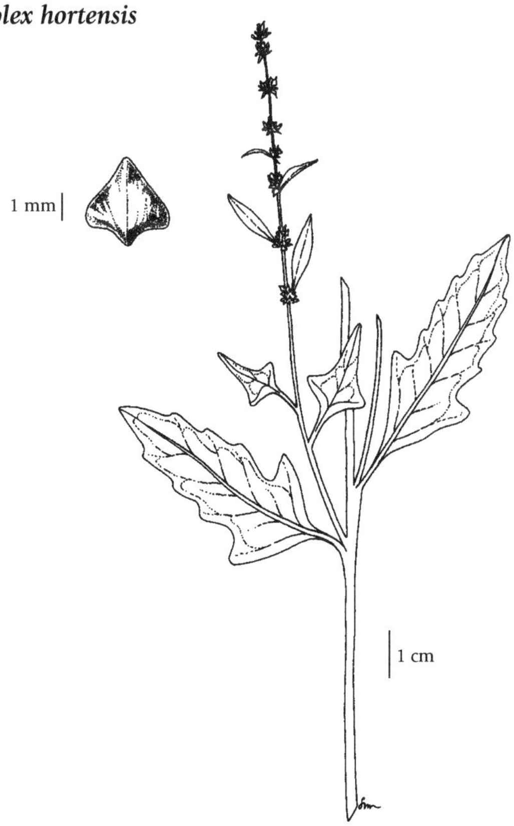

Atriplex nudicaulis

Atriplex nuttallii **S. Wats.**

Nuttall's orache; moundscale

HABITAT/RANGE:

Dry grassy slopes in the montane zone; rare in NE BC, known only from Clayhurst Crossing, Peace River; E to SK and S to ID, CO, UT, NM and AZ.

DESCRIPTION:

General: Perennial semishrub with a woody base; stems erect or spreading, freely branched, 10-60 cm tall.

Leaves: Leaves alternate, narrowly spoon-shaped to oblong, 2-5 cm long, 2-12 mm wide, narrowed gradually to a short stalk or sometimes unstalked, densely greyish-scurfy, usually smooth margined.

Flowers: Inflorescence of axillary or terminal, leafless spikes or panicles; pistillate bracteoles irregularly toothed and smooth to pimply, 3-7 mm long, lanceolate to elliptic or rounded.

Fruits: Membranous pericarps; seeds erect, about 2 mm long.

Atriplex oblongifolia **Waldst. & Kit.**

oblong-leaved orache

HABITAT/RANGE:

Dry roadsides and waste places in the steppe and montane zones; rare in SC and SE BC; introduced from Eurasia.

DESCRIPTION:

General: Annual herb from a taproot; stems erect, branched, 0.2-1.2 m tall.

Leaves: Lower leaves opposite, upper leaves alternate, stalked or stalkless, oblong-lanceolate, smooth or slightly toothed, covered with a whitish-mealy substance below, green above.

Flowers: Inflorescence of terminal or axillary spikes or panicles; pistillate bracteoles of 2 sizes, largest bracteoles egg-shaped or egg-shaped-triangular, up to 13 mm long, enclosing a 3 mm wide yellow-brown fruit, the smaller bracteoles enclosing a black fruit 1.5 mm wide, all bracteoles with margins free to the base and lateral teeth lacking.

Fruits: Membranous pericarps; seeds erect, 1-2 mm long.

Atriplex patula **L.**

common orache

HABITAT/RANGE:

Mesic roadsides, waste places and gardens in the lowland, steppe and montane zones; frequent in S BC; probably introduced from Eurasia.

DESCRIPTION:

General: Annual herb from a taproot; stems erect or ascending, simple or branched, 5-80 cm tall.

Leaves: Lower leaves opposite, upper leaves alternate, stalked or stalkless, 1-10 cm long, lanceolate to broadly lanceolate, 2-4 cm wide, rounded at the base, smooth or toothed, upper leaves green below, others covered with a whitish mealy substance above and below when young, but becoming glabrous and greenish with maturity.

Flowers: Inflorescence of terminal or axillary spikes or panicles; pistillate bracteoles somewhat basally thickened at maturity, the largest bracteoles somewhat diamond-shaped, margins united almost to the middle, lateral teeth present, densely compressed in the inflorescence, more or less uniform in size, 3-6 mm long.

Fruits: Membranous pericarps; seeds erect, 1.5-2.5 mm long.

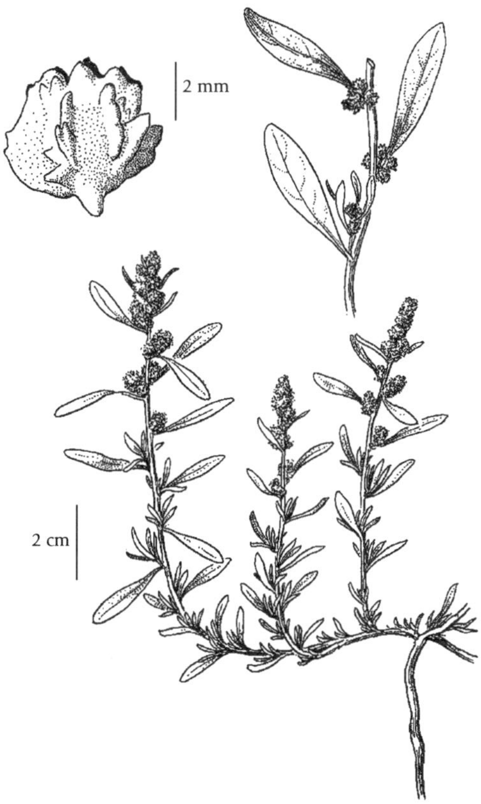

Atriplex nuttallii

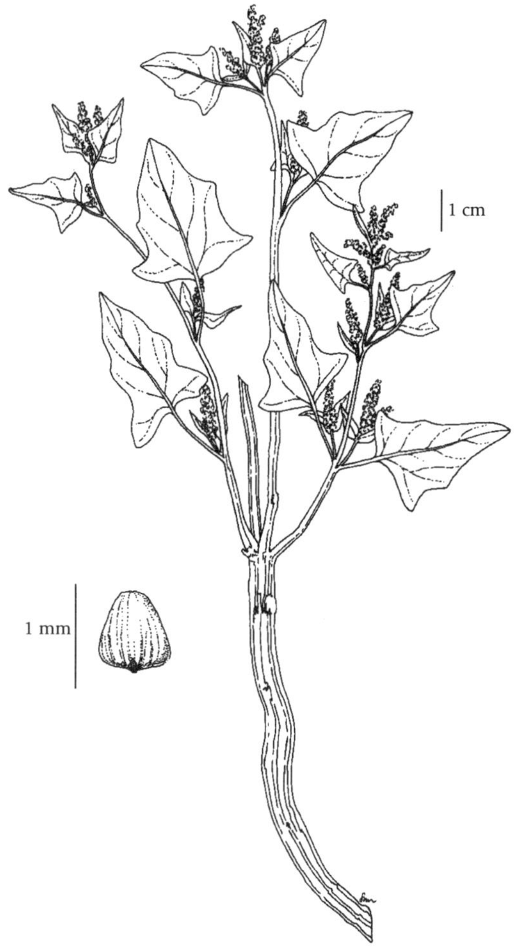

Atriplex oblongifolia

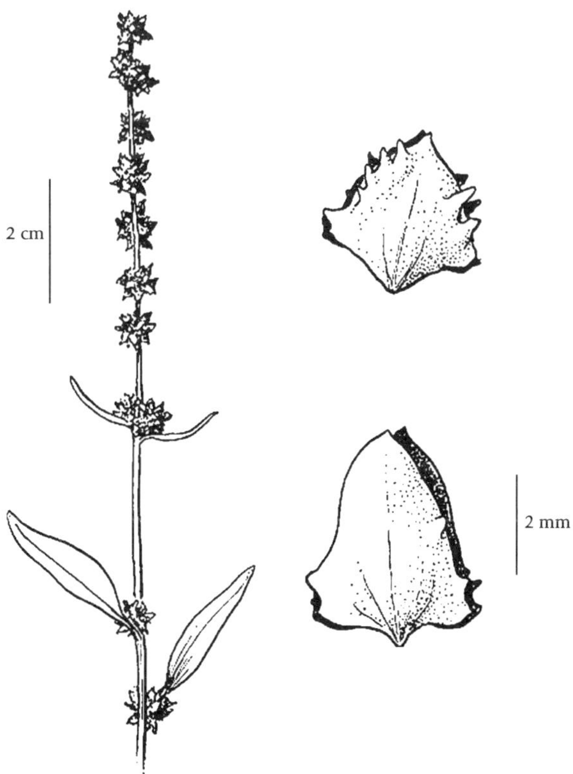

Atriplex patula

***Atriplex rosea* L.**

red orache; tumbling orache

HABITAT/RANGE:

Dry roadsides and disturbed areas in the steppe zone; rare in SC BC; introduced from Europe.

DESCRIPTION:

General: Annual herb from a taproot; stems erect, simple or freely branched, 0.2-1.2 m tall, nearly glabrous.

Leaves: Lower leaves opposite, upper leaves alternate, stalked or stalkless, lanceolate to egg-shaped or nearly diamond-shaped, 2-5 cm long, generally slightly toothed, often blunt to rounded at the base, whitish-grey, covered with a copiously and permanently greyish-mealy substance, especially on the lower leaf surfaces, upper surfaces greenish, fleshy.

Flowers: Inflorescence of terminal or axillary spikes or panicles; pistillate bracteoles lance-shaped to orbicular, strongly flattened, pointed at the tip, margins slightly toothed, the backs sharply pimply.

Fruits: Membranous bladders enclosed by 2 lance-shaped bracts; seeds 2 mm long.

***Atriplex subspicata* (Nutt.) Rydb.** (*A. patula* ssp. *hastata* [L.] Hall & Clem. *pro parte*, *A. patula* ssp. *littoralis auct. non* [L.] Hall. & Clem.)

saline orache

HABITAT/RANGE:

Dry saline and alkaline roadsides and waste places in the lowland, steppe and montane zones; frequent in S BC; E to NF and S to NC and CA.

DESCRIPTION:

General: Annual herb from a taproot; stems erect to ascending, simple to branched, 0.1-1.0 cm tall.

Leaves: Lower leaves opposite, upper leaves alternate, stalked or stalkless, lance-shaped to linear or oblong, rounded or arrowhead-shaped at the base, smooth or toothed, covered with a whitish-mealy substance but becoming glabrous and greenish with maturity, 1-13 cm long.

Flowers: Inflorescence of terminal or axillary spikes or panicles, greenish, tiny, numerous; pistillate bracteoles lance-shaped to orbicular, more or less thickened toward the base by the presence of spongy tissue, the backs pimply; seeds erect, 1.5-2.5 mm long.

Fruits: Membranous bladders enclosed by 2 lance-shaped bracts.

***Atriplex truncata* (Torr. *ex* S. Wats.) A. Gray**

wedgescale orache

HABITAT/RANGE:

Dry roadsides, waste places and alkaline flats in the steppe and montane zones; infrequent in S BC east of the Coast-Cascade Mountains; E to SE SK and S to CO, UT and CA.

DESCRIPTION:

General: Annual herb from a taproot; stems erect or curved, simple or branched, 25-120 cm tall, greyish-mealy above.

Leaves: Lower leaves opposite, upper leaves alternate, short-stalked or stalkless, copiously and permanently greyish-mealy, especially on the lower surfaces, greyish-green above, lance-shaped to lance-linear or oblong, 2-7 cm long, squared-off to arrowhead-shaped at the base, smooth to wavy margins but not toothed.

Flowers: Inflorescence of densely compacted axillary spikes or panicles, greenish, tiny, numerous; pistillate bracteoles wedge-shaped to oblong, 2.0-3.5 mm long, entire to undulating, never toothed.

Fruits: Membranous bladders enclosed by 2 lance-shaped, smooth or wavy-margined bracts; seeds 1.5 mm long.

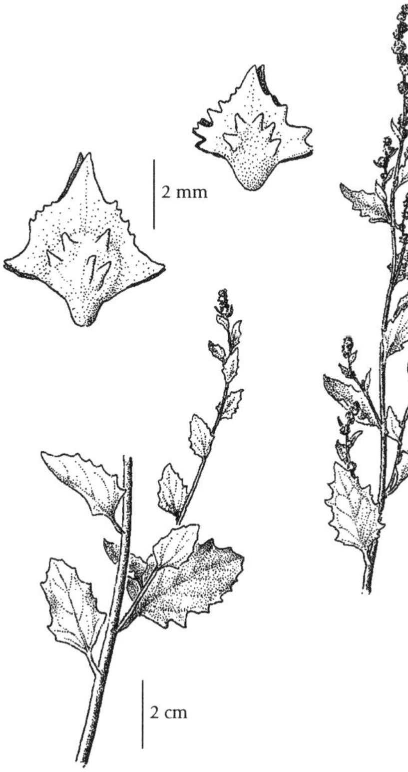

Atriplex rosea

Atriplex subspicata

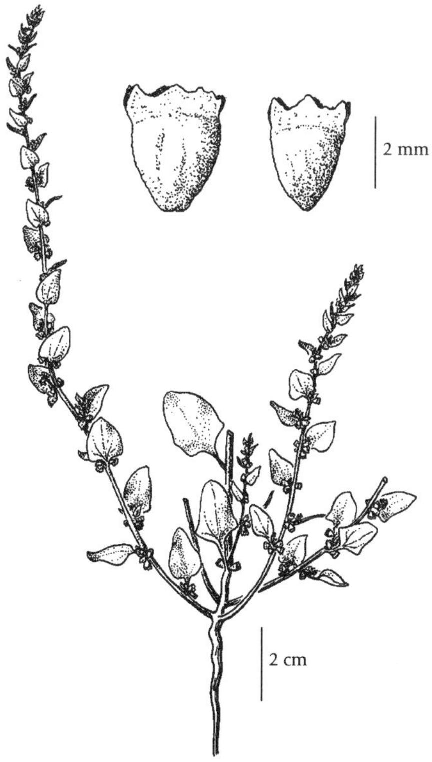

Atriplex truncata

AXYRIS

Axyris amaranthoides L.
Russian pigweed
HABITAT/RANGE:
Dry roadsides, fields and waste places in the steppe zone; infrequent in SC BC; introduced from Eurasia.
DESCRIPTION:
General: Annual herb from a taproot; stems ascending, solitary, branched, with starlike hairs, 0.2-1.2 m tall.

Leaves: Mainly upper leaves, alternate, lanceolate to elliptic or egg-shaped, entire, those of the flowering branches bractlike.

Flowers: Inflorescence of unisexual flowers; staminate flowers in short panicled spikes terminating the branches; pistillate flowers few or single in the axils of the bracts.

Fruits: Thin pericarps, 2-winged at the summit or wingless, the mature bracteoles and calyces becoming whitish and conspicuous; seeds silvery, flattened, egg-shaped, 2.0-2.5 mm long.

BASSIA

Bassia hyssopifolia (Pall.) O. Kuntze
five-hooked bassia
HABITAT/RANGE:
Dry alkaline flats, roadsides and waste places in the steppe and montane zones; infrequent in SC and SE BC; introduced from Eurasia.
DESCRIPTION:
General: Annual herb; stems erect, simple to freely branched, 15-60 cm tall, woolly-hairy, often reddish-tinged.

Leaves: Stem leaves alternate, entire, linear or narrowly lanceolate, 1-4 cm long, without stalks, soft- to woolly-hairy, especially when plants are in flower.

Flowers: Inflorescence of densely clustered terminal and lateral spikes, with smaller axillary spikes below, greenish; bracts reduced.

Fruits: Membranous bladders flattened, 2 mm broad with curved, strongly hooked spines; seeds horizontal, flattened.

CHENOPODIUM [27]

1. Plants usually glandular, strongly aromatic, not mealy-surfaced.
   2. Seeds erect; fruits laterally flattened .... *C. pumilio*
   2. Seeds horizontal; fruits flattened at the top clusters .... *C. botrys*
1. Plants without glands, non-aromatic, often mealy-surfaced.
   3. Seeds usually erect, the fruits laterally flattened; leaves large, triangular, often arrowhead-shaped, greenish on both surfaces, never mealy-surfaced.
      4. Calyces becoming reddish and fleshy in fruit; fruiting clusters more than 4 mm wide at maturity .... *C. capitatum*
      4. Calyces not reddish and fleshy in fruit; fruiting clusters less than 4 mm wide at maturity .... *C. rubrum*

[27] Key adapted from Hitchcock *et al.* (1964).

Axyris amaranthoides L.

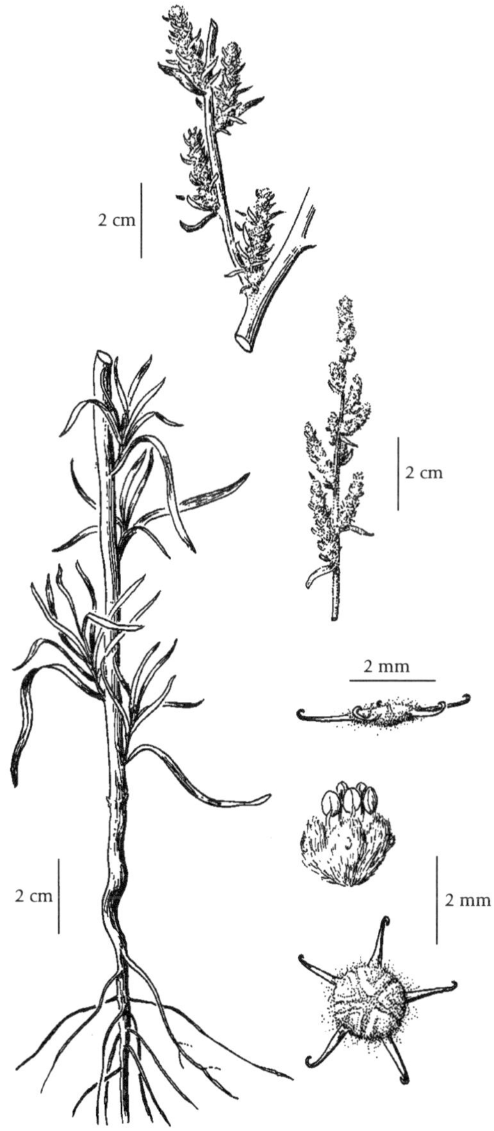

Bassia hyssopifolia

3. Seeds mostly horizontal, the fruits flattened at the top, if seeds erect then the leaves either mealy-surfaced or not arrowhead-shaped.
    5. Plants usually prostrate or low and spreading; leaves green above, greyish-mealy below; at least some of the seeds usually vertical .... *C. glaucum*
    5. Plants usually erect; leaves not greyish-mealy below, or if so, then the seeds all horizontal.
        6. Leaves linear to lanceolate, mostly entire.
            7. Leaves usually white-mealy below .... *C. desiccatum*
            7. Leaves sparsely to moderately greyish-mealy below .... *C. atrovirens*
        6. Leaves broadly lanceolate, elliptic, egg-shaped, diamond-shaped, triangular or deltoid, shallowly to prominently toothed or lobed.
            8. Leaves sparsely to densely mealy-surfaced; sepals of the fruiting plants keeled.
                9. Leaves entire except sometimes with a few teeth at the base .... *C. fremontii*
                9. Leaves toothed throughout .... *C. album*
            8. Leaves glabrous; sepals of the fruiting plants not keeled.
                10. Leaves entire or rarely with a pair of slight teeth at the base; sepals with definite, abruptly sharp-pointed tips .... *C. polyspermum*
                10. Leaves shallowly to deeply toothed above the basal lobes; sepals without abruptly sharp-pointed tips.
                    11. Seed margins rounded, the surface apparently smooth although net-veined .... *C. urbicum*
                    11. Seed margins sharply angled, the surface finely pitted .... *C. hybridum*

***Chenopodium album* L.** (*C. strictum* Roth ssp. *glaucophyllum* [Aellen] Aellen & Just. = ssp. *striatum)*
lamb's-quarters

HABITAT/RANGE:

Mesic to dry fields, waste places and gardens in the lowland, steppe and montane zones; common throughout BC except the Queen Charlotte Islands and adjacent coast; introduced from Eurasia.

DESCRIPTION:

General: Annual herb from a taproot; stems erect, solitary, branched, 0.2-1.0 m tall, greenish to greyish-mealy, often reddish-tinged with age.

Leaves: Stem leaves alternate, somewhat succulent, broadly lanceolate, deltoid, egg-shaped or diamond-shaped, 3-10 cm long, shallowly to prominently toothed or lobed, sparsely to densely mealy.

Flowers: Inflorescence of densely clustered terminal or axillary spikes, greenish; with or without bracts.

Fruits: Thin, membranous envelopes 1-1.5 mm broad; seeds black, shiny, flattened.

Notes: The two subspecies recognized by Tutin (1964) for *C. album* appear to be present in our flora. They may be distinguished as follows:

1. Leaves lanceolate, egg-shaped or diamond-shaped, often abruptly sharp-pointed, margins not always parallel; stems green or red .... ssp. *album*
1. Leaves oblong, obtuse, margins more or less parallel; stems often red striped .... ssp. *striatum* (Krasan) J. Murr in Urban & Graebn.

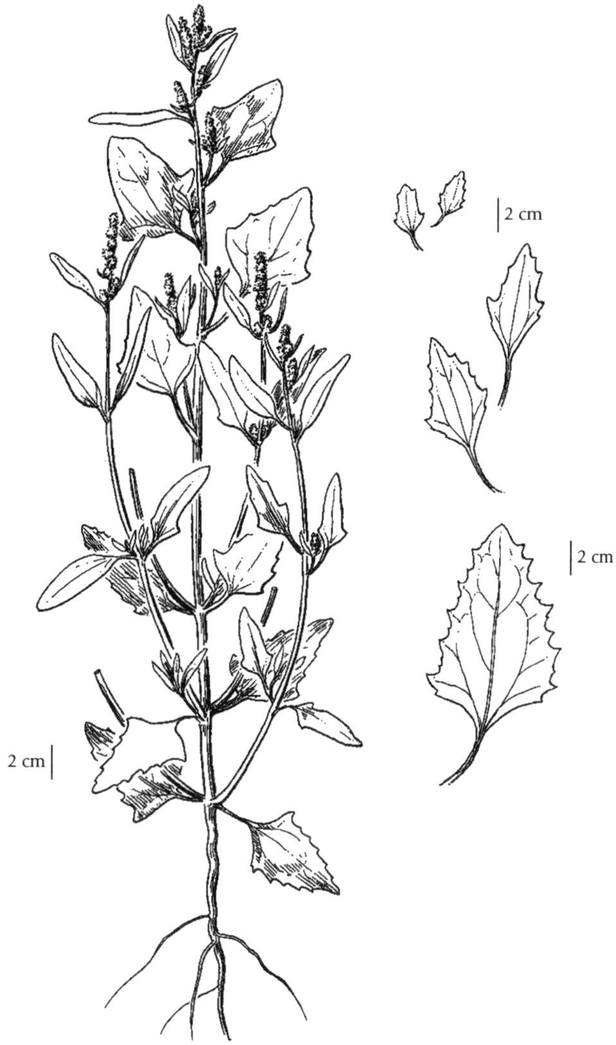

Chenopodium album* ssp. *album

Chenopodium atrovirens **Rydb.** (*C. fremontii* S. Wats. var. *atrovirens* [Rydb.] Fosburg)
dark lamb's-quarters
Habitat/Range:
Dry waste places in the steppe zone; rare, scattered throughout BC east of the Coast-Cascade Mountains S of 55° N; E to AB and S to IA, CO, NV and CA.
Description:
General: Annual herb from a taproot; stems erect, solitary, simple to branched, 10-50 cm tall.
Leaves: Stem leaves lanceolate, usually not arrowhead-shaped, greenish on upper surfaces but sparsely mealy below, stalked, 1-4 cm long, rounded to pointed.
Flowers: Inflorescence of dense clusters in large terminal and smaller lateral spikes, densely mealy, without stalks.
Fruits: Thin, membranous envelopes, 1 mm wide; seeds obtusely margined, slightly roughened.

Chenopodium botrys **L.**
Jerusalem-oak; feather geranium
Habitat/Range:
Dry roadsides, fields and waste places in the steppe and lower montane zones; frequent in SC BC, infrequent in SW BC; introduced from Eurasia.
Description:
General: Annual herb from a taproot; stems ascending to erect, solitary, freely branched from the base, 15-60 cm tall, copiously glandular-hairy, strongly aromatic.
Leaves: Stem leaves alternate, fleshy, stalked, oblong to egg-shaped, 2-4 cm long, with shallowly to deeply toothed lobes, reduced upwards.
Flowers: Inflorescence of numerous, small, greenish, paired, axillary clusters; flowering spikes up to 4/5 the total height of the plant.
Fruits: Thin, whitish, membranous envelopes; seeds black, 0.6 mm wide, horizontal.

Chenopodium capitatum **(L.) Aschers.**
strawberry-blite; Indian paint
Habitat/Range:
Mesic to dry disturbed areas and waste places in the lowland, steppe and montane zones; common throughout most of BC, rare W of the Coast-Cascade Mountains; N to AK, YT and NT, E to PQ and NS and S to PA, MN, NM and CA; also Eurasia; native status uncertain.
Description:
General: Annual herb from a stout taproot; stems erect, several, branched, 20-80 cm tall, glabrous.
Leaves: Leaves alternate, greenish on both surfaces, 2-10 cm long, triangular, arrowhead-shaped at the base.
Flowers: Inflorescence of axillary, dense, spikelike clusters, green, becoming reddish and fleshy in fruit; fruiting clusters more than 4 mm wide at maturity.
Fruits: Thin, membranous envelopes, laterally flattened; seeds erect, oblong, 0.9 mm long.

Chenopodium atrovirens

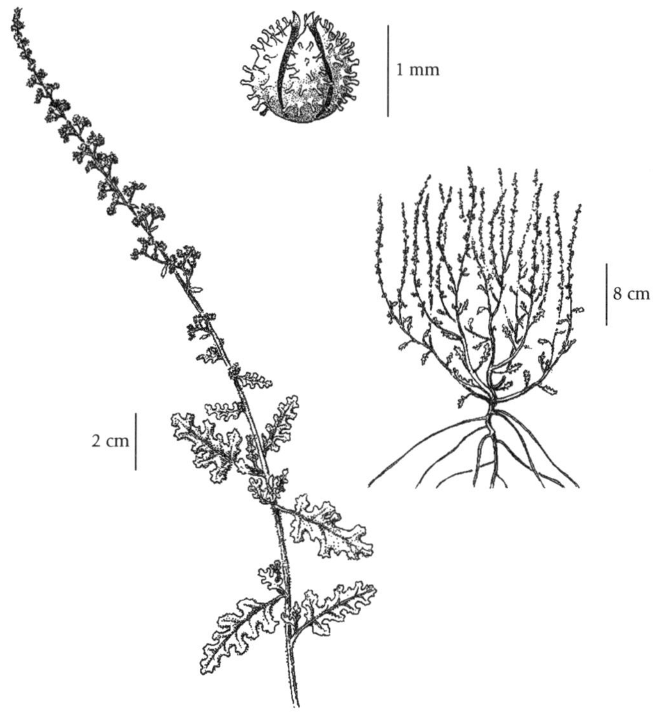

Chenopodium botrys

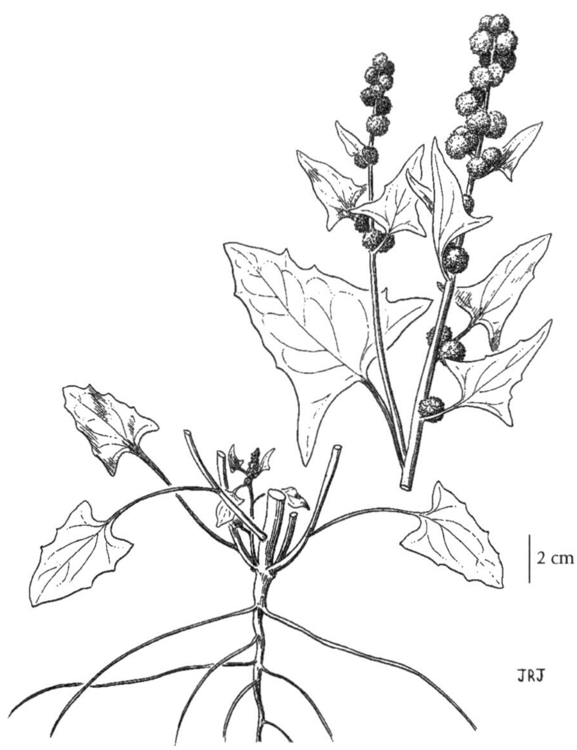

Chenopodium capitatum

***Chenopodium desiccatum* A. Nels.** (*C. leptophyllum* [Moq.] S. Wats. var. *oblongifolium* S. Wats., *C. oblongifolium* [S. Wats.] Rydb.)

narrow-leaved goosefoot

HABITAT/RANGE:

Dry saline or alkaline flats or disturbed areas in the steppe and montane zones; rare in S and NE BC; E to MB and S to MI, TX, MX and CA.

General: Annual herb from a taproot; stems erect, solitary, simple to freely branched, usually greyish-mealy but sometimes greenish, 20-80 cm tall.

Leaves: Stem leaves alternate, linear to lanceolate, mostly entire, white-mealy below, glabrous and pale green above, gradually reduced upward, 1- or 3-nerved, fleshy.

Flowers: Inflorescence of loosely to closely clustered, mostly terminal panicles or short spikes, greyish-mealy, greenish.

Fruits: Thin, membranous envelopes 0.8-1.5 mm wide; seeds horizontal, nearly smooth, shiny.

Chenopodium fremontii* S. Wats. var. *fremontii

Fremont's goosefoot

HABITAT/RANGE:

Dry roadside banks, calcareous cliff bases or open forests in the steppe and montane zones; frequent in S BC east of the Coast-Cascade Mountains; E to AB and S to TX, CA and MX.

DESCRIPTION:

General: Annual herb from a taproot; stems erect, solitary, much branched, 10-50 cm tall.

Leaves: Stem leaves usually triangular with arrowhead-like lobes, entire except sometimes with a few teeth at the base, greenish on upper surfaces, usually greyish-mealy below, stalked, 1-4 cm long, rounded to pointed.

Flowers: Inflorescence of dense clusters in large terminal and smaller lateral spikes, densely mealy, without stalks.

Fruits: Thin, membranous envelopes, 1 mm wide; seeds obtusely margined, slightly roughened.

***Chenopodium glaucum* L.** (*C. glaucum* ssp. *salinum* [Standl.] Aellen, *C. glaucum* var *salinum* [Standl.] Boivin, *C. glaucum* var. *pulchrum* Aellen, *C. salinum* Standl.)

oak-leaved goosefoot

HABITAT/RANGE:

Mesic to dry roadsides, waste places and saline or alkaline flats in the steppe and montane zones; frequent in SC and SE BC, rare in SW BC (Vancouver); introduced from Eurasia.

DESCRIPTION:

General: Annual herb from a taproot; stems usually prostrate or low and spreading, numerous, branched from the base, 10-40 cm long.

Leaves: Stem leaves alternate, pale green on upper surfaces, greyish-mealy below, gradually reduced upwards, oblong egg-shaped or deltoid, 1-3 cm long, coarsely toothed, narrowing to stout stalks, fleshy.

Flowers: Inflorescence of dense clusters in numerous, short, axillary spikes and a terminal, simple or compound spike, greenish.

Fruits: Thin, membranous envelopes flattened at the top, 0.8 mm wide; seeds mostly horizontal, some vertical.

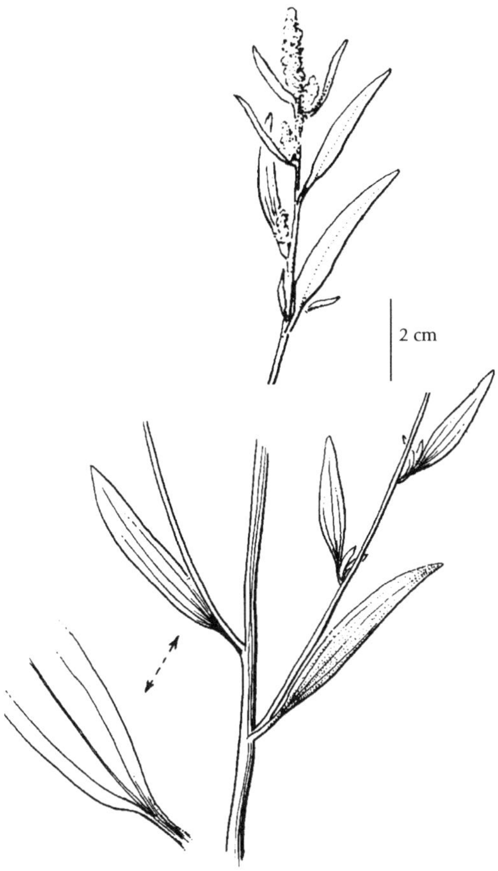

Chenopodium desiccatum

Chenopodium fremontii var. *fremontii*

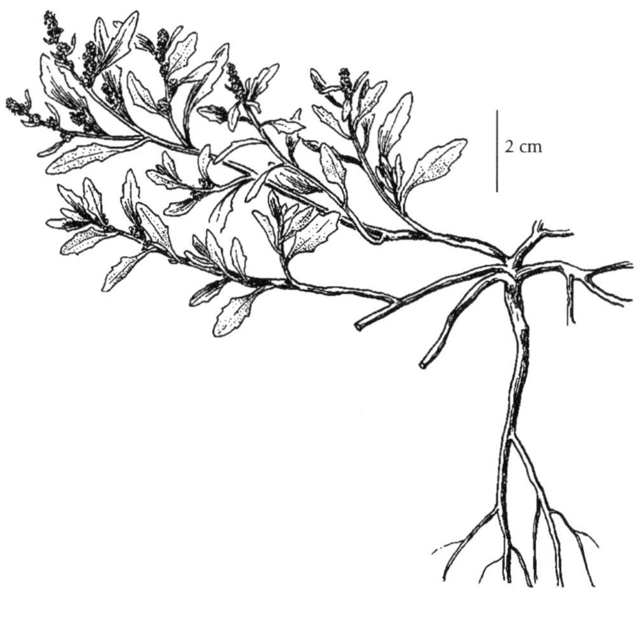

Chenopodium glaucum

Chenopodium hybridum **L.** (*C. gigantospermum* Aellen, *C. hybridum* var. *gigantospermum* [Aellen] Rouleau)
maple-leaved goosefoot

Habitat/Range:

Moist to dry forests and disturbed areas in the montane zone; infrequent in SC and SE BC, rare northward to Dawson Creek; introduced from Eurasia.

Description:

General: Annual herb from a taproot; stems erect, solitary to numerous, simple to freely branched, 0.2-1.5 m tall, glabrous except glandular in the inflorescence.

Leaves: Stem leaves alternate, broadly deltoid to egg-shaped, 5-20 cm long, shallow to prominently toothed or lobed, glabrous.

Flowers: Inflorescence of somewhat dense clusters in large terminal panicles and sometimes in reduced lateral spikes or panicles, greenish; sepals of the fruiting plant not keeled.

Fruits: Thin, wide, membranous envelopes; seeds strongly flattened, margins sharply angled, horizontal, surface finely pitted.

Chenopodium polyspermum **L.**
European goosefoot

Habitat/Range:

Dry roadsides, fields and waste places in the lowland zone; infrequent in SW BC; introduced from Europe.

Description:

General: Annual herb from a taproot; stems decumbent to erect, solitary to numerous, branched, usually squarish, 0.1-1.0 m tall/long, glabrous except glandular in the inflorescence.

Leaves: Stem leaves alternate, broadly elliptic to egg-shaped, 5-8 cm long, entire or rarely with a pair of slight teeth at the base, glabrous, short-stalked.

Flowers: Inflorescence of axillary, branched, spreading clusters; sepals of the fruiting plant abruptly sharp-pointed, not keeled.

Fruits: Thin, wide, membranous envelopes; seeds shiny black, 0.8-1.3 mm long, horizontal.

Chenopodium pumilio **R. Br.**
Australian goosefoot

Habitat/Range:

Mesic to dry beaches, lake shorelines and disturbed areas in the montane zone; rare in SE BC, known only from the Christina Lake area; introduced from Australia.

Description:

General: Annual herb from a taproot; stems prostrate to ascending, solitary, freely branched from the base, 20-40 cm tall/long.

Leaves: Stem leaves alternate, fleshy, stalked, oblong or lanceolate, 1-3 cm long, 2-4 coarse teeth on each side, with numerous yellow glands below, aromatic, reduced upwards.

Flowers: Inflorescence of numerous axillary clusters, greenish.

Fruits: Thin, whitish, membranous envelopes; seeds lens-shaped with a raised margin, laterally flattened, 0.5-0.6 mm wide, erect.

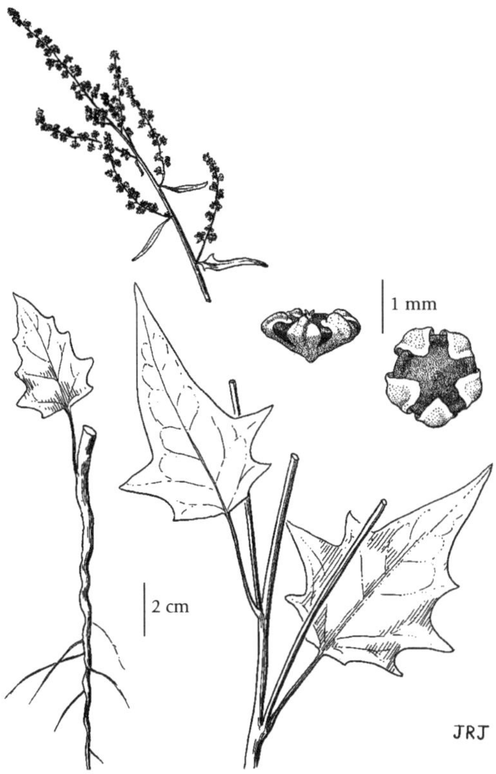

Chenopodium hybridum

Chenopodium polyspermum

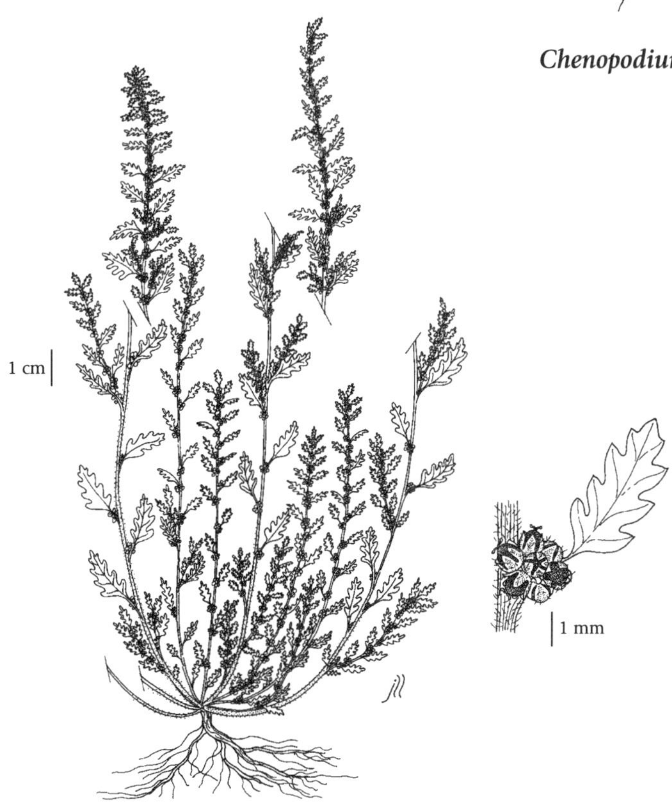

Chenopodium pumilio

***Chenopodium rubrum* L.**

red goosefoot; coast-blite

HABITAT/RANGE:

Wet to moist salt marshes or saline and alkaline flats in the lowland, steppe and lower montane zones; frequent in S BC, rare northward to NE BC; circumpolar, N to YT and NT, E to NF and S to MO, NE, NM, AZ, and CA; Eurasia.

DESCRIPTION:

General: Annual herb from a taproot; stems prostrate to erect, solitary, simple to branched from the base, 0.1-1.0 m tall/long, glabrous to somewhat woolly in the inflorescense.

Leaves: Stem leaves alternate, green on both surfaces, often becoming reddish-tinged, narrowing abruptly to a slender stalk, deltoid, egg-shaped to diamond-shaped, often arrowhead-shaped at the base, 1.5-10 cm long, gradually reduced upwards, fleshy.

Flowers: Inflorescence of unstalked, densely clustered, short, simple to more or less compound axillary spikes and a more freely compounded terminal spike, greenish.

Fruits: Thin, membranous envelopes, egg-shaped, strongly compressed, 0.8-1 mm long; seeds usually vertical.

Notes: Two sometimes poorly defined varieties occur in BC.

1. Stems prostrate, branched at the base, less than 2.5 cm tall .... var. *humile* (Hook.) Wats.
1. Stems, erect, not branched at the base, over 2.5 cm tall .... var. *rubrum*

***Chenopodium urbicum* L.**

upright goosefoot

HABITAT/RANGE:

Dry disturbed areas and waste places in the lowland, steppe and montane zones; rare in S BC; introduced from Eurasia.

DESCRIPTION:

General: Annual herb from a taproot; stems erect, branched, 10-50 cm tall, glabrous.

Leaves: Stem leaves alternate, 3-20 cm long, egg-shaped to deltoid, only slightly or not at all mealy, usually green on both surfaces.

Flowers: Inflorescence of densely clustered terminal and axillary spikes, greenish; sepals of the fruiting plants not strongly keeled along the midvein.

Fruits: Thin, membranous envelopes; seeds smooth but net-veined, margins rounded.

CORISPERMUM

***Corispermum hyssopifolium* L.** (*C. marginale* Rydb.)

bugseed

HABITAT/RANGE:

Dry sandy banks and disturbed areas in the lowland and steppe zones; rare (locally frequent in some years) in SW BC, rare in SC BC; introduced from Eurasia.

DESCRIPTION:

General: Annual herb from a taproot; stems ascending to erect, solitary, freely branching, especially near the base, often reddish-tinged, copiously covered in starlike hairs or nearly glabrous, 10-60 cm tall.

Leaves: Stem leaves alternate, linear to linear-lanceolate, entire, unstalked, primary ones 1-7 cm long, with starlike hairs or glabrous; bracts papery-margined, 5-9 mm long, lanceolate to egg-shaped.

Flowers: Inflorescence of short spikes in the axils of the upper, papery-margined bractlets, numerous, 1-3 cm long; calyx a single delicate sepal on the inner side; bracteoles narrowly to broadly lanceolate, papery-margined, 5-9 mm long.

Fruits: Papery, thin, membranous envelopes, egg-shaped to orbicular, 3-4 mm long, narrowly to broadly winged.

Notes: This species can be especially abundant on dredged sands along the lower Fraser River when the sand remains moist.

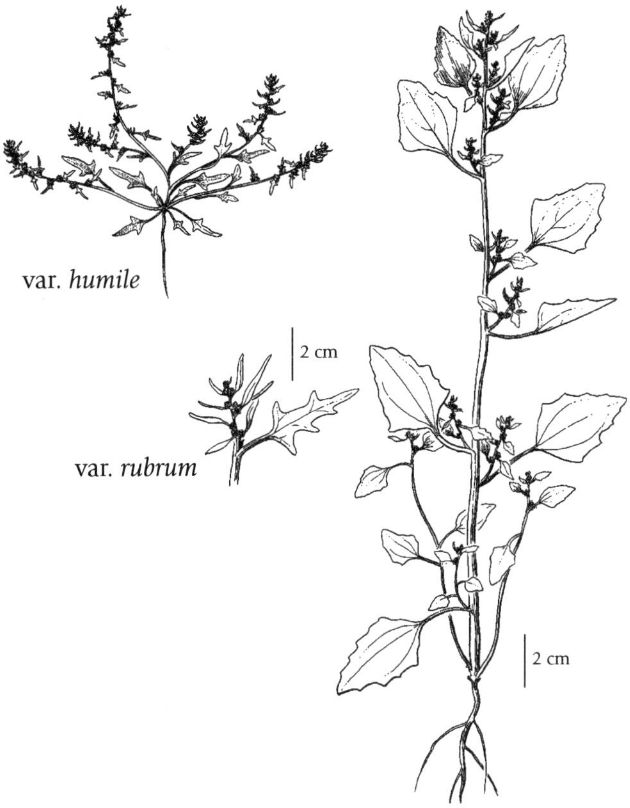

Chenopodium rubrum

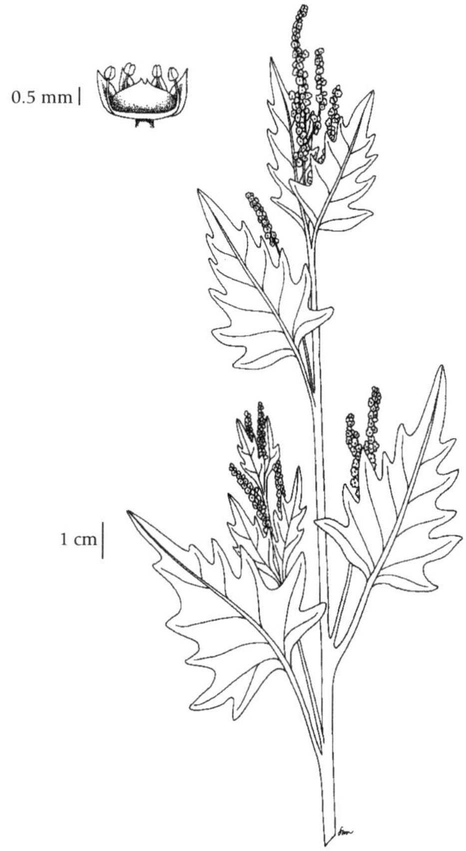

Chenopodium urbicum

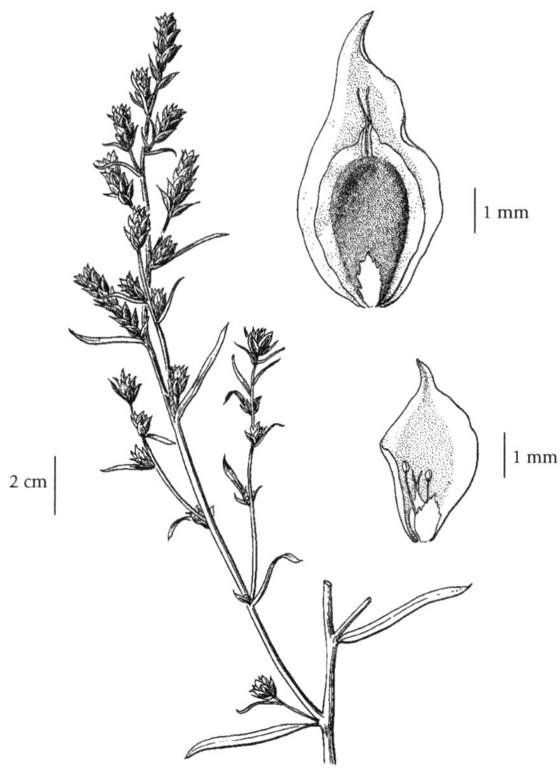

Corispermum hyssopifolium

KOCHIA

Kochia scoparia **(L.) Schrad.**

summer-cypress

HABITAT/RANGE:

Dry waste places in the steppe zone; rare in SW and NE BC, frequent in SC BC; introduced from Eurasia.

DESCRIPTION:

General: Annual herb from a taproot; stems erect, several, freely branching, strongly to finely woolly giving a silvery or rusty appearance, usually reddish-tinged, especially in the fall, 0.3-1.5 m tall.

Leaves: Stem leaves alternate, linear to narrowly lanceolate, 2-6 cm long, 3-5 nerved, softly hairy to rusty-bristled beneath, upper surface usually glabrous but fringed on the margins; bracts 3-10 mm long.

Flowers: Inflorescence of 3-6 flowers in axillary spikes along the stem; calyces rusty and long soft-hairy to woolly long-hairy, prominently flanged in fruit.

Fruits: Papery, thin, membranous envelopes; seeds horizontal.

MONOLEPIS

Monolepis nuttalliana **(J.A. Schultes) Greene**

poverty weed; Nuttall's monolepis

HABITAT/RANGE:

Dry roadsides, waste places, and saline and alkaline flats in the lowland, steppe and montane zones; infrequent in SC and SE BC, rare in SW and NE BC; introduced from the U.S.

DESCRIPTION:

General: Annual herb from a taproot; stems prostrate to ascending, few to much branched, more or less whitish-mealy, 10-30 cm tall/long.

Leaves: Stem leaves alternate, rather succulent, gradually reduced upward to leafy bracts, stalks slender, up to 3 cm long, gradually broadened into an arrowhead-shape, basal lateral lobes spreading, 1-4 mm long.

Flowers: Inflorescence of dense stalkless, axillary clusters, reddish; sepals 1.5-2.5 mm long, oblong to spoon-shaped, more or less abruptly short-pointed.

Fruits: Thin, pitted, membranous envelopes attached to the seeds; seeds 0.9-1.3 mm wide, dark brown.

POLYCNEMON

Polycnemon arvense **L.**

crunch weed

HABITAT/RANGE:

Dry grasslands in the steppe zone; rare in SC BC, known only from the Okanagan Valley; introduced from Eurasia.

DESCRIPTION:

General: Annual herb from a taproot; stems decumbent to erect, solitary, freely branching, the branches spirally twisted, glabrous, covered with wartlike projections, 3-20 cm tall/long.

Leaves: Stem leaves alternate, narrowly spoon-shaped, bases papery-margined, entire, unstalked, 3-12 mm long, softly spine-tipped, glabrous.

Flowers: Inflorescence of solitary flowers in the upper leaf axils, subtended by pairs of spinelike papery, whitish bractlets; calyces deeply 5-lobed, the papery sepals erect, or nearly so; bractlets equalling the perianth.

Fruits: Papery, thin, membranous envelopes; seeds 1-1.5 mm long.

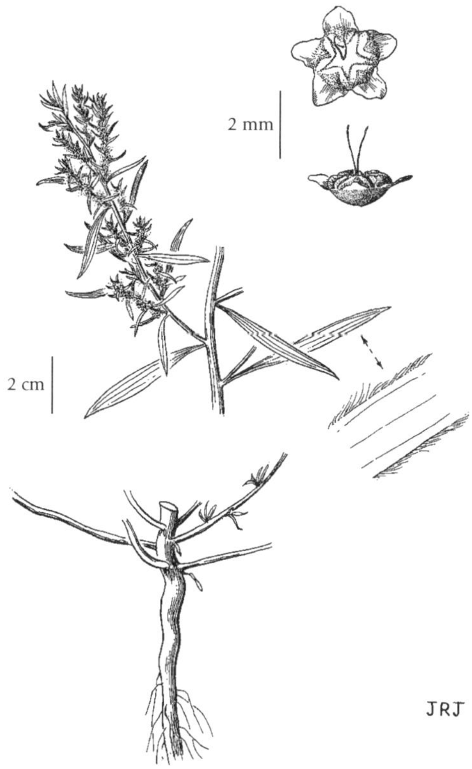

Kochia scoparia

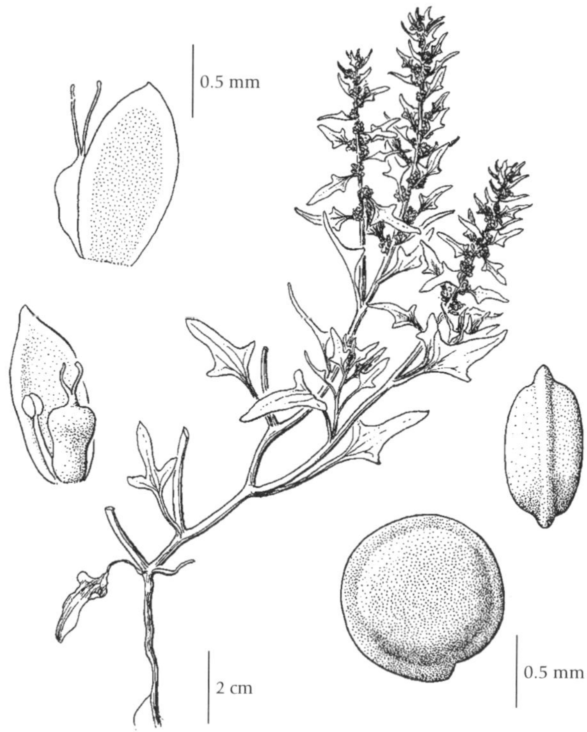

Monolepis nuttalliana

Polycnemon arvense

SALICORNIA

1. Plants perennials with slender rhizomes, usually matted; central flowers in each cluster barely exceeding the lateral ones .... *S. virginica*
1. Plants annuals, usually erect; central flowers in each cluster much exceeding the lateral ones.
    2. Joints of the spikes mostly 2 (rarely 2.5) mm long and as thick or almost as thick as the spikes .... *S. rubra*
    2. Joints of the spikes 2.5-4 (rarely 2) mm long and about half as thick as the spikes .... *S. maritima*

Salicornia maritima **Wolff & Jeffries** (*S. europaea auct. non* L.)

maritime glasswort

HABITAT/RANGE:

Wet salt marshes, shorelines and saline and alkaline flats in the lowland, steppe and montane zones; infrequent in S BC; N to AK, YT and NT, E to NF and S to GA, KS, NV and CA.

DESCRIPTION:

General: Annual herb from a fibrous root; stems decumbent to erect, solitary, simple to freely branched, 5-50 cm tall/long, glabrous, fleshy, often reddish with age.

Leaves: Stem leaves reduced to opposite, scalelike, fused bracts; lower scales slightly spreading, obtuse to rounded.

Flowers: Inflorescence of fleshy spikes, the flowers sunken in depressions of the joints just above the axils and opposite the fused bracts, the joints of the spike 2.5-4 (rarely 2) mm long and half as thick as the spikes; flowers in groups of 3, central flowers in each cluster much exceeding the lateral ones; spikes 2-5 cm long.

Fruits: Thin, papery-walled, membranous bladders enclosed by spongy flower scales, egg- to diamond-shaped; seeds 1 per bladder.

Salicornia rubra **A. Nels.**

red glasswort

HABITAT/RANGE:

Wet salt marshes, shorelines and saline and alkaline flats in the lowland, steppe and montane zones; infrequent in S BC; N to AK, YT and NT, E to NF and S to GA, KS, NV and CA.

DESCRIPTION:

General: Annual herb from a fibrous root; stems decumbent to erect, solitary, simple to freely branched, 10-25 cm tall/long, glabrous, fleshy, often reddish with age.

Leaves: Stem leaves reduced to opposite, scalelike, fused bracts; lower scales slightly spreading, obtuse to rounded.

Flowers: Inflorescence of fleshy spikes, the flowers sunken in depressions of the joints just above the axils and opposite the fused bracts, the joints of the spike mostly 2 (rarely 2.5) mm long and as thick, or almost as thick as the spikes; flowers in groups of 3, central flowers in each cluster much exceeding the lateral ones; spikes 2-5 cm long.

Fruits: Thin, papery-walled, membranous bladders enclosed by spongy flower scales, egg-shaped to diamond-shaped; seeds 1 per bladder.

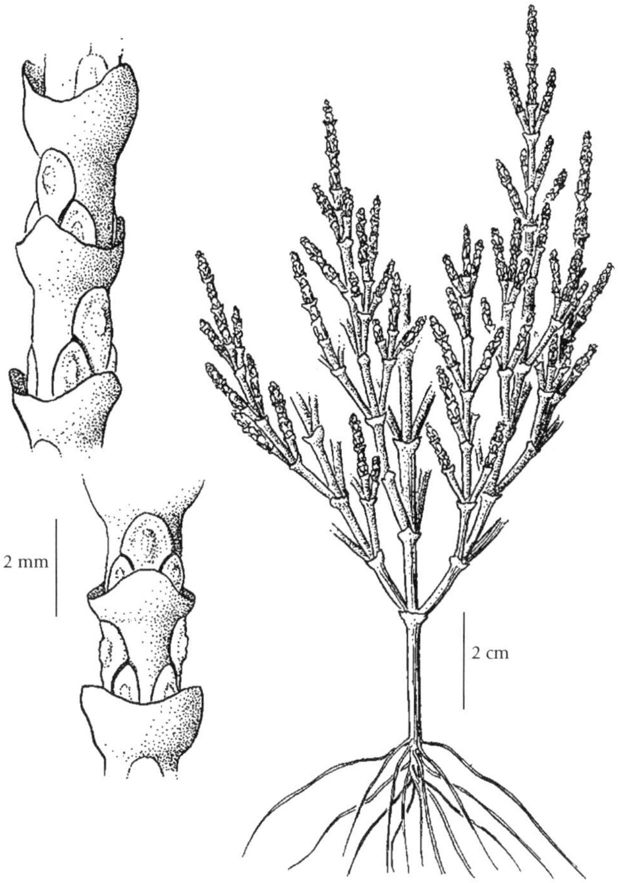

Salicornia maritima

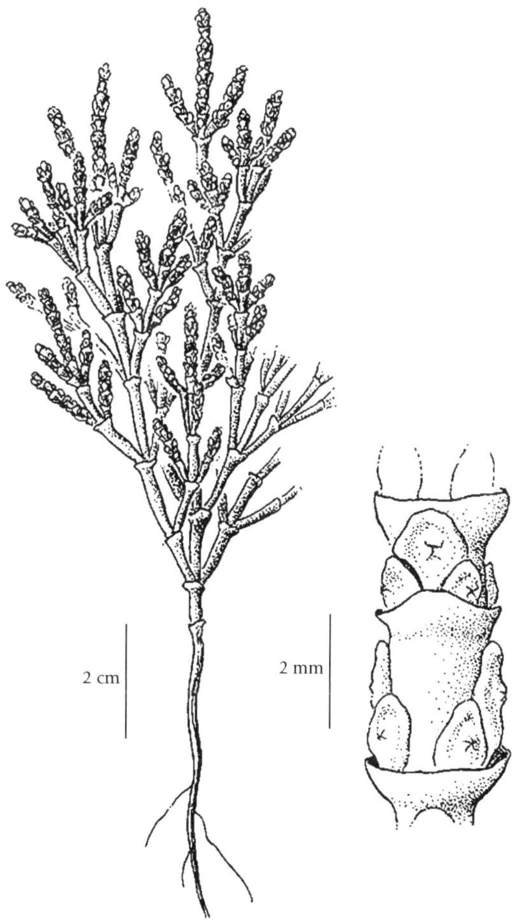

Salicornia rubra

***Salicornia virginica* L.** (*S. pacifica* Standl.)
American glasswort; perennial saltwort
HABITAT/RANGE:
Wet salt marshes and beaches in the lowland zone; common along the coast in BC; N to SE AK and S to CA and MX, also along the Atlantic and Gulf of MX coasts.
DESCRIPTION:
General: Perennial herb from a branching rhizome; matted; stems prostrate, freely branched, up to 1 m long.
Leaves: Stem leaves reduced to opposite, scalelike, fused bracts; lower scales slightly spreading, obtuse to rounded.
Flowers: Inflorescence of fleshy spikes, the flowers sunken in depressions of the joints just above the axils and opposite fused bracts, the joints of the spike mostly 0.5-2.5 mm long and about as thick as the spikes; flowers in groups of 3, central flowers in each cluster barely exceeding the lateral ones; spikes 1-4 cm long.
Fruits: Thin, papery-walled, membranous bladders enclosed by spongy flower scales; seeds 1 per bladder, hairy.

SALSOLA

***Salsola kali* L.**
Russian thistle; tumbleweed
HABITAT/RANGE:
Dry roadsides, fields and waste places in the lowland, steppe and montane zones, frequent in S BC; introduced from Eurasia.
DESCRIPTION:
General: Annual herb from a fibrous root; stems ascending, solitary, freely branched, glabrous to hairy, purplish, 0.1-1.0 m tall, becoming hardened, ridged and rounded in late summer, then breaking at groundlevel to form "tumbleweeds".
Leaves: Stem leaves alternate, linear, entire, succulent, spiny at the tip, reduced upward but broadened at the base, merging with papery-margined bracts.
Flowers: Inflorescence of solitary or clustered flowers in the axils of the spiny bracts; calyces 5-parted, becoming winged or keeled on the back.
Fruits: Thin, papery-walled, membranous envelopes, flattened, purplish, winged with prominent veins, cup-shaped.

SUAEDA

***Suaeda calceoliformis* (Hook.) Moq.** (*S. americana* [Pers.] Fern., *S. depressa auct. non* [Pursh] S. Wats., *S. maritima* [L.] Dumort. var. *americana* [Pers.] Boivin)
seablite; pahute weed
HABITAT/RANGE:
Wet salt marshes, beaches and saline and alkaline habitats in the lowland, steppe and montane zones; frequent in SC and SE BC, rare along the coast; N to YT, E to NF and S to TX and CA.
DESCRIPTION:
General: Annual herb from a taproot; stems decumbent to erect, solitary, freely branched from the base, 20-50 cm tall/long.
Leaves: Stem leaves alternate, fleshy, glaucous, narrow and rounded, 1-3 cm long, reduced gradually during flowering, upper leaves reduced, scarcely 5 mm long.
Flowers: Inflorescence of small, stalkless, greenish flowers, solitary in the axils of the bracts; calyx lobes unequal, 1 or more becoming keeled or crested.
Fruits: Thin, papery-walled, membranous envelopes; seeds horizontal, black, smooth or fine wrinkled, 1.0-1.7 mm wide.

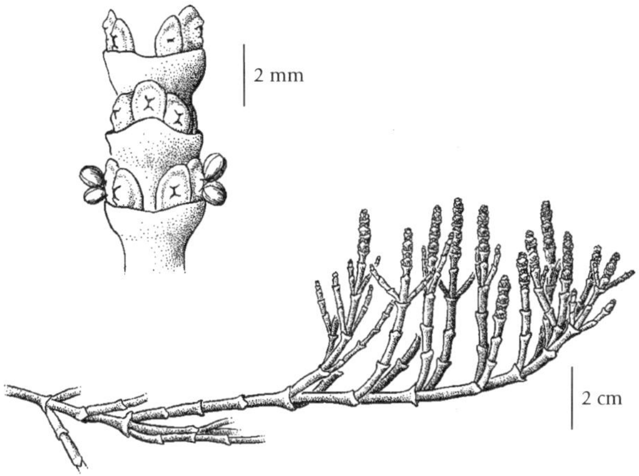

Salicornia virginica

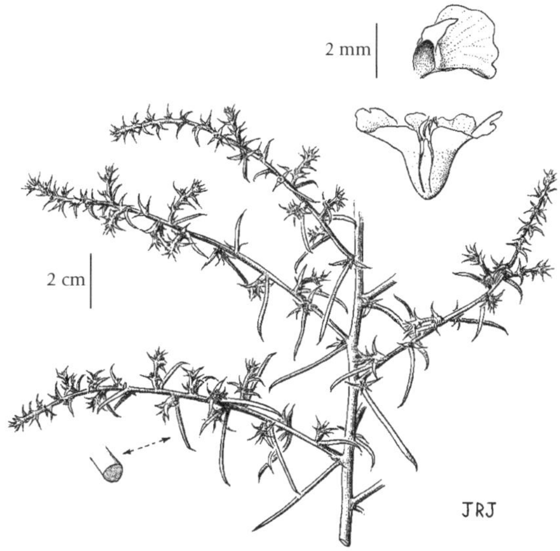

Salsola kali

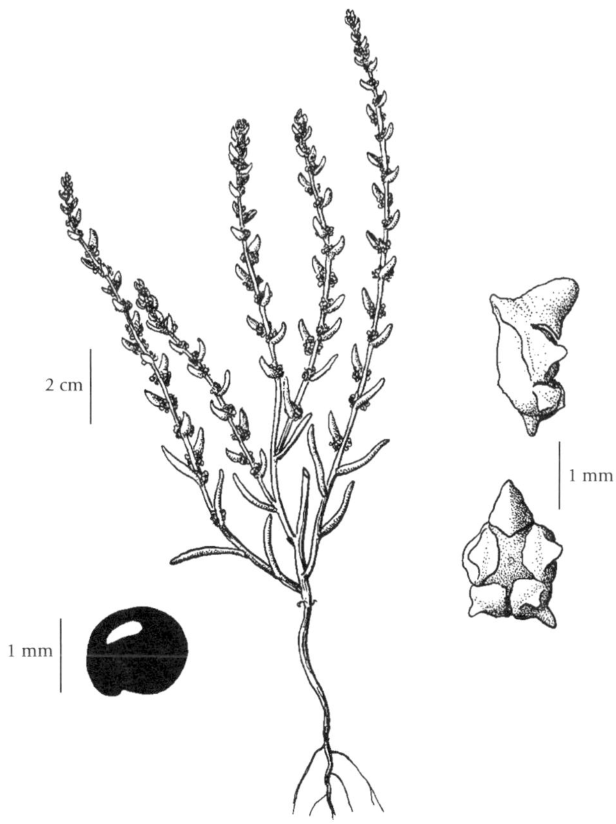

Suaeda calceoliformis

CLUSIACEAE [28]

1. Petals purplish or flesh-coloured; fruits purplish capsules; stamens in three groups, alternating with three large, orange glands .... *Triadenum*
1. Petals yellow; fruits yellowish to brownish capsules or purplish-black berries; stamens lacking intervening glands .... *Hypericum*

HYPERICUM

1. Plants shrubs; leaves 3-15 cm long; fruits purplish-black berries .... *H. androsaemum*
1. Plants herbs; leaves 1-4 cm long; fruits dry capsules.
    2. Flowers conspicuous, the petals 7-12 mm long; stamens more than 35, in 3-5 clusters.
        3. Stems quadrangular; petals with elongated black dots and streaks .... *H. maculatum*
        3. Stems rounded with 2 raised lines; petals bordered with black dots.
            4. Sepals linear-lanceolate to lanceolate, mostly abruptly sharp-pointed; leaves lanceolate to oblanceolate .... *H. perforatum*
            4. Sepals narrowly egg-shaped to triangular, round at the tip; leaves oblong to rounded .... *H. scouleri*
    2. Flowers inconspicuous, the petals 1-4 mm long; stamens less than 30, not in clusters.
        5. Flower bracts leafy, not tapering to a fine point.
            6. Stems usually prostrate and rooting at the lower nodes; flowers 1-15, not spreading on the branches .... *H. anagalloides*
            6. Stems usually erect, not rooting at the lower nodes; flowers 20-150, spreading on the branches .... *H. boreale*
        5. Flower bracts linear and tapering to a fine point.
            7. Principal leaves lanceolate to oblong; inflorescence erect; petals orange-yellow, 2-4 mm long .... *H. majus*
            7. Principal leaves elliptic; inflorescence spreading; petals yellow, 1-2 mm long .... *H. mutilum*

Hypericum anagalloides **Cham. & Schlect.**
bog St. John's-wort

Habitat/Range:
Moist to wet bogs, ditches and lawns in the lowland zone; frequent in SW BC; S to MT, CA and MX.

Description:
General: Perennial herb from freely rooting, matted stolons; stems numerous, prostrate to decumbent, glabrous, 3-25 cm long.

Leaves: Stem leaves egg-shaped to elliptic, 4-15 mm long, 5-to 7-nerved, rounded at the apex, slightly clasping at the base, glabrous.

Flowers: Inflorescence 1- to 15-flowered; petals salmon-yellow, 2-4 mm long; stamens 15-25, longer than the sepals; sepals triangular to egg-shaped, 2-4 mm long, unequal, rounded to abruptly sharp-pointed, margins glabrous; styles 3, 1.5-2 mm long.

Fruits: Membranous capsules, 3 mm long; seeds yellow-brown, about 0.5 mm long, transversely wrinkled and longitudinally finely-lined.

---

[28] Contributed by F. Lomer and G.W. Douglas

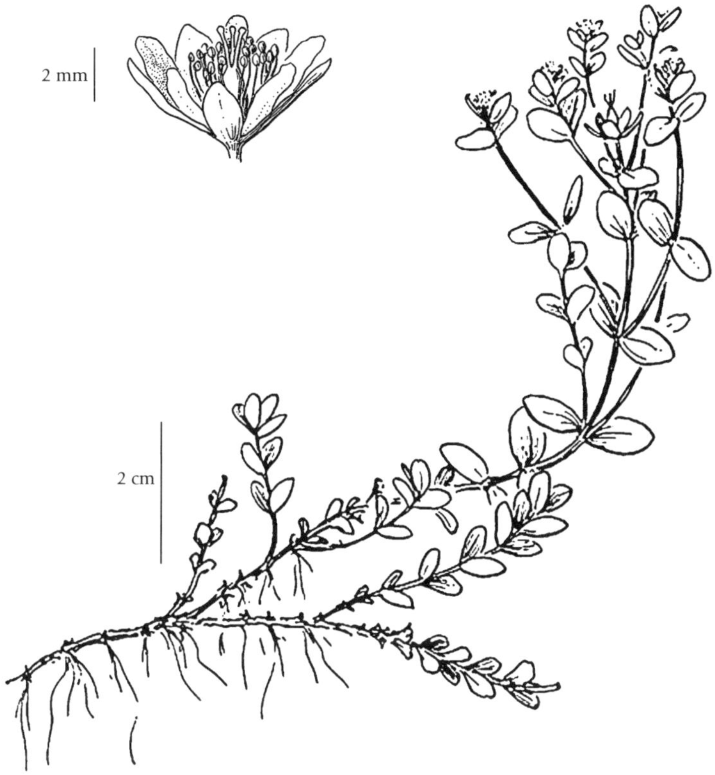

Hypericum anagalloides

Hypericum androsaemum **L.**
tutsan
Habitat/Range:
Moist meadows, ditches and thickets in the lowland zone; rare in SW BC, known only from the lower Fraser Valley; introduced from Europe.
Description:
General: Perennial shrub, partially evergreen; stems erect, few branched, up to 80 cm tall.
Leaves: Stem leaves rounded, unstalked, obtuse, 3-15 cm long, glabrous with minute translucent glands.
Flowers: Inflorescence few flowered, spreading on the branches; petals yellow, 6-10 mm long; sepals rounded, obtuse, unequal, persistent and deflexed in fruit; stamens about equal to the petals; styles 3, shorter than the ovary.
Fruits: Berries, purplish-black, 7-10 mm long.

Hypericum boreale **(Britt.) Bickn.**
northern bog St. John's-wort
Habitat/Range:
Wet to moist ditches, disturbed areas, field and river banks in the lowland zone; frequent in SW BC, known only from the lower Fraser Valley; introduced from E. North America.
Description:
General: Perennial herb from a fibrous root; stems erect, much branched above, glabrous, 10-40 cm tall.
Leaves: Stem leaves elliptic to oval, unstalked, obtuse, glabrous, 1-4 cm long, 0.5-1.0 cm wide.
Flowers: Inflorescence 20-150+ flowered; petals orange-yellow, 2-4 mm long; sepals 3-5 mm long; stamens less than 20; styles 3.
Fruits: Membranous capsules, 5.5-7.5 mm long; seeds yellow-brown, about 0.5 mm long, transversely wrinkled and longitudinally finely-lined.

Hypericum maculatum **Crantz ssp.** ***obtusiusculum*** **(Tourlet) Hayek**
spotted St. John's-wort
Habitat/Range:
Moist waste places, ditches and forest margins in the lowland zone; infrequent in WC (Prince Rupert) and SW BC (Pemberton and the lower Fraser Valley); introduced from Europe.
Description:
General: Perennial herb from a rhizome; stems erect, branched, quadrangular, glabrous, 15-100 cm tall.
Leaves: Stem leaves elliptic, unstalked, obtuse, glabrous with a few scattered glands and black marginal dots, 1.5-4.0 cm long, 1.0-2.5 cm wide.
Flowers: Inflorescence 20-60 flowered; petals golden yellow, with elongated black dots and streaks, up to 12 mm long; sepals rounded, about 5 mm long; stamens in 3 bundles; styles 3.
Fruits: Membranous capsules, up to 8 mm long; seeds brownish, about 1 mm long, net-veined.

Hypericum majus **(A. Gray) Britt.**
large Canadian St. John's-wort
Habitat/Range:
Moist to wet waste places, vernal pools, gravelly shorelines and tidal shores in the lowland zone; rare in SW BC, known only from the lower Fraser Valley; E to NF and S to DE, PA, NE, CO and WA.
Description:
General: Perennial or sometimes annual herb from a fibrous root; stems erect, branched above, glabrous, 10-50 cm tall.
Leaves: Stem leaves lanceolate to oblong, unstalked, obtuse to abruptly sharp-pointed, 5-7 nerved, 1-4 cm long, glabrous.
Flowers: Inflorescence 5-50 flowered, erect; petals orange-yellow, 2-4 mm long; sepals lanceolate, abruptly sharp-pointed, 5-6.5 mm long; stamens less than 30; styles 3.
Fruits: Membranous capsules, 5.5-7.5 mm long; seeds yellowish, 0.5 mm long, longitudinally grooved and finely transversely corrugate.

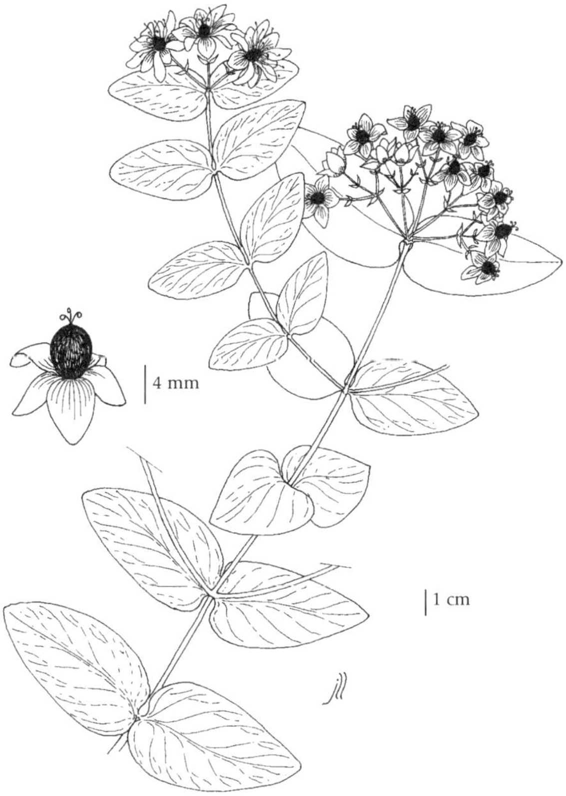

Hypericum androsaemum

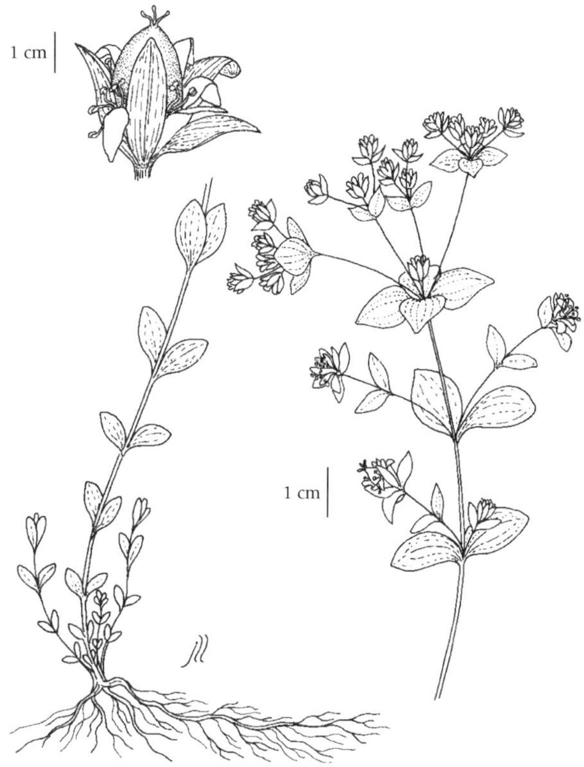

Hypericum boreale

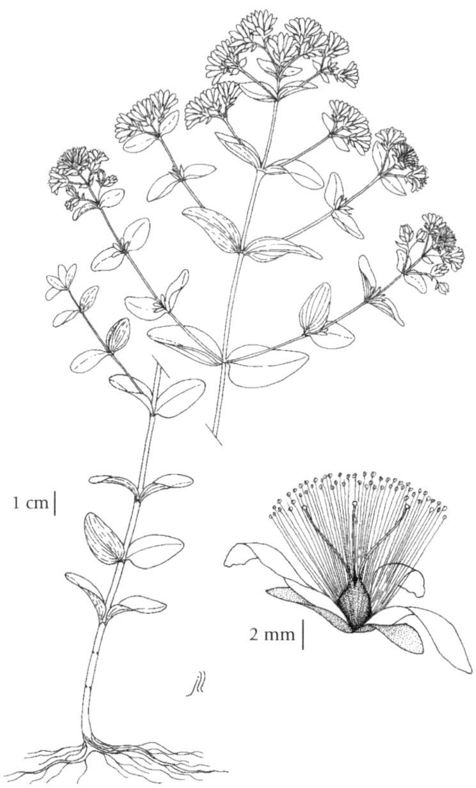

Hypericum maculatum ssp. *obtusiusculum*

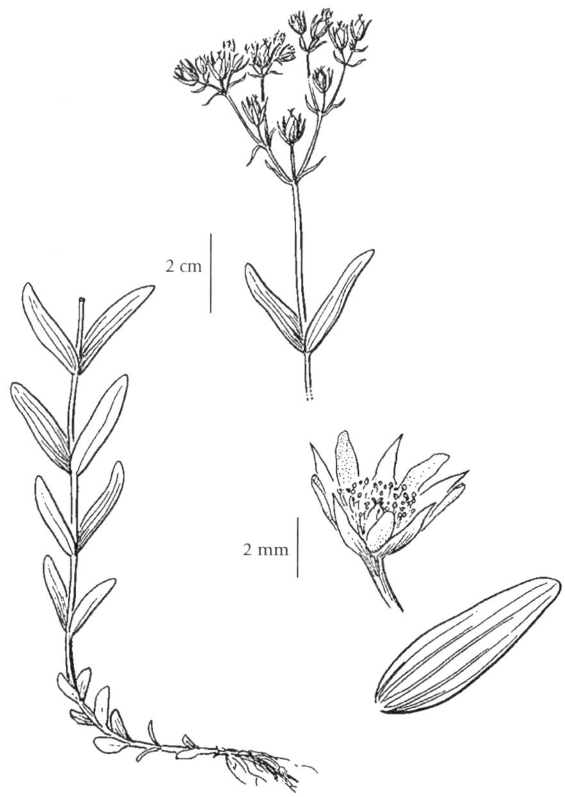

Hypericum majus

Hypericum mutilum **L.**

small Canadian St. John's-wort

HABITAT/RANGE:

Wet to moist river banks and disturbed areas in the lowland zone; rare in SW BC, known only from the lower Fraser Valley; introduced from E North America.

DESCRIPTION:

General: Perennial or sometimes annual herb from a fibrous root; stems erect, much branched above, glabrous, 10-75 cm tall.

Leaves: Stem leaves lanceolate to elliptic or rounded, unstalked, obtuse to abruptly sharp-pointed, 3-5 nerved, 1-3 cm long, 0.5-1.5 cm wide, partly clasping at the base, glabrous.

Flowers: Inflorescence 20-100 flowered, spreading; petals orange-yellow, 1-2 mm long; sepals linear-oblong, abruptly sharp-pointed, 2-3 mm long; stamens less than 20; styles 3.

Fruits: Membranous capsules, 2.5-4.0 mm long; seeds yellowish, 0.5 mm long, longitudinally grooved and finely transversely corrugate.

Hypericum perforatum **L.**

common St. John's-wort

HABITAT/RANGE:

Mesic to dry fields, roadsides and disturbed areas in the lowland, steppe and montane zones; infrequent in SW and WC BC, frequent in SC BC; introduced from Eurasia.

DESCRIPTION:

General: Perennial herb from a rhizome; stems erect, much branched above, glabrous, 0.1-1.0 m tall.

Leaves: Stem leaves lanceolate or oblanceolate, unstalked, obtuse, 1-3 cm long, 0.5-1.5 mm wide, glabrous with translucent glandular dots throughout and black marginal dots.

Flowers: Inflorescence up to 100+ flowered; petals yellow, 8-10 mm long; sepals linear-lanceolate to lanceolate, abruptly sharp-pointed, 4-6 mm long; stamens 75-100, united basally into 3 groups; styles 3.

Fruits: Membranous capsules, 5-8 mm long; seeds dark brownish, 1.0-1.3 mm long, pitted in longitudinal rows.

Hypericum scouleri **Hook.** (*H. formosum auct. non* H.B.K.,*H. formosum* Kunth var. *nortoniae* [M.E. Jones] C.L. Hitchc. = ssp. *nortoniae*; *H. formosum* var. *scouleri* [Hook.] Coult. = ssp. *scouleri*)

western St. John's-wort

HABITAT/RANGE:

Moist to wet streamsides, estuaries, marshes and open slopes in all zones except the alpine zone; infrequent in S BC; S to WY, CA and MX.

DESCRIPTION:

General: Perennial herb from a long stolon and rhizome; stems erect, branched above, glabrous, 5-80 cm tall.

Leaves: Stem leaves oblong to rounded, unstalked, obtuse, 1-3 cm long, 0.5-1.5 cm wide, glabrous with black marginal dots.

Flowers: Inflorescence up to 50+ flowered; petals pale to bright yellow, 7-12 mm long; sepals narrowly egg-shaped to triangular, obtuse, 3-4 mm long; stamens 75-100, united basally into 3 groups; styles 3, 3-5 mm long.

Fruits: Membranous capsules, 6-9 mm long; seeds yellow to brownish, about 0.8 mm long, finely net-veined and longitudinally grooved.

Notes: Two subspecies occur in BC.

1\. Stems few branched in the inflorescence, mostly 5-20 cm tall; leaves rounded; plants infrequent at higher elevations in S BC, most common in SE BC .... ssp. *nortoniae* (M.E. Jones) J. Gillett

1\. Stems branched below the inflorescence, mostly 20-80 cm tall; leaves narrowly egg-shaped; infrequent at lower elevations in S BC, most common in SW BC .... ssp. *scouleri*

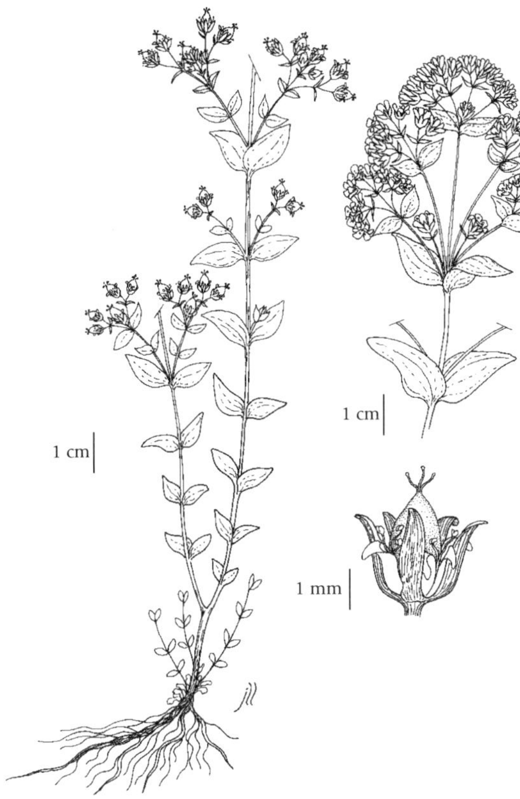

Hypericum mutilum

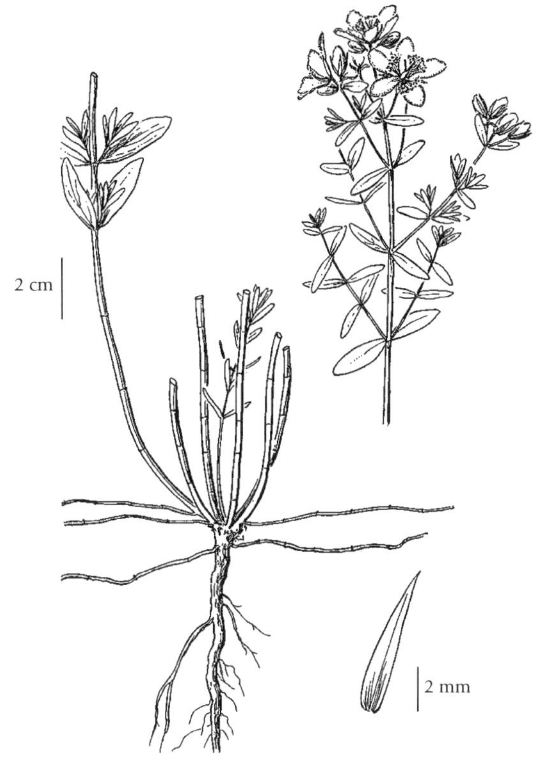

Hypericum perforatum

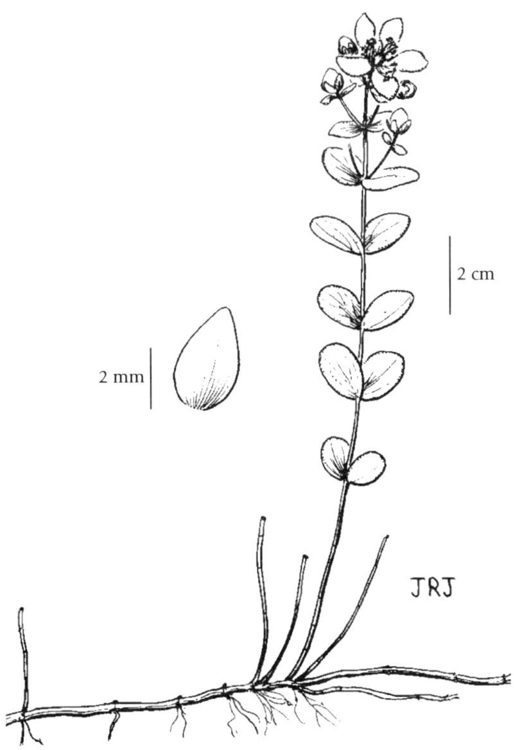

Hypericum scouleri ssp. *nortoniae*

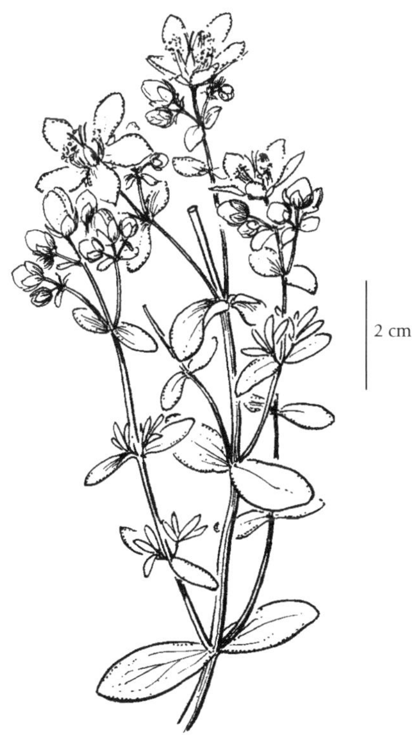

Hypericum scouleri ssp. *scouleri*

TRIADENUM

Triadenum fraseri **(Spach) Gleason** (*Hypericum virginicum* ssp. *fraseri* [Spach] J. Gillett)
Fraser's St. John's-wort

Habitat/Range:

Wet bogs, marshes and shorelines in the lowland zone; rare in SW BC, known only from the lower Fraser Valley; introduced from E North America.

Description:

General: Perennial herb from a taproot; stems erect, few branched, glabrous, 30-60 cm tall.

Leaves: Stem leaves oblong to oblong egg-shaped, unstalked, heart-shaped at the base with rounded or notched tips, 3-6 cm long, 2-3 cm wide, glabrous with translucent glands below.

Flowers: Inflorescence up to 5- to 50-flowered; petals purplish or flesh-coloured, 5-8 mm long; sepals elliptic or oblong, obtuse, 3-5 mm long; stamens in 3 groups of 3 alternating with 3 large, orange glands.

Fruits: Membranous capsules, 7-12 mm long; seeds dark purplish, about 1 mm long, finely corrugated and pimply in lines.

CONVOLVULACEAE

CONVOLVULUS

1. Leaves thick, fleshy, kidney-shaped, glabrous; restricted to coastal sand dunes .... *C. soldanella*
1. Leaves thin, not fleshy, seldom kidney-shaped, glabrous to hairy; seldom, if ever, on coastal sand dunes.
   2. Bracts broad, often heart-shaped, usually borne adjacent to the calyces and largely concealing them .... *C. sepium*
   2. Bracts linear, often borne well below the calyces .... *C. arvensis*

Convolvulus arvensis **L.**
field bindweed

Habitat/Range:

Mesic to dry disturbed areas and waste places in the lowland, steppe and montane zones; common in S BC, rare northward to Dawson Creek; introduced from Eurasia.

Description:

General: Perennial herb from a widespread and deeply descending rhizome; stems trailing to somewhat twining, 0.2-2.0 m long, often forming carpets.

Leaves: Stem leaves alternate, narrowly egg- to arrowhead-shaped at the base, blunt or sharp-pointed at the tip, 2-6 cm long, glabrous to sparsely hairy, stalked.

Flowers: Inflorescence of 1 or 2, funnel-shaped, 15-25 mm wide, wide-mouthed flowers on erect stalks that exceed the leaves; petals white to pinkish-purple, fused, twisted in bud, pleated in flower; linear bracts midway down flowering stalks.

Fruits: Capsules cone-shaped, 5-7 mm long; seeds 4 mm long, smooth.

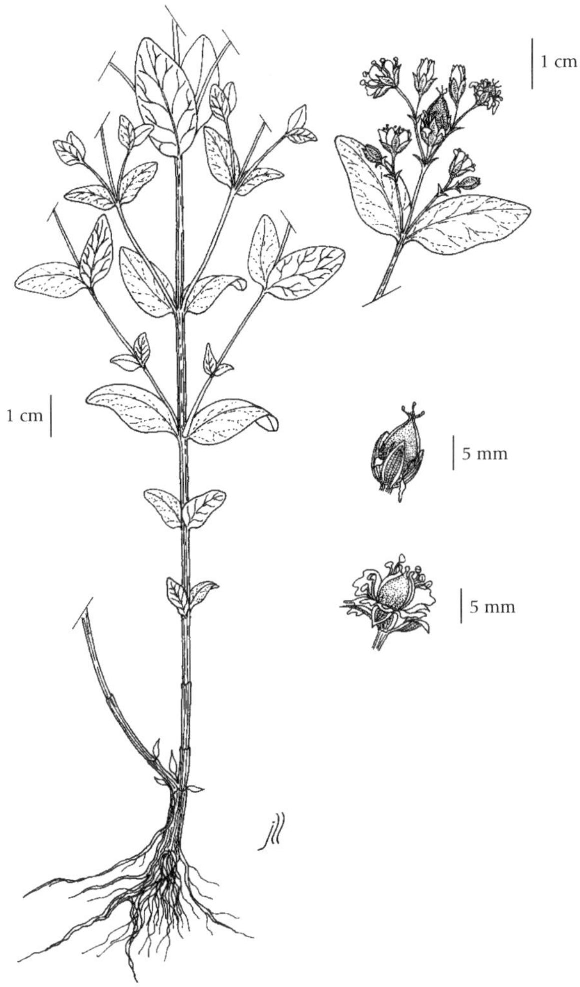

Triadenum fraseri

Convolvulus arvensis

***Convolvulus sepium* L.** (*Calystegia sepium* [L.] R. Br.)
hedge bindweed; wild morning-glory
HABITAT/RANGE:
Moist streamsides, river bottoms and shorelines in the lowland, steppe and montane zones; common in SW BC, infrequent on the Queen Charlotte Islands, rare in NE, SC and SE BC; introduced from E North America.
DESCRIPTION:
General: Perennial herb from an elongate, slender rhizome; stems trailing to climbing, 2-3 m long, glabrous to softly hairy.

Leaves: Stem leaves alternate, 5-12 cm long, arrowhead-shaped with basal lobes rounded to distinctly angled, tips pointed, glabrous to hairy.

Flowers: Inflorescence of single, 4-7 mm wide, wide-mouthed, funnel-shaped flowers on stalks shorter than the leaves; petals white to deep pink, fused, twisted in bud, pleated in flower; bracts broadly heart-shaped to egg-shaped.

Fruits: Capsules egg-shaped to globose, 1 cm long; seeds 5 mm long, finely hairy-warty.

***Convolvulus soldanella* L.** (*Calystegia soldanella* [L.] R. Br.)
beach bindweed; beach morning-glory
HABITAT/RANGE:
Moist to mesic sand dunes in the lowland zone; infrequent along the coast; S to CA, also in the islands of the Pacific Ocean and Europe.
DESCRIPTION:
General: Perennial from a deep rhizome; stems creeping but not twining, 10-30 cm long.

Leaves: Stem leaves alternate, thick, fleshy, broadly kidney-shaped, 1.5-4 cm wide, glabrous, basal lobes somewhat angled.

Flowers: Inflorescence of single, 3-6 mm wide, wide-mouthed, funnel-shaped flowers subtended by or equal to the broad, heart-shaped bracts; petals pinkish-purplish, fused; sepals oblong egg-shaped, bristle-tipped, about 1 cm long.

Fruits: Capsules subglobose, 1.5 cm long; seeds 6-8 mm long, essentially smooth.

CORNACEAE

CORNUS

1. Flowers in open clusters; bracts inconspicuous; mature drupes white to somewhat bluish; plants medium to tall shrubs .... *C. stolonifera*
1. Flowers in dense clusters; bracts large, showy, white, 4 or more; mature drupes red; plants either low herbs or medium-sized trees.
    2. Plants medium-sized trees; flowers unstalked .... *C. nuttallii*
    2. Plants low herbs; flowers short-stalked.
        3. Leaves unstalked, egg-shaped to elliptic, lateral veins arising from the base of the leaf, or nearly so; petals and sepals dark purplish .... *C. suecica*
        3. Leaves short-stalked, oblong egg-shaped to diamond-shaped, lateral veins arising from the midvein in the lower third of the leaf; petals and sepals usually yellowish or greenish, sometimes purplish-tinged, never dark purplish .... *C. canadensis*

Convolvulus sepium

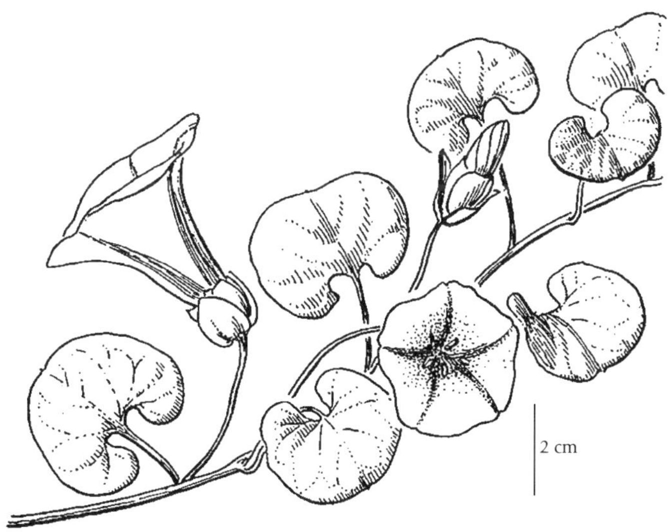

Convolvulus soldanella

***Cornus canadensis* L.**

bunchberry; Canadian bunchberry

Habitat/Range:

Moist to mesic forests and openings in all but the alpine zone; common throughout BC; amphiberingian, N to AK, YT and NT, E to NF and S to MN, PA, NM and CA; E. Asia.

Description:

General: Low, trailing perennial herb from a rhizome with a somewhat woody base; stems erect, solitary, simple, 5-25 cm tall, minutely hairy, greenish to reddish, leafless or with bracts on the lower part.

Leaves: Evergreen more or less, 4-7 in a terminal whorl, oblong egg-shaped to diamond-shaped, short-stalked, 2-8 cm long, green above, whitish beneath, lateral veins arising from the midvein in the lower third of the leaf.

Flowers: Inflorescence of a solitary, stalked, umbrella-shaped cluster subtended by 4 white to pinkish or purplish-tinged, broadly to narrowly egg-shaped bracts; petals 1-1.5 mm long, greenish-white with a purplish tinge.

Fruits: Fleshy drupes, 6-8 mm long, bright coral red when mature, usually several.

Notes: Hybrids between *C. canadensis* and *C. suecica* have been treated as either a variety of *C. canadensis* (var. *intermedia* Farr.) or as a separate species (*C. unalaschkensis* Ledeb.). The latter taxon is a common tetraploid with features between its parents.

***Cornus nuttallii* Aud. *ex* T.& G.**

western flowering dogwood; Pacific dogwood

Habitat/Range:

Mesic forests in the lowland and montane zones; common in SW BC; S to CA, disjunct in ID.

Description:

General: Irregularly branched deciduous tree up to 20 m tall; bark blackish-brown, smooth, becoming finely ridged with age, young branches greyish-purplish.

Leaves: Opposite, deciduous, oval, stalked, pointed at the tip, 4-10 cm long, deep green above, greyish-brown below, turning red in the fall, characteristic veins curving parallel to the leaf edge.

Flowers: Inflorescence of numerous, unstalked, hemispheric clusters 1.5-2 cm wide, subtended by 4-7 conspicuous white or pinkish-tinged, large (2-7 cm long) showy bracts, flowering in spring and often again in the fall; petals greenish-white, usually purplish-tinged.

Fruits: Fleshy drupes, 10 mm long, bright red.

***Cornus stolonifera* Michx.** (*C. sericea* L.)

red-osier dogwood

Habitat/Range:

Wet to mesic streamsides, lakesides, swamps and forests in the lowland, steppe and montane zones; common throughout BC; N to AK, YT and NT, E to NF and S to WV, OH, NE, NM, CA and MX.

Description:

General; Deciduous shrub; stems 1-6 m tall, spreading by layering of prostrate stems that simulate stolons; young branches opposite, bright reddish to reddish-purple, later turning greyish-green.

Leaves: Opposite, deciduous, oval, pointed, 4-12 cm long, greenish above, whitish below, turning reddish in the autumn, 5-7 prominent parallel veins converging at the leaf tips, white threads running through the veins.

Flowers: Inflorescence of flat-topped terminal clusters, 2-4 cm wide; flowers white to greenish-white, 4 petals and stamens, bracts inconspicuous.

Fruits: Berrylike drupes, 7-9 mm long, white sometimes blue-tinged, each with flattened stone.

Cornus canadensis

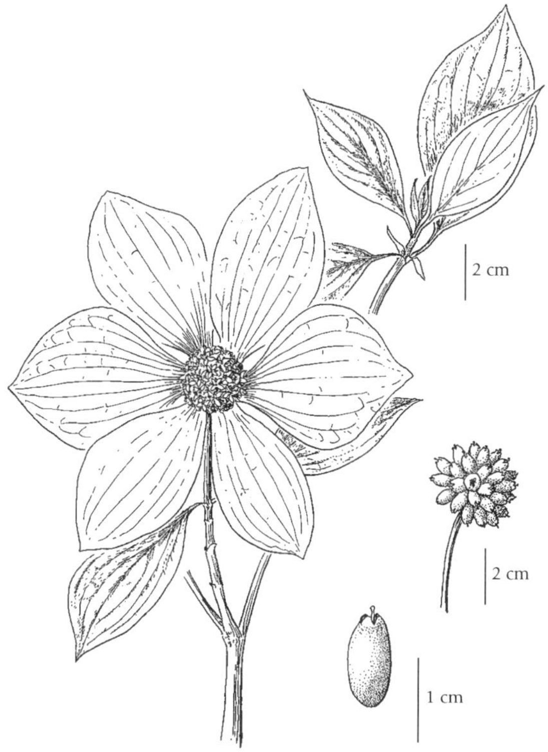

Cornus nuttallii

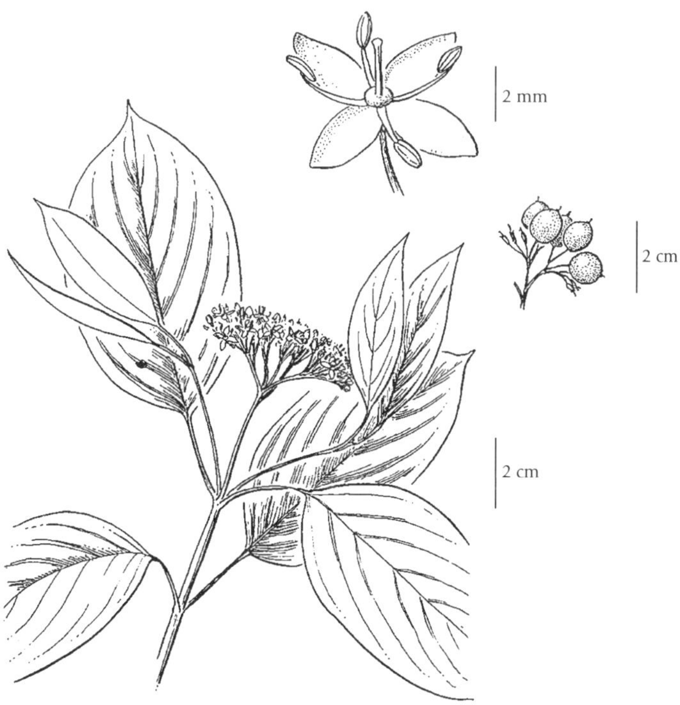

Cornus stolonifera

Cornus suecica **L.**
dwarf bog bunchberry
HABITAT/RANGE:
Moist to mesic forests and meadows in the lowland to alpine zones; rare in coastal BC; circumpolar, N to AK and NT and E to NF; Greenland, Eurasia.
DESCRIPTION:
General: Perennial herb from a rhizome; stems erect, 4-26 cm tall, minutely bristly with 2-pronged hairs.
Leaves: 1-2 pairs of stem leaves below the terminal pair or whorl of 4-6 leaves, egg-shaped to elliptic, unstalked, lateral veins arising from the base of the leaf or nearly so, sparsely hairy above, glabrous and paler below.
Flowers: Inflorescence of solitary, umbrella-shaped clusters subtended by 4 white, yellowish, pinkish or purplish bracts; petals and sepals dark purplish.
Fruits: Fleshy drupes, 7-12 mm long, red.

CRASSULACEAE

1. Plants perennials; flowers usually 5-merous; stamens usually 10 .... *Sedum*
1. Plants annuals; flowers usually 4-merous; stamens 4 .... *Crassula*

CRASSULA

1. Flowers solitary; carpels several-seeded .... *C. aquatica*
1. Flowers clustered; carpels 1-2 seeded .... *C. connata*

Crassula aquatica **(L.) Schoeni.** (*Tillaea aquatica* L.)
pigmyweed
HABITAT/RANGE:
Wet vernal pools and mudflats in the lowland zone; infrequent, scattered through S BC; circumpolar, N to AK and NT and S to CA and MX; Iceland, Eurasia.
DESCRIPTION:
General: Weak annual herb, rooting nodally; stems ascending or erect, freely branching, 2-6 cm tall/long, glabrous.
Leaves: Opposite, entire, succulent, linear to linear-oblanceolate, 3-6 mm long, with heart-shaped, sheathing bases.
Flowers: Inflorescence of single flowers in the leaf axils; flowers usually 4-merous, greenish, short-stalked but lengthening in fruit; calyces cup-shaped; petals exceeding the sepals, whitish; stamens shorter than the petals.
Fruits: Follicles, less than 2 mm long, purplish, 6-12 seeded.

Crassula connata **(Ruiz & Pavon) Berger var.** ***connata*** (*C. erecta* [H.& A.] Berger, *Tillaea erecta* H.& A.)
erect pigmyweed
HABITAT/RANGE:
Wet to moist vernal pools on coastal bluffs in the lowland zone; rare on S Vancouver Island and the Gulf Islands; S to AZ, CA and MX, Chile.
DESCRIPTION:
General: Weak annual herb, rooting nodally; stems ascending or erect, freely branching, 2-6 cm tall, glabrous.
Leaves: Opposite, entire, succulent, oblong egg-shaped, 1.5-3 mm long.
Flowers: Inflorescence of flower clusters in the leaf axils; flowers usually 4-merous, greenish, short-stalked but lengthening in fruit; calyces cup-shaped; petals no longer than the sepals, whitish.
Fruits: Follicles, less than 2 mm long, purplish, 1-2 seeded.

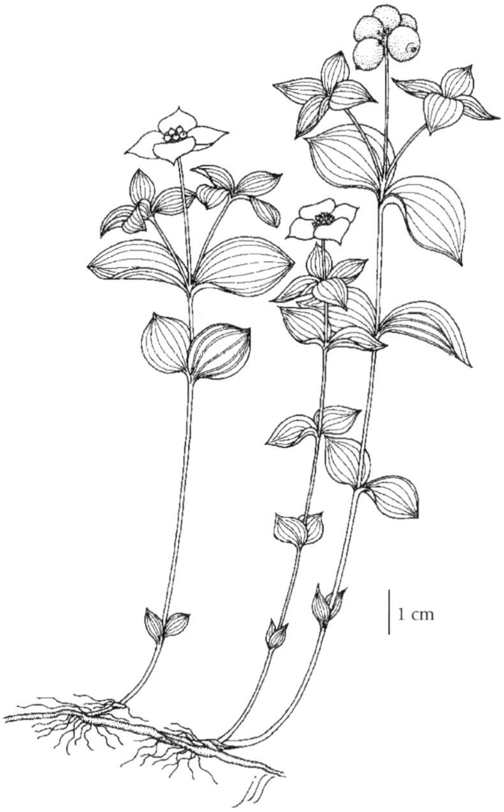

Cornus suecica

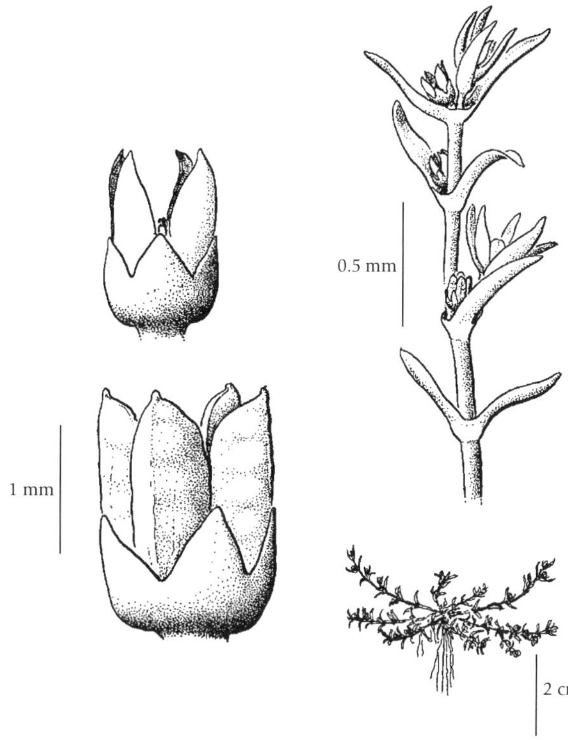

Crassula aquatica

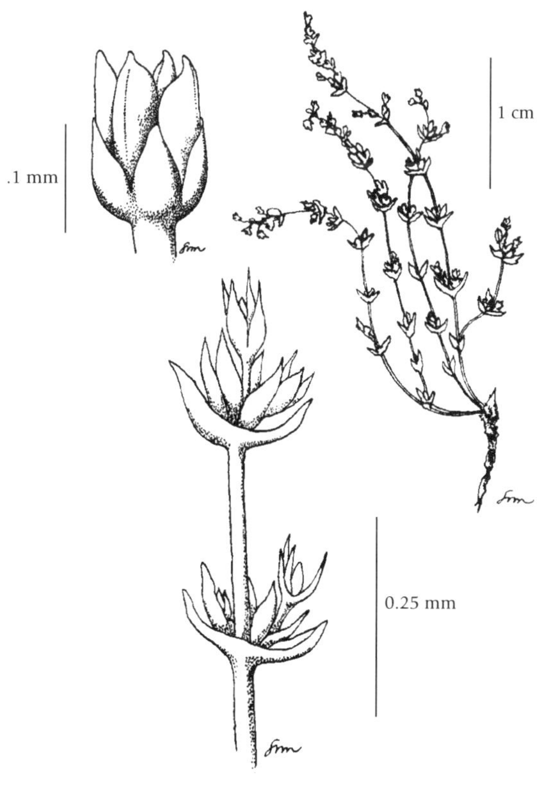

Crassula connata var. *connata*

SEDUM

1. Flowers usually purple .... *S. integrifolium*
1. Flowers usually yellow.
   2. Leaves of the flowering stems mainly opposite .... *S. divergens*
   2. Leaves of the flowering stems alternate.
      3. Leaves usually broadest below midlength, tapering from midlength to tip, or to both ends, either strongly keeled or more or less round in cross section.
         4. Leaves strongly keeled, strongly slender-pointed, usually bearing bulbil-like propagules in the upper leaves .... *S. stenopetalum*
         4. Leaves not strongly keeled, sometimes completely deciduous by flowering time, more or less round in cross section, rarely bearing propagules in the upper leaves.
            5. Follicles spreading; leaves smooth, egg-shaped .... *S. acre*
            5. Follicles erect; leaves usually finely pimply, linear .... *S. lanceolatum*
      3. Leaves usually broadest above midlength and tapered to the base, usually strongly flattened.
         6. Petals white, obtuse, about equal to the stamens .... *S. album*
         6. Petals yellow, abruptly sharp-pointed to long-pointed, subequal to the stamens.
            7. Petals (8) 10-13 mm long, narrowly oblong-lanceolate and long-pointed; follicles erect .. .. *S. oreganum*
            7. Petals 7-10 mm long, narrowly oblong-lanceolate but not long-pointed; follicles more or less spreading .... *S. spathulifolium*

***Sedum acre* L.**

goldmoss stonecrop

Habitat/Range:

Dry rock outcrops in the lowland zone; locally frequent garden escape on S Vancouver Island, the Gulf Islands and adjacent mainland, rare elsewhere in S BC; introduced from Eurasia.

Description:

General: Perennial herb from a stout rhizome; mat-forming; stems ascending to erect, several, branched above, 4-10 cm tall, glabrous.

Leaves: Stem leaves alternate, smooth, evergreen, 3-5 mm long, more or less cylindric, succulent, unstalked, not strongly keeled, sometimes completely deciduous by flowering time, rarely bearing propagules in the upper leaves.

Flowers: Inflorescence of terminal, umbrella-shaped clusters; petals 5, bright yellow; sepals 5, lanceolate.

Fruits: Follicles, spreading.

***Sedum album* L.**

white stonecrop

Habitat/Range:

Dry gravelly areas in the lowland zone; rare garden escape in SW BC; introduced from Eurasia.

Description:

General: Perennial herb from a stout rhizome; mat-forming; stems decumbent to ascending, several, branched above, 10-20 cm tall/long, glabrous.

Leaves: Stem leaves alternate, smooth, evergreen, 6-12 mm long, nearly round in cross section, linear-oblong to ellipsoid, succulent, unstalked, usually broadest above the midlength and tapered to the base, usually somewhat flattened.

Flowers: Inflorescence of terminal, umbrella-shaped clusters; petals white, obtuse, about equal to the stamens.

Fruits: Follicles, spreading.

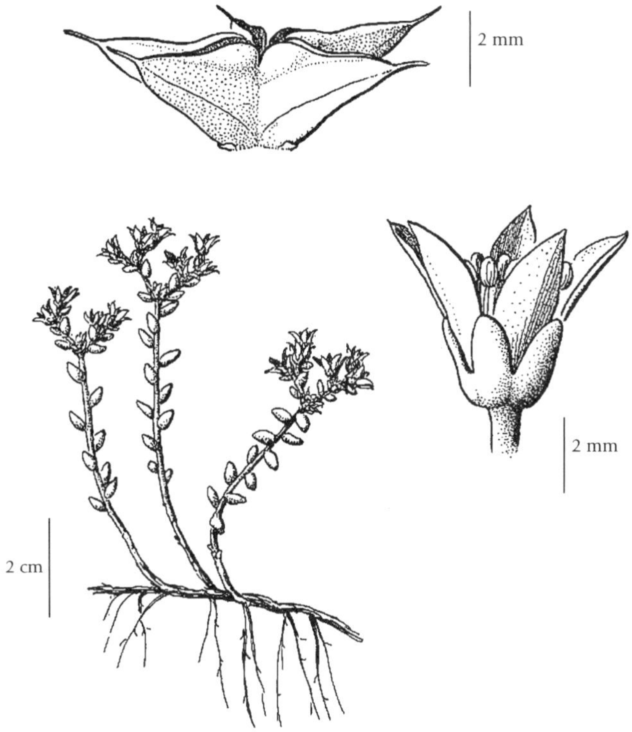

Sedum acre

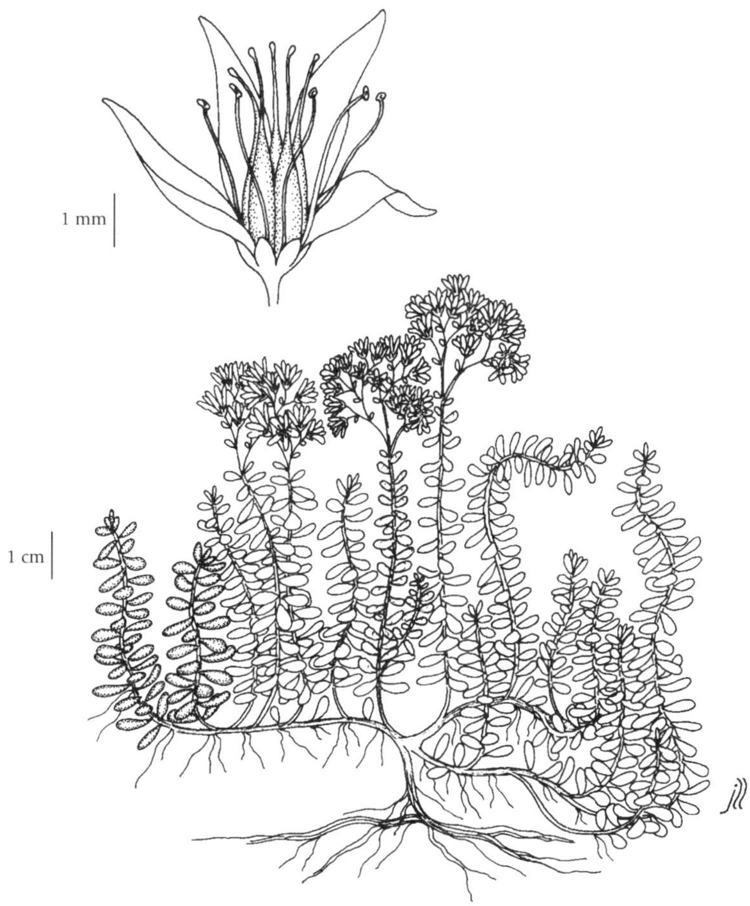

Sedum album

***Sedum divergens* S. Wats.**

spreading stonecrop

HABITAT/RANGE:

Dry rocky cliffs and talus slopes from the lowland to alpine zones; frequent in and W of the Coast-Cascade Mountains, infrequent elsewhere; S to OR.

DESCRIPTION:

General: Perennial herb from a rooting rhizome or stem; mat-forming; stems ascending to erect, branched above, freely rooting, 5-15 cm tall, glabrous.

Leaves: Stem leaves opposite or nearly so, finely hairy, succulent, oval to broadly egg-shaped, mostly 4-8 mm long on sterile shoots, more oblong on flowering stems.

Flowers: Inflorescence of tight, terminal, umbrella-shaped clusters with leafy bracts; petals 5, yellow; stamens slightly shorter than the petals.

Fruits: Follicles, basally fused for 1.5-2 mm then strongly spreading, cream coloured.

Sedum integrifolium* (Raf.) A. Nels. ssp. *integrifolium (*Rhodiola integrifolia* Raf., *Sedum alaskanum* [Rose] Rose *ex* Hutchins., *S. frigidum* Rydb., *S. rosea* ssp. *integrifolium* [Raf] Hult., *S. rosea* var. *integrifolium* [Raf] Berger)

roseroot

HABITAT/RANGE:

Moist to mesic rocky cliffs, talus slopes and meadows in the lowland to alpine zones; common throughout BC; N to AK, YT and NT, E to AB and S to CO, NV and CA.

DESCRIPTION:

General: Perennial herb from a thick, branched rhizome; mat-forming; stems erect, numerous, branched above, clustered, 5-35 cm tall, glabrous.

Leaves: Stem leaves egg-shaped to elliptic or oblanceolate, succulent, reduced and scalelike below, larger further up the stem, 0.5-4 cm long, green to pinkish.

Flowers: Inflorescence of terminal, umbrella-shaped clusters; petals 4, dark purple (female) or yellow (male); stamens longer than the petals.

Fruits: Follicles, erect with spreading tips, red or purple.

***Sedum lanceolatum* Torr.**

lance-leaved stonecrop

HABITAT/RANGE:

Mesic to dry coastal bluffs, rocky slopes, rock outcrops and talus slopes in all zones except the alpine zone; var. *lanceolatum* – common throughout BC, var. *nesioticum* – infrequent; absent from the Queen Charlotte Islands and adjacent coast; N to YT, E to S SK, and S to NE, CO, NM and CA.

DESCRIPTION:

General: Perennial herb from a slender rhizome; mat-forming; stems decumbent to ascending, several, branched above, 5-25 cm tall/long, glabrous.

Leaves: Stem leaves alternate, linear to linear-lanceolate, succulent, rounded in cross section, not strongly keeled, strongly pointed, finely pimply to smooth, upper leaves deciduous by flowering time, lower leaves crowded to form basal rosettes, often reddish-green, sometimes covered with a white waxy powder, turning golden in autumn.

Flowers: Inflorescence of terminal, umbrella-shaped clusters; petals 5, yellow but sometimes reddish-tinged; stamens shorter than the petals.

Fruits: Follicles, erect, fused at the base.

Notes: Two varieties occur in BC.

1. Leaves usually smooth, closely tufted and overlapping on the flowering stems .... var. *nesioticum* (G.N. Jones) C. L. Hitchc.
1. Leaves usually finely pimply, usually not overlapping on the flowering stems .... var. *lanceolatum*

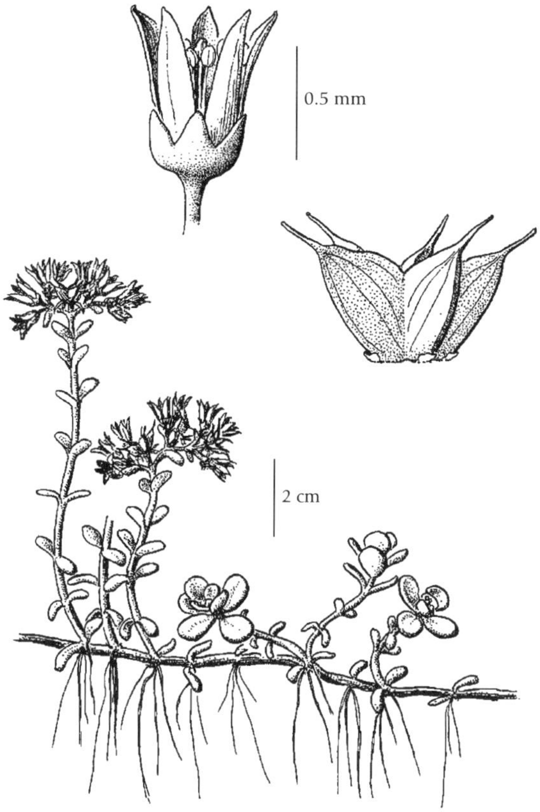

Sedum divergens

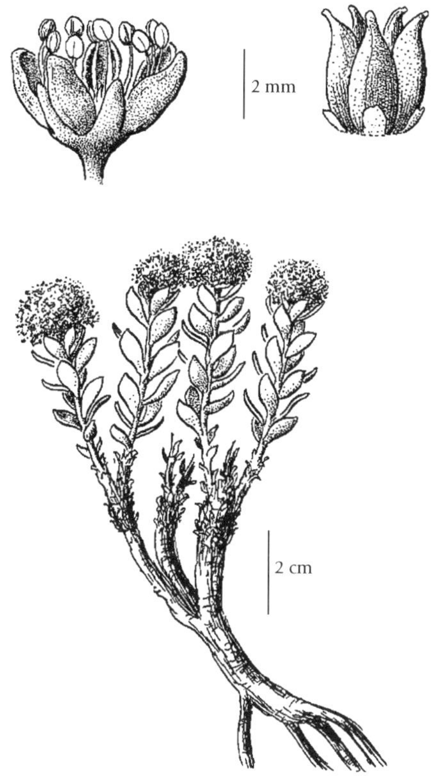

Sedum integrifolium ssp. *integrifolium*

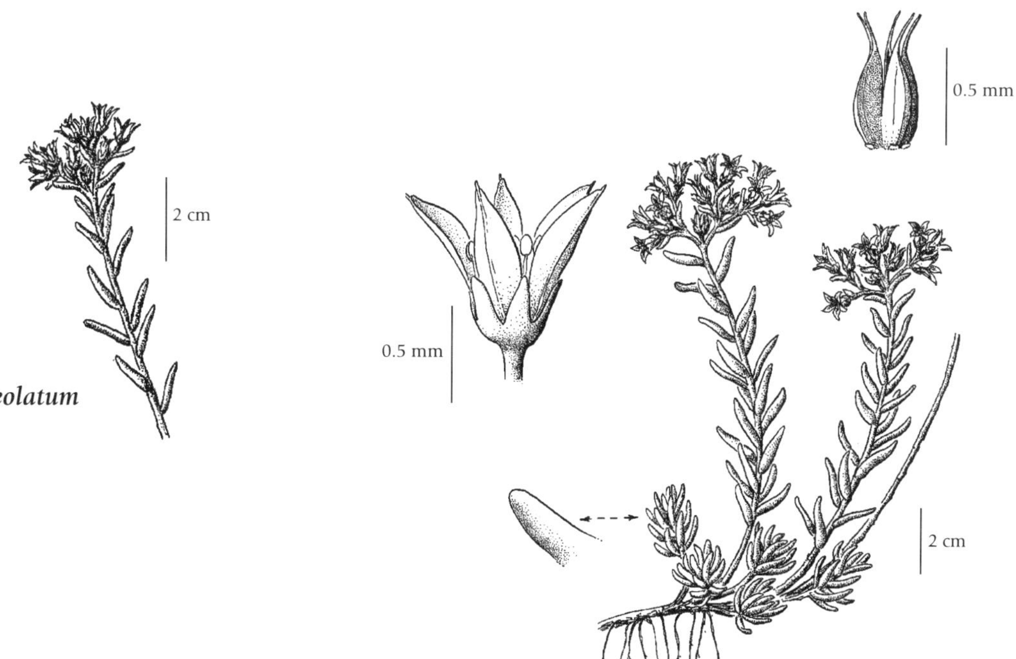

var. *lanceolatum*

var. *nesioticum*

Sedum lanceolatum

Sedum oreganum **Nutt.**
Oregon stonecrop
HABITAT/RANGE:
Mesic to dry rocky cliffs and talus slopes in the lowland and montane zones; common in SW BC; N to AK and S to CA.
DESCRIPTION:
General: Perennial herb from a stout rhizome; stems ascending, several, branched above, 6-15 cm tall, glabrous and glaucous.

Leaves: Stem leaves alternate, crowded and rosette-forming on the sterile shoots, overlapping on the flowering stems, egg- to spoon-shaped, 20-30 mm long, 6-12 mm wide, strongly flattened, broadest above the midlength and tapered to the base, green becoming bronze.

Flowers: Inflorescence of terminal, umbrella-shaped clusters; petals 5, yellow becoming pinkish with maturity, petals pointed, longer than the stamens, (8) 10-13 mm long, narrowly lanceolate and long-pointed.

Fruits: Follicles, erect and only very slightly fused at the base.

Sedum spathulifolium **Hook.** (*S. pruinosum* Britt. and *S. spathulifolium* ssp. *pruinosum* [Britt.] Clausen & Uhl var. *pruinosum*)
broad-leaved stonecrop
HABITAT/RANGE:
Mesic coastal cliffs and forest openings in the lowland and montane zones; common in SW BC, known only from Vancouver Island and the Gulf Islands; S to CA.
DESCRIPTION:
General: Perennial herb from a stout rhizome and rootstock; stems ascending to erect, several, branched above, 5-22 cm tall, glabrous and glaucous.

Leaves: Basal leaves in a rosette; stem leaves alternate, oblong to spoon-shaped or wedge-shaped, 0.5-2.5 cm long, succulent, flattened, broadest above the midlength and tapered to the base, strongly flattened, crowded, pale green to reddish in full sun, glaucous.

Flowers: Inflorescence of terminal, flat-topped umbrella-shaped clusters; petals 5, bright yellow, 7-10 mm long, narrowly oblong-lanceolate but not long-pointed.

Fruits: Follicles, erect but strongly spreading.

Notes: Two weakly separated varieties occur in BC.

1. Rosette leaves lightly glaucous .... var. *spathulifolium*
1. Rosette leaves strongly glaucous .... var. *pruinosum* (Britt.) Boivin

Sedum stenopetalum **Pursh** (*S. douglasii* Hook.)
worm-leaved stonecrop; narrow-petaled stonecrop
HABITAT/RANGE:
Mesic to dry rocky cliffs, talus slopes and meadows from the lowland and steppe to subalpine zones; frequent in S BC, rare northward to 57° N in NE BC; E to SW AB and S to MT and CA.
DESCRIPTION:
General: Perennial from a strong rhizome and rootstock; stems ascending to erect, several, branched above, 5-27 cm tall, glabrous.

Leaves: Stem leaves alternate, pimply roughened at the tips, linear to narrowly lanceolate, strongly keeled or nerved on the dorsal surface, broadest above midlength and tapered to the base, usually strongly flattened; lower leaves becoming deciduous, often leaving their midribs attached to the stem.

Flowers: Inflorescence of terminal, umbrella-shaped clusters; petals 5, yellow, 6-10 mm long, narrowly oblong-lanceolate but not long-pointed, longer than the stamens.

Fruits: Follicles, more or less spreading.

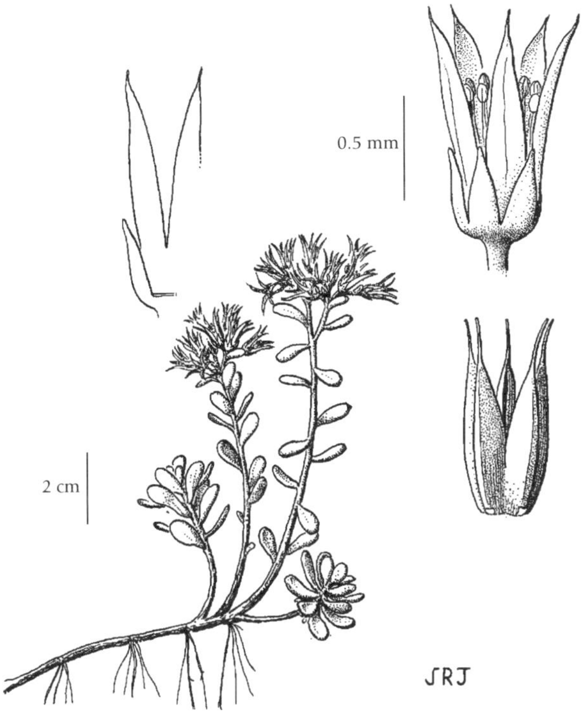

Sedum oreganum

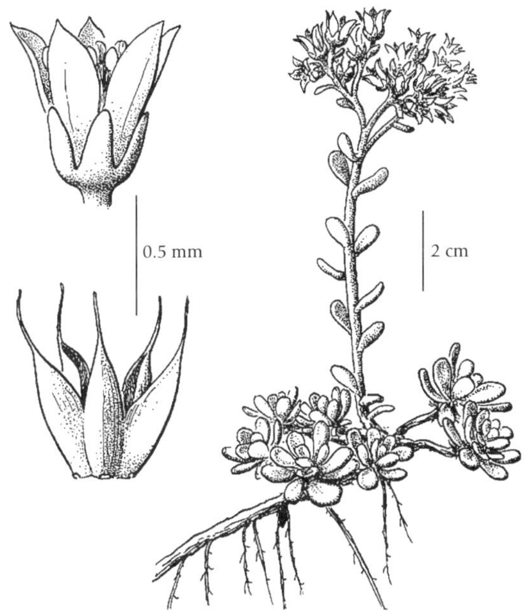

Sedum spathulifolium

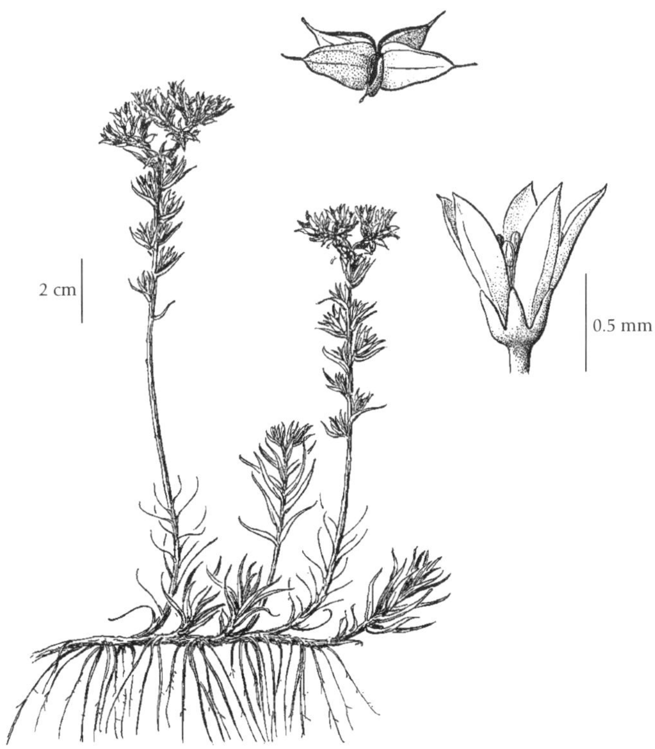

Sedum stenopetalum

CUCURBITACEAE

1. Plants annuals in SC BC; fruits ellipsoid or subglobose; seeds flattened, rough .... *Echinocystis*
1. Plants perennials in SW BC; fruits oblong egg-shaped; seeds turgid, smooth .... *Marah*

ECHINOCYSTIS

Echinocystis lobata **(Michx.) T.& G.**
wild cucumber; prickly cucumber

HABITAT/RANGE:

Moist meadows and thickets in the lowland and steppe zones; rare in SW and SC BC; introduced from E North America.

DESCRIPTION:

General: Climbing annual herb; the vines with tendrils.

Leaves: Stem leaves alternate, palmately 5-lobed, stalked, the stalks up to 15 cm long, short rough-hairy, heart-shaped at the base with the mostly triangular lobes remotely callous-toothed.

Flowers: Inflorescence of axillary panicles; perianth bell-shaped with flaring lobes, greenish-white; corolla lobes spreading, narrowly lanceolate, usually 3-6 mm long, 1 mm wide.

Fruits: Inflated spiny bladders, ellipsoid or suborbicular, fibrous-netted within; seeds flattened, rough.

MARAH

Marah oreganus **(T.& G.) Howell**
manroot; bigroot

HABITAT/RANGE:

Moist fields and open meadows and slopes in the lowland zone; rare in SW BC, known only from Vancouver Island and the Gulf Islands; S to N CA.

DESCRIPTION:

General: Climbing perennial from an enlarged woody root; annual stems climbing or trailing with branched tendrils.

Leaves: Stem leaves alternate, irregularly palmately lobed, stalked, bristly above, sparsely hairy below, heart-shaped at the base, about 20 cm long.

Flowers: Inflorescence of bell-shaped flowers in a panicle; perianths bell-shaped, whitish, with 5 to 8 petals/sepals; male flowers in narrow-stalked clusters from the leaf axils; single female flowers short-stalked from the same axils as the male flowers; corolla lobes flaring, oblong egg-shaped.

Fruits: Inflated, weakly spiny bladders, ellipsoid to oblong egg-shaped; seeds 1 or 2 in each locule, turgid, smooth.

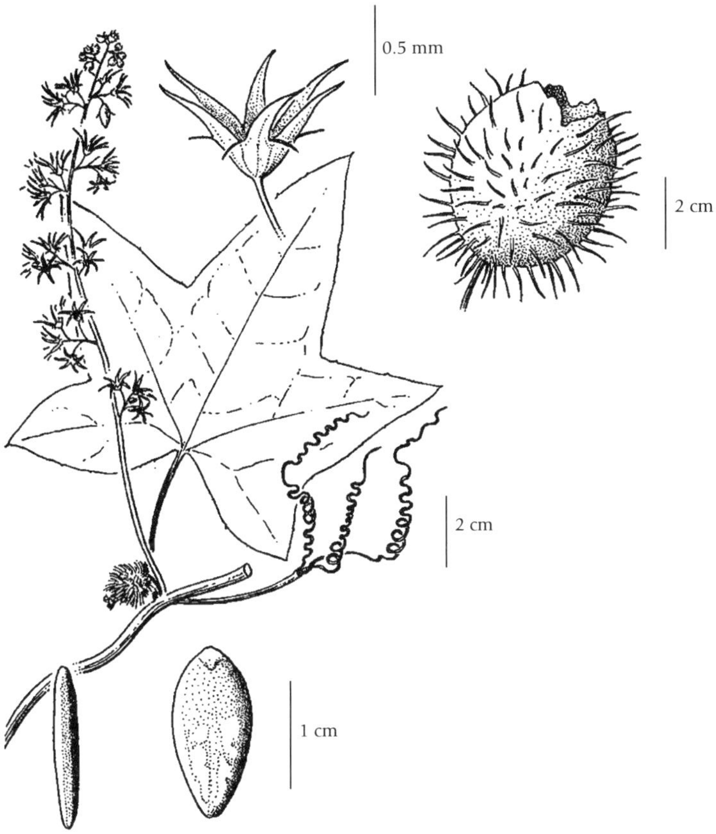

Echinocystis lobata

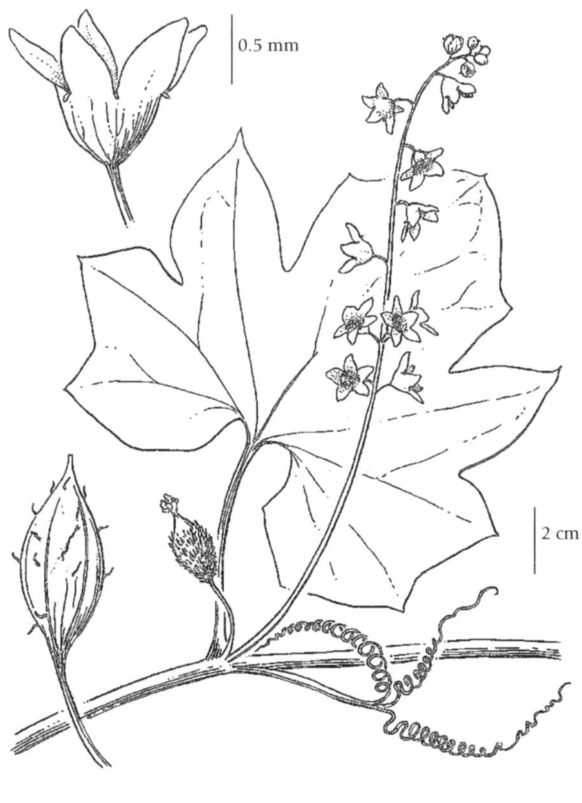

Marah oreganus

CUSCUTACEAE

CUSCUTA

1. Stigmas long-pointed, not headlike; capsules opening in an irregularly circular fashion near the base.
   2. Calyx lobes as wide as long, slightly fleshy; corolla tubes seldom longer than the calyces .... *C. approximata*
   2. Calyx lobes longer than wide, membranous; corolla tubes usually longer than the calyces .... *C. epithymum*
1. Stigmas headlike; capsules not opening in an irregularly circular fashion near the base.
   3. Flowers with 4 (sometimes 3 or 5) petals/sepals, unstalked, or nearly so; sepals free nearly to the base, unequal .... *C. cephalanthi*
   3. Flowers with 5 (sometimes 4) petals/sepals, stalked; sepals free to about halfway to the base, nearly equal.
      4. Capsules depressed-globose, usually not at all beaked; stamens slightly exserted....... *C. pentagona*
      4. Capsules globose, somewhat beaked; stamens not exserted .... *C. salina*

Cuscuta approximata **Bab.**

clustered dodder; alfalfa dodder

Habitat/Range:

Parasitic, especially on legumes, in the steppe and lower montane zones; rare in SC BC; introduced from Eurasia or Africa.

Description:

General: Parasitic, leafless perennial herb; stems threadlike, twining and sticking to other plants by means of suckers.

Leaves: Reduced to tiny scales.

Flowers: Inflorescence of several to many stalkless flowers in clusters or globular masses; flowers 2-3 mm long; calyx lobes yellowish, as long as wide, slightly fleshy; corolla tubes seldom longer than the calyces; stigmas long-pointed.

Fruits: Capsules opening in an irregularly circular fashion near the base.

Cuscuta cephalanthi **Engelm.**

button-bush dodder

Habitat/Range:

Parasitic on herbs, shrubs and subshrubs in the lowland zone; rare in SW BC; S to OR; also Atlantic Coast; possibly introduced.

Description:

General: Parasitic, leafless perennial herb; stems threadlike, twining and sticking to other plants by means of suckers.

Leaves: Reduced to tiny scales.

Flowers: Inflorescence of several to many stalkless flowers in clusters or globular masses; flowers about 2 mm long, with 4 (sometimes 3 or 5) petals/sepals; sepals free nearly to the base, unequal.

Fruits: Capsules globose, not opening in an irregularly circular fashion near the base, 2.5 mm in diameter; seeds 1-4.

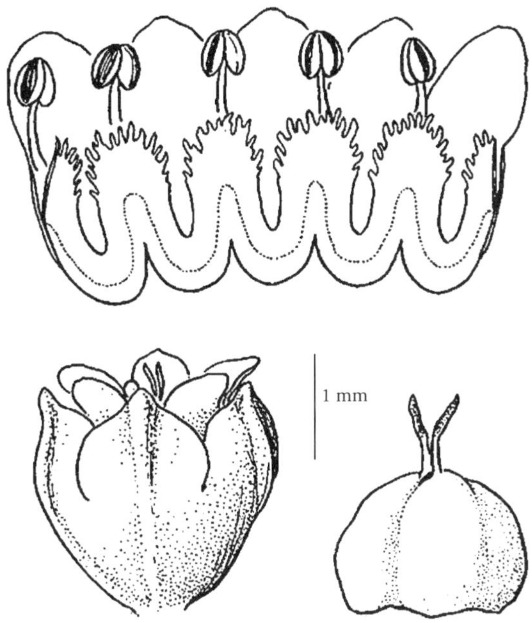

Cuscuta approximata

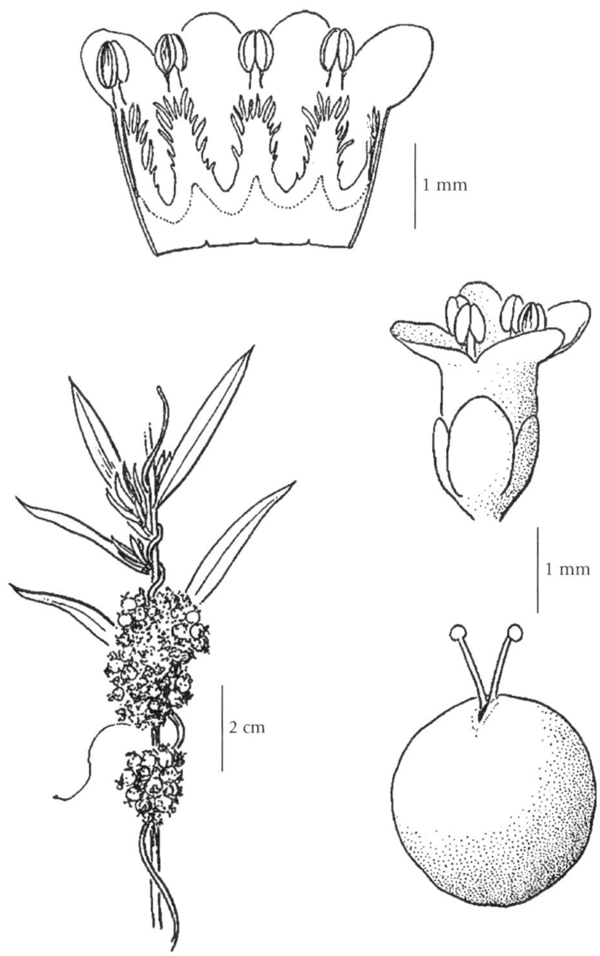

Cuscuta cephalanthi

***Cuscuta epithymum* (L.) L.**

common dodder; thyme dodder

Habitat/Range:

Parasitic, especially on legumes, in the lowland and steppe zones; infrequent in SW and SC BC; introduced from Eurasia.

Description:

General: Parasitic, leafless perennial herb; stems threadlike, twining and sticking to other plants by means of suckers.

Leaves: Reduced to tiny scales.

Flowers: Inflorescence of several to many stalkless, often purplish-tinged flowers in clusters or globular masses; flowers 2.5-3 mm long, with 5 petals/sepals; calyx lobes longer than wide, membranous; corolla tubes longer than the calyces; stigmas long-pointed.

Fruits: Capsules spherical, opening in an irregularly circular fashion near the base; seeds 4.

***Cuscuta pentagona* Engelm.** (*Grammica pentagona* [Engelm.] Weber)

field dodder; five-angled dodder

Habitat/Range:

Parasitic, especially on legumes, in the lowland zone; rare in SW and SC BC; ranges over much of U.S. and S Canada.

Description:

General: Parasitic, leafless perennial herb; stems threadlike, twining and sticking to other plants by means of suckers.

Leaves: Reduced to tiny scales.

Flowers: Inflorescence of several to many stalked flowers in clusters or globular masses; flowers 1.5-3 mm long, with 5 petals/ sepals; sepals free to about halfway to the base, nearly equal; stamens slightly exserted; stigmas headlike.

Fruits: Capsules depressed-globose, 2.5-3 mm wide, usually not beaked; seeds usually 2.

***Cuscuta salina* Engelm.** (*Grammica salina* [Engelm.] Tayl. & MacBryde)

salt marsh dodder; alkali dodder

Habitat/Range:

Parasitic on plants of the Chenopodiaceae and Asteraceae on saline sites in the lowland zone; locally frequent in SW BC, known from S Vancouver Island, the Gulf Islands and the lower Fraser Valley; S to UT, AZ, CA and MX.

Description:

General: Parasitic, leafless perennial herb; stems threadlike, twining and sticking to other plants by means of suckers.

Leaves: Reduced to tiny scales.

Flowers: Inflorescence of several stalked flowers in loose clusters or globular masses; flowers 2.0-4.5 mm long, with 5 petals/sepals; sepals free to about halfway to the base; stamens not exserted; stigmas headlike.

Fruits: Capsules globose, somewhat beaked; seeds usually solitary.

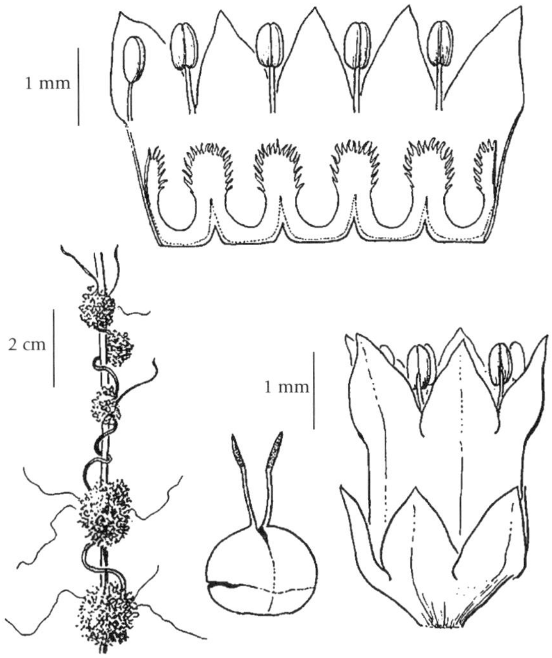

Cuscuta epithymum

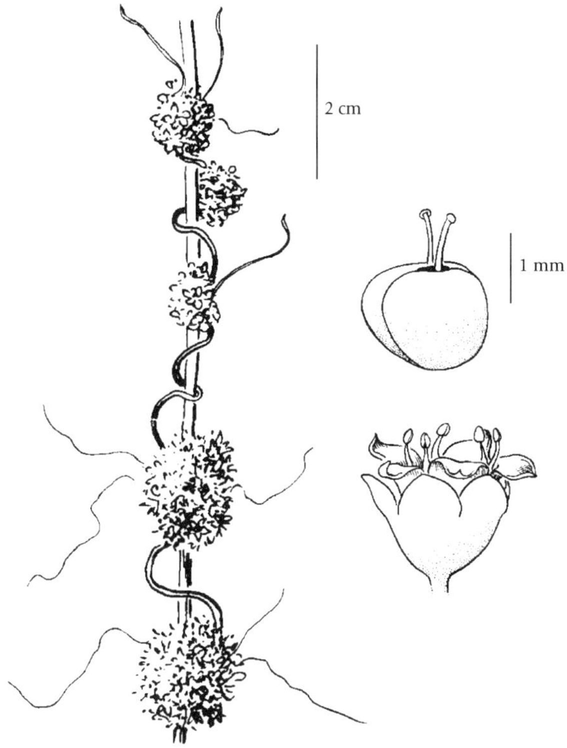

Cuscuta pentagona

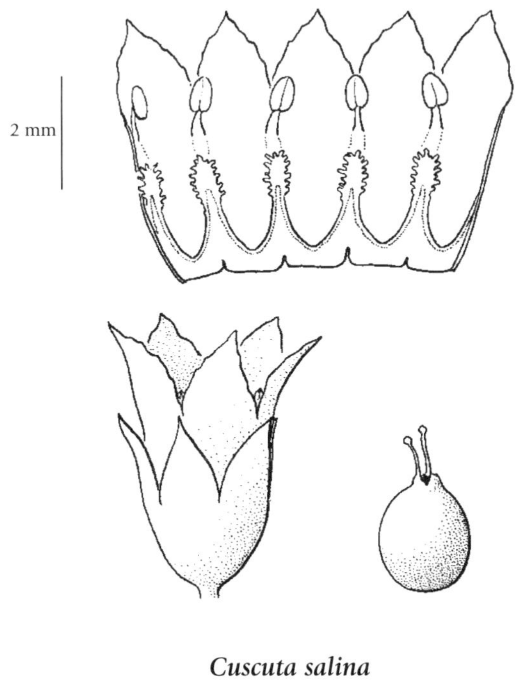

Cuscuta salina

REFERENCES

General

Abrams, L.R. 1923-1951. Illustrated flora of the Pacific States: Washington, Oregon and California. Vols. 1-3. Stanford Univ. Press, Stanford, CT.

Argus, G.W. 1973. The genus *Salix* in Alaska and the Yukon. Natl. Mus. Nat. Sci. Publ. in Bot., No. 2; Natl. Mus. Can., Ottawa, ON. 279pp.

Bailey, L.H. 1949. Manual of cultivated plants. The Macmillan Co., New York, NY.

Boivin, B. 1966-1967. Enumération des plantes du Canada. Provancheria No. 6. Univ. Laval, Québec, PQ.

———. 1968-1969. Flora of the Prairie Provinces. Provancheria No. 3. Univ. Laval, Quebec, PQ.

Brayshaw, T.C. 1985. Pondweeds and bur-reeds, and their relatives: Aquatic families of Monocotyledons in British Columbia. B.C. Prov. Mus. Occas. Pap. Ser. No. 26. Victoria, BC.

———. 1996a. Trees and shrubs of British Columbia. Royal B.C. Mus. Handb. UBC Press, Vancouver, BC.

Calder, J.A., and R.L. Taylor. 1968. Flora of the Queen Charlotte Islands, Part 1. Can. Dep. Agric. Res. Branch Monogr. 4. 659pp.

Campbell, R. 1904. Some conspicuous British Columbia plants. Can. Rec. Sci. 9:176-189.

Clark, L.J. 1973. Wild flowers of British Columbia. Gray's Publ. Ltd., Sidney, BC. 591pp.

———. 1976. Wild flowers of the Pacific Northwest. Gray's Publ. Ltd., Sidney, BC. 604pp.

Cody, W.J. 1996. Flora of the Yukon Territory. NRC Res. Press, Ottawa, ON. 643pp.

Dansereau, P. 1957. Biogeography. Ronald Press, New York, NY.

Daubenmire, R. 1970. Steppe vegetation of Washington. Wash. Agric. Exp. Stn. Tech. Bull. 62.

Davis, R.J. 1952. Flora of Idaho. Wm. C. Brown Co., Dubuque, IA. 828pp.

Douglas, G.W. 1971. The alpine-subalpine flora of the North Cascade Range, Washington. Wasmann J. Biol. 29:129-168.

———. 1972. Subalpine plant communities of the western North Cascades, Washington. Arct. Alp. Res. 4:147-166.

———. 1974. Montane zone vegetation of the Alsek River region, southwestern Yukon. Can. J. Bot. 52:2505-2532.

———. 1996. Endemic vascular plants of British Columbia and immediately adjacent regions. Can. Field-Nat. 110:387-391.

Douglas, G.W., and L.C. Bliss. 1977. Alpine and high subalpine plant communities of the North Cascades Range, Washington and British Columbia. Ecol. Monogr. 47:113-150.

Douglas, G.W., A. Ceska, and G.G. Ruyle. 1983. A floristic bibliography for British Columbia. B.C. Minist. For., Land Manage. Rep. No. 15, Victoria, BC.

Douglas, G.W., F. Lomer, and H.L. Roemer. 1998. New or rediscovered native vascular plant species in British Columbia. Can. Field-Nat. 112:276-279.

Douglas, G.W., J. Pojar, D. Meidinger, and K. McKeown. 1994. Rare vascular plant collections from the St. Elias Mountains, northwestern British Columbia. Can. Field-Nat. 108:391-96.

Douglas, G.W., G.B. Straley, and D. Meidinger. 1989. The Vascular Plants of British Columbia. Part 1 – Gymnosperms and Dicotyledons (Aceraceae through Cucurbitaceae). Spec. Rep. Ser. 1. B.C. Minist. For., Victoria, BC.

———. 1990. The Vascular Plants of British Columbia. Part 2 – Dicotyledons (Diapensiaceae through Portulacaceae). Spec. Rep. Ser. 2. B.C. Minist. For., Victoria, BC.

———. 1991. The Vascular Plants of British Columbia. Part 3 – Dicotyledons (Primulaceae through Zygophyllaceae) and Pteridophytes. Spec. Rep. Ser. 3. B.C. Minist. For., Victoria, BC.

———. 1994. The Vascular Plants of British Columbia. Part 4 – Monocotyledons. Spec. Rep. Ser. 4. B.C. Minist. For., Victoria, BC.

———. 1998. Rare native vascular plants of British Columbia. B.C. Minist. Environ., Lands and Parks, Resourc. Inv. Branch, B.C. Conserv. Data Centre, and Wildlife Branch, Victoria, BC. 430pp.

Eastham, J.W. 1947. Supplement to the Flora of Southern British Columbia. B.C. Prov. Mus. Spec. Publ. No. 1, Victoria, BC. 119pp.

Fernald, M.L. 1950. Gray's Manual of Botany. 8th ed. Am. Book Co., New York, N.Y. 1632pp.

Ferris, R.S. 1960. Illustrated Flora of the Pacific States: Washington, Oregon and California. Vol. IV. Bignoniaceae to Compositae. Stanford Univ. Press, Stanford, CT.

Flora of North America Editorial Committee. 1993. Flora of North America north of Mexico, Vol. 2. Oxford Univ. Press, New York, NY. 475pp.

Franklin, F.J., and C.T. Dyrness. 1973. Natural vegetation of Oregon and Washington. U.S. Dep. Agric. For. Serv. Gen. Tech. Rep. PNW-8.

Frankton, C., and G.A. Mulligan. 1970. Weeds of Canada. Rev. ed. Can. Dep. Agric. Publ. 948. Queen's Printer, Ottawa, ON. 217pp.

Gleason, H.A., and A. Cronquist. 1963. Manual of vascular plants of northeastern United States and adjacent Canada. D. Van Nostrand Co., New York, NY.

Gray, A. 1884. Caprifoliaceae to Compositae. *In* Synoptical flora of North America. Vol. 1. Am. Book Co., New York, NY. 474pp.

Groh, H. 1946. Canadian weed survey. Third Ann. Rep. Can. Dep. Agric., Ottawa, ON.

Henry, J.K. 1915. Flora of Southern British Columbia. W. J. Gage and Co. Ltd., Toronto, ON. 363pp.

Hickman, J.C. 1993. The Jepson manual: higher plants of California. Univ. Calif. Press, Los Angeles, CA. 1400pp.

Hitchcock, C.L., and A. Cronquist. 1973. Flora of the Pacific Northwest. Univ. Wash. Press, Seattle, WA.

Hitchcock, C.L., A. Cronquist, M. Ownbey, and J.W. Thompson. 1955-1969. Vascular plants of the Pacific Northwest. Parts 1-5. Univ. Wash. Press, Seattle, WA.

Hosie, R.C. 1969. Native trees of Canada. 7th ed. Can. For. Ser., Dep. Environ. Queen's Printer, Ottawa, ON. 380pp.

Hulten, E. 1968. Flora of Alaska and neighboring territories. Stanford Univ. Press, Stanford, CT. 1008pp.

Kartesz, J.T. 1994. A synonymized checklist of the vascular flora of the United States, Canada, and Greenland. Vol. I. Checklist. Biota of North America. Univ. Timber Press, Portland, OR. 622pp.

Krajina, V. 1965. Biogeoclimatic zones and classification of British Columbia. Ecol. West. North Am. 1:1-17.

———. 1969 Ecology of forest trees in British Columbia. Ecol. West. North Am. 2:1-146.

Lomer, F., and G.W. Douglas. 1998. Additions to the flora of the Queen Charlotte Islands. Can. Field-Nat. (in press).

Löve, D. 1970. Subarctic and subalpine: where and what? Arct. Alp. Res. 2:63-73.

MacKinnon, A., J. Pojar, and R. Coupé. 1992. Plants of northern British Columbia. B.C. Minist. For., Victoria, BC, and Lone Pine Publishing, Vancouver, BC., B.C. 345pp.

Macoun, J. 1883-1890. Catalogue of Canadian plants. Parts 1-5. Geol. Surv. Can., Ottawa, ON.

Macoun, J.M. 1889. Checklist of Canadian plants. Cunningham and Lindsay, Ottawa, ON.

———. 1894-1906 Contributions to Canadian botany. Parts 1-18. Ottawa Field-Nat., Ottawa, ON.

Meusel, H., E. Jager, and E. Weinert. 1965. Vergleichende chorologie der zentraleuropäischen flora. I. Gustav. Fischer, Jena.

Packer, J.G. 1983. Flora of Alberta. Second ed. Univ. Toronto Press, Toronto, ON.

Parish, R., R. Coupé, and D. Lloyd. 1996. Plants of southern interior British Columbia. B.C. Minist. For., Victoria, BC, and Lone Pine Publishing, Vancouver, BC. 463pp.

Pojar, J., and A. MacKinnon. 1994. Plants of coastal British Columbia. B.C. Minist. For., Victoria, BC, and Lone Pine Publishing, Vancouver, BC. 526pp.

Porsild, A.E., and W.J. Cody. 1980. Vascular plants of continental Northwest Territories, Canada. Natl. Mus. Can., Ottawa, ON. 667pp.

Rowe, J.S. 1959. Forest regions of Canada. Can. Dep. North. Affairs Nat. Resour. For. Branch Bull. 123.

———. 1972. Forest regions of Canada. Can. For. Serv. Publ. 1300.

Rydberg, P.A. 1922. Flora of the Rocky Mountains and adjacent plains. 2nd ed. P.A. Rydberg, New York, NY. 1143pp.

Scoggan, H.J. 1978-79. The flora of Canada. Parts 1-4. Natl. Mus. Nat. Sci. Publ. Bot., No. 7:1-1626.

Straley, G.B., R.L. Taylor, and G.W. Douglas. 1985. The rare vascular plants of British Columbia. Natl. Mus. Can. Syllogeus 59, Ottawa, ON.

Szczawinski, A.F. 1959. The orchids of British Columbia. B.C. Prov. Mus. Handb. 16, Victoria, BC. 124pp.

———. 1962. The heather family (Ericaceae) of British Columbia. B.C. Prov. Mus. Handb. 10. Victoria, BC. 205pp.

Szczawinski, A.F., and A.S. Harrison. 1973. Flora of Saanich Peninsula. B.C. Prov. Mus. Occas. Pap. Ser. No.16, Victoria, BC.

Taylor, R.J., and G.W. Douglas. 1995. Mountain plants of the Pacific Northwest. Mountain Press Publ. Co, Missoula, MT. 437pp.

Taylor, R.L., and B. MacBryde. 1977. Vascular plants of British Columbia. Bot. Gard. Tech. Bull. No. 4. UBC Press, Vancouver, BC. 754pp.

Taylor, T.M.C. 1963. The ferns and fern allies of British Columbia. B.C. Prov. Mus. Handb. 12. Victoria, BC. 172pp.

———. 1966. The lily family (Liliaceae) of British Columbia. B.C. Prov. Mus. Handb. 25, Victoria, BC. 109pp.

———. 1970. Pacific Northwest ferns and their allies. Univ. Toronto Press, Toronto, ON. 247pp.

———. 1973. The rose family (Rosaceae) of British Columbia. B.C. Prov. Mus. Handb. 30. Victoria, BC. 223pp.

———. 1974a. The pea family (Leguminosae) of British Columbia. B.C. Prov. Mus. Handb. 32, Victoria, BC. 251pp.

———. 1974b. The figwort family (Scrophulariaceae) of British Columbia. B.C. Prov. Mus. Handb. 33, Victoria, BC. 237pp.

———. 1983. The sedge family (Cyperaceae). B.C. Prov. Mus. Handb. 43, Victoria, BC. 375pp.

Tisdale, E.W. 1947. The grasslands of the southern interior of British Columbia. Ecology 28:346-382.

Tutin, T.G., V.H. Heywood, N.A. Burges, D.M. Moore, D.H. Valentine, S.M. Walters, and D.A. Webb, eds. 1964-1980. Flora Europaea. Volumes 1-5. Univ. Cambridge Press, Cambridge, MA.

Ulke, T.A. 1935. List of the vascular plants of the Horsethief Creek-Purcell Range, B.C. Can. Field-Nat. 49:49-55, 71-76.

Weber, W.A. 1967. Rocky Mountain Flora. Univ. Colorado Press, CO. 437pp.

Welsh, S.L. 1974. Anderson's flora of Alaska and adjacent parts of Canada. Brigham Young Univ. Press, Provo, UT. 724pp.

DICOTYLEDONS

BALSAMINACEAE

Ornduff, R. 1967. Hybridization and regional variation in Pacific Northwest *Impatiens* (Balsaminaceae). Brittonia 19:122-218.

BERBERIDACEAE

Ahrendt, L.W.A. 1961. *Berberis and Mahonia.* A taxonomic revision. J. Linn. Soc. 57:1-410.

Fukuda, I. 1967. The biosystematics of *Achlys.* Taxon 16:308-316.

Taylor, S.T. 1972. The genus *Mahonia* Nutt. in British Columbia. Davidsonia 3:39-43.

Williams, M.P. 1993. Berberidaceae. *In* J.C. Hickman, ed. The Jepson manual: higher plants of California. Univ. Calif. Press, Los Angeles, CA. 1400pp.

BETULACEAE

Brayshaw, T.C. 1976. Catkin Bearing plants (Amentiferae) of British Columbia. B.C. Prov. Mus., Victoria, BC. Occas. Pap. No. 18. 176pp.

———. 1996b. Catkin-Bearing plants of British Columbia. Royal B.C. Mus., Victoria, BC.

Dugle, J.R. 1966. A taxonomic study of western Canadian species in the genus *Betula.* Can. J. Bot. 44:929-1007.

BORAGINACEAE

Bucknall, C. 1913. A revision of the genus *Symphytum.* J. Linn. Soc. 41: 491-556.

Gentry, J.L. 1972. A new combination and a new name in *Hackelia* (Boraginaceae). Madroño 21:490.

Gentry, J.L., and R.L. Carr. 1976. A revision of the genus *Hackelia* (Boraginaceae) in North America, north of Mexico. Mem. N.Y. Bot. Gard. 26:121-227.

Ganders, F.R. 1993. Amsinckia. *In* J.C. Hickman, ed. The Jepson manual: higher plants of California. Univ. Calif. Press, Los Angeles, CA. 1400pp.

Govoni, D. 1973. The taxonomy of the genus *Lithospermum* L. (Boraginaceae) in the western Great Plains. Ph.D. diss. Univ. Nebraska, Lincoln, NE.

Higgins, L.C. 1971. A revision of *Cryptantha* subgenus *Oreocarya.* Brigham Young Univ. Sci. Bull., Biol. Ser. 13:1-63.

Johnston, I.M. 1923. Studies in the Boraginaceae. 1. Restoration of the genus *Hackelia.* Contr. Gray Herb. New Ser. 68:43-48.

———. 1923. Studies in the Boraginaceae. 4. A synopsis and redefinition of *Plagiobothrys.* Contr. Gray Herb. New Ser. 68:57-80.

———. 1923. Studies in the Boraginaceae. Vol. IX. 1. The *Allocarya* section of *Plagiobothrys* in the western United States. Contr. Arn. Arb. 3:5-82.

———. 1924. Studies in the Boraginaceae. A synopsis of the American native and immigrant borages of the subfamily Boraginoideae. Contr. Gray Herb. 70:1-54.

———. 1924. Studies in the Boraginaceae 11. *Lappula.* Contr. Gray Herb. 70:47-51.

———. 1925. The North American species of *Cryptantha.* Contr. Gray Herb. 74:14-19.

———. 1952. Studies in the Boraginaceae 23. A survey of the genus *Lithospermum.* J. Arn. Arb. 33:299-366.

Macbride, J.F. 1916. The true Mertensias of western North America. Contr. Gray Herb. New Ser. 48:1-20.

———. 1917. A revision of the North American species of *Amsinckia.* Contr. Gray Herb. New Ser. 49:1-16.

Payson, E.B. 1927. A monograph of the section *Oreocarya* of *Cryptantha.* Annals of the Missouri Bot. Gardens 14:211-358.

Ray, P.M., and H.F. Chisaki. 1957. Studies in *Amsinckia*. 1. Am. J. Bot. 44: 529-536.

Williams, L. 1937. A monograph of the genus *Mertensia* in North America. Annals of the Missouri Bot. Gardens 24:17-159.

BRASSICACEAE

Detling, L.E. 1936. The genus *Dentaria* in the Pacific States. Am. J. Bot. 23:570-576.

———. 1937. The Pacific Coast species of *Cardamine*. Am. J. Bot. 24:70-76.

———. 1939. Revision of the North American species of *Descurainia*. Am. Midl. Nat. 22:481-520.

Dudley, T.R. 1968. *Alyssum* (Cruciferae) introduced in North America. Rhodora 70:293-300.

Fernald, M.L. 1909. The North American species of *Barbarea*. Rhodora 11:134-141.

———. 1934. *Draba* in temperate northeastern America. Rhodora 36:241-261.

Harms, V.L., J.H. Hudson, and G.F. Ledingham. 1986. *Rorippa truncata*, the blunt-fruited yellow cress, new for Canada, and *R. tenerrima*, the slender yellow cress, in southern Saskatchewan and Alberta. Can. Field-Nat. 100: 45-51.

Hitchcock, C.L. 1936. The genus *Lepidium* in the United States. Madroño 3:265-320.

———. 1941. A revision of the Drabas of western North America. Univ. Wash. Publ. Biol. 11:1-132.

Hopkins, M. 1937. *Arabis* in Eastern and Central North America. Rhodora 39: 63-98, 106-148, 155-186.

Mulligan, G.A. 1961. The genus Lepidium in Canada. Madroño 16:77-90.

———. 1970a. Cytotaxonomic studies of *Draba glabella* and its close allies in Canada and Alaska. Can. J. Bot. 48:1431-1437.

———. 1970b. A new species of *Draba* in the Kananaskis Range of Southwestern Alberta. Can. J. Bot. 48:1897-1898.

———. 1971a. Cytotaxomomic studies of the closely allied *Draba cana*, *D. cinerea*, and *D. groenlandica* in Canada and Alaska. Can. J. Bot. 49:89-93.

———. 1971b. Cytotaxonomic studies of *Draba* species of Canada and Alaska: *D. ventosa*, *D. ruaxes*, and *D. paysonii*. Can. J. Bot. 49:1455-1460.

———. 1971c. The genus *Lepidium* in Canada. Madroño 16:77-90.

———. 1972. Cytotaxonomic studies of *Draba* species in Canada and Alaska: *D. oligosperma* and *D. incerta*. Can. J. Bot. 50:1763-1766.

———. 1974a. Confusion in the names of three *Draba* species of the arctic: *D. adamsii*, *D. oblongata*, and *D. corymbosa*. Can. J. Bot. 52:791-793.

———. 1974b. Cytotaxonomic studies of *Draba nivalis* and its close allies in Canada and Alaska. Can. J. Bot. 52:1793-1801.

———. 1975. *Draba crassifolia*, *D. albertina*, *D. nemorosa* and *D. stenoloba* in Canada and Alaska. Can. J. Bot. 53:745-751.

———. 1976. The genus *Draba* in Canada and Alaska: key and summary. Can. J. Bot. 54:1386-1393.

———. 1996. Synopsis of the genus *Arabis* (Brassicaceae) in Canada, Alaska and Greenland. Rhodora 97:109-163.

Payson, E.B. 1921. A monograph of the genus *Lesquerella*. Annals of the Missouri Bot. Gardens 8:102-236.

Rollins, R.C. 1941. A monographic study of *Arabis* in Western North America. Rhodora 43:289-325, 348-411, 425-481.

———. 1993. The Cruciferae of continental North America. Stanford Univ. Press, Stanford, CT.

Rossbach, G.B. 1958. The genus *Erysimum* (Cruciferae) in North America north of Mexico - a key to the species and varieties. Madroño 14:261-267.

Stuckey, R.L. 1972. Taxonomy and distribution of the genus *Rorippa* (Cruciferae) in North America. Sida 4:279-430.

Thellung, A. 1906. Die Gattung *Lepidium* (L.) R. Br. Eine monographische Studie. Mitt. Bot. Mus. Univ. Zurich 28:1-340.

CALLITRICHACEAE

Fassett, N.C. 1951. *Callitriche* in the New World. Rhodora 53:137-155, 161-182, 185-194, 209-222.

Rubtzoff, R. 1969. Notes on the *Callitriche* in western North America. Wasmann J. Biol. 27:103-114.

CAMPANULACEAE

Bowden, W.M. 1959. Cytotaxonomy of *Lobelia* L. sect. *Lobelia*, 1. Three diverse species and seven small-flowered species. Can. J. Genet. Cytol. 1:49-64.

McVaugh, R. 1941. A monograph on the genus *Downingia.* Mem. Torr. Bot. Club 194:1-57.

Morin, N. 1983. Systematics of *Githopsis* (Campanulaceae). Syst. Bot. 8:436-468.

Shetler, S.G. 1963. A checklist and key to the species of *Campanula* native or commonly naturalized in North America. Rhodora 65:319-337.

CAPRIFOLIACEAE

Bolli, R. 1994. Revision of the genus *Sambucus*. Dissertationes Botanicae 223:1-227.

Eriksson, T., and M.J. Donoghue. 1997. Phlogenetic relationships of *Sambucus* and *Adoxa* (Adoxoideae, Adoxaceae) based on nuclear ribosomal ITS sequences and preliminary morphological data. Syst. Bot. 22:555-573.

Jones, G.N. 1940. A monograph of the genus *Symphoricarpos.* J. Arn. Arb. 21:201-252.

Taylor, S.T. 1977, The genus *Viburnum* in British Columbia. Davidsonia 8:41-47.

———. 1979. The genus *Lonicera* in British Columbia. Davidsonia 10:54-63.

CARYOPHYLLACEAE

Baad, M. 1969. Biosystematic studies of the North American species of *Arenaria,* subgenus *Eremogone* (Caryophyllaceae). Ph.D. Thesis, Univ. Wash., Seattle, WA.

Baker, H.G. 1950. Dioecious *Melandrium* in Western North America. Madroño 10:218-221.

Bocquet, G. 1967. Physolychnidium olim Gastrolychniclium nomenclaturae fundamentur includens combinationes taxaque nova nonnulla Silenes generis. Candollea 22:1-38.

Ceska, A., and O. Ceska. 1980. Additions to the flora of British Columbia. Can. Field-Nat. 94:69-74.

Chater, A.O., and S.M. Walters. 1964. *Silene* L. *In* T.G. Tutin et al., eds. Flora Europaea. Vol. 1 Univ. Press, Cambridge, MA.

Chinnappa, C.C., and J.K. Morton. 1976. Studies on the *Stellaria longipes* Goldie complex - variation in wild populations. Rhodora 78:488-502.

———. 1984. Studies on the *Stellaria longipes* complex (Caryophyllaceae) - biosystematics. Syst. Bot. 9:60-73.

———. 1991. Studies on the *Stellaria longipes* complex (Caryophyllaceae) - taxonomy. Rhodora 93:31-38.

Chowdhuri, P.K. 1957. Studies in the genus *Silene.* Notes Royal Bot. Gard. Edinburgh, 22:221-278.

Crow, G.E. 1978. A taxonomic revision of *Sagina* (Caryophyllaceae) in North America. Rhodora 80:1-91.

Emery, R.J.N., and C.C. Chinnappa. 1994. Morphological variation among members of the *Stellaria longipes* complex: *S. longipes, S. longifolia,* and *S. porsildii* (Caryophyllaceae). Pl. Syst. Evol. 190:69-78.

Fernald, M.L. 1940. *Stellaria calycantha* Goldie and its allies. Bot. Not. 1943:251-270.

Fernald, M.L., and K.M. Wiegand. 1920. Studies of some boreal American Cerastiums of the section *Orthodon*. Rhodora 22:169-179.

Hickman, J.C. 1971. *Arenaria,* section *Eremogone* (Caryophyllaceae) in the Pacific Northwest: A key and discussion. Madroño 21:201-207.

Hitchcock, C.L., and B. Maquire. 1947. A revision of North American species of *Silene.* Univ. Wash. Publ. Biol. 13:1-73.

Hultén, E. 1956. The *Cerastium alpinum* complex. A case of world-wide introgressive hybridization. Svensk Bot. Tidskr. 50:411-495.

———. 1971. The circumpolar plants. II. Dicotyledons. Kongl. Svenska Vetenskapsakad. Handl., Ser. 4, 13:1-463.

Maguire, B. 1947. Studies in the Caryophyllaceae. III. A synopsis of the North American species of *Arenaria,* section *Eremogone.* Fenzi. Bull. Torrey Bot. Club 74:38-56.

———. 1950. Studies of the Caryophyllaceae. Synopsis of the North American species of the subfamily *Silenoideae.* Rhodora 52:233-245.

———. 1951. Studies in the Caryophyllaceae. V. *Arenaria* in America, north of Mexico. A conspectus. Am. Midl. Nat. 46:493-511.

———. 1958. *Arenaria rossii* and some of its relatives in America. Rhodora 60:44-53.

Mattfield, J. 1922. Geographisch – genetische Untersuchungen ber die Gattung *Minuartia* (L.) Hiern. Rep. Spec. Nov. Regni Veg. Beih. 15.

McNeill, J. 1978. *Silene alba* and *S. dioica* in North America and the generic delimitation of *Lychnis, Melandrium,* and *Silene* (Carophyllaceae). Can. J. Bot. 56:297-308.

———. 1980. The delimitation of *Arenaria* (Caryophyllaceae) and related genera in North America, with 11 new combinations in *Minuartia.* Rhodor 82:495-502.

McNeill, J., and I.J. Bassett. 1974. Pollen morphology and the infrageneric classification of *Minuartia* (Caryophyllaceae). Can. J. Bot. 52:1225-1231.

Morton, J.K., and R.K. Rabeler. 1989. Biosystematic studies on the *Stellaria calycantha* (Caryophyllaceae) complex. 1. Cytology and cytogeography. Can. J. Bot. 67:121-127.

———. 1989. Biosystematic studies on the Stellaria calycantha complex. I. Cytology and cytogeograpy. Can. J. Bot. 67:121-127.

Phillipp, M. 1972. The *Stellaria longipes* group in N.W. Greenland. Cytological and morphological investigations. Bot. Tidsskr. 67:64-75.

Porsild, A.E. 1963. *Stellaria longipes* Goldie and its allies in North America. Natl. Mus. Can. Bull. 1861.

Rossbach, R.P. 1940. *Spergularia* in North and South America. Rhodora 42:57-83, 105-143, 158-193, 203-213.

Wolf, S.J., J.G. Packer, and K.E. Denford. 1979. The taxonomy of *Minuartia rossii* (Caryophyllaceae). Can. J. Bot. 57:1673-1686.

Zajac, A. 1975. The genus *Cerastium* L. in Poland: section *Fugacia* and *Caespitosa*. Monogr. Bot. 47:1-100.

CERATOPHYLLACEAE

Ceska, A., and O. Ceska. 1980. Additions to the flora of British Columbia. Can. Field-Nat. 94:69-74.

Fassett, N.C. 1953. North American *Ceratophyllum.* Communicaciones del Instituto tropical de Investigaciones cientificas, Universidad de El Salvador 2:25-45.

CHENOPODIACEAE

Aellen, P. 1929. Beitrag zur Systematik der *Chenopodium-Arten* Amerikas, voriviegand auf Grund der Sammlung des United States Natl. Mus. in Washington, D.C. Rep. Spec. Novi Reg. Veget 26:31-64, 119-60.

Aellen, P., and T. Just. 1943. Key and synopsis of the American species of the genus *Chenopodium* L. Am. Midl. Nat. 30:47-76.

Bassett, I.J., and C.W. Crompton. 1973. The genus *Atriplex* (Chenopodiaceae) in Canada and Alaska. III. Three hexaploid annuals: *A. subspicata, A. gmelinii* and *A. alaskensis.* Can. J. Bot. 51:1715-1723.

———. 1978a. The biology of Canadian weeds. 32. *Chenopodium album.* Can. J. Plant Sci. 58:1060-1072.

———. 1978b. The genus *Suaeda* (Chenopodiaceae) in Canada. Can. J. Bot. 56:581-591.

Bassett, I.J., and K.W. Spicer. 1970. *Atriplex oblongifolia,* a recently recognized weed now well established in British Columbia and Alberta. Greenhouse Garden Grass 9:1-7.

Bassett, I.J., C.W. Crompton, J. McNeil, and P.M. Taschereau. 1983. The genus *Atriplex* (Chenopodiaceae) in Canada. Agric. Can., Monogr. 31, Ottawa, ON.

Brown, G.D. 1956. Taxonomy of American *Atriplex.* Am. Midl. Nat. 55:199-210.

Crawford, D.J. 1972. An analysis of variation in three species of *Chenopodium.* Brittonia 24:118. (abstract).

———. 1975. Systematic relationships in the narrow-leaved species of *Chenopodium* of the western United States. Brittonia 27:279-288.

Crawford, D.J., and J.F. Reynolds. 1974. A numerical study of the common narrow-leaved taxa of *Chenopodium* occurring in the western United States. Brittonia 26:398-410.

Frankton, C., and I.J. Bassett. 1968. The genus *Atriplex* in Canada. 1. Three introduced species: *A. heterosperma, A. oblongifolia,* and *A. hortensis.* Can. J. Bot. 46:1309-1313.

———. 1970. The genus *Atriplex* in Canada. II. Four native western annuals: *A. argentea, A. truncata, A. powellii,* and *A. dioica.* Can. J. Bot. 48:981-989.

Hall, H.M., and F.E. Clements. 1923. The phylogenetic method in taxonomy. The genus *Atriplex.* Publ. Carnegie Inst. Wash. Bot. 326:235-346.

McNeill, J., I.J. Bassett, and C.W. Crompton. 1977. *Suaeda calceoliformis,* the correct name for *Suaeda depressa* Auct. Rhodora 79:133-138.

Tutin, T.G. 1964. *Chenopodium* L. *In* T.G. Tutin et al., eds. Flora Europaea. Volume 1. Lycopodiaceae to Platanaceae. Cambridge Univ. Press, Cambridge, MA.

Uotila, P., and J. Suominen. 1976. The *Chenopodium* species in Finland, their occurrence and means of migration. Ann. Bot. Fenn. 13:1-25.

van der Meijden, R. 1970. Biosystematic notes on *Atriplex patula* L., *A. hastata* L., and *A. littoralis* L. (Chenopodiaceae). Blumea 18:53-63.

Wahl, H.A. 1954. A preliminary study of the genus *Chenopodium* in North America. Bartonia 27:1-46.

CLUSIACEAE

Talbot, J. 1993. Hypericaceae. *In* J.C. Hickman, ed. The Jepson manual: higher plants of California. Univ. Calif. Press, Berkeley, CA.

CONVOLVULACEAE

Brummitt, R.K. 1965. New combinations in North American *Calystegia.* Annals of the Missouri Bot. Gardens. 52:214-216.

Hadac, E., and J. Chrtek. 1970. Notes on the taxonomy of Cuscutaceae. Folia Geobot. Phytotax. 5:443-445.

———. 1973. Some further notes on the taxonomy of Cuscutaceae. Folia Geobot. Phytotax. 8:219-221.

Lewis, W.H., and R.L. Oliver. 1965. Realignment of *Calystegia* and *Convolvulus* (Convolvulaceae). Annals of the Missouri Bot. Gardens. 52:217-222.

Tryon, R. M., Jr. 1939. The varieties of *Convolvulus spithamaeus* and of *C. septum.* Rhodora 41:415-423.

CORNACEAE

Bain, J.F., and K.E. Denford. 1979. The herbaceous members of the genus *Cornus* in NW North America. Bot. Not. 132:121-129.

Ferguson, I.K. 1966. Notes on the nomenclature of *Cornus.* J. Arn. Arb. 47:100-105.

Fosberg, F.R. 1942. *Cornus sericea* L. (*C. stolonifera* Michx.) Bull. Torrey Bot. Club. 69:583-589.

Rickett, H.W. 1944. *Cornus stolonifera* and *Cornus occidentalis.* Brittonia 5:149-159.

Smithberg, M.H., and C.J. Weiser. 1968. Patterns of variation among climatic races of red-osier dogwood. Ecology 49:495-505.

CRASSULACEAE

Clausen, R.T. 1975. *Sedum* of North America north of the Mexican plateau. Cornell Univ. Press. Ithaca, NY.

CUCURBITACEAE

Stocking, K.M. 1955. Some taxonomic and ecological considerations of the genus *Marah* (Cucurbitaceae). Madroño 13:113-137.

CUSCUTACEAE

Yuncker, T.G. 1932. The genus *Cuscuta.* The taxonomy and nomenclature of Cuscutaceae. Mem. Torr. Bot. Club 18:109-331.

———. 1965. *Cuscuta* Linnaeus. Sp. PI. 124. 2753. North Am. Flora Ser. 11. Part 4:1-51.

APPENDIX I – EXCLUDED SPECIES

BALSAMINACEAE

Impatiens pallida **Nutt.** Reports of this species by Macoun (1895) are based on specimens of *I. noli-tangere.*

BERBERIDACEAE

Vancouveria hexandra **(Hook.) Morr. & Dec.** Listed by Taylor and MacBryde (1977) but all B.C. material seen was acquired from gardens.

BETULACEAE

Alnus rhombifolia **Nutt.** Reports of this species by Henry (1915) were based on misidentifications (Scoggan 1978).

BORAGINACEAE

Amsinckia tesselata **A. Gray.** A specimen in UBC from New Westminster collected in 1886 is this species but no recent collections are known.

Cryptantha circumscissa **(H. & A.) I. M. Johnst.** Reported by Henry (1915) from Spences Bridge, but probably a misidentification.

Cryptantha flaccida **(Dougl.) Greene.** Cited by Henry (1915) from S BC, but probably a misidentification.

Cryptantha leucophaea **(Dougl.) Pays.** Listed by Macoun (1884), but probably a misidentification.

Cynoglossum grande **Dougl.** Listed by Henry (1915) and illustrated in Clark (1973), but no specimens known.

Hackelia arida **(Piper) I. M. Johnst.** Attributed to the province by Boivin (1966-1967) from Rock Creek, but it is likely a misidentification.

Hackelia hispida **(A. Gray) I. M. Johnst.** Reported by Henry (1915) from Spences Bridge, presumably based on a misidentification.

Hackelia patens **(Nutt.) I. M. Johnst.** Cited by Eastham (1947) and presumably based on a misidentification.

Lithospermum officinale **L.** Listed by Taylor & MacBryde (1977), but no specimens seen.

Mertensia oblongifolia **(Nutt.) G. Don.** References to this taxon (e.g., Macoun 1884, Henry 1915, Taylor and MacBryde 1977) in the province all seem to be based on misidentifications of *M. longiflora.*

Nonea vesicaria **(L.) Rchb.** There is a specimen in V, collected at McMurdo in 1936 but no recent collections seen.

Trigonotis peduncularis **(Trev.) Benth.** Known in North America only from a 1893 collection by Macoun from Nanaimo (Scoggan 1979).

BRASSICACEAE

Arabis caucasica **Willd.** A European species that is commonly cultivated as a garden ornamental, but it does not usually produce seed, and therefore is propagated by cuttings. It may persist around old gardens and in waste areas, but does not become a part of our flora.

Arabis divaricarpa **A. Nels. var. *interposita* (Greene) Roll.** Included in Douglas et al. (1989), but apparently does not occur in BC.

Cardamine debilis **D. Don.** A European species recorded from the Peace River District, but probably referrable to *C. hirsuta.*

Cardamine flexuosa **With.** A European species collected once at Chilliwack, needs further collecting and confirmation.

Cardamine impatiens **L.** A European species, probably confused with *C. oligosperma.*

***Caulanthus pilosus* S. Wats.** This species, listed from BC. by Rydberg (1922), occurs only to the south of our region.

***Cochlearia anglica* L.** Reports of this species by Henry (1915) probably refer to specimens of *C. officinalis L.* (Scoggan 1978).

***Lepidium lasiocarpum* Nutt.** Cited by Henry (1915), but no specimens seen.

***Lepidium oxycarpum* T. & G.** This western U.S. species, cited by Mulligan (1961) and based on a single collection, is probably not established in BC.

***Lobularia maritima* (L.) Desv.** A European garden plant, listed by Scoggan (1978), but not seen again during this century.

***Lunaria rediviva* L.** Cited by Taylor and MacBryde (1977), but no B.C. material known.

***Rorippa heterophylla* (Blume) R.O. Williams.** Mapped from a single locality on SE Vancouver Island by Stuckey (1972) but no specimens seen.

***Rorippa microsperma* (D.C.) Bailey.** An Asian species, reported by Taylor and MacBryde (1977), known only from a single collection at Nanaimo in 1892.

CAMPANULACEAE

***Jasione montana* L.** Listed by Eastham (1947), but not persisting in our region.

CARYOPHYLLACEAE

***Agrostemma githago* L.** Cited by Taylor and MacBryde (1977), but apparently represented by only a single 1915 collection from the Okanagan.

***Lychnis chalcedonica* L.** Reported by Taylor and MacBryde (1977), but not persisting in our region.

***Spergularia diandra* (Guss.) Boiss.** This European plant, listed by Eastham (1947), has been collected only once at Kamloops and represents an unknown *Spergularia* spp.

CHENOPODIACEAE

***Beta vulgaris* L.** This species (the common beet) is not established in our region (Scoggan 1978).

***Chenopodium macrospermum* Hook. var. *halophilum* Standley.** Known from a single collection at Anahim Lake, this species is sometimes treated under *C. chenopodioides* but the group is poorly understood.

***Chenopodium murale* L.** Reported from ballast by Macoun (1889) but no longer with us.

***Sarcobatus vermiculatus* (Hook.) Torr.** Listed by Henry (1915), but later dropped by Eastham (1947).

APPENDIX II – NAME CHANGES

The following are name changes that have been made since Douglas et al. (1989).

Alnus tenuifolia = A. incana ssp. tenuifolia
Alnus crispa ssp. crispa = A. viridis ssp. crispa
Alnus crispa ssp. sinuata = A. viridis ssp. sinuata
Betula glandulosa = B. nana
Atriplex heterosperma Bunge = A. micrantha C. A. Mey.
Atriplex longipes = A. nudicaulis
Callitriche anceps = C. heterophylla ssp. heterophylla
Callitriche verna = C. palustris
Cardamine umbellata = C. oligosperma var. kamtschatica
Cardamine pulcherrima var. tenella = C. nuttallii var. nuttallii
Cardaria chalapensis = C. draba var. chalapensis
Chenopodium leptophyllum var. oblongifolium = C. desiccatum
Cochlearia officinalis var. oblongifolia = C. groenlandica
Crassula erecta = Crassula connata var. connata
Descurainea richardsonii = D. incana ssp. incana
Hypericum formosum var. nortoniae = H. scouleri ssp. nortoniae
Hypericum formosum var. scouleri = H. scouleri ssp. scouleri
Lappula echinata = L. squarrosa
Lappula redowskii var. cupulata = L. occidentalis var. cupulata
Legousia perfoliata = Triodanis perfoliata var. perfoliata
Myosotis alpestris = M. asiatica
Nasturtium microphyllum = Rorripa microphylla
Nasturtium officinale = Rorripa nasturtium-aquaticum
Pachistima myrsinites = Paxistima myrsinites
Rorripa curvipes var. integra = R. curvipes var. truncata
Salicornia europaea ssp. europaea = S. maritima
Salicornia europaea ssp. rubra = S. rubra
Silene alba = S. latifolia ssp. alba
Stellaria calycantha var. bongardiana = S. borealis ssp. borealis
Stellaria calycantha var. sitchana = S. borealis ssp. sitchana
Stellaria simcoei = S. calycantha
Symphoricarpos mollis = S. hesperius
Thelluginiella salsuginea = Arabidopsis salsuginea
Thellypodium milleflorum = T. lacinatum var. milleflorum
Viburnum opulus ssp. trilobum = V. opulus var. americanum

GLOSSARY [5]

Acerose — "Needle-pointed," or with a needle-like tip.

Achene — A dry, single-seeded fruit that does not open at maturity.

Acicular — Shaped like the needle of a pine tree "needle-like".

Acorn — A one-seeded fruit of oak trees.

Acuminate — Tapering to a narrow tip or concave point, the sides generally concavely narrowing.

Acute — Gradually tapering to a point, the sides straight (i.e., not convex).

Adaxial — On the side of a structure, turned toward the main axis.

Adnate — Fusion or attachment of unlike parts (e.g., the calyx and corolla).

Adventitious — Development of structures in unusual locations; in the case of roots, generally refers to the development of roots from the stem or leaf.

Alternate — Any arrangement of leaves or other parts not opposite or whorled; arranged singly at different heights on the stem.

Ament — A pendulous spikelike inflorescence with apetalous, unisexual flowers; a catkin.

Amphiberingian — Occurring on both sides of the Bering Straits; a phytogeographic element resulting from past migration across the Bering Land Bridge.

Amplexicaule — Clasping the stem.

Androecium — A collective term referring to male floral parts, that is, the stamens.

Androgynous — A spike bearing staminate (male) and pistillate (female) flowers where the staminate flowers are positioned above the pistillate flowers, closer to the apex of the inflorescence.

Annual — Living for only one growing season.

Anthesis —The period of time when a flower is fully open, "flowering time".

Apex — Tip, point or end.

Aphyllopodic — Having lower leaves that are bladeless or reduced to scales (e.g., *Carex* spp.).

Apical — Located at the apex.

Apomictic — Capable of reproducing asexually.

Apomixis — Any of several types of asexual reproduction.

Aquatic — Living in water.

Arachnoid — Covered with long, delicate, tangled hairs; resembling a cobweb.

Areoles — Small, clearly marked, round areas appearing on the leaves of *Opuntia*.

Aristate — Having an awn or a bristle at the tip, "awl-tipped".

Armed — Possessing prickles, spines, or thorns.

Articulate — Jointed; separating at a joint, by a clean cut scar, upon maturity.

Anther — The pollen producing sac of the stamen.

Attenuate — Gradually tapering to a slender base or apex.

Auct. non — Indicates that a species name is used commonly in the literature but is incorrect and is not the name used to refer to the type specimen.

Auricle — An ear-shaped lobe or appendage.

---

[5] Words or phrases used in lieu of technical terms in this manual appear in quotes in the glossary.

Auriculate — With auricles; often used to describe the base of a leaf blade where there is an auricle on either side of the petiole; with "earlike lobes".

Awn — A slender bristlelike appendage, usually at the apex of a structure.

Axil — The angle between a structure and the axis to which it is attached (e.g., the angle between a leaf and the stem).

Axile placentation — A pattern of placentation where the ovules are attached to the partitions of a compound pistil near the center of the ovary.

Axillary — Used in reference to structures attached in the axil or juncture of stem and leaf stalks.

Banner — The largest upper petal of a papilionaceous flower.

Barb — A short, firm, reflexed process.

Barbellate — Minutely or "finely-barbed"; bearing stiff hairs with barbs.

Basal — At, or forming, the base.

Beak — A long, slender projection on a broader structure such as a fruit or seed.

Bi — A prefix meaning two or twice.

Biennial — Living for two seasons.

Bilabiate — Two-lipped; often with reference to a zygomorphic flower.

Bipinnate — *See* bipinnatifid.

Bipinnatifid — Twice branched.

Biserrate — Doubly serrated.

Bisexual — Bearing both male and female reproductive structures.

Biternate — Two groups of three.

Bloom — *See* glaucous.

Bract — A modified leaf, reduced in size or form, that is associated with a flower or an inflorescence; in conifers, a main appendage arising from the axis of a cone.

Bractlet — A small or secondary bract (*syn.* bracteole).

Bristle — A stiff, hairlike structure.

Bristly — With bristles.

Bulb — A small, stout underground stem bearing modified (scalelike) leaves and fibrous basal roots; generally food storage organs.

Bulbiferous — Producing bulbs.

Bulbil — A small bulb produced above the ground, often on the flowering stem.

Caducous — Falling off the plant early relative to other similar structures.

Caespitose — Growing in dense, low-lying clumps, "tufted".

Calcareous — Rich in calcium carbonate.

Callose — Bearing callosities.

Callosity — *See* callus.

Callus — A hard or firm thickening of a structure; in grasses, the thick, hardened basal portion of the lemma.

Calyx — A collective name referring to the sepals, the outer whorl of the perianth.

Campanulate — "Bell-shaped" to saucer-shaped.

Cane — In *Rubus* spp., lignified shoots that arise each year.

Canescent — Appearing white or gray due to very short, fine hairs.

Capillary — Very slender and "hairlike".

Capitate — In a "headlike" cluster.

Capsule — A dry, dehiscent fruit comprised of more than one carpel; method of splitting further classifies the capsule type (e.g., circumscissile, septicidal, etc.).

Carpel — The basic unit of a pistil, a modified seed-bearing leaf.

Caruncle — An often fleshy appendage at or near the attachment point (hilum) of some seeds.

Caryopsis — The dry, one-seeded fruit found in grasses.

Caudate — Bearing a terminal, long and slender appendage like a tail.

Caudex — The woody, thickened and persistent basal portion of a herbaceous perennial that gives rise to leaves and flowering stems each year, i.e. "stem-base".

Caulescent — Having an obvious leafy stem.

Cauline — Of, or pertaining to, the stem, i.e. "stem leaves".

Cellular — A term used in reference to the cell-shaped network of minutely raised and lignified ridges on epidermis (e.g., on the culm of some *Carex* spp.).

Chaffy — With thin dry scales or bracts; "papery", "papery-bracted".

Chartaceous — Having the texture of stiff paper or parchment; "firm-papery", thickly membranous.

Ciliate — Leaves fringed on the margins,with fine hairs (cilia), "fringed".

Ciliolate — Ciliate but with smaller hairs.

Cinereous — Appearing grey because of short hairs.

Circumboreal — Distributed at the top or the bottom of the world in the boreal zone.

Circumpolar — Distributed around the North or the South Pole.

Circumscissile — A capsule that dehisces transversely in a circular line, like the opening of a lid.

Clavate — "Club-shaped", the wider portion at the apex.

Claw — The narrowed, basal portion of some petals or sepals.

Cleft — Usually applied to leaf margins and sinuses cut sharply and about half-way to the midrib or base; deeply lobed.

Collar — In grasses, the band on the outside area of the leaf at the point where the leaf blade and the sheath converge.

Coma — A tuft of hair usually at the apex of structures like seeds.

Compound — With two or more parts or components.

Confluent — Structures merging into one; fusing at the bases.

Conic — Cone-shaped; "conelike".

Connate — Union or attachment of like parts (e.g., petals), "fused".

Connivent — Convergent but not fused.

Convolute — The arrangement of petals or sepals such that each petal or sepal has one edge exposed while the other is covered; in grasses, a leaf that is rolled longitudinally so that one edge is inside the other.

Cordate — Shaped like a heart, with the notch at the base; "heart-shaped".

Coriaceous — Having a thick and leathery texture.

Corm — A thickened, underground stem with papery leaves; a storage organ.

Corniculate — Bearing small horns.

Corolla — The collective term describing the petals of a flower.

Corymb — A flat-topped, indeterminant inflorescence where pedicels of florets are progressively shorter towards the center of the inflorescence; may be simple or compound.

Corymbose — A flat or convex-topped inflorescence; resembling a corymb.

Costa — A prominent rib or vein.

Cotyledon — The first functional leaf of a seedling.

Crenate — Toothed with rounded, broad teeth, "round-toothed".

Crenulate — Crenate but with smaller teeth.

Crown — The persistent basal portion of a perennial plant (e.g., grass), or the leafy portion of a tree; any appendage situated between the corolla and the stamens.

Cruciform — Having the shape of a cross, "cross-shaped".

Cucullate — Hooded or "hood-shaped."

Culm — The flowering stem of grasses, sedges and rushes.

Cuneate — Triangular or "wedge-shaped" with the acute angle at attachment point.

Cupulate — Cup-shaped; "cuplike".

Cuspidate — Bearing a short, rigid and sharp point, usually in reference to the apex of a leaf, "awl-tipped".

Cyme — An inflorescence in which all floral axes terminate in a single flower with the terminal flower, at the apex of the inflorescence, blooming first.

Cymose — Bearing cymes, or cyme-like.

Deciduous — Seasonal shedding of leaves or other structures after the growth period.

Decumbent — Lying on the ground with the end curved upward.

Decurrent — A structure adnate or confluent to the stem near its insertion point, with an adnate wing or margin extending down the stem or axis below the point of insertion.

Definite — Distinct; precise.

Deflexed — Bent downward or backward.

Dehiscent — Discharge of contents by the opening of pores or slits, most often pertaining to the opening of fruit at maturity.

Deltoid — Shaped like an equilateral triangle.

Dentate — With sharp, spreading teeth, "sharp-toothed."

Dichotomous — Forked with both branches of approximately equal size.

Didymous — Occurring or developing in pairs.

Dimorphic — Having two forms.

Dioecious — Male and female reproductive structures borne on different plants.

Disarticulate — The separation of structures at maturity.

Disciform — With the form of a disk; in the Asteraceae, with the central flowers of a head perfect (or functionally staminate) and the marginal flowers pistillate but without a ligule.

Discoid — Resembling a disk; in the Asteraceae, with the flowers of a head all tubular and perfect (or functionally staminate).

Disk corolla (disk flower) — Tubular corolla or flower of some members of the Asteraceae.

Dissected — Divided into many small, fine and often repeated parts or segments.

Dorsal — Pertaining to, or located on, the back of an organ.

Drupe — A fleshy, one-seeded or one-stoned fruit of some plants of the rose family (e.g., cherries, peaches).

Elliptic — Shaped like an ellipse, the broadest point midway between the ends, and the width about one half the length.

Emarginate — Having a shallow notch at the apex, usually with reference to a leaf or petal.

Endocarp — The innermost layer of the three layers comprising the mature ovary wall of a fruit.

Entire — A continuous margin, without teeth or divisions.

Epigaeous — Growing above ground.

Epiphytic — A plant that grows on another plant but does not receive nourishment from it.

Equitant — With alternate leaves whose basal portions overlap and are flattened lengthwise (e.g., *Iris*).

Erose — An irregular margin, appearing eroded or gnawed, not fringed or toothed.

Ex — With respect to plant authorities, indicates that the first author proposed the name for a plant and the second author published that name.

Exserted — Projecting beyond a surrounding structure, like stamens projecting beyond a corolla.

Extravaginal — Located outside the sheath that surrounds another organ.

Falcate — Sickle-shaped; curved sideways and tapering.

Farinose — Covered with a white mealy substance.

Fascicle — A cluster.

Ferruginous — Rust in colour.

Filament — The slender stalk of the stamen that supports the anther.

Filiform — "Threadlike", slender and cylindrical.

Fimbriate — Fringed with hairs coarser than cilia.

Fistulose — Hollow.

Flabellate — "Fan-shaped".

Flabelliform — *See* flabellate.

Flexuous — Bent in opposite directions; having a more-or-less zig-zag or wavy form.

Floccose — Covered with tufts of soft woolly hairs that usually rub off readily.

Floret — A small flower, particularly one in a floral cluster; pertains also to the flower of a grass spikelet which includes the lemma, palea and reproductive structures.

Foliaceous — "Leaflike".

Follicle — A dry, dehiscent fruit composed of one carpel that splits at the side upon maturing to release its seeds.

Fornix (*pl.* fornices) — A small, scale-like or petaloid appendage in the tube or throat of the corolla, found in many borages.

Free central placentation — A pattern of placentation where the ovules are borne on a single stalk, which is anchored at the base of a one-celled ovary.

Frond — The leaf of a fern.

Fruit — The ripened ovary and any other structures that enclose it.

Funnelform — "Funnel-shaped," usually pertaining to the corolla.

Fusiform — "Spindle-shaped," broadest at the middle and tapering to opposite ends.

Galea — The hooded portion of the perianth in some irregular or bilabiate flowers.

Gametophyte — The sexual reproductive stage of a plant that produces gametes; in ferns it is often the small inconspicuous form of the plant.

Gemma (*pl.* gemmae) — A young bud from which plants vegetatively reproduce.

Geniculate — Abruptly bent.

Gibbous — Swollen on one side.

Glabrate — "Nearly glabrous," or becoming glabrous, usually with age.

Glabrescent — *See* glabrate.

Glabrous — Smooth, without hairs of any type.

Gland — A spot or structure that produces a sticky or greasy substance.

Glandular — Having glands.

Glandular-pubescent — With gland-tipped hairs, "glandular-tipped".

Glaucous — Covered with a whitish or bluish waxy powder that readily rubs off.

Globose — Spherical.

Glochidia — Retrorsely barbed hairs.

Glomerate — Aggregated or condensed into heads.

Glomerule — A dense cluster, usually of flowers.

Glumes — The two small bracts located at the base of a grass spikelet; they do not subtend flowers.

Glutinous — Covered with a sticky, glue-like resin.

Gynaecandrous — A spike bearing staminate (male) and pistillate (female) flowers where the staminate flowers are positioned below the pistillate flowers, thus the pistillate flowers are closest to the spike apex.

Hastate — Shaped like the head of an arrow with the basal lobes flaring outward, "arrowhead-shaped".

Head — A dense indeterminant inflorescence of sessile (or nearly sessile) flowers, situated on an axis that is much compressed vertically and sometimes flattened or saucer-shaped.

Hemispheric — In reference to a head, with a half-sphere shape.

Herb — Plant with stem that dies back to the ground after flowering or at the end of the growing season.

Herbaceous — A non-woody plant.

Heterophyllus — Having two types of leaves.

Heterosporous — Production of two different types of spores, generally differing in size.

Hirsute — Covered with long, stiff and coarse hairs, "stiff-hairy".

Hirtellous — Minutely hirsute.

Hispid — Bearing long, rigid and very sharp bristles or "bristlelike" hairs.

Hispidulous — Slightly hispid.

Hoary — Covered with short and fine white or gray hairs.

Homophyllous — With leaves that are all identical or very similar in form.

Homosporous — Production of one type of spore.

Hyaline — Thin, almost completely transparent tissue, translucent,"wax papery".

Hybrid — A cross between two taxa, usually a cross between two species of the same genus.

Hypanthium — A cup-like structure surrounding the ovary, formed by the fusion of the basal portions of the perianth, calyx, corolla and androecium.

Imbricate — Partly overlapping in a spiral manner, shingle-like, "graduated".

Imperfect — Flowers lacking either female or male reproductive structures, a unisexual flower.

In — With respect to plant authorities, indicates that the first author described the plant in an article edited or published by the second author.

Incised — Cut sharply, deeply and irregularly, as in leaf margins.

Indument — Appendages or outgrowths from the epidermis of a plant.

Indurate — Hard or stiff.

Indusium (*pl.* indusia) — A scale-like outgrowth that covers and protects the cluster of sporangia located on the underside of a fern frond.

Inferior (ovary) — When the perianth and androecium are fused at the top of the ovary wall, the ovary position is described as inferior.

Inflorescence — The flowering cluster, categorized by the arrangement of flowers on the floral axis.

Intergradients — Taxa that interbreed and consequently share many characteristics making it difficult to separate them taxonomically.

Internode — The section of a stem or another structure between two nodes.

Interrupted — Discontinuous.

Involucre — A rosette of bracts subtending an inflorescence or head.

Involute — Margins longitudinally rolled in, on the upper side, toward the center.

Irregular — Pertaining to a flower in which similar parts (e.g., petals) are unequal in shape and size (*syn.* zygomorphic).

Keel — In a papilionaceous flower, the two lowermost connate petals; resembles the keel of a boat; in grasses, the sharp crease or ridge of the leaf blades, the lemma or the glume.

Lacerate — With an irregular or jagged margin, as if torn.

Laciniate — Cut into narrow, often pointed, lobes.

Lamina — The flattened portion (i.e., the blade) of a leaf or petal.

Lanate — Woolly, with long intertwined, curly hairs.

Lanceolate — "Lance-shaped"; longer than broad, widening above the base and tapering to the apex.

Leaflet — One of the segments or divisions of a compound leaf.

Lemma — The outermost of the two bracts that subtend the grass floret (*See* palea).

Lenticular — Biconvex in shape.

Ligulate — With a ligule.

Ligule — Flattened part of the ray corolla in the Asteraceae; in many grasses and some sedges and rushes, a membranous (or hair-like) appendage on the inner side of the leaf at the junction of the leaf blade and the sheath.

Linear — Long and narrow, with parallel margins.

Locule — A compartment in the ovule or anther.

Lustrous — Shiny.

Lyrate — Pinnatifid with the end lobe enlarged and rounded.

Malpighiaceous — Hairs attached to a surface by their centres.

Maritime — Near to, or influenced by, the sea.

Megasporangia — The sporangia of a heterosporous plant that produces megaspores.

Megaspore — On a heterosporous plant, the larger of two spore sizes produced; a spore involved in development of female reproductive organs.

Mericarp — One portion of the fruit (e.g., a schizocarp) that splits into two halves at maturity, found in members of the Apiaceae.

Merous — A suffix referring to the division of floral parts, e.g., a 3-merous flower would have 3 petals, 3 sepals, 3 stamens, and a pistil with 3 carpels.

Mesic — Relating to habitats with a moderate or average amount of moisture.

Microsporangia — The sporangia of a heterosporous plant that produces microspores.

Microspores — On a heterosporous plant, the smaller of two spore sizes produced; a spore involved in the development of male reproductive organs.

Monoecious — Male and female reproductive structures borne on the same plant.

Mottled — Marked with spots.

Mucro — A small, short and abrupt point.

Mucronate — Having mucro.

Muricate — Roughened with short, firm and sharp projections.

Naked — Lacking various organs, hairs or appendages.

Naturalized — A well established plant originating from another area.

Neutral — In reference to a flower lacking functional stamens and pistils.

Nodding — Curving to one side; minutely pendulous.

Node — The location on a stem where stems or leaves usually originate, generally a swollen area, other structures like tendrils and flowers may also originate here.

Nom. superfl. — A plant name that has been superceded by a valid name; a superfluous name.

Nut — A hard, indehiscent fruit, with one seed.

Nutlet — A small nut; also used when referring to each lobe of the 4-lobed nuts found in borages.

Ob — A prefix meaning in the reverse of the typical direction.

Obconical — Conical in shape with attachment point at the narrow end of the cone.

Obcordate — Having an inverted heart shape, "heart-shaped".

Oblanceolate — Reverse of lanceolate, widest above the middle and tapering to the base.

Oblong — Much longer than wide with parallel or nearly parallel sides.

Obovate — The reverse of ovate; attached at the narrow end, "egg-shaped".

Obpyramidal — Having the shape of an inverted pyramid.

Obsolete — Reduced to the point of being entirely, or almost, undetectable.

Obtuse — Blunt, rounded.

Ochroleucous — Yellowish-white.

Olivaceous — Olive-like in colour.

Orbicular — Having a circular outline.

Ovary — The enlarged base of the pistil that contains the developing seeds, matures into the fruit.

Ovate — Shaped like a hen's egg, attached at the broader end, "egg-shaped".

Palea — The innermost of the two bracts that subtend the floret in grasses (*See* lemma).

Palmate — Lobed or divided in a palm- or hand-like fashion, digitate.

Panicle — A branching inflorescence on which younger flowers are borne at the apex.

Paniculiform — Resembling a panicle.

Papillate — Covered with minute rounded protuberances (papillae, "pimples").

Papilionaceous flower — A strongly irregular flower with five petals, the lowermost two petals are connate and form the keel, as in Fabaceae.

Pappus — The modified outer perianth series crowning the ovary and achene in the Asteraceae; usually plumose, bristle-like, or of scales.

Parietal placentation — A pattern of placentation, in compound pistils only, where ovules are attached to the outer wall of the carpel.

Pectinate — "Comb-like", pinnately lobed with narrow segments arranged like the teeth of a comb.

Pedicel — The "stalk" of a single flower in an inflorescence, or of a spikelet in grasses.

Pedicellate — Having a pedicel, a "stalk".

Peduncle — The single supporting stalk of a flower or inflorescence.

Peltate — An umbrella-shaped leaf whose attachment point is on the underside of the leaf rather than at the base or margin, "umbrella-like".

Pendulous — Hanging; resembling a pendulum.

Perennial — Growing for more than two seasons.

Perfect — A flower that has functional stamens and pistils.

Perfoliate — Bearing opposite leaves whose bases are wrapped around the stem and connate, giving the appearance that the stem passes through the leaf.

Perianth — A collective term for the calyx and corolla.

Perigynium — A "pouched" bract surrounding the pistil or achene in *Carex* spp.

Persistent — Remaining attached.

Petiole — Leaf stalk.

Phenotypic plasticity — Ability to adapt, morphologically, to changing environmental conditions.

Phyllode — A somewhat expanded but bladeless petiole taking on the function of a leaf.

Phyllopodic — Having well developed lower leaves instead of scales.

Piliform — Having the shape of a long, sometimes wavy, hair.

Pilose — Shaggy with soft hairs, "soft hairy".

Pinna (*pl.* pinnae) — One of the primary divisions of a pinnately compound leaf or frond.

Pinnate — Compound leaf with the leaflets on both sides of an elongated axis.

Pinnatifid — Pinnately lobed.

Pinnule — A secondary leaflet in a bipinnately compound leaf.

Pistillate — Flowers possessing pistils, but no functional stamens (i.e., a female flower).

Pith — The spongy tissue in the center of a root or stem that surrounds vascular bundles.

Placentation — The pattern of attachment of ovules in the ovary. See axile, parietal or free central placentation.

Plumose — "Feather-like".

Pod — Any type of dry, dehiscent fruit.

Pollen — A general term pertaining to the haploid *(n)* microspores produced in the anther.

Polypetalous — Flowers with all petals distinct, not united.

Polyploidy — Two or more sets of chromosomes (e.g., 3n, 4n, 8n).

Prehensile — Grasping.

Prickle — Small, sharp outgrowth from epidermis, contains no vascular tissue.

Prismatic — Prism-shaped.

Procumbent — Prostrate or trailing stem, does not usually root at nodes.

Propagule — Any part of a plant capable of growing into a new organism; e.g., seed, spore, gemma, cutting.

Pro parte — In part.

Pro parte sed non typus — Resembles, in part, but not completely the type specimen; possessing some but not all characteristics of the type specimen.

Prostrate — Lying flat on the ground, may root at the nodes.

Pruinose — *See* glaucous.

Puberulent — Bearing very short, erect hairs, scarcely visible to the unaided eye, minutely pubescent, "short-hairy".

Pubescent — "Hairy".

Pulverulent — Dusty, powdery.

Pulvinus — Swollen tissue at the insertion point of the leaf petiole, or sometimes at a joint.

Punctate — Marked with pits or dots formed by glands or waxy spots.

Puncticulate — Minutely punctate.

Pyriform — Pear-shaped.

Raceme — An indeterminant, elongated inflorescence where single flowers are borne on pedicels arranged on a single axis; the youngest flowers are at the apex.

Racemiform — Possessing the form of a raceme.

Rachilla — The axis of a spikelet in grasses, refers to a secondary axis in sedges.

Rachis — The main stalk of an inflorescence or compound leaf, the "axis of inflorescence".

Radiate — In the Asteraceae, "heads with ray and disk flowers".

Ray (flower) — Strap-shaped corolla (ligule), as in the Asteraceae.

Receptacle — The expanded end of the stem on which floral parts are borne.

Reclining — Lying on the ground or one plant pressed on another.

Reflexed — Bent sharply downward or backward.

Regular (flowers) — A flower with radial symmetry; petals and sepals that are similar in shape, size, colour and orientation.

Reniform — "Kidney-shaped".

Resinous — Covered with a resin.

Reticulate — Forming a network, netted like the veins of a leaf, "net-veined".

Retrorse — Directed backward or downward.

Retuse — See emarginate.

Revolute — Leaf margins rolled toward the underside of the leaf.

Rhizome — A creeping, underground stem or rootstalk that roots at the nodes.

Rhombic — "Diamond-shaped".

Rib — A prominent vein, usually on a leaf.

Root — The descending axis of a plant that anchors and absorbs moisture from the ground, lacks nodes and internodes.

Rootstock — Used loosely to mean a main root or an underground root-like stem or branch.

Rosette — A dense cluster of flowers or leaves radiating around a central axis, usually basal in the case of leaves.

Rosulate — In the form of a rosette.

Rotate — A saucer-shaped corolla where the corolla tube is short and petals are generally reflexed.

Rotund — Rounded.

Rudimentary — Usually a small, imperfectly developed structure, generally reminiscent of an earlier evolutionary stage.

Rufous — "Reddish-brown".

Rugose — Possessing a "wrinkled" surface.

Rugulose — Slightly rugose.

Saccate — Shaped like a sac or pouch.

Sagittate — Shaped like the head of an arrow, triangular with the basal lobes pointing backward, "arrowhead-shaped".

Salverform — A corolla having a long slender tube that flares abruptly at the end.

Samara — A winged, indehiscent fruit.

Samaroid — Resembling a samara.

Saprophytic — A plant that derives its nutrients from dead organic matter and lacks the chlorophyll that gives most plants their green color.

Scaberulous — Slightly scabrous.

Scabrid — *See* scabrous.

Scabridulous — *See* scaberulous.

Scabrous — Having a roughened surface caused by very short stiff hairs or short sharp projections, "rough short-hairy".

Scabrule — A stiff pointed projection or a short stiff hair.

Scale — A small, flat structure resembling the scale of a fish.

Scape — A flowering stem without leaves.

Scapose — Bearing a scape.

Scarious — Having chaffy, membranous, brown or "wax-papery" edges.

Scrambling — Slightly reclining.

Scree — Relatively small broken rock material characteristic of alpine areas, usually a result of freeze-thaw cycles.

Secund — Having all flowers or branches borne on one side of the axis.

Semi-aquatic — A plant that may tolerate partial submersion in water or a plant that roots in the soil but produces aquatic leaves.

Sensu lato — In the broad sense.

Sepals — The outermost whorl of floral leaves, also called the calyx and is usually green in color.

Seriate — In a series of rows.

Sericeous — A surface with long, soft, appressed hairs, giving a silky appearance and texture, "silky-hairy".

Serrulate — A leaf margin with small serrated teeth.

Sessile — Without a stalk, "unstalked".

Seta (*pl.* setae) — A bristle.

Setose — Bearing setae.

Sheath — An thin tubular organ that partially or wholly surrounds another organ; in grasses, the basal portion of the leaf blade that wraps around the stem.

Shrub — A woody perennial plant, less than 10 m tall, with several basal stems.

Silicle — A dry fruit of plants in the Brassicaceae generally not more than two times longer than wide which opens along the edges leaving the seeds attached to a central partition.

Silique — A long slender fruit of the Brassicaceae family which splits open along the edges leaving seed attached to a central partition.

Simple — A leaf not compounded into leaflets, or an unbranched inflorescence.

Sinuate — A pronouncedly waved leaf margin, "wavy".

Sinus — "Notch", a cavity or a depression between two adjacent structures (e.g., the space between two lobes of a leaf).

Sordid — "Dirty tinged".

Sorus (*pl.* sori) — A cluster of spore producing structures (sporangia) on the underside of a fern frond.

Spadix — A spike with small, densely crowded, unisexual flowers on a thick and fleshy axis, the inflorescence is subtended by a spathe.

Spathe — A large bract surrounding a spadix.

Spatulate — Broad and rounded at the apex; narrowing to the base, "spoon-shaped".

Spiciform — An inflorescence resembling a spike, "spike-like".

Spike — An unbranched inflorescence; flowers are sessile.

Spikelet — A small spike; in grasses, the primary inflorescence composed of two glumes and one or more florets.

Sporangia — Spore producing structures.

Sporangium — Reproductive structure of ferns that produces spores.

Spore — A haploid *(n)* reproductive cell.

Sporocarp — The structure that contains the sporangia in ferns.

Sporophore — A leaf bearing reproductive structures; generally used in reference to the spore bearing leaves of ferns.

Sporophyll — A modified leaf bearing sporangia.

Sporophyte — The generation in the reproductive cycle of plants that produces spores; in ferns generally refers to the conspicuous form of the plant (See gametophyte).

Spur — A slender and hollow floral appendage, generally nectariferous.

Squamiform — "Scalelike".

Squarrose — Having parts, generally at the tip of a structure, that are abruptly curving and spreading.

Stamen — The male sex organ of a plant constructed of an anther and a filament, the anther produces pollen.

Staminate — Flowers bearing stamens but no functional pistils; a male flower.

Staminode (*pl.* staminodia) — A sterile, sometimes modified stamen (*syn.* staminodium).

Stellae — Star-shaped structures.

Stellate — "Star-shaped"; having many rays radiating out from a common center.

Steppe — An extensive area of dry, arid grassland.

Sterile — Infertile flowers.

Stigma — The generally expanded, apical portion of the pistil which is receptive to pollen.

Stipe — A stalk supporting the pistil; also the petiole of the frond of a fern.

Stipitate — Borne on a stipe or short stalk.

Stipule — An appendage on each side of the base of a leaf or petiole at its insertion point on the stem.

Stolon — A reclining above-ground horizontal stem that roots at the nodes.or at the tips.

Stramineous — Having the color of straw, "straw-colored".

Strigillose — Like strigose but with shorter hairs, "short appressed hairy".

Strigose — A surface having sharp, appressed and straight hairs that are stiff and often basally swollen, "appressed hairy".

Style — A generally elongated stalk connecting the ovary and the stigma through which the pollen tube grows.

Styliform — Resembling a long slender stalk, like a style.

Stylopodium — A disk-like enlargement at the base of the style.

Sub — A prefix signifying below, somewhat, slightly or rather.

Subspecies (ssp.) — A category of biological classification ranking immediately below the species level.

Subtend — To be positioned directly or closely beneath something.

Subulate — Narrowly triangular and tapering to a small point like an awl, "awl-like".

Succulent — Fleshy, having a soft and thickened texture.

Sulcate — Longitudinally grooved or furrowed.

Sympatric — Occurring in the same area.

Sympodial — Having an apparent main or central axis that is actually comprised of several short overlapping branches.

Synonym — An equivalent superseded name, a second name to a given taxon.

Talus — An accumulation of rock material on or at the base of a slope, larger material than scree.

Taproot — An underground tapering root from which smaller lateral roots may form (may be thick or thin).

Tendril — A long, slender, generally twisting structure from which a climbing plant acquires its support.

Tepals — A collective term applied to petals and sepals when they are morphologically identical.

Terete — A structure that is circular in cross section but imperfectly cylindrical since the object may taper one or both ways.

Ternate — Arranged in groups of three.

Tetrad — In groups of four.

Tetraploid — Four sets of chromosomes (4n).

Thyrse — An elongated and densely flowered panicle, often contracted and pyramidal in shape.

Thyrsoid — Resembling a thyrse.

Tomentose — Densely pubescent with soft, matted wool-like hairs, "woolly-hairy", "woolly".

Tomentulose — Slightly tomentose or "slightly woolly".

Toothed — Bearing teeth-like protrusions.

Torose — Having successive swellings and contractions, "constricted between the seeds".

Torulose — Slightly torose.

Tri — A prefix meaning three or three times.

Tricuspidate — *See* cuspidate.

Trifid — Separation into three parts.

Trifoliate — A compound leaf with three leaflets.

Trigonous — Having three angles.

Tripinnate — Branched three times.

Triquetrous — With three sharp angles, like a triangle with concave sides.

Trophophore — A vegetative leaf or frond, one that does not have reproductive structures attached.

Truncate — Cut-off at the end; having a squared shape at the apex or base.

Tuber — A thickened, modified underground stem having many buds (e.g., a potato).

Tubercle — A small, rounded swelling.

Tuberculate — Bearing tubercles.

Tuberous — Thickened; producing tubers.

Turbinate — Inversely conical, "top-shaped".

Turgid — Swollen.

Turion — A small, scaly bud-like offset on an underground stem.

Type specimen — A specimen identified by the author of a species that, morphologically, represents the best example of this species (*syn.* holotype).

Umbel — An inflorescence in which all flowers arise from a common point, younger flowers are in the center, may be simple or compound.

Umbellate — Arranged in umbels, "umbel-like".

Umbellet — A secondary umbel in a compound umbel.

Undulate — A slightly wavy margin.

Uniseriate — Arranged in a row or series.

Valve — One of the segments into which a dehiscent fruit splits at maturity.

Variety *(var.)* — A category of biological classification ranking immediately below the species or subspecies level.

Veins — Thin threads of vascular tissue, as in a leaf.

Velutinous — Having a velvety texture.

Vernal — Pertaining to, or appearing in, spring.

Verrucose — Covered with wart-like protrusions, "wartlike".

Verticillate — See whorled.

Verticils — See whorl.

Vestige — The reduction of a structure whose functional importance has decreased with evolutionary time.

Villous — With long and soft, unmatted hairs, "long-hairy".

Viscid — Glutinous or greasy.

Wanting — Lacking.

Whorl — Three or more structures arranged in a circular manner around a common insertion point.

Whorled — Having whorls.

Wing — A thin membranous outgrowth.

Winged — Bearing wings.

INDEX

D

E